WITHDRA

D1285296

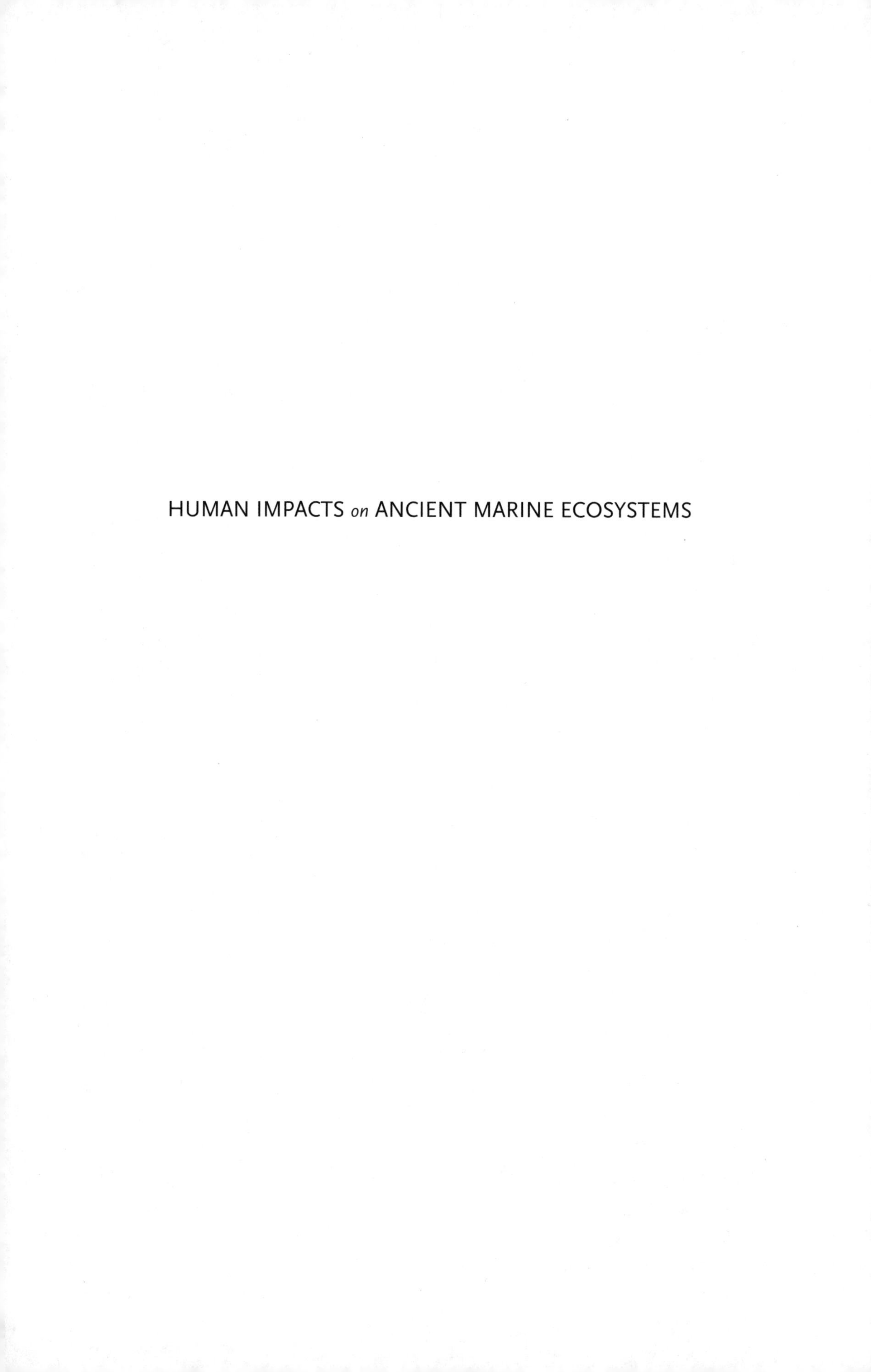

HUMAN IMPACTS *on* ANCIENT MARINE ECOSYSTEMS

HUMAN IMPACTS *on* ANCIENT MARINE ECOSYSTEMS

A Global Perspective

Edited by

Torben C. Rick and Jon M. Erlandson

UNIVERSITY OF CALIFORNIA PRESS
Berkeley Los Angeles London

University of California Press, one of the most distinguished university presses in the United States, enriches lives around the world by advancing scholarship in the humanities, social sciences, and natural sciences. Its activities are supported by the UC Press Foundation and by philanthropic contributions from individuals and institutions. For more information, visit www.ucpress.edu.

University of California Press
Berkeley and Los Angeles, California

University of California Press, Ltd.
London, England

Library of Congress Cataloging-in-Publication Data

Human impacts on ancient marine ecosystems : a global perspective / edited by Torben C. Rick and Jon M. Erlandson.
p. cm.
Includes bibliographical references and index.
ISBN 978-0-520-25343-8 (case : alk. paper)
1. Coastal archaeology—Case studies. 2. Underwater archaeology—Case studies. 3. Prehistoric peoples. 4. Fishing, Prehistoric. 5. Nature—Effect of human beings on. 6. Marine mammals—Effect of human beings on. 7. Marine mammal remains (Archaeology) I. Rick, Torben C. II. Erlandson, Jon.
GN784.H86 2008
930.1028'04—dc22
2007046724

Manufactured in the United States of America
10 09 08
10 9 8 7 6 5 4 3 2 1

The paper used in this publication meets the minimum requirements of ANSI/NISO Z39.48-1992 (R 1997) (*Permanence of Paper*).

Cover photograph: A Middle Holocene red abalone shell midden at San Miguel Island. Photo by Todd Braje.

CONTENTS

CONTRIBUTORS

ATHOLL ANDERSON, Department of Archaeology and Natural History, Research School of Pacific and Asian Studies, Australian National University, Canberra, Australia

C. FRED T. ANDRUS, Department of Geological Sciences, University of Alabama, Tuscaloosa

GEOFF BAILEY, Department of Archaeology, University of York, The King's Manor, York, UK

JAMES BARRETT, McDonald Institute for Archaeological Research, University of Cambridge, UK

BRUCE J. BOURQUE, Department of Anthropology, Bates College, Lewiston, Maine

TODD J. BRAJE, Department of Anthropology, University of Oregon, Eugene

GEORGE M. BRANCH, Department of Zoology, University of Cape Town, South Africa

DOUGLAS CAUSEY, University of Alaska, Anchorage

MARK CLEMENTZ, Smithsonian Institution, Washington, D.C.

DEBRA G. CORBETT, U.S. Fish and Wildlife Service, Anchorage, Alaska

OLIVER CRAIG, Department of Archaeology, University of York, The King's Manor, York, UK

ANGELA DOROFF, U.S. Fish and Wildlife Service, Anchorage, Alaska

JON M. ERLANDSON, Department of Anthropology and Museum of Natural and Cultural History, University of Oregon, Eugene

JAMES A. ESTES, U.S. Geological Survey, Long Marine Laboratory, University of California, Santa Cruz

SCOTT M. FITZPATRICK, Department of Sociology and Anthropology, North Carolina State University, Raleigh

MICHAEL H. GRAHAM, Moss Landing Marine Laboratory, Moss Landing, California

ANTONIETA JERARDINO, Department of Archaeology, University of Cape Town, South Africa

BEVERLY J. JOHNSON, Department of Geology, Bates College, Lewiston, Maine

WILLIAM F. KEEGAN, Florida Museum of Natural History, University of Florida, Gainesville

DOUGLAS J. KENNETT, Department of Anthropology, University of Oregon, Eugene

PAUL L. KOCH, Earth Sciences Department, University of California, Santa Cruz

CHRISTINE LEFÈVRE, Museum National d'Histoire Naturelle, Paris, France

THOMAS H. MCGOVERN, Northern Science and Education Center, Department of Anthropology, Hunter College, New York

NATALIA MARTÍNEZ, Department of Anthropology, University of Arizona, Tucson

NICKY MILNER, Department of Archaeology, University of York, The King's Manor, York, UK

ARTURO MORALES-MUÑIZ, Laboratorio de Arqueozoología, Universidad Autínoma de Madrid, Spain

RENE NAVARRO, Animal Demography Unit, Department of Zoology, University of Cape Town, South Africa

SOPHIA PERDIKARIS, Department of Anthropology, Brooklyn College, New York

ELIZABETH J. REITZ, Georgia Museum of Natural History, Natural History Building, University of Georgia, Athens

TORBEN C. RICK, Department of Anthropology, Southern Methodist University, Dallas, Texas

EUFRASIA ROSELLÓ-IZQUIERDO, Laboratorio de Arqueozoología, Universidad Autínoma de Madrid, Spain

DANIEL H. SANDWEISS, Department of Anthropology and Climate Change Institute, University of Maine, Orono

ROBERT S. STENECK, School of Marine Sciences, Darling Marine Center, University of Maine, Orono

KATHLEEN SULLIVAN SEALEY, Marine Conservation Science Center, University of Miami, Coral Gables, Florida

RENÉ L. VELLANOWETH, Department of Anthropology, Humboldt State University, Arcata, California

BARBARA VOORHIES, Department of Anthropology, University of California, Santa Barbara

THOMAS A. WAKE, Cotsen Institute of Archaeology, University of California, Los Angeles

DIXIE WEST, Natural History Museum and Biodiversity Research Center, University of Kansas, Lawrence

PREFACE

Although their role has been underappreciated until recently, archaeological records from coastlines around the world contain a wealth of information on the history of marine fisheries, human impacts on marine ecosystems, and marine conservation principles. To illustrate the contributions archaeology can make to the study of historical ecology in a variety of marine ecosystems, this volume brings together experts from relatively well studied coastal regions around the world to summarize the history of human coastal occupation, environmental change, and human impacts in their area. The participants, an interdisciplinary group of archaeologists and marine ecologists, are some of the leading researchers involved in reconstructing the historical ecology and human impacts of coastal zones. They provide 11 case studies from the Americas, Pacific Islands, Europe, and Africa, and coverage of diverse marine ecosystems ranging from kelp forests to coral reefs to mangroves.

For this book, we invited contributions from archaeologists and marine ecologists with a firm grasp on the data from particular regions, a deep knowledge of long coastal sequences in their respective areas, and a history of studying human-environmental relationships. Wherever possible, these studies use a multidisciplinary approach to document natural environmental change (sea level history, marine productivity, habitat change, etc.), the antiquity of coastal adaptations, and changes in human demography, technology, social organization, and subsistence through time.

In each case study, we asked the authors to synthesize the evidence for human impacts on marine species or ecosystems across the full range of human occupation, from long prehistoric records, to early historical or colonial periods, to the emergence of increasingly globalized and industrialized fisheries of recent centuries or decades. Finally, these analyses consider the implications of the archaeological, historical, and ecological data from their region for our understanding of the nature of human impacts to marine ecosystems and for the development of fisheries management, conservation, and restoration protocols or policies that are more effective than those that have led to the widespread collapse of aquatic ecosystems and fisheries around the world. This is a large and complex undertaking, but one that we believe can greatly enhance the sustainability of the world's marine ecosystems.

The length, quality, and resolution of archaeological, historical, and ecological records in different coastal areas around the world also vary considerably, so the level of detail and the relative emphasis of individual chapters vary somewhat. Following an introductory chapter by Erlandson and Rick (Chapter 1), the chapters in this volume are organized geographically. In Chapter 2, Anderson provides an in depth analysis of New Zealand and other subtropical Pacific Islands. This is followed by two North American Pacific Coast case studies by Corbett, Causey, Clementz, Koch, Doroff, Lefevre, and West (Aleutians; Chapter 3) and Rick, Erlandson, Braje, Estes, Graham, and Vellanoweth (southern California; Chapter 4). Kennett, Voorhies, Wake, and Martínez (Chapter 5) then provide a detailed analysis of the Pacific Coast of Mexico, which is followed by Reitz, Andrus, and Sandweiss's (Chapter 6) review of human impacts on the fisheries of ancient and modern Peru. In Chapter 7, Fitzpatrick, Keegan, and Sealey provide perspectives from the Caribbean. The next three chapters focus on the North Atlantic, beginning with Bourque, Johnson, and Steneck (Chapter 8) for the Gulf of Maine, Perdikaris and McGovern (Chapter 9) on the Norse, and Bailey, Barrett, Craig, and Milner (Chapter 10) on the North Sea Basin. The focus then shifts to the Mediterranean with Morales and Rosselo's (Chapter 11) 20,000-year analysis of fisheries in Iberia. Jerardino, Branch, and Navarro's (Chapter 12) overview of South Africa is the final regional case study. Finally, Rick and Erlandson (Chapter 13) provide an overview of the volume, pointing out important issues in coastal archaeology and marine conservation as well as areas for future research.

We owe a great deal of gratitude to the contributors of this volume for providing insightful and thought-provoking analyses of many major marine ecosystems around the world, and for graciously meeting our deadlines. David Steadman and Steven James provided a number of important comments on an earlier version of this manuscript that were instrumental in improving and expanding its content. We also thank Blake Edgar, Scott Norton, Matt Winfield, and the staff at the University of California Press, Joanne Bowser of Aptara, and Jake Kawatski of Live Oaks Indexing for helping review, edit, and produce this book. Finally, Jeremy Jackson, Paul Dayton, Bob DeLong, and Daniel Pauly provided much of the inspiration for our voyage into the depths of marine ecology and historical ecology.

Torben C. Rick and Jon M. Erlandson
January 15, 2008

1

Archaeology, Marine Ecology, and Human Impacts on Marine Environments

Jon M. Erlandson and Torben C. Rick

COVERING MORE THAN 70 percent of our blue planet, the oceans dominate the earth in a variety of ways. With an average depth of almost 4 km, they provide over 99 percent of the habitable space for life on earth (Woodard 2000:31). As human populations have grown exponentially over the past century, and with 60 percent of the world's population living within 100 km of the coast, many have looked to the oceans as a source of hope and protein to feed the masses. Once thought to be nearly inexhaustible, many global fisheries have collapsed or are severely depleted (Jackson et al. 2001; Pauly et al. 2002; Roberts 2002; Worm et al. 2006). Pollution, habitat loss, global warming, and the introduction of exotic species also take an increasing toll on coastal and pelagic ecosystems (see Carlton et al. 1999; Earle 1995; Ellis 2003; Vitousek et al. 1997:495; Woodard 2000). We are only beginning to understand the larger ecological consequences of such impacts, including the wholesale collapse of many coral reef, kelp forest, estuarine, arctic, benthic, and other ecosystems—foundations of marine productivity that have nurtured human societies for thousands of years. These impacts are now global in scale, but humans have had the heaviest impact on nearshore and coastal areas (0–50 m in depth), substantial impacts on deeper continental shelf habitats (50–200 m), and comparatively less impact on the deeper oceans (Steele 1998).

In the last few years, two national commissions have issued reports concluding that the world's oceans and fisheries are in a state of crisis (Pew Oceans Commission 2003; U.S. Commission on Ocean Policy 2004). The management of fisheries and our understanding of the broader ocean crisis have been hampered by the shallow historical focus of policy makers and resource managers, who have based many decisions on ecological observations that span 10, 20, or 30 years, or on historic catch records that rarely span more than a few additional decades. Just over a decade ago, Daniel Pauly (1995) referred to this problem as the "shifting baselines syndrome," where fisheries managers use recent historical baselines to manage fisheries that are depleted or collapsed. Such recent historical baselines are often fundamentally flawed because they fail to account for the abundance of key species prior to heavy fishing or hunting by indigenous peoples or early commercial harvests (Dayton et al. 1998; Jackson

2001; Jackson et al. 2001). Roman and Palumbi (2003) analyzed the DNA of living whales, for instance, and suggested that the original population sizes for some whale species were 10 or more times larger than estimated by the historical records currently used as baselines for restoring and conserving whale populations. A growing number of marine scientists are now calling for fundamental changes in the management of marine fisheries and ecosystems, including much deeper historical analyses that incorporate archaeological and other data sets into the development of better fisheries management plans, ecosystem restoration efforts, and a sustainable oceans policy.

An important step forward in this effort was a 2001 article in *Science* called "Historical Overfishing and the Recent Collapse of Marine Ecosystems," named by *Discover* magazine as the top science story of the year. In it, Jackson et al. (2001, p. 630) argued that human impacts on marine fisheries began relatively early, but they also recognized that human fishing has evolved through three general (and often overlapping) historical and geographic stages: (1) aboriginal fisheries confined to "subsistence exploitation of nearshore coastal ecosystems" with "relatively simple watercraft and extractive technologies"; (2) colonial exploitation of coastal and continental shelf ecosystems controlled by "mercantile powers incorporating distant resources into a developing market economy"; and (3) a global stage marked by "more intense and geographically pervasive exploitation of coastal, shelf, and oceanic fisheries integrated into global patterns of resource consumption." Within this framework, human impacts on marine fisheries and ecosystems have accelerated through time and expanded geographically as human populations grew, extraction and distribution technologies improved, and increasingly global markets emerged. Here, we argue that management strategies for fisheries and other ocean resources need to consider not just shifting baselines and the historical ecology of marine ecosystems, but the "shifting timelines" that emerge from the knowledge that the history of boats, maritime migrations, and marine fishing and hunting developed considerably earlier than previously believed in many parts of the world (Erlandson and Fitzpatrick 2006).

Like most archaeologists, we have long thought of ourselves as pretty interdisciplinary guys, with a relatively good grasp on the geology, biology, and ecology of Pacific Coast ecosystems. A few years ago, one of us (JME) was asked to join a working group (Long-Term Ecological Records of Marine Ecosystems) at the National Center for Ecological Analysis and Synthesis (NCEAS). Chaired by Jeremy Jackson, this group of 25 marine scientists met periodically for several years to discuss, debate, and reconstruct regional and global patterns in the historical development of human impacts on marine ecosystems. Several members of that group—Bruce Bourque, Debbie Corbett, Jim Estes, Mike Graham, and Bob Steneck—are participating in one way or another in this volume. After joining the NCEAS group, Erlandson learned that his knowledge of biological and ecological issues was as broad as a river but shallow as a thin sheet of water. It also became clear that most ecologists think of "long-term" ecological records as spanning two or three decades of regular scientific observation, sometimes supplemented with a few decades more of historical catch data. When the archaeologists in the group described coastal archaeological records spanning millennia in California, the Aleutians, and the Gulf of Maine, the marine ecologists were amazed at their potential. When the marine ecologists described the structure and dynamics of kelp forest ecosystems in each area, it was the archaeologists' turn to be astonished. What followed was a remarkable series of meetings with these and other colleagues examining the historical ecology of coastal ecosystems and the changing nature of human impacts to such systems through time.

Growing out of these interdisciplinary sessions, and several years of related research they inspired, we also gained clearer insight into just how much coastal archaeology has to offer current research, debates, and policy related to marine conservation and restoration, fisheries management, and other crucial ocean issues. As Lyman and Cannon (2004) argued, the application of archaeological data to current problems in conservation biology raises a number of thorny issues, but it also represents a tremendous opportunity for archaeologists to better meet the crucial test of relevance for modern society. Already widely perceived in the real world as an interesting but esoteric field, archaeologists cannot afford to miss opportunities to prove our relevance to current and pressing issues, not if we hope to receive continued funding for our work and support for the protection of archaeological sites.

THE ARCHAEOLOGY OF HUMAN IMPACTS

> The archaeological record encodes hundreds of situations in which societies were able to develop long-term sustainable relationships with their environments, and thousands of situations in which the relationships were short-lived and mutually destructive. The archaeological record is "strewn with the wrecks" of communities that obviously had not learned to cope with their environment in a sustainable manner or had found a sustainable path, but veered from it only to face self-destruction.
>
> (Redman 1999:4–5)

In recent decades, there has been tremendous interest in the archaeological study of human impacts on ancient ecosystems, including the role early hunter-gatherers and agriculturalists played in the extinction of animal species, major habitat alterations, and the collapse of complex cultures (e.g., Grayson 2001; Grayson and Meltzer 2002, 2003; Lyman and Cannon 2004; Martin 1967; Martin and Klein 1984; Redman 1999; Redman et al. 2004). Far from "living in harmony" with their environments—if any scientists ever actually believed this notion—the arrival of anatomically modern humans *(Homo sapiens sapiens)* in newly colonized lands often appears to be associated with significant habitat changes and accelerated rates of extinction. As Grayson (1991, 2001; Grayson and Meltzer 2003) and others have shown, it can be difficult to prove that human hunting was the *primary* cause of many animal extinctions. Nonetheless, cases for a significant human contribution to extinction events or other environmental impacts have been made for late Pleistocene Australia (Flannery 1994; Miller et al. 1999; Roberts et al. 2001), the Americas during the terminal Pleistocene (Alroy 2001; Martin 2002), some Caribbean Islands in the Early Holocene (Morgan and Woods 1986; Steadman et al. 2005), many Pacific Islands and Madagascar during the Late Holocene (Anderson 1984, 1989; Kirch and Hunt 1997; Simmons 1999; Steadman 1995, 2006), and other areas around the world. It is not always clear if such changes resulted from human hunting, forest clearance or intentional burning of landscapes, the introduction of exotic animals (dogs, pigs, rats, etc.) and the diseases they carried, or a combination of factors. What is clear, however, is that the initial arrival of behaviorally modern and technologically sophisticated *H. s. sapiens* in any given region appears to have posed significant problems for many endemic species, especially where such species had not been subjected to hominid predation before.

Although both continental and island landscapes have played important roles in studying the impacts of early humans, discussions of human-induced extinctions are limited almost exclusively to vertebrate species that live, nest, feed, or breed on land. In cataloging a global bestiary of large animals potentially driven to extinction by our ancestors, Martin (2002) listed scores of terrestrial genera, but no marine or aquatic species. As Martin (2002:1) noted: "Whatever caused large animal extinctions on land, had no impact on large mammals in the oceans." The differential patterns of prehistoric large animal extinctions in terrestrial versus aquatic ecosystems raise two very interesting questions that require further exploration: (1) why did the expansion of anatomically modern humans differentially affect terrestrial versus marine species?; and (2) does the lack of documented extinctions in marine ecosystems mean that human impacts on those ecosystems were absent or severely limited until very recently?

SHIFTING TIMELINES

In the past, the first question was easily answered with traditional anthropological theory, which held that hominids did not intensively exploit marine or aquatic resources until relatively late in human history—between about 15,000 and 5,000 years ago (e.g., Osborn 1977; Washburn and Lancaster 1968; Yesner 1987). Boats and other relatively sophisticated maritime technologies were also viewed as relatively recent developments, suggesting that most island archipelagoes were not colonized until relatively recently. The supposedly late shift toward intensive aquatic adaptations, in which marine resources were largely ignored for 99 percent of human history, was often explained as a shift toward "marginal" aquatic resources after the "optimal" large land mammals were hunted to extinction or severely depleted. If humans developed intensive marine fisheries relatively late in prehistory, then perhaps the dearth of aquatic extinctions is simply due to the differential antiquity of terrestrial hunting versus aquatic hunting and fishing.

The problem with this scenario, of course, is that we live in an interglacial period of high sea levels unmatched in the last 120,000 years. As global sea levels rose over 100 m between 20,000 and 5,000 years ago, the coastlines of the world changed dramatically. Most coastlines moved laterally tens or even hundreds of kilometers, meaning that most sites located along modern coastlines were far from marine habitats during the late Pleistocene (see Parkington 1981). Coastal erosion also destroyed, dispersed, or severely damaged most coastal sites submerged during this marine transgression (Erlandson 2001; Erlandson and Fitzpatrick 2006). Unfortunately, these dramatic changes in coastal geography coincide with the later development and spread of anatomically modern humans *(H. s. sapiens)*, insuring that records of early coastal adaptations or migrations are severely limited.

Despite such problems, recent research in a series of African sites of Last Interglacial age, other sites in coastal zones where steep bathymetry limited lateral shoreline movements, islands that humans could have colonized only by boat, and freshwater aquatic localities has dramatically altered our view of the antiquity of aquatic adaptations, maritime migrations, and human fishing and hunting in marine ecosystems. Dozens of coastal Middle Stone Age sites in southern Africa appear to contain evidence for the use of shellfish, marine mammals, seabirds, and even fish by anatomically modern humans between about 125,000 and 40,000 years ago (e.g., Brink and Deacon 1982; Henshilwood and Sealy 1997; Klein et al. 2004; Marean et al. 2004; Singer and Wymer 1982). Barbed bone harpoons from Katanda in Zaire suggest that sophisticated freshwater fishing technologies may have existed by 80,000 to 90,000 years ago (Yellen et al. 1995). The Pleistocene peopling of Australia and western Melanesia by maritime migrations through island Southeast Asia between roughly 60,000 and 35,000 years ago (see Allen et al. 1989; Clark 1991; Wickler and Spriggs 1988) suggests that our ancestors may have dispersed out of Africa by land and by sea, some of them following the coastlines of southern Asia (Erlandson 2001, 2002; Stringer 2000). The colonization of the Ryuku Islands and evidence for the use of boats in Japan 35,000 to 25,000 years ago suggests that early maritime peoples had adapted to the relatively cool waters of the Northwest Pacific during the Last Glacial. The settlement of California's Channel Islands by maritime peoples as much as 13,000 to 12,000 years ago (Erlandson et al. 1996; Johnson et al. 2002; Rick et al. 2001, 2005), as well as coastal shell middens on the Andean Coast more than 11,000 years old (Keefer et al. 1998; Richardson 1998; Sandweiss et al. 1998), also demonstrates an earlier and more sophisticated use of maritime or aquatic technologies than previously believed and may support early coastal colonization models. These data, along with the terminal Pleistocene expansion of maritime peoples into recently deglaciated coastal regions of Scandinavia,

FIGURE 1.1. Middle Holocene red abalone midden (SNI-161) eroding out of sand dune on northwestern San Nicolas Island, California (photo by René Vellanoweth).

the Early Holocene settlement of some Caribbean Islands, and the later but equally amazing maritime migrations of Austronesian and Polynesian peoples allow us to view the development of diversified coastal and aquatic economies around the world as the outgrowth of the origins and expansion of intellectually and technologically sophisticated anatomically modern humans in Africa some 150,000 years ago.

The early emergence of maritime peoples in many parts of the world also suggests that it is time to reexamine the archaeological evidence for human impacts in coastal environments. If humans have been gathering, hunting, and fishing in some marine ecosystems much longer than previously believed, then a relatively recent development of coastal economies cannot explain the lack of human-induced extinctions in those marine environments. With greater amounts of time for coastal peoples to expand their populations, increasing evidence for the early development of aquatic and maritime technologies, and a deeper history of colonizing many of the less remote island arcs around the world, we should expect to find earlier and more extensive evidence for prehistoric human impacts to coastal and marine ecosystems than previously believed possible.

LESSONS FROM THE PAST: OVEREXPLOITATION, CONSERVATION, AND SUSTAINABLE ECONOMIES

This brings us back to the question of whether the dearth of prehistoric extinctions in marine ecosystems implies a lack of significant human impacts. By now, the answer to this question should be obvious. We clearly believe that the limited evidence for human involvement in prehistoric aquatic extinctions should not be considered a lack of substantial impacts on ancient marine, estuarine, and freshwater species or ecosystems. In fact, a variety of evidence has been marshaled for a significant impact of early human predation on the size, distribution, or population structure of aquatic species, including shellfish, fish, and some marine mammals (see, among many others, Erlandson et al. 2005; Hildebrandt and Jones 1992; Klein et al. 2004; Reitz 2004; Simenstad et al. 1978; Steneck et al. 2004). The nature and extent of such impacts remain to be documented in various areas around the world, however; and the growing collaboration between coastal archaeologists, marine ecologists, and other scientists is developing new methods for recognizing and measuring the nature of those impacts (Figure 1.1).

Ancient humans did not live in complete harmony with their natural environments—at least not for long. By definition, at least for our purposes, all humans affect their environment, and large human populations generally have larger ecological impacts. Identifying archaeological evidence for such impacts, however, is not necessarily the same as demonstrating that some human societies never developed conservation measures or sustainable economies. If ancient peoples never learned from their mistakes and developed effective conservation strategies, there may be little to be learned from studying the past. While it may be true, as Kay (2002:259–260) has argued, that it is condescending and morally indefensible to claim that indigenous peoples were not capable of significant environmental impacts, it would be equally condescending to suggest that such people were incapable of recognizing the ecological impacts they had or developing practices that encouraged conservation or sustainable yields. As archaeologists and anthropologists, we find it difficult to believe that ecological awareness and conservation strategies are solely the province of literate or modern human societies.

Although some of our participants may touch on such issues, we are not particularly interested in debating theoretical stances about whether ancient peoples engaged in conservation, sustainable practices, or ecologically sound management principles. We have followed this debate closely, particularly as it has played out in relation to Native North Americans (e.g., Alvard 1998; Grayson 2001; Guthrie 1971; Hunn et al. 2003; Kay and Simmons 2002; Krech 1999). Such issues can be quite difficult to address using fragmentary archaeological data, and some of the rhetoric has been unnecessarily polarized and politicized. Native Americans, for instance, are variously portrayed as having either widespread cultural systems for conserving natural resources, or virtually no effective conservation strategies, with little middle ground. Such characterizations ignore the wide range of behaviors that might be expressed by a succession of individuals or groups within a single region over long stretches of archaeological time.

Over more than 10,000 years of Native American occupation along the California Coast, for instance, there is every reason to expect to find evidence for a wide range of harvest strategies, from overexploitation when resources are abundant, to more conservative practices when key resources became depleted. This is clearly the case with historical fishing strategies—where regulation is generally limited or absent until a fishery is significantly depleted. Nor is it correct to conclude that indigenous harvesting of small abalones or female seals or pups—forbidden by most historical regulations—is necessarily the antithesis of conservation or sustainable practices (see Porcasi et al. 2000; Raab 1992). Historical regulations were developed in response to specific harvest levels or practices that may not have applied to or been appropriate for prehistoric peoples. Relatively small hunter-gatherer populations might regularly harvest a limited number of female, juvenile, and infant seals from a large rookery, for instance, without a measurable decline in the local seal population. Under such demographic contexts, the hunting of female seals or their pups may have no bearing on the long-term sustainability of such practices. As populations grow, however, the impact on these resources will inevitably be more severe, perhaps causing the relocation of rookeries and possibly greater impacts on sea mammal population structure.

We are not arguing that prehistoric peoples did not cause resource depression or alter the structure of local ecosystems; in coastal zones around the world the archaeological evidence is overwhelming that they often did (see Anderson 1983; Broughton 1999; Butler 2000, 2001; Grayson 2001; Mannino and Thomas 2002; Nagaoka 2002). Yet the widespread evidence for localized resource depression does not necessarily signal a complete lack of conservation practices or sustainable economies among small-scale societies. As with swidden agriculture under low population densities, shifting residence patterns can cause the serial depletion of marine resources within local foraging territories without a long-term alteration of the larger ecosystem. In such cases, local resource depletion combined with a pattern of "shifting sedentism" might be part of a sustainable settlement and economic strategy that could span hundreds or even thousands of years.

As human populations fill their physical and social landscapes, of course, residential mobility can be constrained by neighboring groups. With the filling of spatial niches and the coalescence of foraging territories ("territorial circumscription") caused by human demographic expansion, impacts in local foraging territories may also coalesce into a regional depression of resources and increasingly anthropogenic landscapes and seascapes. It is here that archaeological investigations for the earliest evidence of significant regional human impacts to marine ecosystems are likely to be most fruitful. At the same time, such serious ecological impacts may sometimes have led to the development of social mechanisms that more effectively managed human harvest practices to encourage sustainable yields. Thus, it is conceivable that widespread resource depression, which some have argued was typical of much of Native North America prior to European contact (see Kay and Simmons 2002), gave rise to the conservation-oriented ecological management principles that many Native American tribes espoused after European contact—beliefs reinforced by the commercial decimation of many animal populations by rapacious colonial exploitation practices under European and Euroamerican regimes (see Ellis 2003; Mowatt 1984).

In our view, recent evolutionary theory often portrays individual humans as overly preoccupied with personal gain and reproductive success, ignoring the fact that human survival and success has most often been accomplished in group settings where people may maximize their success by adhering to communal decisions that benefit the larger group (see Ehrlich 2000:310). In human groups, elaborate cultural mechanisms (e.g., shame, ostracism, banishment, and death) are developed to control or punish those who unacceptably enrich themselves at the expense of the common good. Such strictures, imperfect as they were and are, operated both within and between social groups. The riverine peoples of the Pacific Northwest who blocked streams with weirs each year to harvest prodigious amounts of salmon knew, for instance, that blocking all the salmon from ascending the stream would have disastrous consequences for future fish runs, for their upstream neighbors, and ultimately for their own peace and prosperity. In such cases, the interests of individual tribal members and their larger social group were virtually inseparable from ecologically sustainable practices. This is not to suggest that there were not many difficult lessons learned in the long evolution of subsistence strategies and social relationships among such tribes.

Ultimately, suggesting that conservation did not take place in nonliterate societies implies that we have little to learn from the environmental relationships of smaller-scale cultures. Archaeological and historical records indicate that many human groups were incapable of surmounting the problems caused by their environmental impacts, while others were able to adjust their strategies (see Diamond 2005; Redman 1999). In the past, humans were confronted with countless environmental challenges—some of them of their own making—and responded in a variety of ways, both effectively and ineffectively. Learning what worked for those ancient peoples and what did not holds valuable lessons for us today as we strive to more effectively manage the environmental impacts of our species on both land and sea. Human impacts may be inevitable, but long-term environmental catastrophe and ecosystem collapse are not (see Kirch 1997; Redman 1999).

MEASURING HUMAN IMPACTS TO MARINE ECOSYSTEMS

In the absence of numerous extinctions of marine species linked to prehistoric human colonization, how do we recognize human impacts on ancient marine ecosystems? Other than the standard predictors and proxies of population growth or intensification, how can we use data from coastal archaeological sites to identify significant human impacts in marine environments? How do we compare archaeological data to historical and ecological data that are often collected in very different spatial and temporal scales? The answers to such questions are complex and, in some cases, not fully worked out. There are some general methods archaeologists and marine scientists have used, however, that provide insights into the nature of human interactions with marine ecosystems over long periods of time. At this point, any methodology for measuring human impacts across prehistoric, historic, and recent times must be considered a work in progress, since historical ecology is a very young discipline and interaction between archaeologists and marine scientists is still relatively limited.

Obviously, one of the first and most important issues is to differentiate natural (nonhuman) variations in marine ecosystems from those caused by humans (Redman 1999; Reitz and Wing 1999:252). On the ecological side of things, there are few if any marine ecosystems around the world where we have a comprehensive understanding of the historical ecology with any real time depth. Since careful, "long-term" ecological monitoring records rarely span more than a few decades, we do not yet know the full extent of cyclical fluctuations in most ecosystems that operate on decadal scales or longer. It is only relatively recently, for example, that the notion of decadal regime shifts or the dramatic and far-reaching climatic and oceanic patterns involved in El Niño/La Niña cycles have been fully recognized, and their historical parameters are still being defined. It is also increasingly clear that the historical range and behaviors of many marine species have changed significantly as their populations were devastated by early historical exploitation and, in some cases, have rapidly expanded under protective regulations or legislation. Knowledge of the geographic range and behavior of marine mammals along the Pacific Coast of North America—which ecologists and archaeologists have often extrapolated uncritically into the past—is rapidly changing, for instance, as various species continue to expand and adjust to dynamic oceanic conditions and the alteration of marine ecosystems caused by overfishing, habitat changes, and other human impacts (see Burton et al. 2001, 2002; DeLong and Melin 2002; Estes et al. 1998; Etnier 2004).

Marine ecologists desperately need archaeological colleagues to help expand their historical horizons, and coastal archaeologists desperately need marine ecologists to help interpret the ecological implications of our data and understand the broader import of our work. What are the ecological implications of changes in faunal remains found within archaeological sites? How do we distinguish between natural ecological fluctuations and those caused by humans? How susceptible is a particular marine ecosystem to human interference? These are questions ecologists are much more qualified and capable of answering than archaeologists are. How do we identify evidence for resource depletion in marine species? Intertidal shellfish beds are often considered to be highly susceptible to human overexploitation, but shellfish (and other) populations can also be destroyed or depleted by disease, nonhuman predation, sedimentation, heavy storms, changes in water temperature, and other problems unrelated to humans. For these and other reasons, we need to look for evidence of human impacts not just in individual archaeological sites but in regional records. To most effectively examine the impacts of humans on marine ecosystems, we should focus on long and nearly continuous sequences within relatively small areas, such as a 7,300-year-old sequence at Otter Point on San Miguel Island (Figure 1.2).

Paying particular attention to problems of temporal resolution and geographic scale, we must also find ways to compare archaeological data more effectively—often anchored by radiocarbon chronologies with resolution measured in centuries—to historical and ecological data that are often much more fine grained. For instance, northern elephant seals *(Mirounga angustirostris)*

FIGURE 1.2. A roughly 100 foot high Holocene dune in the Otter Point area on California's San Miguel Island, where at least 10 stratified shell midden components spanning the past 7,300 years have been identified (photo by J. Erlandson).

were abundant along the Baja and Alta California coast before being driven nearly to extinction during the first half of the nineteenth century by hunters who rendered their blubber into a commercial oil. Since Mexico officially protected a tiny relict population on Guadalupe Island in AD 1911, however, elephant seal populations along the Pacific Coast have recolonized much of their historical range and expanded to over 150,000 animals (Ellis 2003:193). Such a dramatic recovery, similar to the story of the California sea otter *(Enhydra lutris)*, is heartening for conservationists but raises fundamental questions about the articulation of archaeological, historical, and ecological records. If California Indians temporarily eradicated sea otters or elephant seals from their hunting territories or even the entire Channel Islands, could we recognize such rapid decline and recovery cycles in the archaeological record? These and other problems are yet to be resolved, but some of the primary methods archaeologists and ecologists are using to identify and understand human impacts on marine ecosystems are briefly summarized below.

Resource Depletion and Depression

For archaeologists, one of the most visible types of evidence for localized human impacts on marine fisheries may be found in cases of resource depression, or other changes in the types of resources people used through time. As predicted by foraging theory, humans entering a given environment will initially focus on harvesting a suite of optimal or "high-ranked" resources that provide relatively high nutritional or other (furs, raw materials for tools, etc.) yields (Grayson 2001). Although high-ranked resources are often assumed to be large animals, this is not always the case in aquatic environments, where relatively small shellfish or fish can often be mass harvested in large quantities. Because human groups often contain a diverse array of individuals (of different ages, sex, social class, etc.), and many coastal ecosystems offer a wide array of foods, the range of resources considered "optimal" by coastal groups may be diverse and related to population size. If intensive harvesting of high-ranked resources reduces their productivity (density, size, accessibility, etc.), people may choose to spend more travel time to access them, switch to lower-ranked alternatives closer to home, or develop a combination of strategies. Eventually, however, a depletion of local resources and the increased travel time invested in making a living may lead to the movement of a village or other residential base.

As mentioned earlier, cases of localized resource depression are relatively common in the archaeological record, especially in coastal areas where people were often relatively sedentary. Changes in the diversity or relative importance of subsistence resources harvested through time, especially within a single occupational component, may provide evidence for localized resource depression. Archaeologists must be careful, of course, to evaluate other possible causes of such changes, as numerous natural processes (storms, sedimentation, disease, other predators, water temperature, etc.) can lead to a reduction in the density of local marine populations. Noncultural variables such as climate change have recently been documented in two interior regions and appear to have affected human hunting and encounter rates (Byers et al. 2005; Wolverton 2005). Nonetheless, when combined with other evidence for human predation pressure, cases of people switching to different resources or resource depression may provide valuable evidence for human impacts on marine communities. We cannot automatically assume, however, that evidence for localized depletion is equivalent to wider degradation of an ecosystem, as heavy local exploitation can be combined with residential mobility in a sustainable economic strategy.

Size or Age Changes in Marine Populations

Historical data suggest that heavy fishing pressure on many fish and shellfish species often reduces the average size or age of local or regional populations. Such changes can have a disproportionate effect on the productivity of a species, since larger and older individuals nearly always are breeding adults and tend to lay more eggs or have more offspring than younger or smaller adults. In the past, such effects may sometimes have been alleviated by technological limitations such as the size or stability of boats, the strength of nets or fishing line, and so forth. The size and age structure of marine populations can also be affected by a variety of noncultural factors alluded to earlier, including predation by other animals, storm events, changes in water temperature, and marine productivity. Even the life histories of certain species, including changes in growth and maturation rates, can be altered by intensive predation (Reitz and Wing 1999:314).

Despite these problems, temporal changes in the average size or age of individuals from a particular fish or shellfish species are one of the simplest, most common, and valuable measures used by archaeologists to reconstruct shifts in human predation pressure and impacts in marine or aquatic ecosystems (see Broughton 2002; Butler 2001; Claassen 1998; Erlandson et al. 2004; Koike 1986; Lightfoot et al. 1993; Mannino and Thomas 2002; Swadling 1976). Sample size is a critical issue in such analyses, so that they are best focused on major prey species that are well represented in a series of strata or sites. Geographic and temporal variability in the size and age of local populations can also be a problem, so such analyses may best be applied to long and relatively continuous sequences within a single site or a relatively small area. However, Klein et al. (2004) recently used variation in average shellfish size from a relatively broad area of the South African Coast to effectively identify changes in predation intensity between Middle and Late Stone Age peoples. One of the great advantages of average size or age studies for zooarchaeological assemblages is that they can be readily compared to paleontological, historical, and recent ecological data sets to construct relatively long and continuous records of change in marine ecosystems (see Steneck and Carlton 2001).

Reductions in Geographic Range

One of the reasons marine animals have historically been more resistant to extinction caused by humans is that they often had geographic refuges where humans were incapable of capturing them, either in deeper offshore waters, remote and inaccessible stretches of coastline, or on offshore islands. As humans expanded around the globe and developed increasingly effective maritime technologies, these refuges gradually shrank and were probably increasingly limited to deeper waters and more remote islands that humans had not yet reached. Remote island populations of pinnipeds, seabirds, sea turtles, and other animals that spent part of their life cycle on land were especially vulnerable to human impact, as they often bred, laid eggs, roosted, or rested in terrestrial landscapes devoid of large predators. When maritime peoples first arrived on such islands, or recolonized them after sustained absences, they often had a heavy impact on such vertebrate populations.

If we can accurately reconstruct the distribution of such animal populations prior to the arrival of humans, the study of faunal assemblages from archaeological sites has great potential for understanding the impacts that both prehistoric and historic peoples had on the distribution of seabird colonies, pinniped rookeries, and other animal aggregations. A number of studies have implicated human hunting as the cause of impacts on the prehistoric distribution of pinniped rookeries and haul-outs in the Pacific (e.g., Anderson 2001; Bryden et al. 1999; Burton et al. 2002; Hildebrandt and Jones 1992, 2002; Jones and Hildebrandt 1995; Lyman 1995), for instance, and archaeological and historical accounts have both contributed to an understanding of the reduction in the geographic range of walrus in the North Atlantic (see Mowatt 1984).

We should be cautious about attributing changes in the distributions of such animals solely to human impacts, however, because a variety of other processes can affect both their density and distributions. Along the Pacific Coast of North America, the geographic range, feeding patterns, and behavior of marine mammals have been altered by the historical decimation of their own populations as well as those of their predators and prey (see Burton et al. 2002; Estes et al. 1998; Etnier 2004). Hildebrandt and Jones (1992) proposed that early hunting by Native Americans eradicated numerous mainland rookeries, for instance, but mainland rookeries were probably always rare because of their vulnerability to grizzly bears and other predators (Erlandson et al. 1998:12). Coastlines are also extremely dynamic and were even more so with rapidly rising postglacial sea levels, so that coastal erosion may also have destroyed many small islets or islands that once contained nesting colonies, rookeries, and haul-outs. Finally, it is not always clear that the naive behavior of some modern animal populations would have persisted under sustained human hunting, and some seabird, pinniped, or other colonies may have been abandoned for more remote locations, without a regional depletion of an animal population.

Trophic Cascades

One of the basic tenets of ecology is that the components of an ecosystem are inextricably linked—one component does not change without affecting others—although the linkages may not always be immediate or easily recognized. Given this fact, we should expect heavy human predation on a particular marine species to have a corresponding effect on the competitors, prey, or predators that the depleted species strongly interacted with (Suchanek 1994). In some cases, including those pinnipeds or seabirds that feed in very deep waters or far offshore, a reduction in local population may have only minor effects on nearshore coastal ecosystems. In others cases, however, a reduction in some "keystone" species can have dramatic effects that set off "trophic cascades" or create "alternate stable states" within coastal ecosystems.

Trophic cascades caused by human overfishing have been documented in North American kelp forest ecosystems from the Aleutians, California, and the Gulf of Maine (Estes et al. 1998; Jackson et al. 2001; Simenstad et al. 1978; Steneck et al. 2002, 2004), where the removal of apex predators such as cod and sea otters caused dramatic regime shifts in local and regional nearshore ecosystems. These case studies demonstrate the diversity of responses to human impacts in similar ecosystems found in various regions. In the North Pacific, the most profound and immediate impacts appear to have occurred in Aleutian kelp forest ecosystems, where species diversity is relatively low and the removal of a single "keystone" predator (sea otters) can have dramatic effects, creating trophic cascades and alternate stable state communities. In the Aleutians, the removal of sea otters by Aleut and Russian hunters allowed a rapid proliferation of sea urchins that overgrazed nearshore kelp forests and created "urchin barrens" that support a much less productive and diverse suite of marine resources. Several cycles of kelp deforestation have been documented in the Aleutians, the most recent being the result of heavy killer whale predation on sea otters (Estes et al. 1998; Steneck et al. 2002).

In the more diverse food webs of the southern California Coast, in contrast, the eradication of sea otters from much of their historical range during the early 1800s had dramatic effects on nearshore ecosystems but never caused the wholesale collapse of kelp forests, probably because other predators such as sheephead and lobsters helped keep urchin populations in check. On San Miguel Island, however, understanding the dynamics of such trophic interactions has allowed us to tentatively identify some ecological changes that may signal localized impacts by Chumash Indians thousands of years ago. One of these is found in the proliferation of large red abalone middens between about 7,500 and 3,000 years ago, an archaeological site type that modern ecological data suggest could not exist unless sea otter populations were held in check, probably by native hunting (Erlandson et al. 2005).

In cases like that of the sea otter, which strongly influence the structure of nearshore biological communities, understanding the ecological consequences of heavy marine fishing or hunting of keystone species in coastal ecosystems can help develop a series of predictions for what related changes might be visible in the archaeological record. In many cases, such predictions can be developed from modern ecological studies and historical fisheries data that can also provide strong support for archaeological inferences about ecological changes in marine ecosystems caused or contributed to by humans. In our experience, however, the active participation of marine ecologists is a crucial component of such modeling and analyses.

Fishing Down Food Webs

A relatively new quantitative method for understanding human impacts on fisheries was introduced by Pauly et al. (1998) in an influential quantitative synthesis of marine fishing practices in recent historical times. Using twentieth-century fisheries and ecological data on the average trophic level of economically important species, Pauly et al. (1998) identified a pattern of declining average trophic level in regional and global fisheries over time. They argued that this pattern reflected an intensive early focus of most commercial fisheries on relatively large and long-lived carnivores (e.g., cod, haddock, tuna, swordfish). When these fisheries declined or collapsed, the emphasis of commercial fishing switched to higher proportions of smaller fish (herring etc.), invertebrates (lobster, shrimp, shellfish, etc.), and other organisms that generally fill the lower trophic levels of ecosystems. Historically, sustained overfishing of some key predatory species (cod etc.) can lead to "ecological extinction" or "ecological ghosts," where a species is still present in an ecosystem but its numbers are so depleted that it no longer fills its normal ecological role. In cases like that of the sea otter, this can lead to the creation of trophic cascades and dramatic phase shifts in marine ecosystems, such as those described above for the Aleutian Islands (Estes et al. 1998; Simenstad et al. 1978; Steneck et al. 2002). Another classic example is the historical overexploitation of the Atlantic cod *(Gadus morhua)* and other large apex predators in the Gulf of Maine and the western North Atlantic (see Jackson et al. 2001; Steneck et al. 2002, 2004). Here, the ecological extinction of large nearshore fish released predatory controls on herbivorous sea urchins, which then greatly reduced the productive three-dimensional kelp forest habitats that appear to have dominated a relatively stable ecosystem for thousands of years. After heavy commercial fishing of sea urchins began in the late 1980s, however, kelp forests returned to the ecosystem, but the apex predator role is now filled by invertebrates (crabs). In this process of "accelerating trophic-level dysfunction," the average trophic level of marine fisheries has declined substantially (Steneck et al. 2004).

It is still not clear how much such historical examples may apply to prehistoric fishing practices, but the use of quantitative trophic level analysis has the potential to help bridge the gap between archaeological, historical, and ecological data on human impacts to marine fisheries (see Morales and Rosello 2004; Reitz 2004:63). For ecologists and fisheries managers, it provides a technique to explore changes in marine fisheries over much greater time depths and reexamine the shallow historical baselines on which fisheries management policy has long been based. As Reitz (2004:63) noted, the technique provides archaeologists a new perspective for understanding changes in archaeological fish faunas and an opportunity to use archaeological data to help restore marine ecosystems. So far, archaeological applications of trophic level analysis have been limited, but work by Reitz (2004) and Steneck et al. (2004) has provided important case studies for coastal archaeologists and marine ecologists to build on. We should not expect, however, that the patterns of the twentieth century will necessarily hold true through long periods of archaeological time. On California's Channel Islands, for instance, one of the secrets to the relative stability of indigenous fisheries over 10,000 years (Erlandson et al. 2005; Rick et al., this volume) may be that the Chumash and their ancestors appear to have focused first on the lower trophic levels, relying heavily on shellfish and smaller nearshore fish during the Early and Middle Holocene.

SUMMARY AND CONCLUSIONS

Throughout human history, the oceans generally appear to have been more resistant to human impact and degradation than terrestrial ecosystems. This fact is probably due to a combination of factors, including the longer history of hominid exploitation of terrestrial landscapes, the susceptibility of terrestrial landscapes to fire, the limited physiological and technological ability of humans (or their dogs, rats, pigs, etc.) to access deeper or more remote aquatic habitats, and the greater resistance of most marine organisms to diseases carried by humans and our domesticated companions. Our hominid ancestors may have used marine and aquatic resources to some extent for millions of years, but archaeological and anthropological data suggest that the intensity, diversity, and technological sophistication of aquatic resource use increased significantly after the appearance of anatomically modern humans roughly 150,000 years ago (Erlandson 2001; Erlandson and Fitzpatrick 2006).

Through the comparative approach of historical ecology, archaeologists have the opportunity to evaluate the evolution of human impacts on marine ecosystems through time and space. In the process, we can contribute valuable insights into one of the most important ecological problems currently facing humanity. In so doing, we can strengthen the relevance of archaeology in the modern world, as well as the arguments for increased protection of archaeological sites and increased research on the archaeology of coastal societies around the world. In the process, however, we should be cautious in how we interpret archaeological evidence for human impacts in marine ecosystems. We should be equally cautious in our use of historical and ecological data on the demography of marine species and the structure of past ecosystems, for in many cases historical or modern patterns have been affected by centuries or millennia of anthropogenic influence. Differences in both temporal and spatial scales should also be carefully considered, and further work needs to be done on the methods and theory required to effectively integrate paleontological, archaeological, historical, and ecological data sets.

For archaeologists and ecologists, evidence for human impacts on marine fisheries may be found in cases of resource depression, depletion, or shifting, changes in abundance and geographic distribution, reductions in size or age profiles for specific populations, in trophic cascades, or in changes in the average trophic level of marine species harvested. Understanding the nature of trophic cascades or other recent impacts documented with historical or ecological data can provide models that can help us understand the potential impacts of humans on marine ecosystems. On California's Channel Islands, for instance, marine ecologists have provided new insights on old archaeological problems by helping us better understand the dynamic historical ecological linkages between humans, sea otters, abalones, and sea urchins, and some previously unsuspected impacts of the Island Chumash on kelp forest ecosystems (Erlandson et al. 2004, 2005; Rick and Erlandson 2003). At the same time, we have learned that the biological diversity of Channel Island kelp forests makes them considerably more resistant to ecological collapse than the less-diverse kelp forest ecosystems of the Aleutian Islands and Gulf of Maine (Jackson et al. 2001; Steneck et al. 2002), helping explain why the Chumash and their ancestors may have had relatively limited impacts during a history of systematic fishing and hunting that spans more than 10,000 years (see Erlandson et al. 2005; Rick et al. this volume). Anderson (2001) and Kirch (1997) have made similar points about variation in the productivity, diversity, and resilience of different Pacific Island ecosystems. Before we indict some human societies for excessive environmental degradation and celebrate others for their sustainability, we would do well to thoroughly understand the ecological and historical underpinnings of their successes and failures (see Rainbird 2002).

Despite tremendous variation in coastal cultures and ecosystems around the world—and what we suspect are a variety of adaptive trajectories that vary widely in their success—it seems reasonably clear that there is a general geographic expansion of human impacts to marine ecosystems over time. These began in supratidal and intertidal zones, expanded to subtidal and nearshore waters, then to pelagic zones not far from land, to island arcs more and more distant from the continents, to the vast and relatively empty expanses of the Atlantic, Pacific, Indian, and other oceans. Although these large oceanic expanses were traversed by Austronesians/Polynesians, Vikings, and others, their resources were largely untouched by humans until the advent of whaling and other industrial fishing technologies of the eighteenth, nineteenth, and twentieth centuries. Archaeology is obviously positioned to illuminate the expansion of marine, estuarine, and freshwater fisheries during prehistoric times, but it may also shed considerable light on some poorly documented fisheries of the historic era, such as the Chinese abalone fishers of Alta and Baja California in the mid- to late 1800s (see Braje et al. 2007). As a variety of studies have shown (see Jackson et al. 2001; Steneck et al. 2002, 2004; and the case studies that follow), archaeological data can play a key role in defining the acceleration of human impacts to marine ecosystems through time, the development of more realistic notions of the abundance of past populations prior to industrialized fishing, and the reconstruction of more effective historical baselines for the future restoration and management of aquatic fisheries and ecosystems. With this volume, we hope to highlight that potential, encourage further interdisciplinary work between coastal archaeologists and marine ecologists, and inspire ecologists, archaeologists, and others to pursue new directions in their research.

REFERENCES CITED

Allen, J., C. Gosden, and J. P. White
1989 Pleistocene New Ireland. *Antiquity* 63: 548–561.

Alroy, J.
2001 A Multispecies Overkill Simulation of the End-Pleistocene Megafaunal Mass Extinction. *Science* 292:1893–1896.

Alvard, M. S.
1998 Evolutionary Ecology and Resource Conservation. *Evolutionary Anthropology* 7:62–74.

Anderson, A. J.
1983 Faunal Depletion and Subsistence Change in the Early Prehistory of Southern New Zealand. *Archaeology in Oceania* 18:1–10.
1984 The Extinction of Moa in Southern New Zealand. In *Quaternary Extinctions*, edited by P. S. Martin and R. G. Klein, pp. 728–740. University of Arizona Press, Tucson.
1989 Mechanics of Overkill in the Extinction of New Zealand Moas. *Journal of Archaeological Science* 16:137–151.
2001 No Meat on that Beautiful Shore: The Prehistoric Abandonment of Subtropical Polynesian Islands. *International Journal of Osteoarchaeology* 11:14–23.

Braje, T. J., J. M. Erlandson, and T. C. Rick
2007 An Historic Chinese Abalone Fishery on San Miguel Island, California. *Historical Archaeology* 41(4): 115–125.

Brink, J. S., and H. J. Deacon
1982 A Study of a Last Interglacial Shell Midden and Bone Accumulation at Herold's Bay, Cape Province, South Africa. *Paleoecology of Africa* 15:31–39.

Broughton, J. M.
1999 *Resource Depression and Intensification during the Late Holocene, San Francisco Bay.* Anthropological Records 32. University of California Press, Berkeley.
2002 Prey Spatial Structure and Behavior Affect Archaeological Tests of Optimal Foraging Models: Examples from the Emeryville Shellmound Vertebrate Fauna. *World Archaeology* 34:60–83.

Bryden, M. M., S. O. O'Connor, and R. Jones
1999 Archaeological Evidence for the Extinction of a Breeding Population of Elephant Seals in Tasmania in Prehistoric Times. *International Journal of Osteoarchaeology* 9:430–437.

Burton, R. K., D. Gifford-Gonzalez, J. J. Snodgrass, and P. L. Koch
2002 Isotopic Tracking of Prehistoric Pinniped Foraging and Distribution along the Central California Coast: Preliminary Results. *International Journal of Osteoarchaeology* 12:4–11.

Burton, R. K., J. J. Snodgrass, D. Gifford-Gonzalez, T. Guilderson, T. Brown, and P. L. Koch
2001 Holocene Changes in the Ecology of Northern Fur Seals: Insights from Stable Isotopes and Archaeofauna. *Oecologia* 128:107–115.

Butler, V. L.
2000 Resource Depression on the Northwest Coast of North America. *Antiquity* 74:649–661.
2001 Changing Fish Use in Mangaia, Southern Cook Islands: Resource Depression and the Prey Choice Model. *International Journal of Osteoarchaeology* 11:88–100.

Byers, D. A., C. S. Smith, and J. Broughton
2005 Historical Artiodactyl Populations and Large Game Hunting in the Wyoming Basin, U.S.A. *Journal of Archaeological Science* 32:125–142.

Carlton, J. T., J. B. Geller, M. L. Reaka-Kudla, and E. A. Norse
1999 Historical Extinctions in the Sea. *Annual Review of Ecology and Systematics* 30:515–536–538.

Claassen, C. L.
1998 *Shells.* Cambridge University Press, Cambridge, U.K.

Clark, J. T.
1991 Early settlement of the Indo-Pacific. *Journal of Anthropological Archaeology* 10:27–53.

Dayton, P. K., M. J. Tegner, P. B. Edwards, and K. L. Riser
1998 Sliding Baselines, Ghosts, and Reduced Expectations in Kelp Forest Communities. *Ecological Applications* 8:309–322.

DeLong, R. L., and S. R. Melin
2002 Thirty Years of Pinniped Research at San Miguel Island. In *Proceedings of the Fifth California Islands Symposium*, edited by D. R. Browne, K. L. Mitchell, and H. W. Chaney, pp. 401–406. Santa Barbara Museum of Natural History, Santa Barbara.

Diamond, J.
2005 *Collapse: How Societies Choose to Fail or Succeed.* Viking, New York.

Earle, S. A.
1995 *Sea Change: A Message of the Oceans.* Fawcett/Columbine, New York.

Ehrlich, P. R.
2000 *Human Natures.* Island Press, Washington, D.C.

Ellis, R.
2003 *The Empty Ocean: Plundering the World's Marine Life.* Island Press, Washington, D.C.

Erlandson, J. M.
2001 The *Archaeology* of Aquatic Adaptations: Paradigms for a New Millennium. *Journal of Archaeological Research* 9:287–350.

2002 Anatomically Modern Humans, Maritime Adaptations, and the Peopling of the New World. In *The First Americans: The Pleistocene Colonization of the New World*, edited by N. Jablonski, pp. 59–92. Memoirs of the California Academy of Sciences, San Francisco.

Erlandson, J. M., and S. M. Fitzpatrick

2006 Oceans, Islands, and Coasts: Current Perspectives on the Role of the Sea in Human Prehistory. *Journal of Island and Coastal Archaeology* 1(1):5–33.

Erlandson, J. M, D. J. Kennett, B. L. Ingram, D. A. Guthrie, D. P. Morris, M. A. Tveskov, G. J. West, and P. L. Walker

1996 An Archaeological and Paleontological Chronology for Daisy Cave (CA-SMI-261), San Miguel Island, California. *Radiocarbon* 38:355–373.

Erlandson, J. M., M. A. Tveskov, and R. S. Byram

1998 The Development of Maritime Adaptations on the Southern Northwest Coast of North America. *Arctic Anthropology* 35(1):6–22.

Erlandson, J. M., T. C. Rick, and R. L. Vellanoweth

2004 Human Impacts on Ancient Environments: A Case Study from California's Northern Channel Islands. In *Voyages of Discovery: The Archaeology of Islands*, edited by S. M. Fitzpatrick, pp. 51–83. Praeger, New York.

Erlandson, J. M., T. C. Rick, J. A. Estes, M. H. Graham, T. J. Braje, and R. L. Vellanoweth

2005 Sea Otters, Shellfish, and Humans: A 10,000 Year Record from San Miguel Island, California. In *Proceedings of the Sixth California Islands Symposium*, edited by D. Garcelon and C. Schwemm, pp. 9–21. National Park Service Technical Publication CHIS-05-01. Institute for Wildlife Studies, Arcata, California.

Estes, J. A., M. T. Tinker, T. M. Williams, and D. F. Doak

1998 Killer Whale Predation on Sea Otters Linking Oceanic and Nearshore Ecosystems. *Science* 282:473–476.

Etnier, M. A.

2004 The Potential of Zooarchaeological Data to Guide Pinniped Management Decisions in the Eastern North Pacific. In *Zooarchaeology and Conservation Biology*, edited by R. L. Lyman and K. P. Cannon, pp. 88–102. University of Utah Press, Salt Lake City.

Flannery, T. F.

1994 *The Future Eaters: An Ecological History of Australasian Lands and People*. George Braziller, New York.

Grayson, D. K.

1991 Late Pleistocene Mammalian Extinctions in North America: Taxonomy, Chronology, and Explanations. *Journal of World Prehistory* 5:193–231.

2001 The Archaeological Record of Human Impacts on Animal Populations. *Journal of World Prehistory* 15:1–68.

Grayson, D. K., and D. J. Meltzer

2002 Clovis Hunting and Large Mammal Extinction: A Critical Review of the Evidence. *Journal of World Prehistory* 16:313–359.

2003 A Requiem for North American Overkill. *Journal of Archaeological Science* 30:585–593.

Guthrie, D. A.

1971 Primitive Man's Relationship to Nature. *Bioscience* 21:721–723.

Henshilwood, C., and J. Sealy

1997 Bone Artefacts from the Middle Stone Age at Blombos Cave, Southern Cape, South Africa. *Current Anthropology* 38:890–895.

Hildebrandt, W. R., and T. L. Jones

1992 Evolution of Marine Mammal Hunting: A View from the California and Oregon Coasts. *Journal of Archaeological Anthropology* 11:360–401.

2002 Depletion of Prehistoric Pinniped Populations along the California and Oregon Coasts: Were Humans the Cause? In *Wilderness and Political Ecology: Aboriginal Influences and the Original State of Nature*, edited by C. F. Kay and R. T. Simmons, pp. 72–110. University of Utah Press, Salt Lake City.

Hunn, E. S., D. Johnson, P. Russell, and T. F. Thornton

2003 Huna Tlingit Traditional Environmental Knowledge and the Management of a "Wilderness Park." *Current Anthropology* 44:79–104.

Jackson, J. B. C.

2001 What Was Natural in the Coastal Oceans? *Proceedings of the National Academy of Sciences* 98:5411–5418.

Jackson, J. B. C., M. X. Kirby, W. H. Berger, K. A. Bjorndal, L. W. Botsford, B. J. Bourque, R. H. Bradbury, R. Cooke, J. Erlandson, J. A. Estes, T. P. Hughes, S. Kidwell, C. B. Lange, H. S. Lenihan, J. M. Pandolfi, C. H. Peterson, R. S. Steneck, M. J. Tegner, and R. R. Warner

2001 Historical Overfishing and the Recent Collapse of Coastal Ecosystems. *Science* 293: 629–638.

Johnson, J. R., T. W. Stafford, H. O. Ajie, and D. P. Morris

2002 Arlington Springs Revisited. In *Proceedings of the Fifth California Islands Symposium*, pp. 541–545. Santa Barbara Museum of Natural History, Santa Barbara.

Jones, T. L., and W. R. Hildebrandt

1995 Reasserting a Prehistoric Tragedy of the Commons: Reply to Lyman. *Journal of Anthropological Archaeology* 14:78–98.

Kay, C. E.
2002 False Gods, Ecological Myths, and Biological Reality. In *Wilderness and Political Ecology: Aboriginal Influences and the Original State of Nature*, edited by C. E. Kay and R. T. Simmons, pp. 238–261. University of Utah Press, Salt Lake City.

Kay, C. E., and R. T. Simmons
2002 Preface. In *Wilderness and Political Ecology: Aboriginal Influences and the Original State of Nature*, edited by C. E. Kay and R. T. Simmons, pp. xi–xix. University of Utah Press, Salt Lake City.

Keefer, D. K., S. D. deFrance, M. E. Moseley, J. B. Richardson, III, D. R. Satterlee, and A. Day-Lewis
1998 Early Maritime Economy and El Niño Events at Quebrada Tacahuay, Peru. *Science* 281:1833–1835.

Kirch, P. V.
1997 Microcosmic Histories: Island Perspectives on "Global" Change. *American Anthropologist* 99:30–42.

Kirch, P. V., and T. L. Hunt (editors)
1997 *Historical Ecology in the Pacific Islands: Prehistoric Environmental and Landscape Change*. Yale University Press, New Haven, Connecticut.

Klein, R. G., G. Avery, K. Cruz-Uribe, D. Halkett, J. E. Parkington, T. Steele, T. P. Volman, and R. Yates
2004 The Ysterfontein 1 Middle Stone Age Site, South Africa, and Early Human Exploitation of Coastal Resources. *Proceedings of the National Academy of Sciences* 101:5708–5715.

Koike, H.
1986 Prehistoric Hunting Pressure and Paleobiomass: An Environmental Reconstruction and Archaeo-zoological Analysis of a Jomon Shellmound Area. In *Prehistoric Hunter-Gatherers in Japan: New Research Methods*, edited by T. Akazawa and C. M. Aikens. *University of Tokyo Museum Bulletin* 27:27–53.

Krech, S.
1999 *The Ecological Indian: Myth and History*. W. W. Norton, New York.

Lightfoot, K. G., R. M. Cerrato, and H. V. E. Wallace
1993 Prehistoric Shellfish-Harvesting Strategies: Implications from the Growth of Soft-Shell Clams *(Mya arenaria)*. *Antiquity* 67:358–369.

Lyman, R. L.
1995 On the Evolution of Marine Mammal Hunting on the West Coast of North America. *Journal of Anthropological Archaeology* 14:45–77.

Lyman, R. L., and K. P. Cannon
2004 Applied Zooarchaeology, Because It Matters. In *Archaeology and Conservation Biology*, edited by R. L. Lyman and K. P. Cannon, pp. 1–24. University of Utah Press, Salt Lake City.

Lyman, R. L., and K. P. Cannon (editors)
2004 *Archaeology and Conservation Biology*. University of Utah Press, Salt Lake City.

Mannino, M. A., and K. D. Thomas
2002 Depletion of a Resource? The Impact of Prehistoric Human Foraging on Intertidal Mollusc Communities and Its Significance for Human Settlement, Mobility and Dispersal. *World Archaeology* 33:452–474.

Marean, C. W., P. J. Nilssen, K. Brown, A. Jerardino, and D. Stynder
2004 Paleoanthropological Investigations of Middle Stone Age Sites at Pinnacle Point, Mossel Bay (South Africa): Archaeology and Hominid Remains from the 2000 field Season. *Paleo-Anthropology* 5(2):14–83.

Martin, P.S.
1967 Pleistocene Overkill. *Natural History* 76(10):32–38.

Martin, P. S.
2002 Prehistoric Extinctions: In the Shadow of Man. In *Wilderness Political Ecology: Aboriginal Influences and the Original State of Nature*, edited by C. E. Kay and R. T. Simmons, pp. 1–27. University of Utah Press, Salt Lake City.

Martin P. S., and R. G. Klein
1984 *Quaternary Extinctions*. University of Arizona Press, Tucson.

Miller, G. H., J. W. Magee, B. J. Johnson, M. L. Fogel, N. A. Spooner, M. T. McCulloch, and L. K. Ayliffe
1999 Pleistocene Extinction of *Genyornis newtoni*: Human Impact on Australian Megafauna. *Science* 283:205–208.

Morales, A., and E. Rosello
2004 Fishing down the Food Web in Iberian Prehistory? A New Look at the Fishes from Cueva de Nerja (Malaga, Spain). In *Pettis Animaux Et Societes Humaines Du Complement Alimentaire Aux Ressources Utilitaires XXIV*, edited by J. Brugal and J. Desse, pp. 111–123. APDCA, Antibes.

Morgan, G. S., and C. A. Woods
1986 Extinction and the Zoogeography of West Indian Land Mammals. *Biological Journal of the Linnean Society* 28:167–203.

Mowatt, F.
1984 *Sea of Slaughter*. Atlantic Monthly Press, Boston.

Nagaoka, L.
2002 The Effects of Resource Depression on Foraging Efficiency, Diet Breadth, and Patch Use in Southern New Zealand. *Journal of Anthropological Archaeology* 21:419–442.

Osborn, A. O.
1977 Strandloopers, Mermaids, and Other Fairy Tales: Ecological Determinants of Resource Utilization—the Peruvian Case. In *For Theory*

Building in Archaeology, edited by L. R. Binford, pp. 157–205. Academic Press, New York.

Parkington, J.
1981 The Effects of Environmental Change on the Scheduling of Visits to the Elands Bay Cave, Cape Province, S.A. In *Patterns of the Past*, edited by I. Hodder, G. Isaac, and N. Hammond, pp. 341–359. Cambridge University Press, Cambridge.

Pauly, D.
1995 Anecdotes and the Shifting Baselines Syndrome of Fisheries. *Trends in Ecology and Evolution* 10(10):430.

Pauly, D., V. Christensen, J. Dalsgaard, R. Froese, and F. Torres, Jr.
1998 Fishing down Marine Food Webs. *Science* 279:860–863.

Pauly, D., V. Christensen, S. Guénette T. J. Pitcher, U. R. Sumaila, C. Walters, R. Watson, and D. Zeller
2002 Toward Sustainability in World Fisheries. *Nature* 418:689–695.

Pew Oceans Commission
2003 *America's Living Oceans: Charting a Course for Sea Change.* Pew Oceans Commission, Arlington, Virginia.

Porcasi, J. F., T. L. Jones, and L. M. Raab
2000 Trans-Holocene Marine Mammal Exploitation on San Clemente Island, California: A Tragedy of the Commons Revisited. *Journal of Anthropological Archaeology* 19:200–220.

Raab, L. M.
1992 An Optimal Foraging Analysis of Prehistoric Shellfish Collecting on San Clemente Island, California. *Journal of Ethnobiology* 12:63–80.

Rainbird, P.
2002 A Message for Our Future? The Rapa Nui (Easter Island) Ecodisaster and Pacific Island Environments. *World Archaeology* 33:436–451.

Redman, C. L.
1999 *Human Impact on Ancient Environments.* University of Arizona Press, Tucson.

Redman, C. L., S. R. James, P. R. Fish, and J. D. Rogers (editors)
2004 *The Archaeology of Global Change: The Impact of Humans on Their Environment.* Smithsonian Books, Washington, D.C.

Reitz, E. J.
2004 "Fishing down the Food Web": A Case Study from St. Augustine, Florida, USA. *American Antiquity* 69:63–83.

Reitz, E. J., and E. S. Wing
1999 *Zooarchaeology.* Cambridge University Press, Cambridge.

Richardson, J. B., III
1998 Looking In the Right Places: Pre-5000 BP Maritime Adaptations in Peru and the Changing Environment. *Revista de Arqueologia Americana* 15:33–56.

Rick, T. C., and J. M. Erlandson
2003 Archeology, Ancient Human Impacts on the Environment, and Cultural Resource Management on Channel Islands National Park, California. *CRM: The Journal of Heritage Stewardship* 1:84–87.

Rick, T. C., J. M. Erlandson, and R. L. Vellanoweth
2001 Paleocoastal Marine Fishing on the Pacific Coast of the Americas: Perspectives from Daisy Cave, California. *American Antiquity* 66: 595–613.

Rick, T. C., J. M. Erlandson, R. L. Vellanoweth, and T. J. Braje
2005 From Pleistocene Mariners to Complex Hunter-gatherers: The Archaeology of the California Channel Islands. *Journal of World Prehistory* 19:169–228.

Roberts, C. M.
2002 Deep Impact: The Rising Toll of Fishing in the Deep Sea. *Trends in Ecology and Evolution* 17(5):242–245.

Roberts, R. G., T. F. Flannery, L. K. Ayliffe, H. Yoshida, J. M. Olley, G. J. Prideaux, G. M. Laslett, A. Baynes, M. A. Smith, R. Jones, and B. L. Smith
2001 New Ages for the Last Australian Megafauna: Continent-Wide Extinction about 46,000 Years Ago. *Science* 292:1888–1892.

Roman, J., and S. R. Palumbi
2003 Whales before Whaling in the North Atlantic. *Science* 301:508–510.

Sandweiss, D. H., H. McInnis, R. L. Burger, A. Cano, B. Ojeda, R. Paredes, M. D. C. Sandweiss, and M. Glascock
1998 Quebrada Jaguay: Early South American Maritime Adaptations. *Science* 281:1830–1833.

Simenstad, C. A., J. A. Estes, and K. W. Kenyon
1978 Aleuts, Sea Otters, and Alternate Stable-state Communities. *Science* 200:403–411.

Simmons, A. H.
1999 *Faunal Extinction in an Island Society: Pygmy Hippopotomus Hunters of Cyprus.* Kluwer Academic/Plenum, New York.

Singer, R., and J. J. Wymer
1982 *The Middle Stone Age at Klasies River Mouth in South Africa.* University of Chicago Press, Chicago.

Smith, E. A., and M. Wishnie
2000 Conservation and Subsistence in Small-scale Societies. *Annual Review of Anthropology* 29: 493–524.

Steadman, D. W.
1995 Prehistoric Extinctions of Pacific Island Birds: Biodiversity Meets Zooarchaeology. *Science* 267:1123–1131.

2006 *Extinction and Biogeography of Tropical Pacific Birds.* University of Chicago Press, Chicago.
Steadman, D. W., P. S. Martin, R. D. E. MacPhee, A. J. T. Jull, H. G. McDonald, C. A. Woods, M. Iturralde-Vinent, and G. W. L. Hodgins
2005 Asynchronous Extinction of Late Quaternary Sloths on Continents and Islands. *Proceedings of the National Academy of Sciences USA* 102: 11763–11768.
Steele, J. H.
1998 Regime Shifts in Marine Ecosystems. *Ecological Applications* 8(1,Supplement): S33–S36.
Steneck, R. S., and J. T. Carlton
2001 Human Alterations of Marine Communities: Students Beware! In *Marine Community Ecology*, edited by M. D. Bertness, S. D. Gaines, and M. E. Hay, pp. 445–468. Sinauer Associates, Sunderland, Massachusetts.
Steneck, R. S., M. H. Graham, B. J. Bourque, D. Corbett, J. M. Erlandson, J. A. Estes, and M. J. Tegner
2002 Kelp Forest Ecosystems: Biodiversity, Stability, Resilience, and Future. *Environmental Conservation* 29:436–459.
Steneck, R. S., J. Vavrinec, and A. V. Leland
2004 Accelerating Trophic-Level Dysfunction in Kelp Forest Ecosystems of the Western North Atlantic. *Ecosystems* 7:323–332.
Stringer, C.
2000 Coasting out of Africa. *Nature* 405:24–26.
Suchanek, T. H.
1994 Temperate Coastal Marine Communities: Biodiversity and Threats. *American Zoologist* 34:100–114.
Swadling, P. A.
1976 Changes Induced by Human Predation in Prehistoric Shellfish Populations. *Mankind* 10:156–162.
U.S. Commission on Ocean Policy
2004 An Ocean Blueprint for the Twenty-first Century. Final Report. U.S. Commission on Ocean Policy, Washington, D.C.
Vitousek, P. M, H. A. Mooney, J. Lubchenco, and J. M. Melillo
1997 Human Domination of Earth's Ecosystems. *Science* 277:494–499.
Washburn, S. L., and C. S. Lancaster
1968 The Evolution of Hunting. In *Man the Hunter*, edited by R. B. Lee and I. DeVore, pp. 293–303. Aldine, Chicago.
Wickler, S., and M. Spriggs
1988 Pleistocene Human Occupation of the Solomon Islands, Melanesia. *Antiquity* 62: 703–707.
Wolverton, S.
2005 The Effects of the Hypsithermal on Prehistoric Foraging Efficiency in Missouri. *American Antiquity* 70:91–106.
Woodard, C. S.
2000 *Ocean's End: Travels through Endangered Seas.* Basic Books, New York.
Worm, B., E. B. Barbier, N. Beaumont, J. E. Duffy, C. Folke, B. S. Halpern, J. B. C. Jackson, H. K. Lotze, F. Micheli, S. R. Palumbi, E. Sala, K. A. Selkoe, J. J. Stachowicz, and R. Watson
2006 Impacts of Biodiversity Loss on Ocean Ecosystem Services. *Science* 314:787–790.
Yellen, J. E., A. S. Brooks, E. Cornelissen, M. J. Mehlman, and K. Stewart
1995 A Middle Stone Age Worked Bone Industry from Katanda, Upper Semliki Valley, Zaire. *Science* 268:553–556.
Yesner, D. R.
1987 Life in the "Garden of Eden": Constraints of Marine Diets for Human Societies. In *Food and Evolution*, edited by M. Harris and E. Ross, pp. 285–310. Temple University Press, Philadelphia.

2

Short and Sometimes Sharp

HUMAN IMPACTS ON MARINE RESOURCES IN THE ARCHAEOLOGY AND HISTORY OF SOUTH POLYNESIA

Atholl Anderson

THE POLYNESIAN ISLANDS LIE in the central Pacific Ocean where sheer distance from continental margins, and prevailing easterly winds and currents severely restricted biotic diversity in native terrestrial taxa, as exemplified by the complete absence of mammals other than bats. Human colonists, their own inventory of cultigens and domesticates slimmed in much the same way, had to look substantially to marine resources, especially in the distinctive region of South Polynesia, which extends across a 6,500,000-km^2 area of the Pacific Ocean. South Polynesia encloses New Zealand and the outlying island groups that surround it at a 500- to 800-km distance (Figure 2.1).

The historical native populations of South Polynesia, Maori in New Zealand, and Moriori in the Chatham Islands, were closely related in lineage, language, and society to those of East Polynesia, but, for environmental reasons primarily, their historical experiences were significantly different. East Polynesian islands generally are small (Hawaii is an exception), of geologically young basalts and coral, and lie entirely within the tropics. South Polynesia is dominated by the relatively immense New Zealand islands (Table 2.1), which contain over 90 percent of the total Polynesian land area. Lying in the geologically continental southwest Pacific, and extending from subtropical (29° S) to subpolar (51° S) zones, South Polynesia has a broadly temperate environment with relatively ancient Gondwanan elements in its geology and biota.

In Polynesia as a whole, the recent advent of human colonization postdates all significant sea-level change. A high stand of 1 to 2 m above modern levels peaked about 4,000 years ago and was still receding when people reached West Polynesia (Samoa, Tonga, and nearby islands) about 3,000 years ago, so the earliest sites are found on relict beach ridges. The rest of Polynesia was colonized after sea levels became very close to modern (Dickinson 2001). Prehistoric settlement in declining or modern sea levels has the considerable archaeological advantage that the surviving sites are likely to sample the full chronological and geographical ranges of those originally deposited.

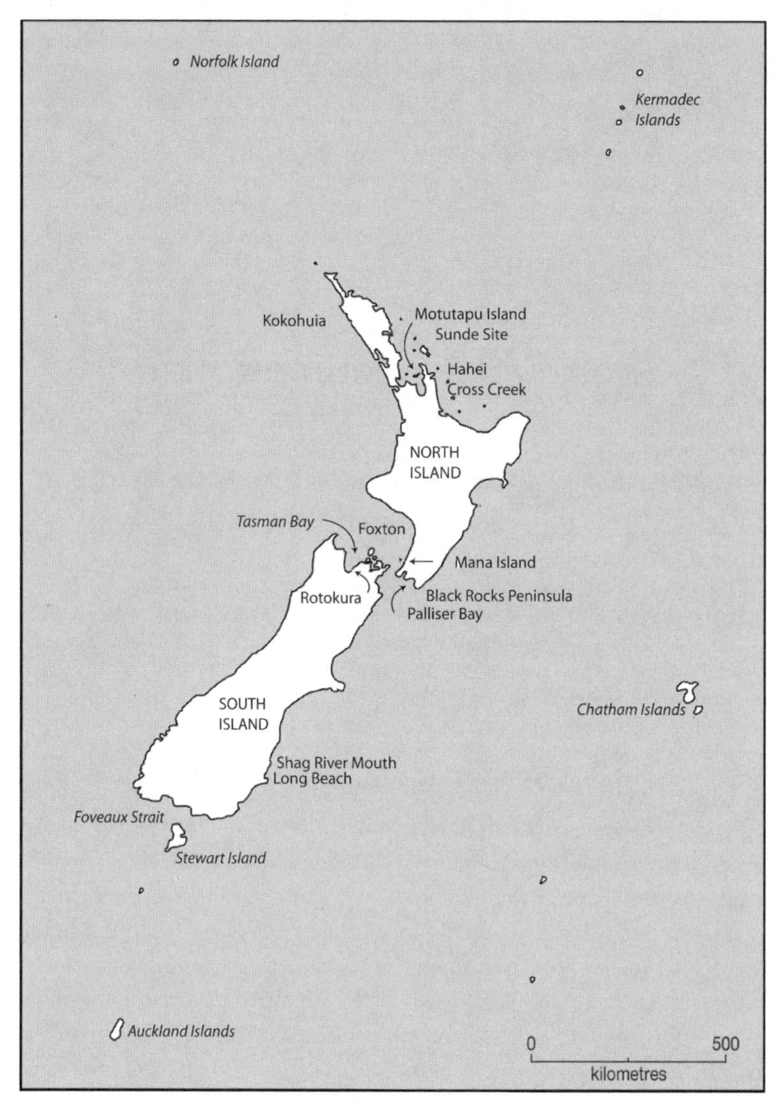

FIGURE 2.1. South Polynesia, showing islands and sites mentioned in the text. Lord Howe Island (not shown) lies southwest of Norfolk Island.

East Polynesia was probably settled initially between AD 700 and 900, but South Polynesia not until about AD 1100–1300, making New Zealand the last substantial landmass lying outside the polar zones to be colonized by people (Anderson 1991; Anderson and Sinoto 2002; Higham and Hogg 1997; McGlone and Wilmshurst 1999). The outlying archipelagos were reached in an early, and probably one-way, radial dispersal from New Zealand during the colonization era (Anderson 2000a, 2003, 2006). The Chatham Islands, like New Zealand, were occupied continuously, but colonization failed in the Norfolk and Kermadec Islands after perhaps 200 years, and it was confined to a very brief period in the Subantarctic Islands. Lord Howe Island, which is geographically within South Polynesia, was not reached until AD 1788 by European expeditions (Anderson 1980a, 2004, 2005; Anderson and White 2001).

At the arrival of Europeans in the eighteenth century, South Polynesia was culturally distinct. New Zealand had a comparatively low population

TABLE 2.1

South Polynesian Island Size, Sea Temperatures, and Number of Breeding Marine Species

GROUP	AREA (KM^2)	ANN. RANGE (°C) MEAN SEA TEMP.	SEALS	SEABIRDS	SEA FISH
Aucklands	626	6–10	3	23	<30
Chathams	1980	10–15	1	20	<50
New Zealand	267,000	9–20	3	37	650
Kermadecs	35	18–23	–	10	145
Norfolk	33	19–24	–	8	254

NOTE: Data from Falla et al. (1979), Paulin et al. (1989).

density (average .4 people per km^2, varying from .1 to 2.5, south to north), and there were about two people per square kilometer in the Chathams. This was more than an order of magnitude lower than the 30 or more people per square kilometer in East Polynesia generally (Anderson 1980b; McGlone et al. 1994). Additional characteristics such as mobile foraging systems, which also reflected the seasonality of temperate climates, comparatively unstratified chiefdoms, and an absence of the monumental religious architecture common in East Polynesia were also partly connected to low population density.

Few plants and animals had been introduced successfully to New Zealand. They included the Polynesian dog *(Canis familiaris)* and rat *(Rattus exulans)*, but not the pig or chicken. Horticulture was based on the sweet potato *(Ipomoea batatas)*, with only minor contributions from a restricted range of tropical cultigens that lacked breadfruit, banana, coconut, and other important Polynesian crops. Horticulture occurred in lowland areas in the north, but conditions were too cold for cultivation on most of the South Island, the Chathams, and the Subantarctic Islands. Consequently, South Polynesian prehistoric subsistence had to rely substantially on foraging, and marine resources were of major importance.

There are no coral reef and lagoon systems in South Polynesia, but subtidal coral occurs in the Kermadec and Norfolk groups where, despite the very small island size, it supports a typically diverse Indo-Pacific coral reef fauna. In these islands, and elsewhere in South Polynesia, coasts are generally exposed and comprise a mixture of sand beaches and rocky shores. Sheltered estuaries and harbors occur in New Zealand, and glaciated fjord coasts in southern New Zealand and the Auckland islands. Sea temperatures vary from subtropical to subpolar and marine communities covariantly (Table 2.1). Breeding seals (especially New Zealand fur seal, *Arctocephalus forsteri*) and seabirds are predominantly of southern distribution. Among the latter, four penguin species breed in southern New Zealand and the Aucklands, along with up to 20 species of Procellaridae, including four species of albatross. In South Polynesia as a whole, there are more than 200 families and 1,000 species of marine fish, but only about 300 of these are of reasonable size (>300 mm length) and found inshore (<100 m depth). The most speciose of those families are Squalidae, Moridae, Labridae, Serranidae, Scorpaenidae, Carangidae, Pleuronectidae, and Scombridae, all with 10 species or more (Paulin et al. 1989).

In New Zealand, the Chathams, and the Auckland islands, rocky coasts often support kelp communities of seals, fish (e.g., Chironemidae, Cheilodactylidae, Latrididae, Labridae, Odacidae), shellfish (e.g., Haliotidae, Mytilidae, Echinoidea), and rock lobster (*Jasus* sp.). Sand coasts and estuaries, which are more prominent in central and northern New Zealand, support substantial shellfish beds of economic importance (notably of Mesodesmatidae and Veneridae), together with schooling fish such as Sparidae (especially snapper, *Pagrus auratus*), Moridae, and Triglidae. Other important food species throughout New Zealand included the large gropers (Percichthyidae) and the barracouta *(Thyrsites atun)*.

Marine resources in South Polynesia were differentially vulnerable to human predation. The highest risks were for seals near the edges of their ranges because of their predictable use of the same breeding localities, their high visibility in islands bereft of terrestrial mammals, and the desirability of fat-rich meat. Colony-breeding seabirds were in a similar position, and one aggravated by the human introduction of rats and dogs; but for a time at least, their vulnerability in New Zealand was muted by the human focus on large-bodied flightless birds, especially moas (Dinornithiformes). In the broadly temperate region of South Polynesia, fish taxa open to human predation generally existed in populations sufficiently large and mobile to avoid permanently deleterious affects from prehistoric exploitation by small human populations. However, both they and the larger coastal invertebrates, such as abalone and rock lobster, were vulnerable to decline under modern commercial harvesting practices.

The prehistoric significance of marine resources is described here, first for the outlying groups, and then for New Zealand. The potential human impact on various taxa is considered, then the evidence of historical exploitation of marine resources. Lastly, I discuss the nature of resource exploitation inferred from these data. It is worth noting at the outset that human impact upon marine resources occurred within broader anthropogenic influences on island ecosystems (Allen 2003). In particular, frequent extinction of Polynesian avifaunas was associated strongly with colonization of each archipelago (Anderson 1989; Steadman 1995; Worthy 1997). As it mostly preceded the extensive exploitation of marine ecosystems, it provides a useful context to understanding the subsequent movement of resource exploitation down island trophic levels.

THE OUTLYING ARCHIPELAGOS

Archaeological data on resource exploitation in South Polynesia are, as in most regions, variable in terms of sampling by sieve sizes, excavation areas, sample sizes, and site sequences (Allen 2003). However, all recent excavations, including in the outlying archipelagos, have employed 2- or 3-mm screens, and remains are quantified as the minimum number of individuals (MNI).

Test excavations of thirteenth-century sites at Sandy Bay on the Auckland Islands disclosed sparse shellfish and ice-cod (*Paranotothenia* sp.) remains, but abundant bird and mammal bones. Birds (MNI = 124), 98 percent of them of marine origin, included sooty shearwater *(Puffinus griseus)*, yellow-eyed penguin *(Megadyptes antipodes)*, white-chinned petrel *(Procellaria aequinoctinialis)*, and southern royal albatross *(Diomedea epomophora)*. Mammals (MNI = 14), were split equally between Hooker's sea lion *(Phocarctos hookeri)* and New Zealand fur seal. As fur seal remains occur mainly in upper layers, it is possible that human disturbance had driven away much of the sea lion colony that bred (and still breeds) at Sandy Bay, but if so the effect was likely only temporary. Given the scarcity of edible plants on these islands, and therefore a diet that was composed almost entirely of marine fauna, nutritional problems such as ketonuria (Anderson 1981a; Leach et al. 1999a), coupled with the cold, wet, and sunless climate, probably caused the departure or demise of the settlers within a few years of arrival (Anderson 2005).

Zooarchaeological data for the Chatham Islands are confined largely to the Waihora village site and nearby areas, with deposits dated between about AD 1500 and 1700 (Sutton 1982, 1989). Excavation of about 25 percent of a large seal bone midden produced remains of 135 New Zealand fur seals, 12 southern elephant seals *(Mirounga leonina)*, 8 leopard seals *(Hydrurga leptonyx)* and 2 sea lions. Other excavations produced about 10,000 fish MNI, of which over 90 percent were from inshore taxa: blue cod

(Parapercis colias), butterfish *(Odax pullus)*, and Labridae. Sutton (1989:127) estimated that 85 percent of the food energy represented by the archaeological data came from seals, and 10 percent from marine fish, with about 5 percent from marine birds (mostly petrels and shearwaters), shellfish, and terrestrial birds. Terrestrial resources were more prominent in the Chathams than in the Subantarctic islands. These included landbirds and freshwater eels (*Anguilla* sp.), both represented consistently in the Waihora middens. Bracken fern *(Pteridium esculentum)* tubers and kernels from one of the common trees, kopi *(Corynocarpus laevigatus)*, may also have been harvested, but these are not represented archaeologically. The contribution of terrestrial resources was potentially substantial, although stable isotope analyses of prehistoric human bone from the Chathams indicate that marine resources remained wholly dominant (Leach et al. 2003). Moriori harvesting and deforestation appears to have caused the extinction of several landbirds, but there are no indications of any such impact on the marine environment.

Excavations in a fourteenth-century settlement site at Low Flat on Raoul Island in the Kermadecs produced small quantities of 115-mm-long giant limpet *(Scutellastra kermadecensis)* shells, and a few fish bones, but the main taxa were procellarid birds (MNI = 43), mostly Kermadec petrel *(Pterodroma neglecta)* and black-winged petrel *(Pterodroma nigripennis)*. There were also some bones from dog, fur seal, and a phocid seal, probably southern elephant seal. Taro *(Colocasia esculenta)* and ti *(Cordyline fruticosa)* were growing on Raoul Island in the twentieth century, but whether they and other Polynesian domestic plants were introduced prehistorically is unknown. No horticultural features or artifacts have been found among the recorded archaeological sites, and as various Polynesian groups lived on Raoul Island in the nineteenth century (Anderson 1980a, 2006; Johnson 1995), these traditional food plants might have been introduced at that time.

Extensive excavations at Emily Bay on Norfolk Island uncovered a small village site occupied during the thirteenth to fifteenth centuries AD (Anderson and White 2001). Associated middens contained fish bones (MNI = 153), of which at least 63 percent were from Lethrinids. However, most bones were from marine birds (MNI = 1115), predominantly petrels and shearwaters (87 percent). By estimated biomass, leaving aside the cranium of a southern elephant seal, marine birds contributed 62 percent, fish 30 percent, turtle 6 percent, and shellfish 1 percent. Terrestrial birds make up the balance. When Norfolk Island was colonized by Europeans in AD 1788, a discrete patch of bananas was found, but if bananas had survived since Polynesian settlement ceased more than 300 years earlier, they would probably have spread more widely. The historical patch might represent a brief eighteenth-century visit, as early European discoveries of wooden artifacts on the beaches also suggest. At any rate, there is no archaeological evidence to suggest the former existence of Polynesian horticulture on Norfolk Island.

The dog, often used as food, was introduced to the Kermadecs, Subantarctic, and possibly Norfolk islands, and the edible rat to the Kermadecs, Norfolk, and Chathams, but as Polynesian food plants would not grow in the Subantarctic or Chathams and may not have been introduced prehistorically to Norfolk or the Kermadecs, subsistence in the outlying archipelagos relied heavily on native resources. Among these, terrestrial birds were relatively scarce, and cursorial taxa were vulnerable to rapid depletion and extinction. Native food plants were also scarce, except in the Chathams. Prehistoric subsistence in the outlying archipelagos of South Polynesia was closely aligned, therefore, to the abundance and accessibility of marine resources. Clumped resources, such as breeding colonies of seals and seabirds were harvested substantially everywhere, and the main fish taxa were those that occur in concentrations around rocky coasts or other inshore localities.

The existence of seals in early Norfolk Island and Kermadec sites, and even in some tropical islands of East Polynesia at the advent of colonization (Walter and Smith 1998), and their apparent disappearance thereafter invite consideration of the cause. As climatic cooling by AD 1500, and perhaps earlier, would, if anything, favor an increase in seal use of tropical islands, it was probably the advent of a human presence either as disturbance or by predation that was enough to deter continued use of islands at the edge of seal ranges. Otherwise there is no evidence of the local disappearance of marine taxa from the outlying archipelagos in the pre-European era, and there is little evidence of any significant depletion until after European settlement (below). Nevertheless, it is probable that faunal depletion was a factor in abandoned or failed settlement in Norfolk and the Kermadecs. The reliance in these groups on ground-breeding petrels and shearwaters was necessitated by few alternatives and none that produced sufficient amounts of the fats and carbohydrates essential to a diet otherwise dominated by animal protein (Davidson and Leach 2001; Leach et al. 1999a). The bird colonies, however, were vulnerable to overexploitation by humans and introduced rats. Once these birds were gone or severely depleted, there was not much left to sustain long-term human settlement (Anderson 2001). Such a situation occurred historically on Norfolk Island when the initial European colony, facing starvation with the loss of its supply vessel in AD 1790, turned to harvesting the highly abundant Providence petrel (*Pterodroma solandri*). At least 172,000 birds were taken in the first winter and by AD 1792 the species was very scarce and soon became locally extinct. There is an analogous suggestion (Holdaway and Anderson 2001) that *P. pycrofti*, very abundant in the archaeological bird remains, but extremely scarce historically, were severely depleted prehistorically and never recovered.

Aggravating the low diversity of the potential food supply was the small size of Norfolk and the Kermadec islands. The Chathams are 30 times larger, had greater diversity and abundance of faunal resources, and had access to significant native food plants that, even if exploited fairly lightly, might have been the critical factor in producing a diet that could underwrite long-term occupation in that group but not elsewhere (Anderson 2006).

NEW ZEALAND

New Zealand is 270 times the size of the Chathams, and, spread from 34° to 47° S, it is correspondingly more diverse environmentally. Most importantly, when New Zealand was discovered by Polynesians, it contained a range and abundance of terrestrial subsistence resources suitable, alone, to sustaining long-term settlement. The best known of these were the giant wingless moas or Dinornithiformes (Anderson 1989), which ranged up to several hundred kilograms in body weight, and there were other large, flightless or poor-flying birds including geese, rails, parrots, and ducks (Tennyson and Martinson 2006; Worthy 1997). Flighted forest, grassland, and water birds were also abundant and diverse. There were freshwater fish, including native trout (Galaxidae), grayling, and eels. Plant foods were few but relatively abundant. They included the root and stem of young *Cordyline* trees, pith from tree ferns and bracken fern (*Pteridium* sp.) root, as well as numerous kinds of berries and kernels.

Of marine resources, the fur seal was most abundant, followed by Hooker's sea lion and the southern elephant seal. Dolphins were hunted by harpoon, and whales, which beached commonly in New Zealand, were important sources of scavenged food, bone, and teeth. Petrels and shearwaters nested in huge colonies on offshore islands, and probably on the mainland initially. In the absence of cursorial terrestrial mammals, there was little threat to ground-nesting birds, but once the Pacific rat was introduced by Maori, the eggs of most of the smaller procellarids were vulnerable. Several species of schooling fish, notably barracouta, red cod *(Pseudophycis bachus)*, and snapper, were particularly abundant, the former two in the south, the latter in the north. Shellfish included several species of abalone (*Haliotis* spp.) and other large gastropods. There were clams in great abundance in the large northern estuaries, and crayfish *(Jasus edwardsii)* around all the rocky coasts. Various species of seaweed, notably karengo *(Porphyra columbina)* were used for food (Anderson and McGlone 1991; Beattie 1994).

Although viewed predominantly as food for both people and their dogs, marine resources also provided important raw materials. Giant kelp *(D'Urvillea antarctica)* fronds were opened out to make airtight bags, iridescent abalone shell was used in ornaments and fishing lures, shell from other large gastropods was used to make bait hooks, and mako shark teeth were used in earrings. From marine bird bone came needles, awls, fishhook points, and beads, and albatross long bones were favored for toggles and flutes. Worked bone or ivory from toothed whales is common in archaeological sites. Sperm whale and dolphin teeth were carved, drilled, and strung as various kinds of neck and breast ornaments, and whale bone was fashioned into imitation whale teeth, hooks, harpoons, beads, and fighting clubs. Whale tooth and bone artifacts are predominantly early, probably because the initial settlers came from East Polynesian societies where whale teeth were a prestige item, and also because they were able to harvest the prehuman accumulation of cetacean remains on New Zealand beaches.

The use of marine resources as food also varied through time and space. In economic terms, New Zealand's archaeology can be divided broadly into two phases. Dating to the early phase, about AD 1200–1500, and especially up to AD 1400, are numerous sites of moa hunting, which are concentrated in the south where populations of moas and other large cursorial birds reached their highest concentrations in relatively dry forests (Anderson 1996a). The largest sites are coastal, and they were often located adjacent to seal colonies, which were also most frequent in this region. Dating to the late phase, AD 1500–1800, are numerous fortified sites, and many unfortified sites as well, which are concentrated in the north where conditions were most favorable to the extensive cultivation of food crops (Anderson 2002). Horticulture had almost certainly begun with the first arrival of people in New Zealand, but as hunting and foraging was, for a time, highly productive, cultivation flourished later in conjunction with substantial growth in population density, especially in the north. Relative stagnation to the south, reflecting the scarcity of horticultural production, was probably exacerbated by the impact of cooling temperatures during a climatic phase similar to the Northern Hemisphere Little Ice Age, after about AD 1500, which placed a premium on frost-free sites and moved the southern margin of effective horticulture northward (Leach 1981). The later northern sites were often coastal and concentrated around sheltered gulfs and estuaries.

A survey of 49 archaeological assemblages of bone and shell distributed around the New Zealand Coast has begun to quantify regional variation in exploitation of fish, marine mammals, and other resources (Smith 2004). In these data, reworked in Table 2.2, "terrestrial mammals" refers to the introduced dog and rat, mainly the former, which was bred for food. The early moa contribution is underweighted, particularly in the south, because the data exclude numerous inland moa-hunting sites. Most of the sizeable and reasonably representative assemblages lie toward the early end of occupation chronologies ($n = 29$), and the sample is weighted toward the southern region (south of 45° S, $n = 21$; central region assemblages, $n = 15$, extend up to 39° S, and northern region assemblages, $n = 13$, up to 34° S).

In New Zealand generally, fish contributed more than 20 percent of the total meat weight (MNI × average body weight × usable meat proportion of .7) in 28 assemblages, and marine mammals in 26, while by estimated energy yields, the main sources were marine

TABLE 2.2

Usable Meat Weight Percentage by Region, Age, and Faunal Class in 29 Early and 20 Late Archaeological Assemblages

	MARINE MAMMAL	TERRESTRIAL MAMMAL	MOA	SMALL LAND BIRD	SEABIRD	FISH	SHELLFISH
North							
Early	36.6	5	9	0.5	1.5	47	0.1
Late	0	30.3	0	0	2.7	34.4	32.6
Central							
Early	20	5.4	10	1.8	2.9	12.9	46.8
Late	10.3	6.2	0.4	0.2	2.1	42.1	38.6
South							
Early	35.4	4.1	24.8	1.3	4.2	28.2	1.7
Late	22.1	6.2	2.1	0.6	10.7	56.4	1.8
Overall							
Early	32.6	4.5	18.4	1.2	3.3	29.3	10.4
Late	15.4	7.2	1.1	0.4	6.2	48.6	21

NOTE: Data calculated from Smith (2004:14, Table 5).

mammals, predominant in 21 assemblages, and fish in 13. Looking at the geographical and temporal distribution of proportional meat weights (Table 2.2), it is apparent that moas disappeared relatively early, and marine mammals declined substantially and disappeared in the north. In compensation, there was increased use of marine birds, of fish in central and southern regions, and of shellfish in the north. The overall exploitation of faunal resources represents a shift from the coastal and inland megafauna (marine mammals and large flightless birds) toward the smaller colonial taxa of offshore islands, the intertidal zone, and the open sea. Human impacts on marine populations were varied and are best followed by discussing each group separately.

Marine Invertebrates

There is no archaeological evidence of the extinction of shellfish or of widespread, sustained depression in the mean size of any species, but it should be emphasized that quantitative data are localized and mostly early. At the local level, variations in taxonomic representation and trends toward lower mean sizes can be documented, but they are not easily interpreted. Many late period sites in northern New Zealand are associated with extensive middens dominated by estuarine clams, notably *Paphies australe* and *Chione stutchburyi*, and casual inspection suggests relatively small specimens are more common than in early sites. Samples that indicate a downsizing trend have been measured from several sites (e.g., Swadling 1977), but whether they reflect depletion of larger specimens by harvesting pressure or other factors has yet to be investigated satisfactorily. Increasing frequency of smaller specimens in estuarine colonies could be caused by crowding (Reyment 1971) as nutrient levels rose through runoff. Those, in turn, could be related to either or both expanding deforestation and horticultural activity, which, along with human population growth, occurred around the northern estuaries in the late phase, or to increased stream flow during cooler and stormier conditions that set in with the Little Ice Age. Alternatively, shellfish growing conditions might have deteriorated in northern estuaries through lowered water temperatures or increased silt loads and turbidity. Until more data exist on archaeological trends in estuarine molluskan shell-size frequencies, however, it cannot be assumed that they were usually toward downsizing or that explanations are validly anthropogenic.

Open-coast shellfish, and especially those on exposed rocky shores, are usually less affected by runoff, and as they often occur in smaller populations of discontinuous linear distribution, rather than in the extensive areal accumulations of large estuaries, they were potentially more vulnerable to harvesting impacts. The limpet *Cellana denticulata* occurs in early sites in northern New Zealand but not later, and Rowland (1976) suggested that its range became restricted southward, where it is abundant, by overexploitation. In South Polynesia, as elsewhere in Oceania, there is a general perception that large individuals of rocky shore shellfish are found predominantly in the earliest sites or levels. This has been documented for green mussels *(Perna canaliculus)* and sea urchins *(Evechinus chloroticus)* at the early-phase Sunde site by Nichol (1988) and for various early sites in Palliser Bay (Leach and Leach 1979).

At Black Rocks Peninsula there was a program of midden research, which included a quantitative survey of intertidal shellfish taxa, the results of which were compared with the distribution of molluskan taxa in excavated middens spanning the greater part of the prehistoric era. This suggested that the archaeological pattern could be obtained by harvesting according to selection for individual size or yield, rather than by random collection or taxonomic preference—an early archaeological illustration of the principles of optimal foraging (Allen and Nagaoka 2004; Anderson 1973, 1979, 1981b). A similar case was advanced for shellfishing at Hahei (Nichol 1986), and Leach and Anderson (1979a) argued that the early, marked, and sustained depression in crayfish size through the Palliser Bay middens sequence represents human impact (Figure 2.2), but they also observed that reports elsewhere of large crayfish traded by Maori to eighteenth-century European explorers show that the local trend cannot be extrapolated into a general case.

Climatic change might have been involved in the temporary disappearance of the warm-water snail, *Nerita atramentosa*, from shell middens on Motutapu Island. It occurs commonly up to the fourteenth century and again in late, probably eighteenth-century, middens, but it is missing in between. As it lives on the upper shore and is particularly vulnerable to changes in atmospheric temperatures, and as it is near the southern margin of its range on Motutapu Island, it might have disappeared during the Little Ice Age (Szabo 2001a, 2001b). McFadgen and Goff (2001) argued that it disappeared through overharvesting. More precise radiocarbon dates and greater clarity in the sequence of climatic changes are needed to resolve the issue.

Marine Fish

Variation in the taxonomic representation or mean size of fish is even more problematic, because fish often move between different depths or follow moving fronts of current, salinity, or water temperature. Whereas shellfish gatherers are likely to sample the same populations repeatedly, fishers must often have sampled different age-graded or taxonomically diversified populations moving across the accessibility range from a particular site, and frequently will have complicated the issue further by using watercraft to fish in different areas, and different technologies to fish at varied times or places. Maori fishing technology was particularly diverse, involving double- and single-hulled canoes to fish with baited hooks and trolling lures offshore, and hook-and-line fishing from the shore. Seines, sometimes a kilometer or more long, and various movable and fixed fish traps and nets were also used.

Most variation in archaeological fish catches between regions can be ascribed to the relative natural abundance of taxa available (Anderson 1997; Leach and Boocock 1993), but not in the case of Labridae. These small, territorial taxa, which are not generally sought after, are at least as abundant in the north as the south, yet occur predominantly in southern archaeological samples, where they can make up 30 to 60 percent of the total MNI. Leach and Anderson (1979b) suggested that poorer sea conditions toward the

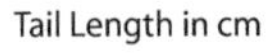

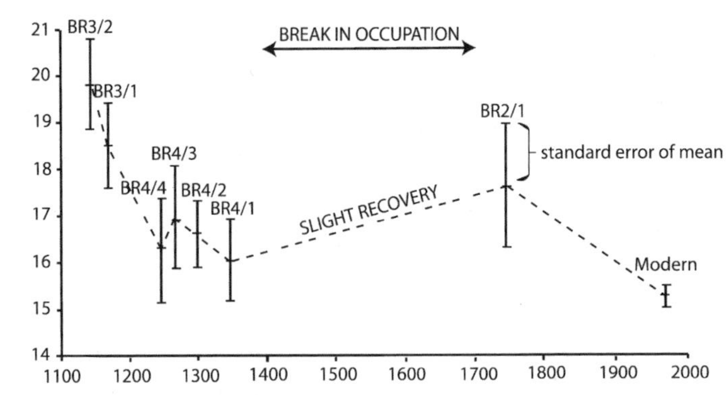

FIGURE 2.2. Variation in the estimated length of crayfish tails across a sequence of middens at Black Rocks Peninsula (after Leach and Anderson 1979a: Figure 10.6).

south, which inhibited offshore fishing, and marginal or absent horticulture, which placed a stronger emphasis on fish resources, required more use of reliable, though otherwise unfavored, taxa that could be taken directly from the shore.

Several time trends in taxonomic representation have been documented at the regional level. In the southern South Island, large snapper *(Pagrus auratus)* occur sparsely in early sites, but snapper are absent later (Anderson and McGlone 1991). It seems improbable that this is due to human impact and more plausible that it reflects a range shift northward, perhaps in the face of cooling sea temperatures as the Little Ice Age set in. A similar case may be apparent in relation to the separate snapper population in Tasman Bay, northern South Island. The first regional review of archaeological data (Leach and Boocock 1993) showed that snapper represented about 22 percent of the overall catch, but most of the data came from early sites. Subsequent data from later sites (Barber 1994) showed snapper as less than 3 percent of the overall catch (Anderson 1997). While variation in fishing strategies is possible, the trend might also reflect a northward withdrawal of the Tasman Bay population as water temperatures declined in the Little Ice Age. A similar proposition has been advanced by Davidson et al. (2000) to account for declining snapper numbers in the Foxton site, southern North Island.

Turning to size-frequency data, the Rotokura site in Tasman Bay, where data span most of the prehistoric era, showed an increasing mean size of snapper. This, too, could reflect climatic variation, since greater fluctuations in snapper recruitment, and therefore in secular movements of mean size, occur in conditions of suboptimal water temperature (Leach and Boocock 1994). A northward movement of the snapper population in cooling water might also leave relatively more of the larger individuals on the southern margins of the population range.

Another possible explanation suggested by Leach and Boocock (1994) is that overexploitation of inshore snapper, which are generally in the smaller size ranges, resulted in a shift of the fishing range into deeper water, where larger individuals lived. This is one of a number of older propositions that relate change in the mean size of snapper, as reconstructed from archaeological and icthyological data, to possible affects of overexploitation. Examples include Leach et al. (1997) on decreasing snapper size in Northland, Nichol (1986, 1988) on decreasing snapper size in the Sunde site and on snapper and other fish at Hahei (both are early sites), and Shawcross (1967) on the larger size of archaeological than modern snapper. Leach (2006), however, rejected Anderson and McGlone's (1991) suggestion that snapper overharvesting was possible in northern Northland, but he mistook the area at issue for the entire province, invalidating his conclusions. Allen's (2005) broader review of the catch data from several northern North Island sites has taken into account taphonomic and sampling issues and concluded that there is evidence of localized snapper depletion.

A patchy, localized record of data pertinent to resource depression in fish stocks is also suggested by recent research on size-frequency changes. There is a possible slight increase in snapper size at Houhora, and a slight decrease in size at Cross Creek, but no significant trend in either (Leach and Davidson 2000). Similar research on blue cod showed that mean size increased significantly over time at Mana Island and varied in Chatham Islands sites (Leach et al. 2000). Labrid sizes also increased over time in the Black Rocks sites, at Mana Island, and at Kokohuia. As the catches include several species of different size that cannot be distinguished osteologically, it is possible that these data represent a change from the early, inshore net or trap fishing that was so well developed among Polynesians generally, to a later general emphasis in New Zealand upon bait-hook fishing, which probably selected for the larger species. On the other hand, perhaps it was simply that the Polynesian culinary preference for small fish also declined in New Zealand (Leach et al. 1999a). Very large barracouta catches from sites spanning about 500 years at Long Beach and 40 to 200 years at Shag River Mouth, in the southern South Island, both show a declining mean size. However, inshore barracouta shoals are drawn from immense, mobile populations lying beyond the Maori fishing range. It would have been impossible for traditional harvesting to have made any significant impact on the latter, so the archaeological data more probably represent either variability in the age grades represented by the archaeological samples from each layer, or the effect of changes in sea temperatures on barracouta growth rates (Leach et al. 1999b).

Leach and Davidson (2001) reviewed research on size-frequency data and concluded that there is no clear support for the hypothesis that prehistoric fishing by Maori and Moriori had a significant impact on fish size or abundance. They suggest that the data are complex and conceal various potential causes of variability, including changes in fishing technology. Use of nets of different mesh size and of different sizes and types of hooks affect modern catch characteristics substantially and must have done so prehistorically as well (Anderson and McGlone 1991; Nichol 1988). Regional variation in the capacity to use offshore fishing canoes (Anderson 1986) and changes in fishing grounds with population growth and changing settlement patterns probably were also implicated, but whatever the factors involved in fish catch and size-frequency variation, Leach and Davidson (2001) argued that abundance and sizes overall remained essentially unaffected by prehistoric exploitation. Prehistoric fishing, which relied considerably on hook and line, tended to take fish in larger size ranges than many modern nets, accounting for some of the difference, but decline in some fish sizes has probably also occurred through overfishing in the European era. An additional factor, emerging from the current discussion, is the potential role of slight changes in sea temperatures over the last 800 years in altering the distribution and growth rates of some prominent target species such as snapper and barracouta (Leach and Davidson 2001).

Marine Birds

The heavy harvesting of marine birds in South (and East) Polynesia is so archaeologically striking as to be regarded as a marker of the colonization era (cf. Moniz 1997). Albatrosses, penguins, cormorants, gannets, and other seabirds were commonly taken, but only petrels, prions, and shearwaters (grouped collectively as "muttonbirds"; see Anderson 1995, 1996b) have been studied in any detail in South Polynesia. In subtropical areas, as in the Kermadecs and Norfolk Island, common muttonbird taxa (e.g., *Pterodroma solandri, P. neglecta, Puffinus pacificus*) are weakly migratory and range over seas with relatively sparse and dispersed food resources. The muttonbird colonies thus represent the entire local populations, and they are, and were, vulnerable to predation. In New Zealand, however, especially in the southern region, and in the Chathams, the main species (e.g., *Puffinus griseus, P. tenuirostris*) are strongly migratory, and often the bulk of their populations bred south of the areas to which Polynesians had access, so they were resilient in the face of even quite severe predation at accessible colonies.

Muttonbirding occurred widely but at low levels throughout most of New Zealand during prehistory, but it became a systematic, seasonal industry in the far south, especially on the offshore islands around Stewart Island. Muttonbird squabs were harvested in large numbers and preserved in fat, inside kelp bags, as a prestige food that was traded to northern communities. There is no archaeological evidence of depletion or extinction among muttonbirds in New Zealand, and catches of 250,000 birds a year have been a regular occurrence in a traditional industry that continues to this day with both subsistence and commercial objectives.

Marine Mammals

Marine mammal bone occurs widely in New Zealand sites, but there is considerable variability in the representation of taxa. Smith's (1989) analysis of 180 archaeological sites with marine mammal remains shows cetacean bone in 51 sites (28 percent). Pilot whale *(Globicephala melaena)* was identified at 10 sites, and all in areas with historical records of pilot whale strandings. Dolphins (mostly common dolphin, *Delphinus delphis*) were identified in 13 sites, and they may have been hunted using the barbed and toggling harpoons often found in the same sites. Seal remains occurred in 173 (96 percent) of the sites. Fur seals were represented in 167 sites (93 percent), and other taxa less frequently: sea lions in 38 percent, southern elephant seal in 26 percent, and leopard seal in 6 percent of sites. Although seal remains occur in sites along high-energy coasts throughout New Zealand, they are concentrated toward the south, and there is evidence of a significant human impact on seal distribution through the prehistoric era.

In modern populations of fur seals, adult females and pups occur only within the breeding range of the species. Smith (1989, 2005) found that bone from these groups occurred in 14 sites, half of which were located north of the modern breeding range, which extends to Cook Strait. Examining the chronology of the sites, Smith (1989, 2002, 2005) showed that there had been a progressive southward withdrawal in breeding range to the southern South Island, from which it has subsequently rebounded, at least in part (Figure 2.3).

In New Zealand both regular cropping and opportunistic capture of seals are documented, but analysis of element distribution and butchery patterns shows that regular cropping was the main strategy. In fact, some 75 percent of all fur seal remains in New Zealand sites came from deposits in which the existence of whole animals and the predominance of low-yield elements indicate that hunting occurred in close proximity to seal colonies from sites located so as to extract a regular harvest, a proposition supported by seasonal analysis of seal teeth (Smith 2005).

In northern North Island sites, regular cropping is evident only in the earliest sites, and opportunistic seal hunting, marked by few and generally high yield elements, persisted up to about AD 1500. No seal remains are found in later northern archaeological sites. In southern New Zealand, regular cropping and opportunistic hunting occurred throughout the prehistoric era. At the local level, rapid seal decline can be documented repeatedly. At the early village site at Shag River Mouth, located beside a modern seal

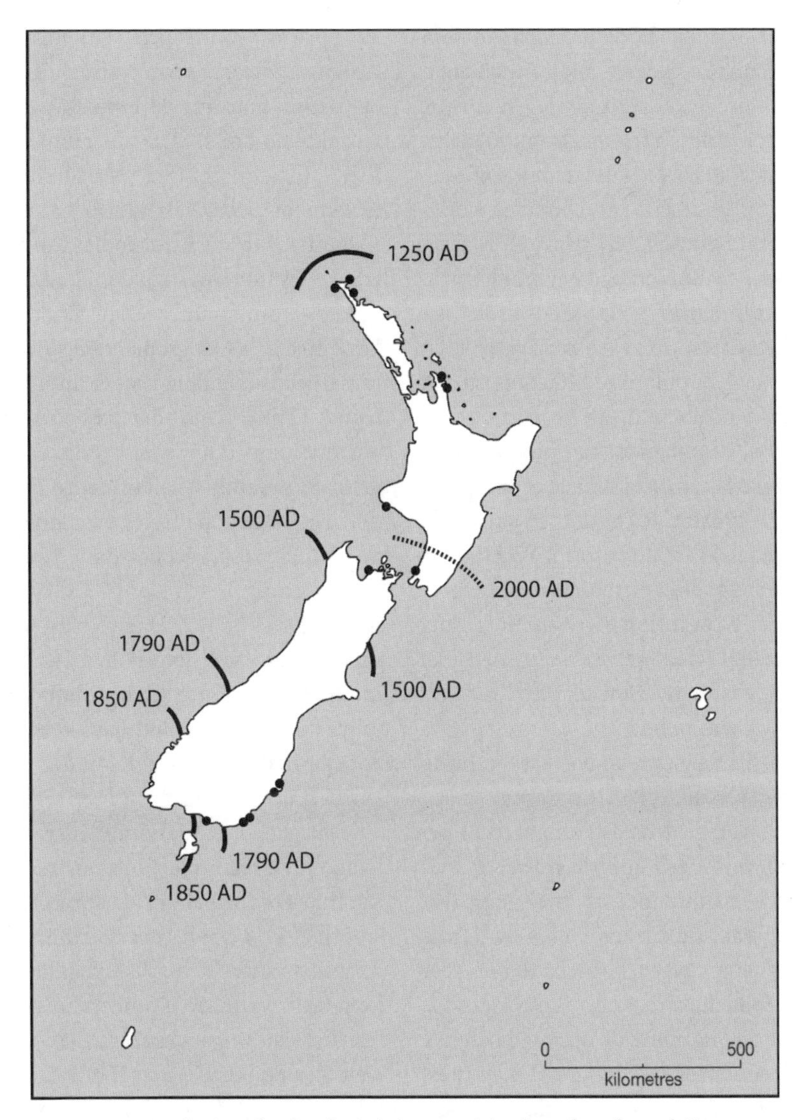

FIGURE 2.3. Distribution of archaeological sites containing bone from fur seal (*Arctocephalus forsteri*) females and pups (black dots), and estimated changes in the breeding range (after Smith 2005: Figures 2 and 4).

colony that probably also existed 700 years ago, the proportional yield of seal meat, among all meat sources represented, fell from about 40 percent to about 10 percent in less than a century, and the site was abandoned soon afterward. The successive, short-term occupation of villages, each abandoned as local moa and seal reserves declined, is a form of shifting sedentism common to the early period in New Zealand prehistory, especially in the south (Anderson et al. 1996; Anderson and Smith 1992, 1996; Nagaoka 2002).

It is difficult, nonetheless, to derive an overall demonstration of prehistoric overkill from the relative archaeological abundance of seal remains and local cases of rapid decline, except by elimination of alternatives. Substantial seal decline between AD 1200 and 1500 cannot be attributed plausibly to climatic or any other aspect of seal habitat change, since any early onset of Little Ice Age was in the direction of cooling sea temperatures, which were more favorable to seal breeding and probably also to increases in fish populations. Human impact on seal habitat, for example, by deforestation increasing ocean turbidity, may have had a small effect inshore, but fur seals hunt preferentially in deeper offshore waters (Smith 2002). Whether the accumulated archaeological remains of seals in northern sites are sufficiently abundant to support the validity of overkill depends on how large the original seal population had been, and that is very difficult to determine. Two kinds of fur seal population estimates are available. Data on the modern fur seal population indicates that it ranged from 20,000 in 1948 to 100,000 in 1996. However, estimates of the pre–AD 1800 population, based on historical data of seal hunting, are much greater; Richards (1994) calculated it at 1,500,000 to 2,000,000. Whether all the estimates are approximately correct, or there is much greater error in some, is very uncertain; but Lalas and Bradshaw (2001) suggested that the decline of mainland New Zealand fur seal colonies during the prehistoric era could have occurred at an average annual decrease in numbers of only .5 percent, or if it occurred within 200 years, an average decrease of 2 percent per year. Fur seals can breed quickly, however, with modern rates of natural increase averaging over 5 percent per year and annual increases in pup numbers reaching over 30 percent per year (Lalas and Bradshaw 2001). Smith (2002) observed that culling rates of up to 35 percent, especially concentrated on young males, are sustainable in various managed populations elsewhere in the world, so the demographic reaction of fur seals to prehistoric exploitation is difficult to model. Nevertheless the New Zealand fur seals were more vulnerable than some others for several reasons: they are nonmigratory, they are strongly faithful to continuing use of the same breeding and nonbreeding sites, and colonies are small—none exceed 8,000 individuals, and most are much smaller, especially toward the north.

HISTORICAL RESOURCE DEPLETION

Fish and shellfish populations do not seem to have declined immediately after the arrival of Europeans in South Polynesia, probably because Maori soon joined Europeans in regarding introduced crops such as potatoes and wheat, and pigs, cattle, and sheep as staple foods. In addition, there was almost no external market for New Zealand fish, except for a small quantity of dried fish exported to Sydney, Australia (Anderson 1998). Graham's (1956) records from southern New Zealand show that large fish such as groper *(Polyprion oxygeneios)* remained in shallow waters until the late nineteenth century and that schools of barracouta, red cod, and other inshore taxa continued in great abundance. Such evidence is mostly anecdotal, of course, but it is common to both European and Maori recollections (Anderson 1988; Beattie 1994; Leach 2003). On Norfolk Island, the large, sought-after sweetlip *(Lethrinus lineata)* is also said locally to have declined dramatically during the last half-century, and groper in New Zealand have continued to disappear inshore through modern overfishing (Annala et al. 2002; Paul 1986).

Paua (abalone) landings increased strongly in 1970, rising rapidly to 15 to 20 times earlier catch levels (Annala et al. 2002). The resource is depleted around most of the New Zealand coastline and is being fished at levels where its long-term viability is uncertain (Leach 2003). Crayfish landings were less than 1,000 tons in New Zealand until 1950, then began to rise rapidly to a peak of 11,000 tons in 1969 as populations were heavily exploited successively in southern New Zealand and the Chathams. Since 1990 landings have declined to about 3,000 tons per annum (Leach 2003). It is difficult to say whether these declines represent serious threats to long-term sustainability. One of the problems is that there are too few data of sufficient age to estimate the preexploitation levels of most taxa, and there is a tendency for management models to work from data compiled during periods when exploitation levels were already high—the "shifting baseline syndrome" (Jackson et al. 2001; Pauly et al. 1998).

The historical sealing industry in New Zealand focused on fur seal skins, and most of its activity occurred AD 1792–1833. Seals were clubbed on shore and skins salted and stored for eventual transport. Sealing occurred around the southern coasts of the South Island and the islands of Foveaux Strait and the Subantarctic. Southern Maori regarded seals as their property, and there were skirmishes and killings between Maori and European sealers in the early nineteenth century (Anderson 1998).

There were at least 113 sealing voyages to mainland New Zealand, and they took 167,000 to 372,000 skins, according to cargo records (Smith 2005). Richards (2003), however, estimates that from the southern fur seal populations generally, a minimum of 7,000,000 skins had reached the London and Canton markets by 1833. Undoubtedly, this assault severely depleted most populations, including in New Zealand. By the mid-nineteenth century, fur seals were scarce and did not breed north of the southern South Island. Protection, in various forms, began in 1875, and it has been total since 1947, with the result that fur seal populations are now expanding vigorously. Breeding colonies now reach to the southern North Island and, in time, will probably extend throughout New Zealand again (Lalas and Bradshaw 2001).

There is no evidence that prehistoric Maori or Moriori hunted large cetaceans, but dolphin hunting by Maori, using traditional harpoons, continued into the early European era, as did scavenging of beached whale carcasses for meat, bone, and teeth (Anderson 1998). Whaling ships began to arrive in New Zealand in AD 1792, at the same time as sealing vessels, but whaling occurred mostly after 1820, when sealing was in decline. Whaling ships in New Zealand waters averaged about 20 per year until 1830, then rose to 150 per year until 1840, when a steep decline set in (Richards 2002). Maori and Moriori were often recruited as crew. Whaling from shore stations began in 1827 in New Zealand, and Maori were commonly engaged as boat crew and harpooners and to supply vegetables and other victuals. About 80 shore whaling stations, often lasting only a season or two, operated around the New Zealand Coast from AD 1830 to 1850 (Richards 2002).

The main target species from 1820 to 1845 was the sperm whale *(Physeter macrocephalus)*. About AD 1800 this species had a total population estimated at 1,110,000, and although that declined 15 to 20 percent by the mid-nineteenth century, the species was in no imminent danger (Baker and Clapham 2004). The human impact was much greater on the southern right whale *(Eubalaena australis)*. As sperm whales became scarcer and more wary in the New Zealand grounds in the early 1830s, attention turned to the slower baleen right whale. It often bred in bays and harbors along the east coast of New Zealand. Over 14,000 adult right whales, plus juveniles and infants, were killed between 1832 and 1842, by which time the New Zealand industry was in decline; 4,000 were killed from 1843 to 1851 and then only about 10 per year up to 1927 (Richards 2002). From a mid-eighteenth-century total population of about 60,000 to 100,000, catch records show that about 150,000 were taken during the first half of the nineteenth century, and the total population may have reached as low as 60 females by 1920 (Baker and Clapham 2004). In New Zealand, right whales had effectively disappeared, with none being observed off the South Island from 1927 to 1963 (Richards 2002). Today, there is a small population (<100) that breeds mainly at Port Ross in the Auckland Islands.

DISCUSSION

As might be expected on islands, the use of marine resources in South Polynesia was very substantial. In comparison with the rest of Polynesia, and the Oceanic islands generally, it was enhanced by relative isolation and climatic conditions. The very long and difficult ocean passage prevented some domestic plants and animals from being introduced, and it made any subsequent contact with East Polynesia unlikely (none has been documented), while the temperate climate reduced both the range and productivity of horticultural practices that did reach New Zealand. Similarly, in southern New Zealand, the Chathams, and the Subantarctic, cool conditions suited extensive colonization by several species of seals and by numerous marine birds. In addition, the extent and diversity of New Zealand coastlines ensured immense reserves of fish and shellfish.

In archaeological data on the exploitation of these resources, it is difficult to perceive any system of conservation management. There is no evidence to indicate that any kind of "temporal discounting" (Smith and Wishnie 2000), that is, deliberate constraint in the short term to ensure sustainability in the long term, occurred in prehistoric South Polynesia. The evidence from individual prehistoric sites in the early phase in southern New Zealand suggests that there was rapid local depletion of the larger but scarcer taxa, such as moas and seals, with small birds and fish becoming more prominent. Such sites were often abandoned at that point, and it is surmised that at a regional level there was serial occupancy of high-quality localities as each became exhausted of its preferred big-game resources (Anderson and Smith 1996). Maximization of harvesting efficiency seems to have been the primary, perhaps unconscious, objective. After about 300 years of settlement, moas and many other landbirds were extinct, and fur seal breeding had ceased in the North Island and southern colonies were substantially depleted.

Yet, there is a theoretical rationale, at least, to such focused resource exploitation. The biggest danger to initial human colonists was their own extinction, especially in isolated island situations where colonizing groups were small and unlikely to receive additional members (Williamson and Sabath 1983). An overriding concern had to be to get beyond the demographic danger point as quickly as possible. Where colonizing populations were very small, as probably was the case in outlying island groups, the matter was of particular urgency. But even in New Zealand, where the colonizing population might have been larger (Murray-McIntosh et al. 1998), the size of the islands probably imposed considerable internal isolation, so that each local group faced its own fitness imperative. Resource management, in other words, seems to have focused strongly and necessarily on conservation of the consumers. It required continual attention to maintaining optimal foraging efficiency among the resources of highest food value and bulk return for effort. On outlying islands the choice of patches was highly limited and soon exhausted, which, for all except the Chathams, elicited settlement extinction. In New Zealand, the choice of high-quality patches was much greater, and settlement mobility could be used to avoid, as long as possible, the consequences of localized depletion.

It is impossible to show that consumer-prioritized foraging was a deliberate strategy, except to note that in the way in which it proceeded, Maori must have seen and understood repeatedly at the local level, if not regionally, the results of overexploitation of cursorial landbirds and seals. That they continued to press these resources so hard might suggest that they understood where their priorities needed to lie. As populations grew, social pressures probably also came into play. In Maori tradition (Buck 1938), and by the uncertain nature of prehistoric voyaging (Anderson 2000b), the colonizing population was not socially homogeneous. As it increased, lineages diverged still further and grew increasingly territorial, so that competitive forces between chiefdoms continued to emphasize the wisdom of localized population growth and the reinforcement of vigorous exploitation of resources.

The disappearance or severe depletion of the big-game taxa can be attributed substantially to overharvesting, even if the process was more extended (for ecological reasons; Anderson 2000c; Brook and Bowman 2004) for moas than Holdaway and Jacomb (2000) contend. In the case of marine birds in New Zealand and the Chathams, and especially in regard to fish and shellfish, the evidence of a human impact on population distribution and size-frequency is more equivocal. The few changes of species distribution are slight and may not have a cultural explanation. There are some changes in the average size of rocky shore shellfish, which are in accord with localized depletion. Similarly, some sites suggest localized depletion of snapper stocks. Overall, however, there are no widespread patterns of resource depression among fish and shellfish, probably because both were abundant relative to human population density, even at European contact. The prehistory of South Polynesia was simply too brief for consumer impact to make much of a mark among the more numerous taxa, especially since they were to some extent protected by their marine habitats, unlike seals that bred ashore.

The impact of people can be seen, nevertheless, in a general sequence of switching from terrestrial birds and seals to fish and shellfish, which exemplifies foraging down the food web (cf. Pauly et al. 1998), and it is possible that this process, which included depletion and localized extinction of an apex predator, had significant effects on some fish taxa that were common prey for seals, for example, barracouta and red cod. These may have become more abundant in proportion to the effects of sealing (cf. Raffaelli 2004).

To what extent climatic change during the prehistoric era was implicated in some of the changes observed in distribution or size of taxa remains an open but important question. There is abundant historical evidence worldwide of decadal correlations between climatic variation and the composition of pelagic fish catches (see Alheit and Hagen 2001), and it can be assumed that climatic variation also operated in similar ways over longer periods. One set of climatic changes, already noted above as potentially influential in this regard, is the onset of the Little Ice Age. That began before AD 1600 (McKinzey et al. 2004), and it was preceded by slighter temperature fluctuations over several centuries (Burrows 1982). Its effect on seawater temperatures and currents is still uncertain in relation to the frequency distribution of inshore fish taxa.

Another climatic effect, noticed by Thresher (2002), is that midlatitude South Pacific recruitment rates of fish, and some other faunal cycles, are matched to the impact of quasi-decadal sunspot cycles on climatic parameters of the strongly zonal west winds. The extent of such changes, which have been documented widely and linked to El Niño–Southern Oscillation (ENSO) variability (Sharp 1999), is sufficient, hypothetically, to account for most of the variability seen in archaeological data. In addition, ENSO variability had, potentially, much greater impact at centennial scales since South Polynesian prehistory occurred during the millennium with the highest intercentennial variability of El Niño frequency (Moy et al. 2002). How that affected the maritime environment or human access to it, has yet to be explored, except in matters of seafaring (e.g., Anderson et al. 2006), although some case studies are emerging (e.g., Glynn 2004).

Climatic issues are likely to loom much larger in future analyses of many kinds of Pacific cultural changes (Allen 2006), including variation in prehistoric fish catch records, for they are plausible in principle and demonstrable in modern cases, if yet difficult to document in prehistoric cases. Increasingly high resolution in climatic records is bringing the prospect closer.

In the historical European era, a rapid increase in population density, the arrival of new marine technologies, such as ships and metal harpoons and hooks, which improved resource accessibility, and changes in commodity value through the connection of New Zealand to external markets led to severe depletion of seals and southern right whales. Conservation of marine birds and mammals became accepted by the mid-twentieth century, but fish and shellfish populations continued to be exploited with relatively little restraint, as commercial imperatives strengthened their claim to guide the management of marine resources.

CONCLUSIONS

In the long-term history of marine foraging, a subject still in its infancy, a broad pattern is emerging of relative stability in coastal marine ecosystems through the prehistoric Holocene, followed in the historical and modern eras by progressive trophic level dysfunction as commercial hunting and fishing eroded ecosystem diversity and resilience (Worm et al. 2006) and removed or severely depleted apex-level predators, with effects including trophic cascades and fishing down food webs (e.g., Jackson et al. 2001; Pauly et al. 1998; Steneck et al. 2004). South Polynesia, especially New Zealand, adds another case, but with some interesting variations. An early focus upon high-ranked resources caused terrestrial extinctions and severe range reduction—probably regional extirpation—of seals, followed by a shift toward fishing and shellfishing. This represents foraging behavior, typical of initial entry into previously uninhabited coasts and islands, which skimmed off the most desirable and often most vulnerable resources, producing a finer-grained resource environment in which exploitation patterns became relatively more stable or cyclic. South Polynesian prehistory was simply too brief to document that latter consequence, and the sequence jumps, therefore, from a prehistoric colonizing pattern almost directly to the colonial-commercial phases, in which focused, overexploitative behaviors recurred.

While the human impact seems significant, the role of climatic variation needs more attention at various levels of analysis. The transition from prehistoric to colonial exploitation throughout the Indo-Pacific and the Americas occurred during a period of rapid climatic change involving millennial-scale shifts in ENSO frequency and the strongest impact of the Little Ice Age. The consequence of these for change in coastal ecosystems needs to be better understood before any potential role of overharvesting can be perceived more clearly.

ACKNOWLEDGMENTS

I thank Foss Leach and Ian Smith for providing offprints and other information for this chapter, and Melinda Allen, Ian Smith, Jon Erlandson, and Torben Rick for their useful comments.

REFERENCES CITED

Alheit, J., and E. Hagen
  2001 The Effect of Climatic Variation on Pelagic Fish and Fisheries. In *History and Climate: Memories of the Future?* edited by P. D. Jones, A. E. J. Ogilvie, T. D. Davies, and K. R. Briffa, pp. 247–265. Kluwer Academic, New York.

Allen, M. S.
  2003 Human Impact on Pacific Nearshore Marine Ecosystems. In *Pacific Archaeology: Assessments and Prospects,* edited by C. Sand, pp. 317–325. Le Cahiers de l'Archaeologie en Nouvelle-Caledonie 15. Noumea, New Caledonia.
  2005 Temporal Trends in Harataonga Fishing Strategies. In Unpublished manuscript, Anthropology Department, University of Auckland, Auckland, New Zealand, in press.
  2006 New Ideas about Late Holocene Climate Variability in the Central Pacific. *Current Anthropology* 47:521–535.

Allen, M. S., and L. A. Nagaoka
  2004 "In the Footsteps of von Haast . . . the Discoveries Something Grand": The Emergence of Zooarchaeology in New Zealand. In *Change through Time: 50 years of New Zealand Archaeology,* edited by L. Furey and S. Holdaway, pp. 193–214. NZAA Monograph 26. New Zealand Archaeological Association, Auckland.

Anderson, A. J.
  1973 Archaeology and Behavior: Prehistoric Exploitation of Marine Resources at Black Rocks Peninsula, Palliser Bay. Unpublished Master's thesis, Anthropology Department, University of Otago, New Zealand.
  1979 Prehistoric Exploitation of Marine Resources at Black Rocks Point, Palliser Bay. In *Prehistoric Man in Palliser Bay,* edited by F. Leach and H. Leach, pp. 49–65. NM Bulletin 21. National Museum, Wellington, New Zealand.
  1980a The Archaeology of Raoul Island and Its Place in the Settlement History of Polynesia. *Archaeology and Physical Anthropology in Oceania* 15:131–141.
  1980b Towards an Explanation of Protohistoric Social Organization and Settlement Patterns amongst the Southern Ngai Tahu. *New Zealand Journal of Archaeology* 2:3–23.

1981a The Value of High Latitude Models in South Pacific Archaeology: A Critique. *New Zealand Journal of Archaeology* 3:146–160.

1981b A Model of Prehistoric Collecting on the Rocky Shore. *Journal of Archaeological Science* 8:109–120.

1986 *Mahinga ika o te moana:* Selection in the Pre-European Fish Catch of Southern New Zealand. In *Traditional Fishing in the Pacific,* edited by A. Anderson, pp. 151–165. Pacific Anthropological Records 37. Bishop Museum Press, Honolulu.

1988 *Wahi Mahinga Kai o Ngai Tahu.* Evidence for the Waitangi Tribunal, Wai 27. The Waitangi Tribunal, Department of Justice, Wellington, New Zealand.

1989 *Prodigious Birds: Moas and Moahunting in prehistoric New Zealand.* Cambridge University Press, Cambridge.

1991 The Chronology of Colonization in New Zealand. *Antiquity* 65:767–795.

1995 Historical and Archaeological Aspects of Muttonbirding in New Zealand. *New Zealand Journal of Archaeology* 17:35–55.

1996a *Te Whenua hou:* Prehistoric Polynesian Colonization of New Zealand and Its Impact on the Environment. *Historical Ecology in the Pacific Islands,* edited by P.V. Kirch and T. Hunt, pp. 271–283. Yale University Press, New Haven, Connecticut.

1996b Origins of Procellariidae Hunting in the Southwest Pacific. *International Journal of Osteoarchaeology* 6:1–8.

1997 Uniformity and Regional Variation in Marine Fish Catches from Prehistoric New Zealand. *Asian Perspectives* 36:1–26.

1998 *The Welcome of Strangers: An Ethnohistory of Southern Maori AD 1650–1850.* University of Otago Press, Dunedin.

2000a The Advent Chronology of South Polynesia. In *Essays in honour of Arne Skjolsvold 75 years,* edited by P. Wallin and H. Martinsson-Wallin, pp. 73–82. Occasional Papers 5. Kon-Tiki Musem, Oslo.

2000b Slow Boats from China: Issues in the Maritime Prehistory of the Indo-Pacific Region. In *East of Wallace's Line: Studies of Past and Present Maritime Cultures of the Indo-Pacific Region,* edited by S. O'Connor and P. Veth, pp. 13–50. Modern Quaternary Research in Southeast Asia 16. Balkema, Rotterdam.

2000c Defining the Period of Moa Extinction. *Archaeology in New Zealand* 43:195–200.

2001 No Meat on that Beautiful Shore: The Prehistoric Abandonment of Subtropical Polynesian Islands. *International Journal of Osteoarchaeology* 11:14–23.

2002 A Fragile Plenty: Pre-European Maori and the New Zealand Environment. In *Environmental Histories of New Zealand,* edited by E. Pawson and T. Brooking, pp. 19–34. Oxford University Press, Melbourne.

2003 Initial Human Dispersal in Remote Oceania: Pattern and Explanation. In *Pacific Archaeology: Assessments and Prospects,* edited by C. Sand, pp. 71–84. Le Cahiers de l'Archaeologie en Nouvelle-Caledonie 15. Noumea, New Caledonia.

2004 Investigating Early Settlement on Lord Howe Island. *Australian Archaeology* 57:98–102.

2005 Subpolar Settlement in South Polynesia. *Antiquity* 79:791–800.

2006 Retrievable Time: Prehistoric Colonization of South Polynesia from the Outside In and the Inside Out. In *Disputed Histories: Essays in Honour of Erik Olssen,* edited by T. Ballantyne and B. Moloughney, pp. 25–41. University of Otago Press, Dunedin.

Anderson, A., and M. McGlone

1991 Living on the Edge: Prehistoric Land and People in New Zealand. In *The Naive Lands: Human-Environmental Interactions in Australia and Oceania,* edited by J. Dodson, pp. 199–241. Longman Cheshire, Sydney.

Anderson, A., and Y. Sinoto

2002 New Radiocarbon Ages of Colonization Sites in East Polynesia. *Asian Perspectives* 41:242–257.

Anderson, A., and I. Smith

1992 The Papatowai Site: New Evidence and Interpretations. *Journal of the Polynesian Society* 101: 129–158.

1996 The Transient Village in Southern New Zealand. *World Archaeology* 27:359–371.

Anderson, A., and P. White (editors)

2001. *The Prehistoric Archaeology of Norfolk Island, Southwest Pacific.* Records of the Australian Museum, Supplement 27. Australian Museum, Sydney.

Anderson, A., B. Allingham, and I. Smith (editors)

1996 *Shag River Mouth: The Archaeology of an Early Southern Maori village.* ANH Publications, Canberra.

Anderson, A., J. Chappell, M. Gagan, and R. Grove

2006 Prehistoric Maritime Migration in the Pacific Islands: An Hypothesis of ENSO Forcing. *The Holocene* 16:1–6.

Annala, J.H., K.J. Sullivan, N.W. M. Smith, and S.J.A. Varian.

2002 *Report from the Fishery Assessment Plenary, May 2002: Stock Assessments and Yield Estimates, Part 1.* Ministry of Fisheries, Wellington.

Baker, C.S., and P.J. Clapham

2004 Modelling the Past and Future of Whales and Whaling. *Trends in Ecology and Evolution* 19:365–371.

Barber, I.
1994 Culture Change in Northern Te Wai Pounamu. Unpublished Ph.D. dissertation, Anthropology Department, University of Otago, New Zealand.

Beattie, J. H.
1994 *Traditional Lifeways of the Southern Maori*, edited by A. Anderson. University of Otago Press, Dunedin.

Brook, B. W., and D. M. J. S. Bowman
2004 The Uncertain Blitzkrieg of Pleistocene Megafauna. *Journal of Biogeography* 31:517–523.

Buck, P. H.
1938 *Vikings of the Sunrise*. Frederick Stokes, New York.

Burrows, C.
1982 On New Zealand Climate within the Last 1000 years. *New Zealand Journal of Archaeology* 4:157–167.

Davidson, J., and F. Leach
2001 The Strandlooper Concept and Economic Naivety. In *The Archaeology of Lapita Dispersal in Oceania*, edited by G. Clark, A. Anderson, and T. Vunidilo, pp. 115–123. Terra Australia 17. Pandus Books, Canberra.

Davidson, J., F. Leach, K. Greig, and P. Leach
2000 Pre-European Maori Fishing at Foxton, Manawatu, New Zealand. *New Zealand Journal of Archaeology* 22:75–90.

Dickinson, W.
2000 Paleoshoreline Record of Relative Holocene Sea levels on Pacific Islands. *Earth-Science Reviews* 55:191–234.

Falla, R. A., R. B. Sibson, and E. G. Turbott
1979 *The New Guide to the Birds of New Zealand*. Collins, Auckland.

Glynn, P. W.
2004 High Complexity Food Webs in Low-Diversity Eastern Pacific Reef-Coral Communities. *Ecosystems* 7:358–367.

Graham, D. H.
1956 *A Treasury of New Zealand Fishes*. Reed, Dunedin.

Higham, T. G., and A. G. Hogg
1997 Evidence for Late Polynesian Colonization of New Zealand: University of Waikato Radiocarbon Measurements. *Radiocarbon* 39:149–192.

Holdaway, R. N., and A. Anderson
2001 Avifauna from the Emily Bay Settlement Site, Norfolk Island: A Preliminary Account. In *The Prehistoric Archaeology of Norfolk Island, Southwest Pacific*, edited by A. Anderson and P. White, pp. 85–100. Records of the Australian Museum, Supplement 27. Australian Museum, Sydney.

Holdaway, R. N., and C. Jacomb
2000 Rapid Extinction of the Moas (Dinornithiformes): Model, Test and Implications. *Science* 287:2250–2254.

Jackson, J. B. C., M. X. Kirby, W. H. Berger, K. A. Bjorndal, L. W. Botsford, B. J. Bourque, R. H. Bradbury, R. Cooke, J. Erlandson, J. A. Estes, T. P. Hughes, S. Kidwell, C. B. Lange, H. S. Lenihan, J. M. Pandolfi, C. H. Peterson, R. S. Steneck, M. J. Tegner, and R. R. Warner
2001 Historical Overfishing and the Recent Collapse of Coastal Ecosystems. *Science* 293: 629–638.

Johnson, L.
1995 *In the Midst of a Prodigious Ocean: Archaeological Investigation of Polynesian Settlement of the Kermadec Islands*. DOC Resource Series 11. Department of Conversation, Auckland.

Lalas, C., and C. J. A. Bradshaw
2001 Folklore and Chimerical Numbers: Review of a Millennium of Interaction between Fur Seals and Humans in the New Zealand Region. *New Zealand Journal of Marine and Freshwater Research* 35:477–497.

Leach, F.
1981 The Prehistory of the Southern Wairarapa. *Journal of the Royal Society of New Zealand* 11:11–33.
2003 *Depletion and Loss of the Customary Fishery of Ngati Hinewaka*. Waitangi Tribunal Report Wai 863 #A71. The Waitangi Tribunal, Department of Justice, Wellington, New Zealand.
2006 *Fishing in Pre-European New Zealand*. *New Zealand Journal of Archaeology* special publication and *Archaeofauna* Vol. 15. New Zealand Archaeological Association, Dunedin.

Leach, F., and A. Anderson
1979a Prehistoric Exploitation of Crayfish in New Zealand. In *Birds of a Feather*, edited by A. Anderson, pp. 141–164. BAR S62. British Archaeological Reports, Oxford.
1979b The Role of Labrid Fish in the Prehistoric Economies of New Zealand. *Journal of Archaeological Science* 6:1–15.

Leach, F., and A. Boocock
1993 *Prehistoric Fish Catches in New Zealand*. BAR 584. British Archaeological Reports, Oxford.
1994 The Impact of Pre-European Maori Fishermen on the New Zealand Snapper, *Pagrus auratus*, in the Vicinity of Rotokura, Tasman Bay. *New Zealand Journal of Archaeology* 16: 69–84.

Leach, F., and J. Davidson
2000 Pre-European Catches of Snapper *(Pagrus auratus)* in Northern New Zealand. *Journal of Archaeological Science* 27:509–522.

2001 The Use of Size-Frequency Diagrams to Characterize Prehistoric Fish Catches and to Assess Human Impact on Inshore Fisheries. *International Journal of Osteoarchaeology* 11: 150–162.

Leach, F., and H. Leach
1979 *Prehistoric Man in Palliser Bay*. NMNZ Bulletin 21. National Museum of New Zealand, Wellington.

Leach, F., J. Davidson, and M. Horwood
1997 Prehistoric Maori Fishermen at Kokohuia, Hokianga Harbour, Northland, New Zealand. *Man and Culture in Oceania* 13:99–116.

Leach, F., J. Davidson, and K. Fraser
1999a Pre-European Catches of Labrid Fish in the Chatham Islands and Cook Strait, New Zealand. *Man and Culture in Oceania* 15:113–144.
2000 Pre-European catches of Blue Cod *(Parapercis colias)* in the Chatham Islands and Cook Strait, New Zealand. *New Zealand Journal of Archaeology* 21:119–138.

Leach, F., J. Davidson, K. Fraser, and A. Anderson
1999b Pre-European Catches of Barracouta, *Thyrsites atun*, at Long Beach and Shag River, Otago, New Zealand. *Archaeofauna* 8:11–30.

Leach, F., C. Quinn, J. Morrison, and G. Lyon
2003 The Use of Multiple Isotope Signatures in Reconstructing Prehistoric Human Diet from Archaeological Bone. *New Zealand Journal of Archaeology* 23:31–98.

McFadgen, B., and J. Goff
2001 A Discussion of Molluskan Evidence for Late Holocene Climate Change on Motutapu Island, Hauraki Gulf. *Journal of the Polynesian Society* 110:313–316.

McGlone, M., and J. Wilmshurst
1999 Dating Initial Maori Environmental Impact in New Zealand. *Quaternary International* 59: 5–16.

McGlone, M., A. Anderson, and R. N. Holdaway
1994 An ecological approach to the Polynesian settlement of New Zealand. In *The Origins of the First New Zealanders*, edited by Douglas Sutton, pp. 136–163. Auckland University Press, Auckland.

McKinzey, K., M., W. Lawson, D. Kelly, and A. Hubbard
2004A Revised Little Ice Age Chronology of the Franz Josef Glacier, Westland, New Zealand. *Journal of the Royal Society of New Zealand* 34:381–394.

Moniz, J.
1997 The Role of Seabirds in Hawaiian Subsistence: Implications for Interpreting Avian Extinction and Extirpation in Polynesia. *Asian Perspectives* 36:27–50.

Moy, C. M., G. O. Seltzer, D. T. Rodbell, and D. M. Anderson
2002 Variability of El Niño/Southern Oscillation Activity at Millennial Timescales during the Holocene Epoch. *Nature* 420:162–165.

Murray-McIntosh, R. P., B. J. Scrimshaw, P. J. Hatfield, and D. Penny
1998 Testing Migration Patterns and Estimating Founding Population Size in Polynesia by Using Human mtDNA Sequences. *Proceedings of the National Academy of Sciences* 95: 9047–9052.

Nagaoka, L.
2002 Explaining Subsistence Change in Southern New Zealand Using Foraging Theory Models. *World Archaeology* 34:84–102.

Nichol, R.
1986 Analysis of Midden from N44/215: Hard Times at Hahei? In *Traditional Fishing in the Pacific*, edited by A. Anderson, pp. 179–188. Pacific Anthropological Records 37. Bishop Museum, Honolulu.
1988 Tipping the Feather against the Scale: Archaeozoology from the Tail of the Fish. Unpublished Ph.D. dissertation, Anthropology Department, University of Auckland, New Zealand.

Paul, L.
1986 *New Zealand Fishes: An Identification Guide*. Reed Methuen, Auckland.

Paulin, C., A. Stewart, C. Roberts, and P. McMillan
1989 *New Zealand Fish: A Complete Guide*. National Museum of New Zealand, Wellington.

Pauly, D., V. Christensen, J. Dalsgaard, R. Froese, and F. Torres, Jr.
1998 Fishing down Marine Food Webs. *Science* 279:860–863.

Raffaelli, D.
2004 How Extinction Patterns Affect Ecosystems. *Science* 306:1141–1142.

Reyment, R. A.
1971 *Introduction to Quantitative Paleoecology*. Elsevier, Rotterdam.

Richards, R.
1994 The Upland Seal of the Antipodes and Macquarie Islands: A Historian's Perspective. *Journal of the Royal Society of New Zealand* 24: 289–295.
2002 Southern Right Whales: A Reassessment of Their Former Distribution and Migration Routes in New Zealand Waters, Including on the Kermadec Grounds. *Journal of the Royal Society of New Zealand* 32:355–377.
2003 New Market Evidence on the Depletion of Southern Fur Seals: 1788–1833. *New Zealand Journal of Zoology* 30:1–9.

Rowland, M. J.
1976 *Cellana denticulata* in Middens on the Coromandel Coast, New Zealand: Possibilities of a Temporal Horizon. *Journal of the Royal Society of New Zealand* 6:1–15.
Sharp, G. D.
1999 Fishery Catch Records, El Niño/Southern Oscillation, and Longer-Term Climate Change as Inferred from Fish Remains in Marine Sediments. In *El Nino: Historical and Paleoclimate Aspects of the Southern Oscillation*, edited by H. F. Diaz and V. Margraf, pp. 380–417. Cambridge University Press, New York.
Shawcross, W.
1967 An Investigation of Prehistoric Diet and Economy on a Coastal Site at Galatea Bay, New Zealand. *Proceedings of the Prehistoric Society* 33:107–131.
Smith, E. A., and M. Wishnie
2000 Conservation and Subsistence in Small-Scale Societies. *Annual Review of Anthropology* 29:493–524.
Smith, I.
1989 Maori Impact on the Marine Megafauna: Pre-European Distribution of New Zealand Sea Mammals. In *Saying So Doesn't Make It So: Papers in Honour of B. F. Leach*, edited by D. Sutton, pp. 76–108. NZAA Monograph 17. New Zealand Archaeological Association, Dunedin.
2002 *The New Zealand Sealing Industry*. Department of Conservation, Wellington.
2004 Nutritional Perspectives on Prehistoric Marine Fishing in New Zealand. *New Zealand Journal of Archaeology* 24:5–31.
2005 Retreat and Resilience: Fur Seals and Human Settlement in New Zealand. In *The Exploitation and Cultural Importance of Sea Mammals*, edited by Gregory Monks, pp. 6–18. Oxbow Books, Oxford.
Steadman, D. W.
1995 Prehistoric Extinctions of Pacific Islands Birds: Biodiversity Meets Zooarchaeology. *Science* 267:1123–1131.
Steneck, R. S., J. Vavrinec, and A. V. Leland
2004 Accelerating Trophic-Level Dysfunction in Kelp Forest Ecosystems of the Western North Atlantic. *Ecosystems* 7:323–332.
Sutton, D. G.
1982 Towards the Recognition of Convergent Cultural Adaptation in the Subantarctic Zone. *Current Anthropology* 23:77–97.
1989 Moriori Fishing: Intensive Exploitation of the Inshore Zone. In *Saying So Doesn't Make It So: Papers in Honour of B. F. Leach*, edited by D. Sutton, pp. 116–131. NZAA Monograph 17. New Zealand Archaeological Association, Dunedin.
Swadling, P.
1977 The Implications of Shellfish Exploitation for New Zealand Prehistory. *Mankind* 11:11–18.
Szabó, K.
2001a Molluskan Evidence for Late Holocene Climate Change on Motutapu Island, Hauraki Gulf. *Journal of the Polynesian Society* 110:79–87.
2001b Letter to the Editor. *Journal of the Polynesian Society* 110:436–439.
Tennyson, A., and P. Martinson
2006 *Extinct Birds of New Zealand*. Te Papa Press, Wellington.
Thresher, R. E.
2002 Solar Correlates of Southern Hemisphere Mid-Latitude Climate Variability. *International Journal of Climatology* 22:901–915.
Walter, R., and I. Smith
1998 Identification of a New Zealand Fur Seal *(Arctocephalus forsteri)* Bone in a Cook Island Archaeological Site. *New Zealand Journal of Marine and Freshwater Research* 32:483–487.
Williamson, I., and M. D. Sabath
1982 Small Population Instability and Island Settlement Patterns. *Human Ecology* 12:21–34.
Worm, B., E. B. Barbier, N. Beaumont, J. E. Duffy, C. Folke, B. S. Halpern, J. B. C. Jackson, H. K, Lotze, F. Micheli, S. R. Palumbi, E. Sala, K. A. Selkoe, J. J. Stachowicz, and R. Watson.
2006 Impacts of Biodiversity Loss on Ocean Ecosystem Services. *Science* 314:787–790.
Worthy, T. H.
1997 What Was on the Menu? Avian Extinction in New Zealand. *New Zealand Journal of Archaeology* 19:125–160.

3

Aleut Hunters, Sea Otters, and Sea Cows

THREE THOUSAND YEARS OF INTERACTIONS IN THE WESTERN ALEUTIAN ISLANDS, ALASKA

Debra G. Corbett, Douglas Causey, Mark Clementz, Paul L. Koch, Angela Doroff, Christine Lefèvre, and Dixie West

AMERICAN ARCHAEOLOGISTS have long considered environmental reconstruction integral to interpreting ancient cultures. Interdisciplinary studies involving geologists, soil scientists, palynologists, faunal analysts, and tree ring specialists all contribute to the interpretation of past human cultures and provide a framework for interpreting cultural change (Butzer 1982; Watson et al. 1971). In recent years, environmental reconstructions attracted the interest of wider audiences as biologists and environmentalists have sought information to restore damaged ecosystems. As our understanding of ecosystem processes increases, it is clear historical conditions and records are inadequate to characterize ecosystem variability and function. Paleoenvironmental studies are gaining greater urgency due to concern over global climate change. Marine ecosystems are also receiving increased scrutiny due to the collapse of virtually every major fishery around the world in recent decades (Amorosi et al. 1996; Jackson et al. 2001; Lyman 1996; Redman 1999; Reitz 2004).

Among other findings is the recognition that human impacts are far more pervasive and subtle than previously recognized. Preindustrial human impacts have shaped the modern world in ways we are only beginning to comprehend (Flannery 1994, 2001; Jackson et al. 2001; Mann 2005; Nicholson and O'Connor 2000).

Bering Sea fisheries are among the most productive on earth. Presently 25 species of fish, crustaceans, and mollusks are commercially harvested (National Research Council 1996). In 2003, 1.5 million metric tons of pollack alone were worth, after processing, $900 million (Heath et al. 2004). In 1977, the sea suddenly increased in temperature by 2°C, and fishermen and researchers alike have noticed ominous changes in the region's fish, bird, and sea mammal populations. Several species, including nearly all of the mammals, have undergone large and sudden population fluctuations, and fish catches are declining (National Research Council 1996).

This chapter presents preliminary research into the functioning of the nearshore marine

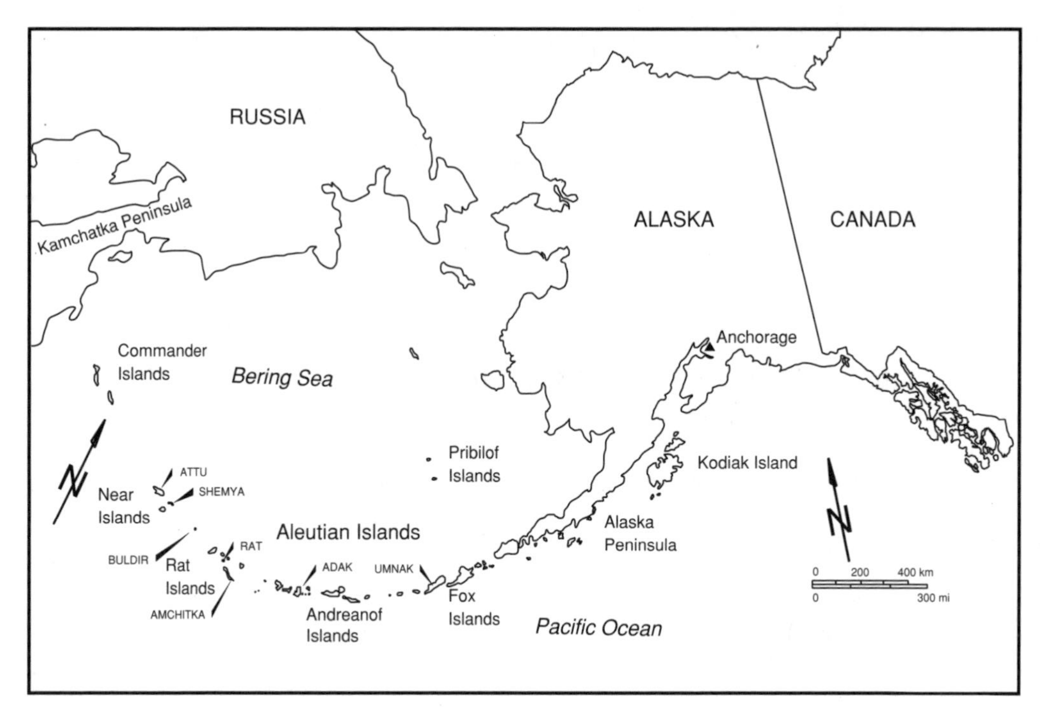

FIGURE 3.1. Aleutian Islands Project area.

ecosystem in the western Aleutian Islands over the last 3,000–5,000 years. Coastal resources were critically important to prehistoric Aleuts, and they intensively exploited shellfish, fish, birds, and mammals, especially in nearshore environments. Specifically, we look at interactions between Aleut hunters and sea otters, a keystone species in the nearshore environment. Human exploitation of these animals affected the structure and function of the nearshore kelp forest ecosystem, with corresponding effects on other organisms dependent on this zone. As one example, the presence of sea cow bones indicates that these animals coexisted with humans in the Aleutians until the late eighteenth century. We believe human hunting of sea otters and the resulting effects on the nearshore environment limited the habitat available for sea cows and kept their populations low.

BACKGROUND

Extending 1,800 km between the American and Asian continents, the Aleutian Islands mark the boundary between the north Pacific Ocean and the Bering Sea (Figure 3.1). This treeless, fog-bound chain of volcanic mountains consists of over 200 islands divided into six clusters separated by wide and rough ocean passes. The climate is subarctic maritime, with cool temperatures, frequent fogs and rain, and violent winds. Biotic communities demonstrate that the chain is an extension of both Asia and North America. Currents flowing through the interisland passes force tongues of cold, nutrient-rich waters into warm, oxygen-rich waters. This functions as an estuarine system, creating greater biological productivity than either ocean possesses alone (Favorite et al. 1976). Massive spring plankton blooms form the base of a rich, diverse food chain. The nutrient-rich oceans support abundant marine animal life: whales, sea lions, seals, sea otters, seabirds, fish, and invertebrates.

The marine environment provided substantial resources for the human occupants of the archipelago. Prehistoric Aleuts lived in large, stable villages, with complex social and political structures supported entirely by an economy based on exploitation of marine and coastal resources.

Sea otters are another top carnivore in the nearshore ecosystem. They prefer sea urchins but prey on a wide variety of benthic invertebrates and bottom fishes (Burgner and Nakatani 1972; Kenyon 1969; Lensink 1962). Palmisano and Estes (1977) first suggested that sea otter predation could affect the structure of the nearshore community, and the presence or absence of otters was the prime cause for differences observed between nearshore communities at Amchitka and in the Near Islands. They hypothesized that dense populations of sea otters control sea urchins, which graze on marine algae, releasing the vegetation from grazing pressure. This resulted in large intertidal kelp beds (Jones and Kain 1967; Paine and Vadas 1969) that support a wide range and density of marine animals. Conversely, the lack of otter predation allows sea urchins to overgraze the vegetation, leaving behind bare rock. The absence of kelp beds increases exposure to the intertidal area because all wave energy reaches shore. Increased herbivory and exposure create open patches in kelp beds that permit establishment of sessile invertebrate populations. The resulting nearshore and intertidal communities support a less-diverse community of marine animals.

Laughlin (1980) suggested Aleuts influenced sea otter populations directly by hunting and indirectly by competing for sea urchins as food. Palmisano and Estes (1977) doubted Aleuts could effectively compete for sea urchins but agreed that they were certainly capable of reducing otter populations. Evolutionarily, humans are recent additions to the Aleutian ecosystem, and their role as predators is unclear. In particular, biologists are trying to reconcile archaeological evidence for low sea otter populations with historic harvest records indicating abundant otter populations (Estes 1990).

History of the Project

The Western Aleutians Archaeological and Paleobiological Project (WAAPP) grew out of biologically oriented research on archaeological collections from the Aleutian Islands. Problems associated with poor excavation techniques limited the usefulness of existing collections, and it became apparent new collections were needed (Lefèvre and Siegel-Causey 1993; Lefèvre et al. 1997; Siegel-Causey et al. 1991a). Fieldwork was launched in 1991 with an archaeological reconnaissance on Buldir Island. From that beginning the project quickly expanded into an interdisciplinary research program aimed at understanding the human and environmental history of the western Aleutians. The team eventually included archaeologists, zooarchaeologists, biologists, a taphonomist, geologist, soils expert, and botanist. Fieldwork began on Buldir (1991, 1993) and Shemya (1994). In 1997, with funding from the National Science Foundation and the Institut Français pour la Recherche et la Technologie Polaires, work resumed on Buldir (1997, 2001) and Shemya (1999) and expanded to Attu (1998, 2000, 2002, and 2003), Adak (1999), and Rat islands (2001, 2003).

From the beginning the WAAPP had a dual focus. First was to identify, document, and define the characteristics and development of the distinctive western Aleut culture. The second aimed to document Holocene environmental processes and determine to what extent observed changes can be ascribed to natural and/or anthropogenic factors. Archaeological research is one of the few ways to obtain information on ecosystem function and process over hundreds or thousands of years. Over the 12 years of the project, paleoenvironmental work became increasingly biologically oriented. With the assistance of ecologists and biologists, the WAAPP team refined its methodologies to address specifically biological issues. Although analysis is in the early stages, this chapter presents a portion of this biologically focused work.

A Brief Outline of Aleutian Culture History

Archaeological work has been conducted in the Aleutian Islands for more than 120 years. Most early work was poorly executed and remains poorly published. In 1874, Dall (1877), a member of the U.S. Coast Survey, tested shell middens on Attu, Agattu, and Amchitka, as well as islands further east. He was the first to recognize the Aleutians had a large prehistoric population, were colonized from the Alaskan mainland, and had been occupied for a long time.

Jochelson (1925, 1933) spent 19 months in 1907–1908 excavating sites on Unalaska, Umnak, Atka, and Attu. He excavated 57 features in 13 sites and six caves. Jochelson's team of anthropologists, geologists, meteorologists, and zoologists inaugurated multidisciplinary studies in the Aleutians. His work remains a benchmark for Aleutian archaeology.

Hrdlicka (1945), a physical anthropologist, spent three seasons in the Aleutians excavating sites on Attu, Agattu, Kiska, Little Kiska, and Amchitka, as well as further east. His work was crude and poorly documented, and he was primarily interested in the collection of human remains.

The only fully reported Near Island excavation is that of Spaulding (1962) at Krugloi Point, Agattu. Spaulding's goals included dating the earliest occupation, obtaining data on regional cultural variability and the nature of temporal change, and recovering human skeletal material.

In the late 1960s, the Atomic Energy Commission sponsored a series of archaeological investigations on Amchitka prior to a series of underground nuclear tests. A descriptive monograph resulted (Desautels et al. 1971), one of the few from the Aleutians. In all, Desautels et al. (1971) excavated seven sites. Quantitative faunal samples were recovered and analyzed from RAT-031. A related project on Amchitka resulted in a second short monograph (Cook et al. 1972).

Other excavations were small scale and remain unpublished. Turner (1886, 1891) excavated near the historic village in Chichagof Harbor, Attu, and the National Park Service tested a site in Massacre Bay, Attu (U.S. National Park Service 1968). Corbett (1990) tested several sites on Shemya Island. More recent work has consisted of site surveys with little excavation (Frohlich and Kopjansky 1975; McCartney 1974; U.S. Bureau of Indian Affairs [US BIA] 1985, 1986, 1988, 1989a, 1989b, 1990, 1994, 1995).

The earliest evidence of human occupation in the Aleutian Islands is from Anangula Island, off the southwestern end of Umnak Island, and dated to around 8000 uncalibrated radiocarbon years BP (Laughlin and Marsh 1951, 1954; McCartney and Turner 1966). Additional Anangula style sites have been located in Unalaska Bay. Called the Early Anangula Phase (Knecht and Davis 2001:269–288), it is characterized by a stone-tool assemblage of prismatic blades struck from polyhedral cores, transverse burins, and end and side scrapers. No bifacial tools have been found at Anangula. No faunal remains have been found, but the people almost certainly had a maritime adaptation.

A Late Anangula Phase (7000–4000 BP) follows Early Anangula, differing from it by the addition of stemmed points, bifaces, and semi-subterranean houses to the assemblage (Knecht and Davis 2001). Dumond (1987a, 2001) has suggested this phase shares similarities with the Ocean Bay tradition of Kodiak.

The Margaret Bay Phase (4000–3000 BP) follows and is notable for houses with coursed stone walls, clay floors, interior fireplaces, and side entries (Denniston 1966; Dumond 1987b; Knecht and Davis 2001:271). Elements of the lithic technology are reminiscent of the Arctic Small Tool tradition (Knecht et al. 2001).

By 3000 BP, the Aleutian tradition was characterized by an elaborate, sea-hunting adaptation (McCartney 1984). Knecht and Davis (2001) divide the Aleutian tradition into two phases: Amaknak (3000–1000 BP) and Late Aleutian (1000–200 BP). Lithic technology is based on irregular cores and bifacial reduction. Harpoon foreshafts, socket pieces, and elaborately barbed harpoon and lance points characterize the bone industry. Sites are marked by thick midden accumulations made up of sea urchin, shellfish, and fish remains (McCartney 1984:124).

The timing of the Aleutian Tradition expansion through the chain is subject to dispute, but

TABLE 3.1
Aleut Procurement Systems

PROCUREMENT SYSTEM	OPEN-SEA HABITAT	LITTORAL HABITAT	TERRESTRIAL HABITAT
Offshore sea mammal hunting	X	X	
Onshore sea mammal hunting		X	
Bird hunting on water	X	X	
Onshore bird hunting		X	X
Offshore fishing	X	X	
Onshore fishing		X	X
Intertidal collecting		X	
Onshore collecting			X

recent dates suggest a rapid spread by 6000 BP to Adak, and between 4700 and 4300 BP to Amchitka, indicating the central and western islands were occupied during the Margaret Bay Phase. Dates from the Near Islands are at least 1,200 years younger than those from the Rat Islands (Corbett 1990, 1991; West et al. 1998b).

The Aleut Maritime Economy

The Aleuts depended almost exclusively on the sea for their material needs (Table 3.1). Limited terrestrial resources were insufficient to sustain human populations. However, the rich marine environment provided substantial resources for the human occupants of the archipelago. Aleuts hunted sea mammals and seabirds, fished, and collected invertebrates from seashores and reefs. They used skins from sea otters, pinnipeds, birds, and even fish for clothing, bedding, containers, and boat covers. Tools and weapons were made from driftwood, bone, and ivory. The region's natural riches allowed people to live in large, stable villages and develop complex social and political structures.

Coastal environments were of critical importance to the Aleuts. McCartney (1977) identified three "habitats" exploited by the Aleuts—open sea, littoral, and terrestrial—and quantified their contribution to the economy. He emphasized the importance of the littoral zone. The number of settlements, and thus the population density of an island, was highest in areas with the largest proportion of shallow (less than 60 m) water. McCartney (1977) identified eight procurement systems, defined as the habitat, resources, tools, and social and technological processes for extracting and processing of resources, all of which were involved in exploiting the three habitats. Although habitat and procurement systems overlap, the critical importance of the littoral zone is clear.

Aleut settlements reflect this reliance. Virtually all sites are located near modern shorelines, with easy access to beaches, but above storm tides. In the Near Islands, 70 percent of sites are in small protected bays. Sites were also located near freshwater and had some degree of protection from the elements (Corbett 1991; Corbett et al. 2007).

Veniaminov (1984) reported that all villages were self-sufficient and relied on resources located proximally to the settlement. Clark (1990) noted a correlation with shellfish resources. Yesner (1977) and Haggarty et al. (1991) predicted that the earliest occupations, and all permanent settlements, would be located in the area with the greatest variety of resources. Faunal remains from Attu and Shemya indicate resources were intensively exploited within about 4 km of villages. The average spacing of settlements in the Near Islands reflects this territoriality (Corbett et al. 2007).

TABLE 3.2
Excavations and Units

ISLAND GROUP	ISLAND	SITE NUMBER	UNITS	UNIT SIZE (METER)	DEPTH (METER)	DEPOSIT TYPE
Andreanov	Adak	ADK-011	1	2 × 2	0.6	Clam shell midden
Andreanov	Adak	ADK-011	1	2 × 2	0.5	Ashy soil
Rat	Amchitka	RAT-031	4	2 × 2 ea	2.5	Sea urchin midden
Rat	Rat	RAT-081	1	2 × 2	0.8	House floor
Rat	Rat	RAT-081	1	2 × 2	1.1	Sea urchin midden
Buldir	Buldir	KIS-008	1	1 × 0.5	0.5	Sea mammal bone
Buldir	Buldir	KIS-008	2	2 × 6 total	0.9	Anaerobic midden
Buldir	Buldir	KIS-008	1	4 × 7	0.6	House deposits
Buldir	Buldir	KIS-008	2	2 × 2 ea	1.5	Anaerobic midden
Near	Shemya	ATU-003	3	1 × 2 ea	1.2	Sea urchin midden
Near	Shemya	ATU-021	1	2 × 2	0.8	Sea urchin midden
Near	Shemya	ATU-022	3	2 × 2 each	1.8	Sea urchin midden
Near	Shemya	ATU-061	1	2 × 2	2.1	Sea urchin midden
Near	Attu	ATU-014	4	2 × 2 each	0.6	House deposits
Near	Attu	ATU-014	1	1 × 1	0.6	House deposits
Near	Attu	ATU-014	1	1.5 × 1.5	1.2	Sea urchin/ Periwinkle midden
Near	Attu	ATU-019	2	2 × 2	0.8	House deposits
Near	Attu	ATU-019	2	2 × 2	1.2	Sea urchin midden
Near	Attu	ATU-019	1	2 × 2	0.9	Storage pits
Near	Attu	ATU-019	1	2 × 2	0.9	Midden and house

SITES, EXCAVATIONS, AND ANALYSIS

Ten sites from six islands in the Near, Rat, and Andreanov groups of the Aleutian Islands (Figure 3.1, Table 3.2) were examined to characterize the structure of the nearshore marine ecosystem over the last 3,000 to 5,000 years. We examine top level predator-prey relationships, specifically interactions among Aleut hunters, sea otters, and sea urchins. ADK-011 is the only central Aleutian Island site included in this analysis. It is a large site on the east coast of Zeto Point on the north coast of Adak. The excavations at this site lacked the typical sea urchin midden of most sites (West et al. 1999). Two sites from the Rat Islands are included. RAT-031, located on the Pacific Coast of Amchitka Island, was excavated by Desautels et al. (1971) in the late 1960s. RAT-081 is a small site on the north shore of Rat Island excavated by WAAPP in 2003 (West et al. 2003). Both sites are predominantly sea urchin middens. The only known site on Buldir is on the north coast at the mouth of a small valley. Archaeologists excavated on the island in 1991, 1993, 1997, and 2001 (Corbett et al. 1997; Siegel-Causey et al. 1991b; West et al. 1997, 2001). Six of the seven units were in a water-saturated, anaerobic midden filled with wooden artifacts and other organic debris. The seventh pit sampled a structure built of whalebones. The island coastline is steep and mountainous, lacking shallow inshore waters and rocky reefs; fish and shellfish remains are therefore rare in the excavations.

Four sites were excavated on Shemya Island during field work in 1994 and 1999 (Corbett et al. 2007; Siegel-Causey et al. 1994; West et al. 1999). All four sites consist primarily of sea urchin shell midden. Site ATU-021 is on the northwest coast and originally consisted of two midden mounds. ATU-022 at the northeast tip of Shemya Island originally stretched nearly a kilometer along the beach and had at least

60 house depressions. ATU-003 is near the western end of the runway, on the south coast. ATU-061 is also on the south coast, west of Laundry Lake, and sits well back from the present shoreline. It is the smallest site on Shemya and is one of the oldest known from the Near Islands. Shemya is one of a group of three tiny islands in the Near Island group, often lumped together as the Semichi Islands. Shemya has been a military base since World War II, and every site has been extensively damaged by construction.

Two large midden sites were excavated on Attu Island in 1998, 2002, and 2003 (West et al. 1998a, 2002, 2003). ATU-014 is located on Murder Point in Massacre Bay on the southeast coast. One excavation focused on a whalebone house feature. A second pit was placed in a deep, nearly pure midden section at the south end of the site. ATU-019 is in Austin Cove on the north coast. This large midden was mapped and tested in 1998, and a large house partially excavated in 2002. Four pits excavated adjacent to the house sampled associated middens and features.

Dates

Every site has been dated with multiple ^{14}C dates on clearly defined and identified stratigraphic units. Generally speaking, there are six old site components (800 BC to AD 200), five early components or sites (AD 0–600), five early/middle components or sites (AD 0–1300), three middle components (AD 600–1300), and five late components and sites (AD 1200–1800) (Table 3.3). A variety of materials were dated, which increases uncertainty as bone and shell from marine animals can yield dates that appear to be significantly older than they actually are. The marine reservoir for the western Aleutians is up to 1,000 years (Dumond and Griffin 2002), but the central Aleutians reservoir has not been tested or defined.

Methods

Bird and mammal remains were collected from all levels in all excavations. For analysis of sea urchin volumes and sizes, bulk samples from four sites were used (ATU-014, ATU-019, RAT-031, and RAT-081). To provide the greatest comparability among studies, we utilized similar methodologies among sites, and between our project and Desautels et al.'s (1971). Bulk samples from both projects were collected from measured areas (volume) and sorted using quarter-inch- and eighth-inch-mesh screens. Weights and volumes of sea urchin material in each fraction were measured separately, but related units were combined for analysis. Desautels et al. (1971) excavated 20-×-30-cm-column samples following the natural stratigraphy exposed by their large block excavation. The sample was water screened through quarter-inch, eighth-inch, sixteenth-inch, and twenty-second-inch mesh. The quarter-inch and eighth-inch fractions were sorted into bird, mammal, and fish bone, sea urchin shell, other shell, artifacts, charcoal, and debris. Each component was weighed, tabulated, and analyzed (Desautels et al. 1971).

Samples from ATU-014, ATU-019, and RAT-081 were taken from 25-cm-square blocks collected from each 10-cm-thick level in each 2-×-2-m excavation unit. The samples were weighed, wet screened in nested quarter-inch and eighth-inch screens, and bagged. After drying, they were sorted into sea urchin shell, Aristotle's lanterns from sea urchins, other shell, bird, fish and mammal bone, stone and bone artifacts, and charcoal. Each component was weighed and bagged separately. At RAT-081, each natural layer within each 10-cm arbitrary level was sampled. Up to four bulk samples were recovered from each level. They were the same size and processed the same way as previously described. For this analysis, like units were combined, and the total volume of each sample calculated for each natural level (Table 3.4).

RESULTS

Sea Otters

Analysis of mammal remains is currently underway. Results presented here are from ADK-011, RAT-031, KIS-008, and the four Shemya sites

TABLE 3.3

Radiocarbon Date List

LAB NO.[a]	AHRS NO.	PIT/FEATURE	LEVEL	RCYBP	CAL BC AD (2 SIGMA, 95% PROBABILITY)	PERIOD
B132882	ADK-011	F-2	Level 5	180 ± 60	AD 1635–1955	Late
B132880	ADK-011	F-1	Unit 1 bottom	2,150 ± 40	BC 360–280, BC 240–60	Old
B132881	ADK-011	F-2	Level 3	220 ± 50	AD 1525–1560, AD 1630–1695, AD 1725–1815	Late
B132879	ADK-011	F-1	Unit 2 Level 5	2,490 ± 50	BC 795–410	Old
B132878	ADK-011	F-1	Unit 1 Level 3	440 ± 40	AD 1420–1500	Late
B133745	ADK-011	F-1	Burial	840 ± 40	AD 995–1160	Middle
I-4737	RAT-031	Unit IV	Stratum IV-H	1,980 ± 95	AD 60	Early
I-4735	RAT-031	Unit I	Stratum I-J	2,550 ± 95	BC 600	Old
I-4736	RAT-031	Unit IV	Stratum III-	890 ± 90	1060 AD	Middle
B187487	ATU-014	SMound	Level 5	880 ± 40	AD 1035–1250	Middle
B187485	ATU-014	N 30	Unit E Level 3/4	1,580 ± 90	AD 250–650	Early
B187482	ATU-014	N 30	Unit C Level 3	1,740 ± 80	AD 95–445	Early
B187480	ATU-014	N30	Unit A Level 4	1,690 ± 60	AD 230–460, AD 480–520	Early
B187486	ATU-014	N30, Pit F	Level 5	1,170 ± 50	AD 720–740, AD 760–990	Middle
B187476	ATU-019	F-73	Unit F Level 7	1,990 ± 70	BC 170–AD 140	Old
B187474	ATU-019	F-73	Unit C Level 8	1,850 ± 60	AD 45–330	Early
B187468	ATU-019	F-73	Unit B, Level 9	1,640 ± 60	AD 250–550	Early
B187469	ATU-019	F-73	Unit C Level 2	890 ± 70	AD 1010–1270	Middle
B187470	ATU-019	F-73	Unit C Level 3	1,400 ± 60	AD 550–710	Early
B187472	ATU-019	F-73	Unit C Level 6	1,770 ± 110	AD 20–530	Early
B187473	ATU-019	F-73	Unit C Level 7	1,760 ± 90	AD 70–450	Early
B187479	ATU-019	F-73	Unit G Level 8	1,720 ± 100	AD 80–550	Early
IE1224	KIS-008	Pit 7	355–358 cm	2,347 ± 84		Early
B54256	KIS-008	Pit 2	7	1,160 ± 50	AD 735–995	Middle
B108969	KIS-008	Pit 4	Roof, Unit H	220 ± 60	AD 1515–1590, AD 1620–1705, AD 1715–1885	Late
B2	KIS-008	Pit 3	Level 4 Cut 8	240 ± 60	AD 1500–1695, AD 1725–1815	Late

B108970	KIS-008	Pit 4	Cut 3 Unit L	250 ± 70	AD 1470–1700, AD 1720–1820, AD 1835–1880	Late
B54254	KIS-008	Pit 1	6	280 ± 50	AD 1480–1675, AD 1775–1800	Late
B71567	KIS-008	Pit 5		320 ± 60	AD 1445–1665	Late
B71566	KIS-008	Pit 5	Level 2 (Cut 1)	350 ± 80	AD 1420–1670, AD 1780–1795	Late
B108972	KIS-008	Pit 4	Cut 4 Unit H	390 ± 90	AD 1410–1660	Late
B54253	KIS-008	Pit 1	1	460 ± 50	AD 1405–1500	Late
B54255	KIS-008	Pit 2	3	530 ± 60	AD 1300–1455	Late
B1	KIS-008	Pit 3	Level 4 Cut 2	530 ± 60	AD 1300–1455	Late
B108965	KIS-008	Pit 6	Level 10	630 ± 60	AD 1275–1420	Middle
B108966	KIS-008	Pit 6	Level 10 Unit C	760 ± 60	AD 1175–1305	Middle
B135539	KIS-008	Pit 7	9	1,240 ± 60	AD1240–1460	Late
IE1228	KIS-008	Pit 7	4 A	1,611 ± 67		Early
IE1292	KIS-008	Pit 7	4 A	1,238 ± 58		Late
IE1278	KIS-008	Pit 7	6 B	1,034 ± 50		Middle
IE1241	KIS-008	Pit 7	10 B	1,185 ± 51		Middle
IE1238	KIS-008	Pit 7	2 B	1,527 ± 41		Middle
IE1225	KIS-008		14 B	1,258 ± 64		Middle
IE1223	KIS-008		248-258 cm	1,888 ± 75		Early
B200550	KIS-008	Pit 4	Unit E	330 ± 40	AD 1460–1650	Late
B200551	KIS-008	Pit 4	Unit E	1,240 ± 40	AD 680–890	Middle
B200552	KIS-008	Pit 4	Unit E	410 ± 50	AD 1420–1530, AD 1550–1630	Late
B200553	KIS-008	Pit 3		710 ± 50	AD 1240–1320, AD 1350–1390	Late
B200554	KIS-008	Pit 3		1,230 ± 40	AD 690–890	Middle
B200555	KIS-008	Pit 3		430 ± 40	AD 1420–1510, AD 1600–1620	Late
B160585	AA-11939	Small feature		1,300 ± 60	AD 1000–1240	Middle
B160586	AA-11939	Large house		2,270 ± 60	BC 50–AD 220	Early
B40420	ATU-003	3	1 (middle)	1,720 ± 70		Early
B39092	ATU-003	2	1 (middle)	1,770 ± 120	BC 5–AD 545	Early
B39090	ATU-003	1	4 (middle)	1,790 ± 160	BC 155–AD 610	Early
B40422	ATU-003	3	3 (bottom)	1,810 ± 60		Early
B39091	ATU-003	1	6 (top)	1,860 ± 90	BC 45–AD 390	Early
IE1177	ATU-003	4	7	2,340 ± 125	BC 50–AD 565	Early
B40421	ATU-003	3	3 (top)	2,030 ± 70		Early

(*continued*)

TABLE 3.3

(Continued)

LAB NO.[a]	AHRS NO.	PIT/FEATURE	LEVEL	RCYBP	CAL BC AD (2 SIGMA, 95% PROBABILITY)	PERIOD
IE1172	ATU-003	4	2	1,935 ± 70	AD 535–860	Early
B131312	ATU-021	1	4	1,700 ± 70	AD 155–530	Early
B131313	ATU-021	1	2	1,980 ± 60	BC 115–AD 135	Old
IE1231	ATU-022	1	Cut 9 Sample 1	1,640 ± 70	AD 790–1170	Middle
B110114	ATU-022	1	Cut 8	2,020 ± 110	BC 365–AD 235	Old
IE1229	ATU-022	1 east	Cut 10 Sample 3	1,375 ± 60	AD 1065–1390	Middle
IE1226	ATU-022	2	Cut 9 Sample 6	1,830 ± 80	AD 620–1000	Middle
IE1175	ATU-061	1	110–120 cm	2,880 ± 155	BC 800–AD 15	Old
IE1206	ATU-061	1	Level 2	2,570 ± 140		Old
IE1205	ATU-061	1	Level 7	3,080 ± 110		Old
B131312	ATU-021	1	4	1,700 ± 70	AD 155–530	Early
B131313	ATU-021	1	2	1,980 ± 60	BC 115–AD 135	Early

[a]Laboratory prefix: B, Beta Analytic, Miami, Florida; IE, Institute of Evolutionary Morphology and Animal Ecology, Moscow; I, Isotopes, Westwood Laboratories, Westwood, New Jersey.

TABLE 3.4

Sea Urchin Volumes

SITE	LEVEL OR UNIT	DATE RANGE	TIME PERIOD	BULK SAMPLE DIMENSIONS	SAMPLE VOLUME (LITERS)	DRY WEIGHT OF URCHIN	GRAMS/LITER OF URCHIN
RAT-031	A	<AD 1060	middle	20 × 20 × 15	6	9.0	1.5
RAT-031	B	AD 1060	middle	20 × 20 × 12	4.8	929.5	193.7
RAT-031	C	AD 60–1060	Early-middle	20 × 20 × 25	10	2,342.5	234.3
RAT-031	D	AD 60–1060	Early-middle	20 × 20 × 10	4	305.8	76.5
RAT-031	E	AD 60–1060	Early-middle	20 × 20 × 70	28	11,349.5	405.4
RAT-031	H	AD 60	Early	20 × 20 × 14	5.6	251.9	45
RAT-031	I	>AD 60	Early	20 × 20 × 19	7.6	2,160.5	284.3
RAT-031	J	600 BC[a]	Old (Early?)	20 × 20 × 17	6.8	66.5	9.8
RAT-081	F-2, Midden A	AD 1470–1810	Late	25 × 25 × 30	18.75	0.1	.005
RAT-081	F-2, Midden B	±AD 1150–1500	Late	25 × 25 × 20	12.5	453.3	36.3
RAT-081	F-2, Midden D	AD 1160–1500	Late	25 × 25 × 30	18.75	3107	165.7
RAT-081	F-2, Midden E	AD 1160–1500	Late	25 × 25 × 10	6.25	10.4	1.7
RAT-081	F-6, Midden B	AD 1420–1630	Late	25 × 25 × 20	12.5	509.1	40.7
RAT-081	F-6, Midden D	±AD 900–1300	Middle	25 × 25 × 20	12.5	314	25.2
RAT-081	F-6, Midden E	AD 670–880	Middle	25 × 25 × 10	6.25	21.4	3.4
ATU-014	F-N30, house, Units A–E	AD 95–520	Early	25 × 25 × 170	106.25	105.7	1.0
ATU-014	F-N30, midden, Unit F	AD 720–990	Middle	25 × 25 × 60	37.5	1025.4	27.3
ATU-014	South Mound	AD 1035–1250	Middle	25 × 25 × 60	37.5	729.9	19.5
ATU-019	House	AD 20–710	Early	25 × 25 × 15	93.75	38.2	0.4
ATU-019	Midden	AD 250–550	Early	25 × 25 × 16	100.0	11,881.7	118.8
ATU-019	Unit F	BC 170–140 AD	Old	25 × 25 × 3	18.75	250	13.3
ATU-019	Unit G	AD 80–550	Early	25 × 25 × 6	37.5	2,290	61

[a]Date on sea mammal bone may be 500 to 1,000 years older than the calendar date.

TABLE 3.5

Otter Remains from Archaeological Sites

SITE	LEVELS	TIME PERIOD	DATE RANGE	NISP	MNI
ADK-011	F-1 bottom	Old	800–100 BC	0	0
ATU-061	Pit 1 Levels 3–7	Old	800 BC–200 AD	1	1
ATU-022	Pit 1 Cuts 8 and 9	Old	BC 400– 300 AD	15	3
ATU-003	Pits 1 and 3	Old	BC 50–AD 500	4	2
ATU-021	Pit 1, Level 2-4	Old	BC–100–AD 500	1	1
RAT-081a	F-2, test pit	Old	BC 50–AD 220	2	1
ATU-019 a	F-73, house	Early	AD 20–1000	0	0
RAT-031	Levels A-J (all units)	Early/Middle	AD 60–1060	744	77
RAT-031	Level A (whole unit)	Early	<AD 1060	17	3
RAT-031	Level B (whole unit)	Middle	AD 1060	213	23
RAT-031	Level C (whole unit)	Early/Middle	AD 60–1060	83	8
RAT-031	Level E (whole unit)	Early/Middle	AD 60–1060	28	3
RAT-031	Level G (whole unit)	Early/Middle	AD 60–1060	23	2
RAT-031	Level K (whole unit)	Early/Middle	AD 60–1060	4	1
RAT-031	Level H (whole unit)	Early	AD 60	50	6
RAT-031	Level I (whole unit)	Early	>AD 60	291	26
RAT-031	Level J (whole unit)	Early	600 BC (date on bone)[a]	35	5
KIS-008	Pit 2	Middle	AD 700–1000	0	0
ADK-011	F-1 L-3	Late	AD 1250–1700	2	2
ADK-011	F-2 L-3, 5	Late	AD 1250–1700	9	2
KIS-008	Pits 6 and 7	Late	AD 1200–1450	0	0
KIS-008 a	Pits 1, 3, 4, 5	Late	AD 1400–1700	0	0
KIS-008 a	Pits 3 and 5	Late	AD 1700–1800	0	0

[a]Phalanges, vertebrae, and ribs are not completely identified, but easily identifiable long bones and skull fragments have been identified.

(Table 3.5). Desautels et al. (1971:315) recovered 11,000 mammal bones from RAT-031. They selected 1,300 complete bones for analysis. Of these, 744 were sea otters, while harbor seals (n = 418 bones), sea lions (n = 76), and northern fur seals (n = 30) were far less common. Fifty percent of the otter remains were from mature animals. Because a large number of sea otter ribs, vertebrae, and phalanges were recovered, and because the bones were generally complete, Desautels et al. (1971) concluded that whole animals had been brought to the sites and processed there.

Only 17 sea otter bones, representing six animals, were recovered from four sites on Shemya, spanning 1,300 years. Mammal remains of any kind were rare in the Shemya excavations (Corbett et al. 2007). Buldir Island is poor sea otter habitat. Nevertheless, the island forms a vital link for otters moving between the Rat and Near Islands. Excavations on Buldir sampled four time periods: AD 700–1000, 1200–1450 AD, AD 1400–1700, and AD 1700–1850+. No otter bones were recovered from any context on Buldir.

The lack of sea otters in our excavations could be the result of sampling. However, all excavation units locally yielded large quantities of sea lion, fur seal, harbor seal, and whale bones. The excavation units at KIS-008 and ADK-011, with few or no sea otter bones, were as large as those by Desautels et al. (1971) which yielded dozens. Excavations on Shemya that yielded sea otter bones were smaller than the units lacking them. We do not have enough bones to determine the butchering patterns as

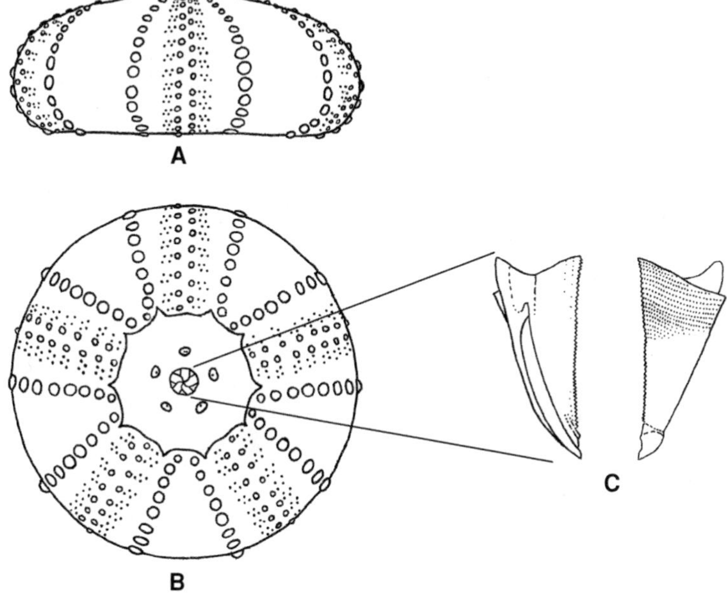

FIGURE 3.2. Sea urchin demipyramids and mouth structure. (A) Side view diagram of a sea urchin (regular echinoid), much simplified for clarity. (B) Base view with mouth central, showing location of the five demipyramids, arranged in a circle around the mouth opening. (C) Sea urchin jaw or Aristotle's lantern demipyramid.

Desautels et al. (1971) did. Hunters could have processed sea otter carcasses away from the sites, but the bones of larger prey such as sea lions would seem to belie this explanation. We conclude that the lack of sea otter remains in Near Island archaeological sites represents a real lack of animals available to human hunters. More abundant sea otter remains from Amchitka suggest the animals were more common or easier to capture on that island.

Sea Urchins

We estimated the sizes and abundance of harvested sea urchins by measuring sea urchin demipyramids, a distinctive articulating structure of the urchin mouth and a common component of sea urchin remains found in sea otter scat and midden deposits (Figure 3.2). The demipyramid is a useful proxy in ecological studies as it is distinctive and durable, and the length relates directly to size and weight in the living animal (Desautels et al. 1971; Simenstad et al. 1978). Furthermore, ecologists and zooarcheologists use this structure in similar ways, which facilitates comparison among approaches.

Not all excavated levels in our studies or in previous studies contained recoverable amounts of sea urchin shell or measurable demipyramids. In all, we recovered sea urchin remains from three sites (ATU-014, ATU-019, RAT-081) and 45 units or levels, 40 of them with nontrivial ($N > 3$) numbers of specimens; in total, we measured 3,216 intact demipyramids. By comparison, the material recovered by Desautels's team was obtained from a single site (RAT-031), seven levels, and included 1,403 intact demipyramids. A graphical summary of the numbers and length distribution of sea urchin demipyramids by site and level is shown in Figure 3.3 and includes our data and those published by Desautels et al. (1971:335).

The urchin demipyramids recovered from three levels excavated in the house feature at site ATU-014-D ranged between 6 mm and 14.5 mm; means ranged between 10.91 mm and 12.14 mm (Figure 3.3). There were no significant differences in size distributions among levels (post-hoc Scheffé F-test, not significant).

Mean lengths of urchin demipyramids recovered from site ATU-014-F, outside the

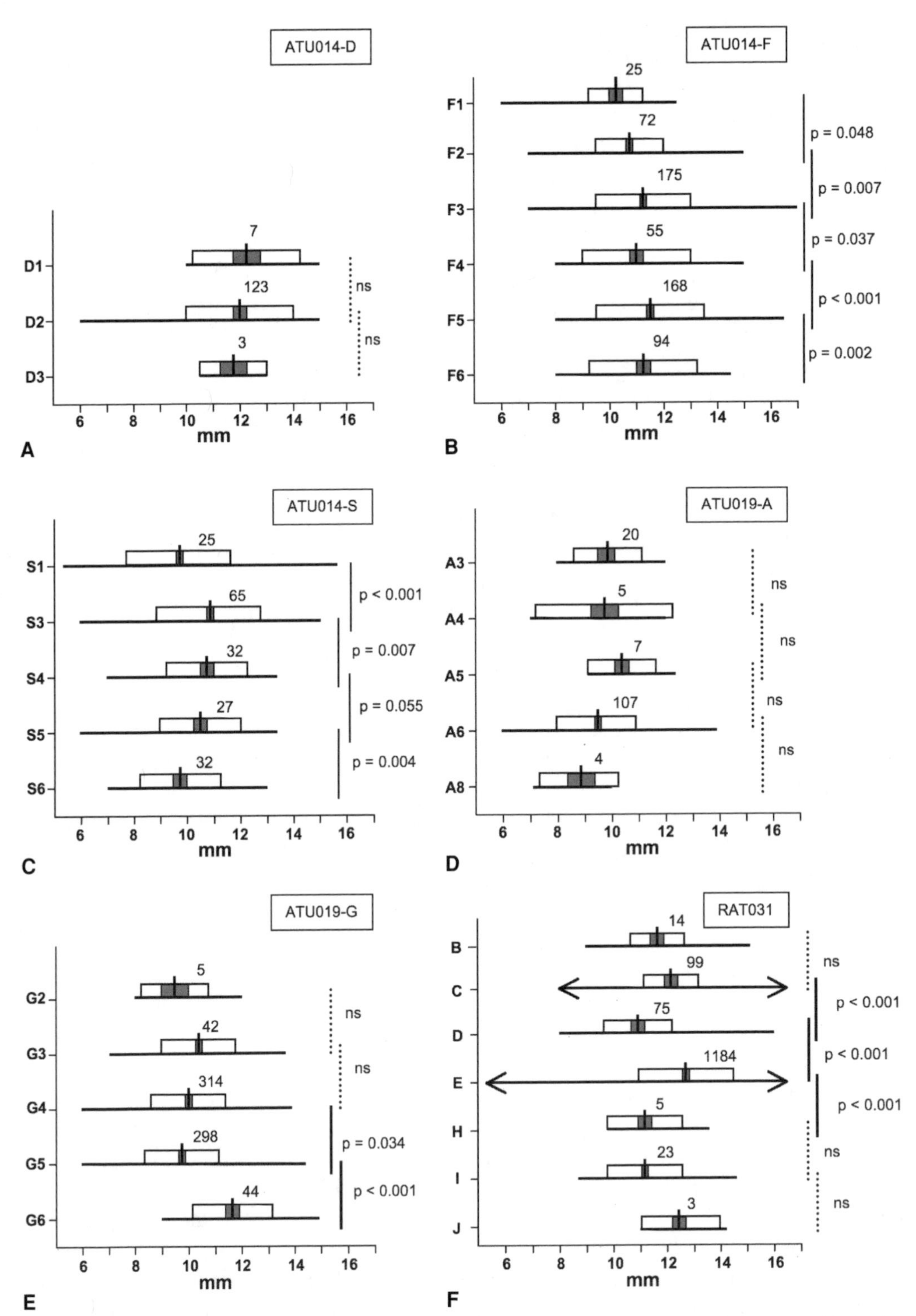

FIGURE 3.3. (A) Size distribution of sea urchin demipyramids recovered from Site ATU-014-D, Levels D1, D2, D3. The horizontal line represents the range of measurements (mm) recovered from each level; the vertical line represents the mean; the open box, one standard deviation from the mean (± 1 Sd); the shaded box, the standard error of the mean (± SEM). The sample size is given adjacent to the mean. Vertical bars at the left indicate pairwise comparisons between distributions; probabilities were determined by student's *t*-test, see text for explanation. (B) Size distribution of sea urchin demipyramids recovered from Site ATU-014-F, Levels F1, F2, F3, F4, F5, F6. (C) Size distribution of sea urchin demipyramids recovered from Site ATU-014-S, Levels F1, F3, F4, F5, F6. (D) Size distribution of sea urchin demipyramids recovered from Site ATU-019-A, Levels A3, A4, A5, A6, A8. (E) Size distribution of sea urchin demipyramids recovered from Site ATU-019-G, Levels G2, G3, G4, G5, G6. (F) Size distribution of sea urchin demipyramids recovered from Site RAT-031, Levels B, C, D, E, H, I, J (from Desautels et al. 1971).

house, ranged from 10.63 mm (level F1) to 11.48 mm (level F5); the length range was 6 mm (level F1) to 17 mm (level F3). The length distributions showed significant differences among all six levels (MANOVA: $F = 53.63$, $df = 5, 1{,}150$, $P < .001$), and pairwise comparisons were also highly significant (*t*-test, Figure 3.3). Mean lengths significantly increased through time from earliest deposits to most recent, but with a pronounced stepped decrease in the upper three levels. Variances tracked mean length, with the greatest dispersion of measurements recovered from earliest levels and the smallest variances in most recent layers.

Mean lengths of urchin demipyramids recovered from site ATU-014-S, a midden at the south end of the site, ranged from 9.75 mm (levels S1, S6) to 10.75 mm (level S3). The length distributions showed significant differences among all levels (MANOVA: $F = 46.47$, $df = 4, 655$, $P < .001$), and pairwise comparisons were also highly significant (*t*-test, Figure 3.3). Mean lengths increased in time from earliest deposits (S6, S5, S4) but leveled and then decreased in the most recent levels (i.e., S1). Variances decreased from earliest levels to most recent times.

The urchin demipyramids recovered from five sea urchin midden levels excavated at site ATU-019-A ranged between 6 mm and 14 mm; means ranged between 9.70 mm and 10.31 mm (Figure 3.3). There were no significant differences in size distributions among levels (post-hoc Scheffé *F*-test, not significant).

Mean lengths of urchin demipyramids recovered from site ATU-019-G, a mix of house and midden deposits, ranged from 9.40 mm (level G2) to 11.61 mm (level G6). The length distributions showed significant differences only among the lower three levels (MANOVA: $F = 21.42$, $df = 2, 244$, $P < .02$), and pairwise comparisons among these levels were significant (*t*-tests, Figure 3.3). Mean lengths decreased in time from earliest deposits (G6–G5) but increased slightly or remained stable after that. Variances decreased from earliest levels to most recent times, with the exception of the uppermost level.

Figure 3.3 displays a similar analysis of Desautels et al.'s (1971) results from RAT-031 on Amchitka Island. Mean lengths of urchin demipyramids ranged from 10.9 mm (level D) to 12.4 (level E). The length distributions showed significant differences only among the middle three levels (MANOVA: $F = 26.98$, $df = 2, 877$, $P < .01$), and pairwise comparisons among these levels were significant (*t*-tests, Figure 3.3). Mean lengths varied complexly, with the lowest levels insignificantly decreasing through time, increasing in length by level E, and then decreasing in most recent levels. The pattern of change in variances was equally complex.

The relationship between length of demipyramid and sea urchin size has been investigated by two independent studies (Desautels et al. 1971; Simenstad et al. 1978), but their results differ considerably. Desautels et al. (1971) found that urchin test diameters were related directly to demipyramid lengths by $Y = -19.248 + 6.330X$, while Simenstad et al. (1978) found the relation to be $Y = -5.948 + 5.7132X$, where Y is the urchin test diameter (mm) and X is the demipyramid length. Minimum and maximum demipyramid lengths, and computed urchin test diameters for each site and unit are given in Figure 3.3.

The temporal trends in size distribution of urchin demipyramids based on results given in Figure 3.3 reveal several complex patterns. For two sites—ATU-014-F and ATU-019-G—mean lengths, variances, and minimum size of demipyramids decreased through time, moving to the present, and more or less tracked together. That is, for these two sites, the distributions became smaller, less variable, and skewed toward small sizes as the sites were occupied longer over time. For site ATU-014-S, these parameters decreased through time as in the other two sites, but these were complex patterns. Here, mean values increased, then decreased through time, variances remained stable, then decreased, and the minimum size decreased but insignificantly (Table 3.6).

Our results from the Near Islands (i.e., ATU-014 and ATU-019) do not compare easily with Desautels et al.'s study on Amchitka Island

TABLE 3.6

Temporal Trends in Sea Urchin Demipyramid Lengths

PARAMETER	ATU-014-F	ATU-014-S	ATU-019-G	RAT-031
Mean	0 –	+ –	–	+ –
Variance	–	0 –	–	+ –
Min. size	–	–	– +	– +

NOTE: 0, stable; +, increase; –, decrease.

(i.e., RAT-031). Their results show that means and variances of harvested sea urchins increased then decreased through time. However, the pattern of minimum size varied inversely, with minimum sizes decreasing early on but gradually increasing in more recent times.

These results imply a strong harvesting influence on sea urchin size by early Aleuts, particularly as indicated by the dramatic decrease in harvested sea urchin sizes, minimum sizes, and decreased variances. Desautels et al. (1971:336) concluded that the presence of larger sea urchins on the intertidal bench at the time of these cultures "has some serious ecological implications." The greatest of these is that harvesting of sea mammals, especially sea otters, initiated a recolonization of the intertidal area by the sea urchin in the absence of its major predator. Faced by the reduced availability of the mammals, the natives turned to the new resources and began systematically harvesting the urchin as a food supply.

HUMAN DEMOGRAPHICS

The distribution and density of human settlement reflects human interactions with their environment. Population estimates for the Aleutian chain range from 7,000 to over 30,000 people (Lantis 1970, 1984; Laughlin 1980; Liapunova 1996). Population estimates based on archaeological data are riddled with uncontrollable assumptions, and censuses are unreliable due to the massive population declines following European contact in the mid-1700s. Laughlin (1980:10) estimated the Near Islands could support 1,000 people, based on a population density of 1.25 Aleuts per square kilometer of land area. The earliest Russian figures are by explorer, M. Nevodchikov, who reported over 100 men on Agattu in 1745, and three villages on Attu with 5 to 15 men each (Black 1984). If each man supported five dependents, Agattu had at least 500 and Attu at least 175 people. These minimum population figures generally support Laughlin's estimates (Black 1984; Corbett 1991). There is less evidence for the early historic population of the Rat Islands, but McLean (Alaska Office of History and Archaeology, personal communication 2005) has estimated a population of 750 for that group. Recent DNA research suggests an archipelago-wide population on the order of 25,000 to 30,000 people (Hayes 2002; O'Rourke, University of Utah, personal communication 2004).

The Early Periods: 800 BC to AD 600

The oldest midden sites on Amchitka appear between 600 and 105 BC (Desautels et al. 1971), around 900 BC on Shemya, and by about 760 BC on Agattu (Spaulding 1962). The population of the islands was apparently small and scattered for several hundred years. Around 400 BC the number of settlements and population began rapidly increasing (Figure 3.4). By AD 500 every dated site on Shemya was occupied. The situation in the Rat Islands is less clear but the information available supports a similar expansion in number of settlements (Figure 3.4). Sites are also not evenly distributed around the island shorelines. Distances between sites on Shemya, Alaid, and Nizki range between .5 and 3.8 km, with an average of 1.9 km (Corbett et al.

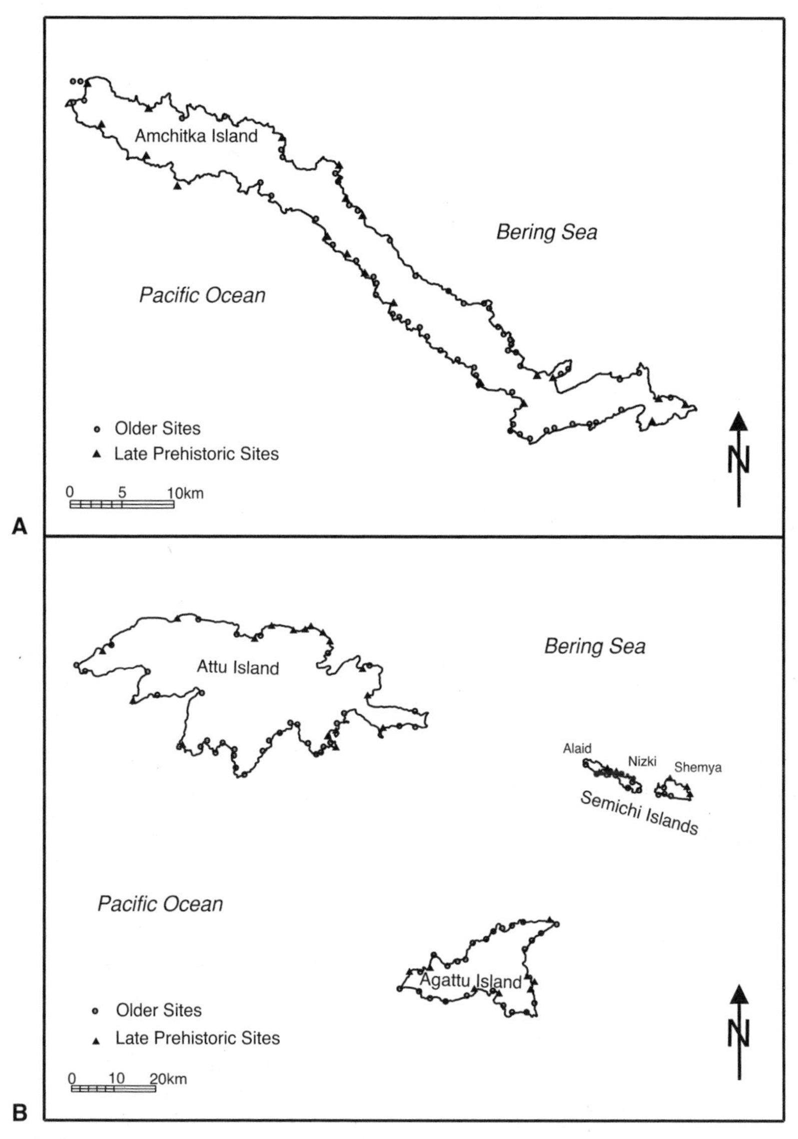

FIGURE 3.4. Settlement locations on (A) Amchitka and in the (B) Near Islands for both early (800 BC–AD 600) and late periods (AD 1200–1800). (A) Base map modified from U.S. Geological Survey Rat Islands, Alaska and (B) base map modified from U.S. Geological Survey Attu, Alaska; 1:250,000 scale quadrangle.

2007). This pattern is repeated on the two areas of Attu Island that have been intensively surveyed (Figure 3.4).

Amchitka Island is one of the most thoroughly surveyed islands in the Aleutian chain, with 83 midden sites recorded (US BIA 1986). They are irregularly distributed around the coast, ranging from 400 m apart in Constantine Harbor to 13 km apart on the southwest coast, and with an average of 4.25 km between sites. Population density on Amchitka was about half that in the Near Islands.

Evidence from Attu suggests these early communities were small, with perhaps 20 to 40 people in each (Corbett et al. 2001). They were sedentary and stable; some attest to occupations spanning 1,500 years (Corbett et al. 2007; Turner 1970). Artifacts suggest all sites shared a similar range of functions, and there is no evidence of seasonal mobility. It appears that

TABLE 3.7

Population Estimates for the Early Period in the Near and Rat Islands

AUTHOR	ISLAND/ GROUP	NO. SITES	NO. HOUSES PER SITE	PEOPLE/ HOUSE	% IN USE	POPULATION ESTIMATE
McCartney (1977	Rat Islands	110	7	6–8	50%	3,300–4,400
McCartney (1977)	Rat Islands	110	7	6–8	25%	1,650–2,200
This paper	Amchitka	83	8	6–8	50%	2,016–2,688
This paper	Amchitka	83	8	6–8	25%	1,008–1,344
This paper	Near Islands	103	10	6	50%	3,000
This paper	Near Islands	103	10	6	25%	1,500

people intensively exploited resources within 2 to 3 km surrounding each village. The Aleutian environment is highly redundant in resource distribution and density—especially intertidal resources. Given this redundancy, a pattern of dispersed settlements may have been more efficient than centralized base camps. The entire coastline of the Semichi Islands and the eastern halves of Attu and Amchitka would have fallen within the economic sphere of a human community (Corbett et al. 2007).

Population Estimates

McCartney (1977) estimated the population of the Rat Islands at 1,650 to 2,200 based on a best fit of archaeological, ecological, and historic information. He based his estimate on an average number of houses per site, multiplied by a number of people per house. Assuming that not every house would have been occupied simultaneously he calculated figures based on 50 percent and 25 percent occupancy. Using McCartneys assumptions we calculated figures for the Near Islands and for Amchitka (Table 3.7). Updated information on site sizes was available from the U.S. Bureau of Indian Affairs (1986) investigations. Early sites in the Near Islands average 10 houses, but the houses are smaller so we estimated six people occupied each one (Corbett et al. 2001).

The Late Prehistoric Near Islands: AD 1200–1800

Human populations are rarely stable. The western Aleutian population grew over time, increasing pressure on available resources. Humans adapt to scarcity in a variety of economic, social, and technological ways. Late prehistoric sites on Shemya show an increase in the use of nearshore fish and invertebrates. The sites also have a larger number and variety of cetaceans, especially small dolphins. Aleuts also developed new technology, such as new styles of fishhooks to catch nearshore and reef fish (Corbett et al. 2007).

Social institutions are critical human adaptations. Later prehistoric sites in the Near Islands show changes in house size and village organization, indicating a period of social transformation. People began aggregating in larger settlements by AD 1200 (Corbett et al. 2007). On Shemya, evidence suggests the population coalesced into about six villages. The shoreline distances between these sites increased slightly to 2.9 km. Sites on all the islands grew in size and share several features, including storage pits and prominent midden mounds (Figure 3.4). The average number of houses increased to 36 on Attu, and 16 on Agattu. In 75 percent of the sites at least one substantially larger dwelling up to 13 m long appears. Early Russian explorers reported that Near Island chiefs lived in large houses. Chiefs had more wives, succored orphans and widows, and entertained foreign guests in these dwellings. The chiefs' houses also had ceremonial functions (Black 1984).

On Amchitka these changes are far less obvious. Only 25 percent of the sites possess

TABLE 3.8

Population Estimates for the Late Period in the Near and Rat Islands

ISLAND/GROUP	TOTAL NO. SITES	NO. HOUSES PER SITE	PEOPLE/ HOUSE	% IN USE	POPULATION ESTIMATE	TOTAL POPULATION
Attu/Semichis	42	36	6	50%	4,536	50% = 5,688
Attu/Semichis	42	36	6	25%	2,376	25% = 2,952
Agattu	23	16	6	50%	1,152	—
Agattu	23	16	6	25%	576	—
Amchitka	20	9	6–8	50%	540–720	50% = 1,842–2,456
Amchitka	20	9	6–8	25%	270–360	25% = 942–1,256
Amchitka	62	7	6–8	50%	1,302–1,736	—
Amchitka	62	7	6–8	25%	672–896	—

chiefs' houses, but in those sites the number of houses increase to nine (Corbett et al. 2001). However, settlements without chiefs' houses continued to be occupied into the historic period. The timing of the arrival of these social and political changes in the Rat Islands is unknown but certainly occurred after their appearance in the Near Islands.

Late Prehistoric Population Estimates

Based on these revised parameters, new population estimates were generated for the late prehistoric period. A minimum of 42 sites on Attu and the Semichi Islands, and 23 sites on Agattu were occupied in the late prehistoric period. On Amchitka, 25 percent of the sites with chiefs' houses averaged nine houses each. Settlements without chiefs houses continued to be occupied into the Historic period.

The 25 percent occupancy rates for Agattu yields an estimate that closely matches Nevodchikovs report for that island (Black 1984), but the total for the island group is considerably higher than any previous estimates and represents an 80 percent population increase over the earlier period. This estimate for Amchitka is also far higher than before (Table 3.8).

Two factors complicate the picture for the Near Islands. At Russian contact in 1745, the Near Islanders were under military pressure from more easterly Aleuts. They suffered from slave raids and deaths in skirmishing for an unknown period of time prior to the Russian arrival (Berkh 1974:30; Black 1984:65). This could have reduced the estimated population. Secondly, in 1880 Turner (1891) recorded the names of 30 villages on Agattu that had been inhabited in the immediate pre-Russian past. This suggests that the number of occupied villages may have been higher than estimated above. All of this emphasizes the uncertainty of population estimates for the prehistoric period. These estimates, and the number and concentration of known sites, suggest that the human population of these islands was likely far higher that previously estimated.

Implications for Environmental Impacts

Virtually the entire coastline of the Semichi Islands and Agattu, and the east ends of Attu and Amchitka were fully exploited by concentrated and dense populations of human hunters and fishermen. Human predation near villages would have been intense, and hunters may have had to travel to western Amchitka and Attu, or along the reef system east of Shemya to find sea mammals. Faunal evidence and recent isotope work presented here support this scenario. Sea otters were rare in most sites, and this, coupled with evidence of numerous and large sea urchins in the middens, indicates a lack of sea otters near settlements.

On Amchitka and Attu, the rugged and remote areas of the western ends of the islands limited the number of human settlements and provided refuges for mammal populations. On Amchitka, this refuge area was closer to the majority of the settlements, allowing hunters easy access to marine mammals and resulting in the larger number of sea otter remains in the site midden there. Still, the sea urchin evidence indicates that otters were scarce near the settlements, and that sea urchins dominated the nearshore environment.

The lack of otters near settlements had profound effects on nearshore ecosystems, which were very different from those described by biologists over the last 50 years. With extensive areas of kelp barrens surrounding human settlements, the lightly populated coastlines would have provided valuable refuges for sea mammals. Overall, the islands would have presented a mosaic of kelp forests and kelp barrens. As in terrestrial habitats, this mosaic was highly diverse and productive. However, it is likely that overall populations of sea otters, seals, and sea lions were much lower than the historic high levels recorded in the 1970s. This ecosystem was stable over the entire prehistoric span of human occupation; we see no major shifts, declines, increases, or species extinctions for nearly 3,000 years (Causey et al. 2005; Corbett et al. 2007). Furthermore, the Near and Rat island ecosystem probably supported over 3,000 people with a high standard of living until the Russians arrived.

STELLER'S SEA COWS

One important difference from the modern fauna was the presence of Steller's sea cows *(Hydrodamalis gigas)*. Sea cows were the largest sirenian, measuring up to 5 m long and weighing up to 3.5 tons, and were the only species inhabiting temperate to subarctic waters. In 1998, a complete sea cow rib was recovered from a beach on Kiska Island, apparently eroded from a nearby midden. In 1999, the WAAPP identified a Steller's sea cow bone from the archaeological site on Adak Island, in the central Aleutians. This was the first confirmed occurrence of this animal in the archaeological record (West et al. 1999).

In 1741, one of two ships of Vitus Bering's expedition wrecked on the Commander Islands. The survivors wintered there, eventually capturing sea cows for food. Stejneger (1887) estimated the population in 1741 at between 1,500 and 2,000 animals, and Domning (1978:133) agrees. The animals traveled in herds and fed exclusively on seaweeds in estuaries around the Commander Islands (Myers 1999; Steller 1988).

Bering's explorations opened the Aleutian Islands to Russian hunting and trading expeditions. Ships traveling to the Aleutians provisioned themselves with sea cow meat. The meat was highly prized, tasting like veal with fat like "sweet almond oil" (Steller 1988:163). A mining engineer, Jakovlev, in 1754, reported the wasteful use of sea cows to Irkutsk authorities and urged a ban on hunting (Stejneger 1887:1049). No action was taken, and Stejneger (1887:1053) reported the Popof party in 1767 or 1768 had killed the last one. Stejneger (1887) estimated 495 animals had been retrieved by hunters, while up to 2,475 were struck and lost. He speculated the Commander Island animals were the survivors of a once more numerous and widely spread species, which had been driven to the remote and unoccupied islands by human hunters (Stejneger 1898:28). Domning (1972) believed predation by humans was the sole cause of the animal's extinction. Guthrie (1984:290) used sea cows as an example of a species exterminated "immediately" by prehistoric seafaring people on the Northwest Coast of North America. Krupnik (1993:235) stated that long before European contact, the Aleuts had exterminated the sea cow from the subarctic fringe. Sea cows are a classic example of Martin's (1984) "prehistoric overkill" hypothesis that postulates human hunting as the primary cause of megafaunal extinctions at the end of the Pleistocene.

The population history of sea cows is virtually unknown. Savinetsky (1995) dated bones from raised beach ridges in the Commanders, spanning 2,250 years. He found that during colder periods (2200–1800 BP and after 800 BP), populations were lower than historic levels. The Little Climatic Optimum or Medieval Warm Period (1100–800 BP) saw populations two to three times the historic levels. Savinetsky notes that Steller reported the animals were emaciated from lack of food during the winters and occasionally drowned beneath the ice (Savinetsky 1995:403). These data support Domning's (1972) contention the Commanders were marginal sea cow habitat.

Sea cows have long been presumed to have occupied only the Commander Islands. Stejneger (1898:28) asserted that there is no evidence the animals inhabited other coasts, though he admitted the discovery of a rib on Attu Island made it "not improbable" the animal once lived there as well. However, Steller (1988:64) reported they were "known from Cape Kronotski to Avacha Bay and was sometimes thrown ashore dead." The Kamchadals called them "kelp-eaters" (Steller 1988). Russians reported they were rare visitors in the Near Islands in the 1760s (Liapunova 1979). Turner (1891) recorded Aleut traditions that sea cows had once been abundant around Attu and had been hunted. One elderly woman reported that her grandfather had seen sea cows in the Near Islands, probably as late as the early 1800s, and they had died out later in the Aleutians than in the Commander Islands (Turner 1886). He also recovered bones identified as sea cow from a midden near the Chichagof Harbor village (Turner 1891). The actual range of the sea cow in the 1700s extended from Kamchatka to the western Aleutians.

Human Hunting of Steller's Sea Cow

Much of the speculation about human caused extinction of sea cows assumes the animals were easy prey. Although slow and defenseless, sea cows were large, incredibly strong, thick skinned, and hard to kill. The Aleuts were accomplished hunters, possessing a highly developed and deadly technology of bone and wood with kelp ropes. Their boats were fragile masterpieces of wood and hide. With their technology, sea cows were formidable prey. Sea cow hides regularly turned aside Russian iron weapons. The animals straightened metal hooks and broke ropes. They swam away from men in large boats trying to tow them to shore, often taking the grappling hooks and ropes with them (Steller 1988:159–160). Steller described the animals coming to the assistance of stricken companions by pressing on ropes or pulling hooks and harpoons out of wounded animals. They also bumped or pushed the boats of their tormentors (Haley 1980:9). They were only successfully captured by Russian or Aleut when caught alone and driven into shallow waters to be tethered there and butchered at low tide (Steller 1988; Turner 1891; Yakovlev in Stejneger 1887:1051).

Sea Cow Dates

The WAAPP identified possible archaeological sea cow bones during campaigns on Buldir and Adak Islands. Identifications were made in the field by Savinetsky (West et al. 1997, 1999). In both sites, the bone fragments had been cut and shaped for tool manufacture. Identification of the bone as sea cow caused heated debate among team members. Given the fragmentary nature of the bone, one potential criticism was that the remains were misidentified. However, sea cow bone is morphologically distinct, being dense and hard, ivory-like in appearance, lacking porous cancellous tissue. Isotope analyses were performed to corroborate the field identification. In 1998, a complete rib recently eroded from archaeological site KIS-009 was collected by National Marine Fisheries biologist James Thomasson from the beach the southern shore of Kiska Island and is one of the few known sea cow bones.

Assuming all of the samples were sea cow, they were directly dated to demonstrate that they are from contemporaneous prehistoric cultural contexts and, thus, represent animals killed or scavenged by Aleuts. An unidentified bone from Laughlin's Chaluka excavations was also submitted for testing. The results are presented in Table 3.9. In none of the cases were the

TABLE 3.9

Steller's Sea Cow Bone Dates

LAB NO.	ISLAND	SITE	PIT AND LEVEL	MATERIAL	CONVENTIONAL AGE (RCYBP)	ISOTOPE RATIO	CALIBRATED AGE AD/BC
IEMAE-1241	Buldir	KIS-008	Pit 7, Level 11	Bird bone	1185 ± 50	N/A	765 AD
Beta-135539	Buldir	KIS-008	Pit 7, Level 11	Sea cow bone	1240 ± 60	0/00 –14.0	710 AD
Beta-33327	Kiska	KIS-009	Test Pit 1	Charcoal	2270 ± 80	0/00 –25.0	320 BC
Beta-124353	Kiska	KIS-009	Beach	Sea cow bone	1040 ± 40	0/00 –14.5	910 AD
Beta-135540	Umnak	Chaluka	Section 10, Depth 7	Sea cow bone	3930 ± 40	0/00 –14.0	1675 BC
P-1086	Umnak	Chaluka		Charcoal	NA	NA	1156 ± 59 BC
Beta-135537	Adak	ADK-009	Pit 1, Level 1	Sea cow bone	1710 ± 70	0/00 –15.6	240 AD
Beta-141258	Adak	ADK-009	Pit 1, Level 1	Marine shell	1260 ±70	0/00 0.20	690 AD

bones salvaged from Pleistocene deposits. Nor were they recently imported by Russian traders. The discovery of Steller's sea cow bones in at least one prehistoric context demonstrates the animals coexisted with human populations. Aleuts had at least occasional access to sea cows, either as hunted prey or as scavenged resources. Prehistoric overhunting now appears to be a simplistic explanation for the extinction of the Steller's sea cow, and other explanations may better account for their restricted range and low populations during the Holocene.

Isotope Results

The stable isotope composition of biogenic materials (e.g., bone, shell, hair, etc.) is a valuable tool for reconstructing the diet and habitat preferences of extinct species (Koch et al. 1994). Both the mineral (bioapatite) and the organic (collagen) components of vertebrate remains can be exploited for isotopic paleodietary information. Three primary elements commonly used in these analyses included carbon ($\delta^{13}C$), nitrogen ($\delta^{15}N$), and oxygen ($\delta^{18}O$). The stable isotope ratios of these elements are largely dictated by the food an animal eats ($\delta^{13}C$, $\delta^{15}N$), the fluids contained within its food ($\delta^{18}O$), and/or the water it drinks ($\delta^{18}O$) (Clementz and Koch 2001; Koch et al. 1994). Based on the unique kelp diet of the Steller's sea cow, bone from this species was expected to be isotopically distinct from that of other marine mammals, thus providing a novel means for validating the identification of archaeological remains.

Eight bone fragments visually identified as Steller's sea cow were selected from midden sites for isotope analyses (Table 3.10). The remains of other marine mammals at these sites, and bones from extant populations of marine mammals in Alaska were also sampled for comparison. In addition, confirmed specimens of *H. gigas* collected from Bering Island (~.2 ka), central California (~18 ka), and Amchitka Island (~127 ka) were analyzed to provide an estimate of expected isotope values for this species before and after human habitation of the region (Jones 1967; Whitmore and Gard 1977). From each specimen, subsamples of bone bioapatite (~10 mg) and bone collagen (~100 mg) were analyzed.

Carbon and oxygen isotope values were obtained from the bioapatite portion of marine mammal bones (Figure 3.5). For recent marine mammal bones and confirmed specimens of *H. gigas*, there was a clear separation between consumers foraging in kelp ecosystems or close to shore (i.e., *H. gigas*, sea otters, harbor seals) and those foraging further offshore (i.e., Steller's sea lions, northern fur seals). Consumer $\delta^{13}C$ values are strongly controlled by those of the primary producers at the base of a food web, which are mainly phytoplankton and macroalgae within nearshore marine ecosystems. The fast growth rate and high productivity of kelp causes its tissues to be significantly enriched in ^{13}C relative to that of nearshore phytoplankton, creating a handy label for identifying consumers that forage within these ecosystems (Duggins et al. 1989). Oxygen isotope values also appear to differ between kelp foraging species and other marine mammals (Figure 3.5), but the underlying mechanism to explain this is unclear at present. Clementz and Koch (2001) noted similar differences among cetaceans, pinnipeds, and sea otters, so this offset may in part reflect phylogenetic differences in osmoregulatory capacity and physiology rather than ecological differences among species.

Significant differences in bioapatite $\delta^{13}C$ and $\delta^{18}O$ values are also observed among the archaeological specimens. For most sea otters and pinnipeds, $\delta^{13}C$ and $\delta^{18}O$ values closely match those observed for modern populations of these species. Sea otter specimens from Shemya Island, however, have much lower $\delta^{13}C$ and $\delta^{18}O$ values than sea otters from other islands or from modern populations. These low values suggest that these sea otters may have foraged outside the kelp ecosystem or that kelp beds were less abundant locally (Simenstad et al. 1993). For sea cows, only one specimen (Kiska Island, Beta-124353) had $\delta^{13}C$ and $\delta^{18}O$ values

TABLE 3.10

Isotope Samples

LOCALITY	AGE	SPECIES	ELEMENT	BIOAPATITE	COLLAGEN
Bering Island	AD 1750	*Hydrodamalis gigas*	Ribs	X	X
Shemya Island					
ATU-003	BC 50–AD 500	*Enhydra lutris*	Rib	X	X
		Eumatopias jubatus	Unidentified	X	X
ATU-021	BC 100–AD 500	*Phoca vitulina*	Limb	X	X
ATU-022	BC 400–AD 200	*Enhydra lutris*	Humerus	X	X
		Pinniped	Phalanges	X	X
		Phoca vitulina	Limb	X	X
ATU-061	BC 800–AD 200	*Enhydra lutris*	Vertebra	X	X
		Pinniped	Metapodial?	X	X
		Callorhinus ursinus	Phalanges	X	X
		Eumatopias jubatus	Limb	X	X
Umnak Island	BC 1675–1156	*Hydrodamalis gigas*	Rib	X	X
		Enhydra lutris	Rib	X	X
		Pinniped, phocid	Unidentified	X	X
		Pinniped, otaroid	Unidentified	X	X
Adak Island	AD 1250–1700	*Hydrodamalis gigas*	Rib	X	X
		Pinniped	Unidentified	X	X
Buldir Island	BC 1410–BC 900	*Hydrodamalis gigas*	Rib	X	X
		Eumatopias jubatus	Rib,	X	X
			metapoidals	X	X
Kiska Island	AD 1200–1700	*Hydrodamalis gigas*	Rib	X	X
Amchitka Island	127 ka ± 8 ka	*Hydrodamalis gigas*	Rib	X	–
Monterey Bay, CA	18 ka	*Hydrodamalis gigas*	Skull	X	X
San Luis Obispo, CA	>10 ka	*Hydrodamalis gigas*	Rib	X	X

similar to those of *H. gigas* sampled from Bering Island (historic), Amchitka Island, and California (both Late Pleistocene). All other specimens had significantly lower $\delta^{13}C$ values than those reported for *H. gigas* and higher $\delta^{18}O$ values than associated marine mammal material. Although it is possible that these bones came from sea cows that foraged on vegetation other than kelp, the fact that sea cow remains spanning over 100,000 years from Bering Island, Amchitka Island, and California show very similar isotope values is compelling evidence that *H. gigas* was highly specialized with little dietary variation across a broad geographic range.

Additional information is provided from bone collagen $\delta^{13}C$ and $\delta^{15}N$ values. As with carbon, the $\delta^{15}N$ composition of bone collagen is

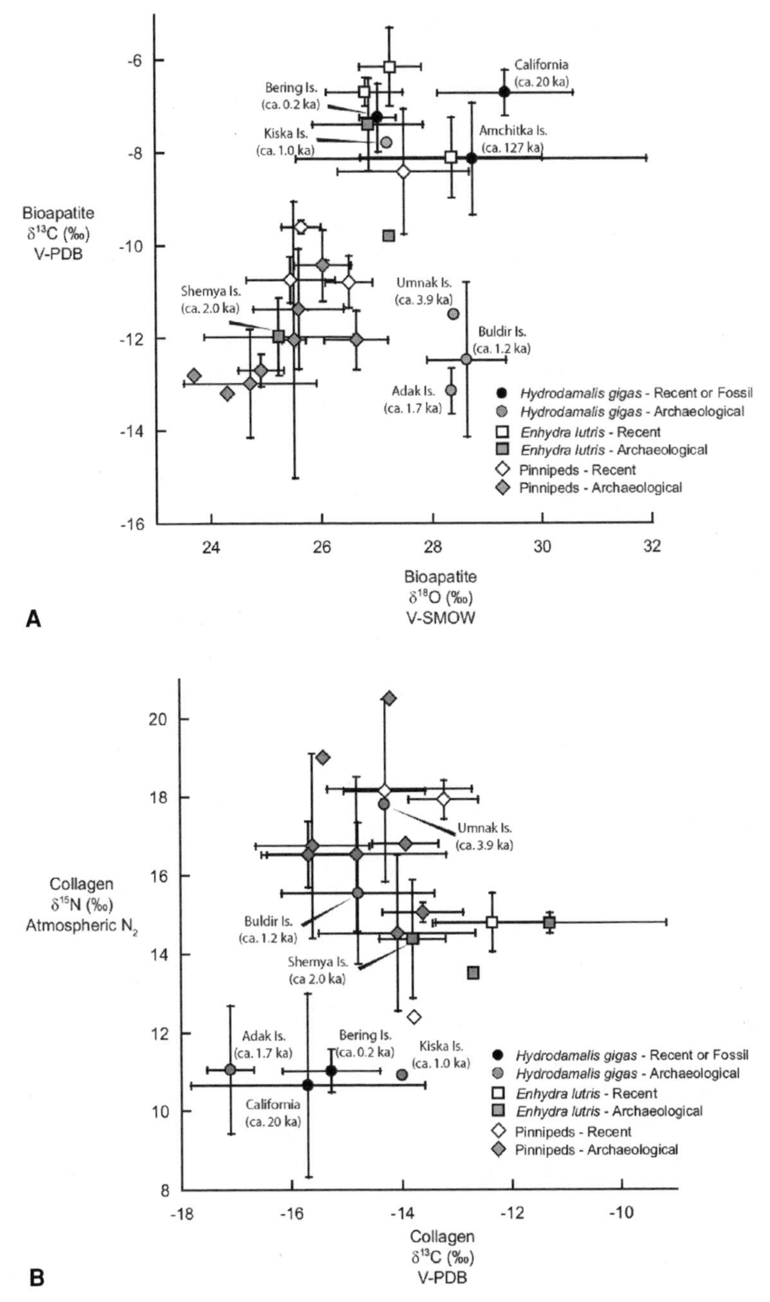

FIGURE 3.5. Isotope values from sea mammal bones.

directly related to that of the diet. Most tissues including collagen are enriched in ^{15}N by ~3‰ relative to diet, and this enrichment is passed up with each step in trophic level, resulting in extremely high $\delta^{15}N$ values for top-level consumers (Minagawa and Wada 1984). These trophic level differences are clearly reflected in the $\delta^{15}N$ values of *H. gigas*, an herbivore, and those of modern marine carnivores (Figure 3.5). Mean $\delta^{15}N$ values for *H. gigas* from Bering Island and California are between 10 and 11‰, whereas those of sea otters and pinnipeds are much higher, typically greater than 14‰. Mean $\delta^{15}N$ values for sea otters at all sites are similar to those reported for modern populations, are lower than those for co-occurring species of pinnipeds, and are consistent with a diet of marine invertebrates and other lower trophic organisms. As with bone bioapatite, collagen $\delta^{13}C$ values for Shemya otters are lower than those for sea otters from other islands. This suggests that even though the sea otters from all islands were foraging at the same trophic level, the $\delta^{13}C$ values of prey species at Shemya Island were much lower than those at other sites. As mentioned above, these low values could be the result of a lower abundance of kelp at Shemya than at other islands, as kelp species growing under the pressure of heavy grazing by sea urchins typically have lower $\delta^{13}C$ values than kelp growing in areas where urchin populations are low and kelp forests are well established (Simenstad et al. 1993). Low $\delta^{13}C$ values for sea otters from Shemya supports archaeological evidence that this population was significantly reduced by human hunting and, consequently, grazing by urchins had significantly reduced kelp abundance at this site.

Archaeological sea cow specimens from Kiska Island and Adak Island have collagen $\delta^{13}C$ and $\delta^{15}N$ values that closely match those reported for other specimens of *H. gigas*, but $\delta^{15}N$ values for the specimens from Buldir and Umnak Islands are extremely high, closer to those of consumers foraging at higher trophic levels. Combined with results from analysis of the bone bioapatite, these collagen $\delta^{13}C$ and $\delta^{15}N$ values raise concerns about the identification of these bones as sea cow. Only the Kiska Island specimen has yielded isotope values from both bone materials that are consistent with those obtained from fossil and recent specimens of *H. gigas*. As for the identities of the remaining material, there are two possible scenarios to account for these isotope values. The most plausible one is that these specimens were misidentified and are actually the remains of other marine mammals. The large differences in collagen $\delta^{13}C$ and $\delta^{15}N$ values between the Adak Island specimens and those from Buldir and Umnak indicate that the remains represent at least two different species. Furthermore, the bioapatite $\delta^{18}O$ values are distinctly higher than those from associated pinniped remains, a finding that is consistent with bones coming from different cetacean species (Clementz and Koch 2001). If true, then the specimens from Adak Island with low $\delta^{15}N$ values could be from a species of baleen whale, whereas those from Buldir and Umnak may represent different species of toothed whales. A less likely, but still possible, scenario is that these bone samples are in fact sea cow remains. Modern manatees and dugongs have been observed to consume marine invertebrates and fish on occasion, although sightings are extremely rare, and these prey items are not thought to contribute significantly to the diets of these species. If these sea cows were foraging at a higher trophic level, then the diets of these individuals were unique. Since these individuals were living in close association with human populations, it is possible that the environmental conditions around these islands had deteriorated to a point that there was no longer a sufficient quantity of kelp available to support local sea cow populations, forcing these animals to consume other resources (i.e., marine invertebrates).

THE SYSTEM DISRUPTED

In 1745, the western Aleutians were densely populated by thriving hunting and fishing cultures dependent on the rich maritime environment. Populations may have been declining due to warfare with stronger neighbors, but nearly 2,000 people were reported by Russian traders. Russian contact precipitated a catastrophic population decline. By 1762, Cherepanov reported 40 males and more females. Fellow crewmember Kul'kov estimated a total population of 100 for the Near Islands (Black 1984). Potap Zaikov, commanding an independent trading vessel, found 27 men in the Near Islands in 1774 (Khlebnikov 1994). By 1805, the Rat Islands were almost totally depopulated, with only a few devastated families remaining in their homes (Andreev 1948; Khlebnikov 1994; Liapunova 1979).

The full impact of this catastrophic population decline on the ecosystem may never be known. The ecosystem's top carnivore, human beings, had been all but eliminated. The population decline, coupled with abandonment of most of the settlements, must have had profound effects on marine mammals through the sudden release of hunting pressure and the opening of vast areas of new habitat. Sea otter populations exploded. Human hunting continued, now for a global market, but there were fewer hunters, and the hunts were seasonal and geographically circumscribed. Overall hunting pressure on sea otters actually declined between 1745 and 1867.

The average annual catch of sea otters between 1745 and 1765, from the Near and Commander islands, was 2,492 (Black 1984; Black and Desson 1986). From 1745 to 1799, the Russians exported 48,216 sea otter furs, 35,887 tails (bits and pieces), and 7,922 medvedkis or pups from the Aleutians, Alaska Peninsula, and Kodiak (Berkh 1974; Black 1984; Black and Desson 1986). After 1799, the Russian American Company enacted and enforced conservation measures to ensure a steady, profitable supply of otter furs. Hunting pressure on sea otters continued to decline. Between 1842 and 1861, only 2,421 otter furs were exported from the Near Islands, an average of 127 per year (Khlebnikov 1994).

In 1867, Russia sold Alaska to the United States, and unregulated slaughter ensued. In the first four years of American management, 12,208 sea otters were shipped from Alaska, more than in the last 20 years of Russian harvest (Corbett 1991; Lensink 1962). By 1895, otters were extinct in the Near Islands. In 1911, when hunting was banned, a few thousand animals still existed in 13 small enclaves. Most of the survivors lived in the Rat Islands (Kenyon 1969; Lensink 1962; U.S. Fish and Wildlife Service 1994). Although no one was watching, the hunting ban led to population increases. By the 1930s, otters had recolonized the Near Islands (Murie 1959), and by the 1990s the population had reached 3,000 animals (Burn et al. 2005).

For 3,000 years in the Near Islands, and 5,000 years in the Rat Islands, the human-dominated ecosystem operated in remarkable equilibrium. This is not to say that there were no changes in the numbers or distributions of species, or that there were no local extinctions. The disappearance of breeding rookeries of fur seals from Shemya demonstrates just one of the changes that took place (Corbett et al. 2007). For the most part, however, shifts in mammal or bird populations were localized, and no species went extinct. Even Steller's sea cows existed until the arrival of the Russians, although the lack of extensive kelp forests may have limited their numbers. The system successfully weathered climatic fluctuations, including two warming and two cooling events within the last 1,800 years (Causey et al. 2005; Savinetsky et al. 2004).

The structure and function of this ecosystem were very different from that documented and monitored by biologists over the last 50 years. Areas near human settlements would have been largely free of extensive kelp forests. Large mammals would have been very rare in populated areas. Kelp forests and their associated faunal assemblages would have been found along coastlines that were relatively inaccessible to hunters in small skin boats. The regional picture would have presented a mosaic of kelp forests and kelp barrens, and fauna would likewise have presented patchy distributions. As in terrestrial habitats, this mosaic was highly diverse and productive. Populations of sea otters, seals, and sea lions were likely well below the historic high levels recorded in the 1970s.

After World War II, biological research increased dramatically and began to describe the ecosystem and its function. The importance of otters in structuring the environment has been exhaustively analyzed (Estes 1990, 1996; Estes and Duggins 1995; Jones and Kain 1967; Paine and Vadas 1969; Palmisano and Estes 1977). The abundance of sea otters and expansive kelp forests visible after the 1950s were heralded as a return to the pristine conditions prevailing before commercial hunting began.

In reality, the Aleutian ecosystem has been heavily influenced by human activity for thousands of years. In spite of this influence, the system was remarkably stable and resilient in the face of climate change and increasing human exploitation. However, for the last 250 years, the Aleutian Islands and Bering Sea region have been characterized by instability. The latest series of population crashes among sea lions, sea otters, several species of birds, and other organisms is only the most recent manifestation of this instability. The apparently pristine conditions of the post World War II period, through the 1970s, are a human artifact of politics, global economics, and recent legislation, rather than a restoration of natural ecosystem processes.

A National Research Council report (1996) concludes that it will be "difficult for human management to cause a large complex marine ecosystem to achieve and maintain a desirable balance." Rather than trying to restore a mythological "pristine" ecosystem, managers and policy makers need to focus on reaching a desirable balance. This will involve researchers striving to understand complex processes operating in the region, including how those processes have fluctuated through the millennia The role of human agency cannot be ignored nor be assumed to be of very recent origin. Finally, managers, policy makers, and researchers will need to accept that decisions made today may have affects that will not be observable within one or several human lifetimes.

ACKNOWLEDGMENTS

First and foremost we owe thanks and appreciation to the Aleut Corporation, especially Melvin Smith, the Aleutian/Pribilof Islands Association, and the Aleut people for their interest in and support of archaeological research into their past. Without the generous support and interest of the Aleut people we would never have even started. The growth of the Western Aleutians Archaeological and Paleoecological Project from relatively simple beginnings to the far more complex end product was stimulated by many friends and colleagues. Biologists from the Alaska Maritime NWR, the Biological Research Division of the U.S. Geological Survey, and a number of universities very early in our work asked complex and probing questions. In particular Vernon Byrd, James Estes, Ian Jones, Jeff Williams, and Douglas Causey spurred our inquiries into the history of the environment of the Aleutian Islands. Their questions inspired much of our research and directed the development of methodologies to address complicated biological questions. Likewise special thanks go to archaeological colleagues Brian Hoffman, L. Lewis Johnson, Rick Knecht, Don Dumond, Roger Powers, Diane Hanson, and Herb Maschner for their comments, suggestions, and ideas. Field research from 1990 through 1994 was funded by grants from the National Geographic Society (#4943-92, #5252-94), and with support from the U.S. Fish and Wildlife Service, the Natural History Museum at the University of Kansas, the Smithsonian Institution Arctic Studies Program, the Museum National d'Histoire Naturelle, and Centre National de la Recherche Scientifique in Paris. From 1997 through 1999, field research for WAAPP was funded by grants from the National Science Foundation (OPP-9314472) and the Institut Français pour la Recherche et la Technologie Polaires (Program 301).

REFERENCES CITED

Amorosi, T., J. Woollett, S. Perdikaris, and T. McGovern
1996 Regional Zooarchaeology and Global Change: Problems and Potentials. *World Archaeology* 28:16–157

Andreev, A. I. (editor)
1948 Report of the Tot'ma Merchant Stepan Cherepanov about his Sojourn on the Aleutian Islands, 1759–1762. In *Russkie Otkrytiia V Tikhom Okeane iv Severnoi Amerike v XVIII–XIX Vekakh* [Russian Discoveries in the North Pacific Ocean and in North America in the Eighteenth and Nineteenth Centuries], pp. 113–120. Akademiia Nauk, Moscow.

Berkh, V. N.
1974 *A Chronological History of the Discovery of the Aleutian Islands,* translated by D. Krenov. Originally published 1923, St. Petersburg, Russia. The Limestone Press, Kingston, Ontario.

Black, L. T.
1984 *Atkha: An Ethnohistory of the Western Aleutians.* The Limestone Press, Kingston, Ontario.

Black, L. T., and D. Desson
1986 *Early Russian Contact.* AHC Studies in History 191. Alaska Historical Commission, Anchorage.

Burgner, R. L., and R. E. Nakatani
1972 Research Program on Marine Ecology, Amchitka Island, Alaska. Annual Progress

Report, July 1, 1970–June 30, 1971 No. BMI-171–144. U.S. Atomic Energy Commission, Amchitka Bioenvironmental Program, Batelle Memorial Institute, Columbus, Ohio.

Burn, D. M., A. M. Doroff, and M. T. Tinker
2005 Carrying Capacity and Pre-decline Abundance of Sea Otters *(Enhydra lutris kenyoni)* in the Aleutian Islands. *Northwestern Naturalist* 84:145–148.

Butzer, K. W.
1982 *Archaeology as Human Ecology.* Cambridge University Press, New York.

Campbell, G.
2008 Sorry, Wrong Phylum: A Neophyte Archaeomalacologist's Experiences in Analyzing a European Atlantic Sea Urchin Assemblage. In *The status of Archaeomalacology in the Twenty-first Century: Proceedings of the First meeting of the Archaeomalacology Working Group,* edited by K. Szabo and I. Quitmyer. ICAZ (Archaeofauna 13). in press.

Causey, D., D. G. Corbett, C. Lefevre, D. West, A. B. Savinetsky, and B. F. Khassanov
2005 The Paleoenvironment of Humans and Marine Birds of the Aleutian Islands: Three Millenia of Change. *Fisheries Oceanography* 14(Supplement 1):259–276.

Clark, F.
1990 At the End of the Chain: Recent Archaeological Reconnaissance and Implications for Settlement Pattern Studies in the Western Aleutians. Presented at the 17th Annual Meeting, Alaska Anthropological Association, Fairbanks, Alaska.

Clementz, M. T., and P. L. Koch
2001 Differentiating Aquatic Mammal Habitat and Foraging Ecology with Stable Isotopes in Tooth Enamel. *Oecologia* 129:461–472.

Cook, J. P., E. J. Dixon, and C. E. Holmes
1972 *Archaeological Site Report, Site 49 Rat 32 Amchitka Island, Alaska.* Atomic Energy Commission Contract AT(29-2)-20. Submitted to Holmes and Narver, Inc., Las Vegas, Nevada.

Corbett, D. G.
1990 Archaeological [MSOffice5]Survey and Testing on Shemya Island, Western Aleutians, Alaska, in June 1990. Final Report to the Geist Fund. University of Alaska Museum, Fairbanks.
1991 *Aleut Settlement Patterns in the Western Aleutian Islands, Alaska.* Unpublished Master's thesis, University of Alaska, Fairbanks.

Corbett, D. G., C. Lefèvre, T. J. Corbett, D. L. West, and D. Siegel-Causey
1997 Excavations at KIS-008, Buldir Island: Evaluation and Potential. *Arctic Anthropology* 34(2):100–117.

Corbett, D. G., D. L. West, and C. Lefèvre
2001 Prehistoric Village Organization in the Western Aleutians. In *Archaeology in the Aleut Zone of Alaska: Some Recent Research,* edited by D. Dumond, pp. 251–266. UO Anthropological Paper 58. Department of Anthropology and Museum of Natural History, University of Oregon, Eugene.
2008 *People at the End of the World: The Archaeology of Shemya Island.* AAA Monograph Series. Alaska Anthropological Association, Aurora, in review.

Dall, W. H.
1877 On Succession in the Shell Heaps in the Aleutian Islands. In *Tribes of the Extreme Northwest,* Vol. 1, Contributions to North American Ethnology. Government Printing Office, Washington, D.C.

Denniston, G. B.
1966 Cultural Change at Chaluka, Umnak Island: Stone Artifacts and Features. *Arctic Anthropology* 3(2):84–124.

Desautels, R. J., A. J. McCurdy, J. D. Flynn, and R. R. Ellis
1971 *Archaeological Report: Amchitka Island 1969–1970.* U.S. Atomic Energy Commission Report TID-25481. Submitted to Holmes and Narver, Inc., Las Vegas, Nevada.

Domning, D. P.
1972 Steller's Sea Cow and the Origin of North Pacific Whaling. *Syesis* 5:187–189.
1978 Sirenian Evolution in the North Pacific Ocean. *University of California Publications in Geological Science* 118:1–178.

Duggins, D. O., C. A. Simenstad, and J. A. Estes
1989 Magnification of Secondary Production by Kelp Detritus in Coastal Marine Ecosystems. *Science* 245:170–173.

Dumond, D. E.
1987a *Prehistoric Human Occupation of Southwestern Alaska: A Study of Resource Distribution and Site Location.* UO Anthropology Paper 36. University of Oregon, Eugene.
1987b A Re-examination of Eskimo-Aleut Prehistory. *American Anthropologist* 89:32–56.
2001 Toward a (Yet) Newer View of the (Pre) History of the Aleutians. In *Archaeology in the Aleut Zone of Alaska: Some Recent Research.,* edited by D. Dumond, pp. 289–309. UO Anthropological Paper 58. University of Oregon, Eugene.

Dumond, D. E., and D. Griffin
2002 Measurements of the Marine Reservoir Effect on Radiocarbon Ages in the Eastern Bering Sea. *Arctic* 55(2):77–86.

Estes, J. A.
1990 Growth and Equilibrium in Sea Otter Populations. *Journal of Animal Ecology* 59:385–401.

1996 The Influence of Large, Mobile Predators in Aquatic Food Webs: Examples for Sea Otters and Kelp Forests. *Aquatic Predators and Their Prey,* edited by S. P. R. Greenstreet and M. L. Tasker, pp. 65–72. Blackwell Scientific, Oxford.

Estes, J. A., and D. O. Duggins

1995 Sea Otters and Kelp Forests in Alaska: Generality and Variation in a Community Ecological Paradigm. *Ecological Monographs* 1:75–100.

Favorite, F. A., J. Dodimead, and K. Nasu

1976 Oceanography of the Subarctic Pacific Region, 1960–1971. *International North Pacific Fisheries Commission Bulletin* 33:1–187.

Flannery, T.

1994 *The Future Eaters: An Ecological History of the Australasian Lands and People.* Grove Press, New York.

2001 *The Eternal Frontier: An Ecological History of North America and its Peoples.* Grove Press, New York.

Frohlich, B., and D. Kopjansky

1975 Aleutian Site Surveys 1975: Preliminary Report to the Aleut Corporation, Anchorage, Alaska. Report prepared by Laboratory of Biological Anthropology, University of Connecticut, Storrs.

Guthrie, R. D.

1984 Mosaics, Allelochemics, and Nutrients: An Ecological Theory of Late Pleistocene Megafaunal Extinctions. In *Quaternary Extinctions: A Prehistoric Revolution,* edited by P. S. Martin, R. Klein, pp. 259–298. University of Arizona Press, Tucson.

Haggerty, J. C., C. B. Wooley, J. M. Erlandson, and A. Grewell

1991 *The 1990 Exxon Cultural Resource Program: Site Protection and Maritime Cultural Ecology in Prince William Sound and the Gulf of Alaska.* Exxon Shipping Company and Exxon Company, Anchorage, Alaska.

Haley, D.

1980 The Great Northern Sea Cow. *Oceans* 13 September, pp. 7–11. The Oceanic Society.

Hayes, M.

2002 Paleogenetic Assessments of Human Migration and Population Replacement in North American Prehistory. Unpublished Ph.D. dissertation, Department of Anthropology, University of Utah.

Heath, T., R. Felthoven, C. Seung, and J. Terry

2004 *Stock Assessment and Fishery Evaluation Report for the Groundfish Fisheries of the Gulf of Alaska and Bering Sea/Aleutian Islands Area: Economic Status of the Groundfish Fisheries off Alaska, 2003.* Alaska Fisheries Science Center, National Marine Fisheries Service, Seattle, Washington.

Hrdlicka, A.

1945 *The Aleutian and Commander Islands and Their Inhabitants.* Wistar Institute, Philadelphia.

Jackson, J. B. C., M. X. Kirby, W. H. Berger, K. A. Bjorndal, L. W. Botsford, B. J. Bourque, R. H. Bradbury, R. Cooke, J. Erlandson, J. A. Estes, T. P. Hughes, S. Kidwell, C. B. Lange, H. S. Lenihan, J. M. Pandolfi, C. H. Peterson, R. S. Steneck, M. J. Tegner, and R. R. Warner

2001 Historical Overfishing and the Recent Collapse of Coastal Ecosystems. *Science* 293: 629–638.

Jochelson, W. I.

1925 *Archaeological Investigations in the Aleutian Islands.* Publication 367. Carnegie Institution, Washington, D.C.

1933 *History, Ethnology and Anthropology of the Aleut.* Publication 432. Carnegie Institution, Washington, D. C. Reprinted in 1968 by Anthropological Publications, Oosterhout, N.B., Netherlands.

Jones, N. S., and J. M. Kain

1967 Subtidal Algal Colonization Following the Removal of Echinus. *Helicolander Wissenschaft Meeresunters* 15:460–466.

Jones, R. E.

1967 A Hydrodamalis Skull Fragment from Monterey Bay, California. *Journal of Mammology* 48:143–144.

Kenyon, K. W.

1969 The Sea Otter in the Northeastern Pacific Ocean. *North American Fauna* 68:1–352.

Khlebnikov, K. T.

1994 *Notes on Russian America,* Parts II-V: *Kad'iak, Unalaska, Atkha, the Pribylovs,* compiled by R. G. Liapunova and S. G. Fedorova, translated by M. Ramsay, edited by R. Pierce. Alaska History Series 42. The Limestone Press, Fairbanks, Alaska.

Knecht, R. A., and R. S. Davis

2001 A Prehistoric Sequence for the Eastern Aleutian Islands. In *Archaeology in the Aleut Zone of Alaska: Some Recent Research,* edited by D. Dumond, pp. 269–288. UO Anthropological Paper 58. University of Oregon, Eugene.

Knecht, R. A., R. S. Davis, and G. A. Carver

2001 The Margaret Bay site and Eastern Aleutian Prehistory. In *Archaeology in the Aleut Zone of Alaska: Some Recent Research,* edited by D. Dumond, pp. 35–70. UO Anthropological Paper 58. University of Oregon, Eugene.

Koch, P. L., M. L. Fogel, and N. Tuross

1994 Tracing the Diets of Fossil Animals Using Stable Isotopes. In *Stable Isotopes in Ecology and Environmental Science,* edited by K. Lajtha and R. H. Michener, pp. 63–92. Blackwell, Oxford.

Krupnik, I.
1993 *Arctic Adaptations: Native Whalers and Reindeer Herders in Northern Eurasia*. University Press of New England, Hanover.

Lantis, M.
1970 *Ethnohistory in Southwestern Alaska and the Southern Yukon*. University Press of Kentucky, Lexington.
1984 Aleut. In *Handbook of North American Indians*, Vol. 5, *Arctic*, edited by D. Damas, pp. 161–184. Smithsonian Institution Press, Washington, D.C.

Laughlin, W. S.
1980 *Aleuts: Survivors of the Bering Land Bridge*. Holt, Rinehart, and Winston, New York.

Laughlin, W. S., and G. H. Marsh
1951 A New View of the History of the Aleutians. *Arctic* 4:75–88.
1954 The Lamellar Flake Manufacturing Site on Anangula Island in the Aleutians. *American Antiquity* 20:27–30.

Lefèvre, C., and D. Siegel-Causey
1993 First Report on Bird Remains from Buldir Island, Aleutian Islands, Alaska. Proceedings of the First Meeting of the ICAZ Bird Working Group, Madrid, Spain, 1993. *Archaeofauna* 2:83–96.

Lefèvre, C., D. G. Corbett, D. L. West, and D. Siegel-Causey
1997 A Zooarchaeological Study at Buldir Island, Western Aleutians, Alaska. *Arctic Anthropology* 34(2):118–131.

Lensink, C. J.
1962 *The History and Status of Sea Otters in Alaska*. Ph.D. thesis, Purdue University, West Lafayette, Indiana. University Microfilms, Ann Arbor.

Liapunova, R. G.
1979 Novyi Dokument o Rannikh Plavaniiakh Na Aleutskii Ostrova ("Izvestiia" Fedora Afanas'evicha Kulkova 1764g). [New Documents about the Early Voyages to the Aleutian Islands (News of Fedor A. Kulkov 1764).] In *Stranyi I Narody Vostoka* [Lands and Peoples of the East] 20(4): 97–105.
1996 *Essays on the Ethnology of the Aleuts (At the end of the Eighteenth and First Half of the Nineteenth Centuries)*. In Rasmuson Library Historical Translation Series, Vol. IX, edited by J. Shelest with assistance by W. B. Workman and L. T. Black, translators. University of Alaska Press, Fairbanks.

Lyman, R. L.
1996 Applied Zooarchaeology: the Relevance of Faunal Analysis to Wildlife Management. *World Archaeology* 28:110–125.

Mann, C. C.
2005 *1491: New Revelations of the Americas Before Columbus*. Alfred A. Knopf, New York.

McCartney, A. P.
1974 *1972 Archaeological Site Survey in the Aleutian Islands, Alaska*. Conference on the Prehistory and Paleoecology of the Western North American Arctic and Subarctic. Archeological Association, University of Calgary, Calgary.
1977 Prehistoric Human Occupation of the Rat Islands. In *The Environment of Amchitka Island, Alaska*, edited by M. L. Merrit and R. G. Fuller, pp. 59–114. U.S. Department of Commerce, Washington, D.C.
1984 Prehistory of the Aleutian Region. In *Handbook of North American Indians*, Vol. 5: *Arctic*, edited by D. Damas, pp. 119–135. Smithsonian Institution Press, Washington, D.C.

McCartney, A. P., and C. G. Turner, II
1966 Stratigraphy of the Anangula Unifacial Core and Blade Site. *Arctic Anthropology* 3(2):28–40

Martin, P. S.
1984 Prehistoric Overkill: The Global Model. In *Quaternary Extinctions: A Prehistoric Revolution*, edited by P. S. Martin, and R. Klein, pp. 354–403. University of Arizona Press, Tucson.

Minagawa, M., and E. Wada
1984 Stepwise Enrichment of 15N along Food Chains: Further Evidence and the Relation between d15N and Animal Age. *Geochimica et Cosmochimica Acta* 48:1135–1140.

Murie, O. J.
1959 *Fauna of the Aleutian Islands and Alaska Peninsula*. North American Fauna, No. 61. U.S. Fish and Wildlife Service, Washington, D.C.

Myers, P.
1999 Animal Diversity: Order Sirenia. Museum of Zoology, University of Michigan. Electronic document, http://animaldiversity.ummz.umich.edu/site/accounts/information/Sirenia.html, accessed January 2007.

National Research Council
1996 The Bering Sea Ecosystem. Report prepared for the National Research Council by the Committee on the Bering Sea Ecosystem, Polar Research Board, and the Commission on Geosciences, Environment and Resources of the National Research Council National Academy Press. National Research Council, Washington, D.C. Available at http://books.nap.edu/catalog/5039.html, accessed January 2007.

Nicholson, R. A., and T. P. O'Connor (editors)
2000 *People as an Agent of Environmental Change*. Oxbow Books, Oxford.

Paine, R. T., and R. L. Vadas
1969 The Effects of Grazing by Sea Urchins, Strongylocentrotus spp. on Benthic Algal Populations. *Limnology and Oceanography* 14: 710–719.

Palmisano, J. F., and J. A. Estes
1977 Ecological Interactions Involving the Sea Otter. In *The Environment of Amchitka Island, Alaska,* edited by M. L. Merritt and R. G. Fuller, pp. 527–567. U.S. Energy Research and Development Administration, Washington, D.C.

Redman, C. L.
1999 *Human Impact on Ancient Environments.* University of Arizona Press, Tucson.

Reitz, E. J.
2004 "Fishing down the Food Web": A Case Study from St. Augustine, Florida, U.S.A. *American Antiquity* 69:63–83.

Savinetsky, A. B.
1995 Ancient Population Dynamics of the Sea Cow (*Hydrodamalis gigas,* Zimm. 1780) in the Late Holocene. *Doklady: Biological Sciences* 320(1–6): 403–405.

Savinetsky, A. B., N. K. Kiseleva, and B. F. Khassanov
2004 Dynamics of Sea Mammal and Bird Populations of the Bering Sea Region over the Last Several Millennia. *Palaeogeography, Palaeoclimatology, Palaeoecology* 209:335–352.

Siegel-Causey, D., C. Lefèvre, and A. B. Savinetsky
1991a Historical Diversity of Cormorants and Shags from Amchitka Island, Alaska. *The Condor* 93:840–852.

Siegel-Causey, D., C. Lefèvre, and D. Corbett
1991b Buldir Island Expedition: Preliminary Excavation of the Aleut Midden Site, July–August 1991. Report on file, U.S. Fish and Wildlife Service, Anchorage, Alaska.

Siegel-Causey, D., D. G. Corbett, C. Lefèvre, S. Loring, and D. West
1994 Report of the Western Aleutian Archaeological Project Shemya Island Investigations. Report on file, U.S. Fish and Wildlife Service, Anchorage, Alaska.

Simenstad, C. A., J. A. Estes, and K. Kenyon
1978 Aleuts, Sea Otters, and Alternative Stable State Communities. *Science* 200:403–411

Simenstad, C. A., D. O. Duggins, and P. D. Quay
1993 High Turnover of Inorganic Carbon in Kelp Habitats as a Cause of 13C Variability in Marine Food Webs. *Marine Biology* 116:147–160.

Spaulding, A.
1962 *Archaeological Investigations on Agattu, Aleutian Island.* Museum of Anthropology, Anthropological Papers 18:1–79. University of Michigan, Ann Arbor.

Stejneger, L.
1887 How the Great Northern Sea Cow Became Exterminated. *American Naturalist* 21:1047–1054.
1898 *The Asiatic Fur-Seal Islands and the Fur-Seal Industry.* The Fur Seals and Fur-Seal Islands of the North Pacific Ocean, Part IV. Government Printing Office, Washington, D.C.

Steller, G. W.
1988 *Journal of a Voyage with Bering, 1741–1742,* edited by O. W. Frost, translated by A. Engel and O. W. Frost. Stanford University Press, Stanford, California.

Turner, C. G.
1970 Archaeological Reconnaissance of Amchitka Island, Alaska. *Arctic Anthropology* 7(2): 118–128.

Turner, L. M.
1886 *Contributions to the Natural History of Alaska.* U.S. Army Signal Service, Washington, D.C.
1891 Descriptive Catalogue of Ethnological Specimens collected by Lucien M, Turner in Alaska. Manuscript at the National Anthropological Archives, Smithsonian Institution, Washington, D.C.

U.S. Bureau of Indian Affairs
1985 Radiocarbon dates for sites AA-11960, 11967, 12011, 12013. U.S. Bureau of Indian Affairs, Alaska Native Claims Settlement Act Office, Anchorage.
1986 Reports of Investigation for Amchitka Island, AA-11945 to AA-12019. U.S. Bureau of Indian Affairs, Alaska Native Claims Settlement Act Office, Anchorage.
1988 Report of Investigation for Attu Village Site Report by D. Corbett and T. Fifield. U.S. Bureau of Indian Affairs, Alaska Native Claims Settlement Act Office, Anchorage.
1989a Reports of Investigation for Shemya Island, AA-11923 to AA-11926. Report by D. Corbett and T. Fifield. U.S. Bureau of Indian Affairs, Alaska Native Claims Settlement Act Office, Anchorage.
1989b Reports of Investigation for Sites AA-11918 to AA-11922. Report by F. Clark. U.S. Bureau of Indian Affairs, Alaska Native Claims Settlement Act Office, Anchorage.
1990 Field Notes and Maps from Investigations on Agattu and in the Semichi Island. On file at U.S. Bureau of Indian Affairs, Alaska Native Claims Settlement Act Office, Anchorage.
1994 Reports of Investigation for Shemya Island, AA-11924 and 11925. Report by J. Bartolini. U.S. On file at Bureau of Indian Affairs, Alaska Native Claims Settlement Act Office, Anchorage.
1995 Reports of Investigation for Sites AA-11908 to AA-11917. Report by D. R. Cooper and B. Hoffman. U.S. Bureau of Indian Affairs, Alaska Native Claims Settlement Act Office, Anchorage.

U.S. Fish and Wildlife Service
1994 *Conservation Plan for the Sea Otter in Alaska.* U.S. Fish and Wildlife Service, Marine Mammals Management, Anchorage, Alaska.

U. S. National Park Service
1968 The Island of Attu, Alaska: A Study of Alternatives. U. S. Department of Interior, National Park Service, Anchorage, Alaska.

Veniaminov, I.
1984 *Notes on the Island of the Unalashka District,* edited by R. A. Pierce, translated by L. Black. The Limestone Press, Kingston, Ontario.

Watson, P., S. LeBlanc, and C. L. Redman
1971 *Explanation in Archeology: An Explicitly Scientific Approach.* Columbia University Press, New York.

West, D. L., D. Corbett, and C. Lefèvre
1997 1997 NSF Grant Report: The Western Aleutians Archaeological and Paleobiological Project. National Science Foundation, Washington, D.C.
1998a 1998 NFS Grant Report: The Western Aleutians Archaeological and Paleobiological Project. National Science Foundation, Washington, D.C.
1999 1999 NSF Grant Report: The Western Aleutians Archaeological and Paleobiological Project. National Science Foundation, Washington, D.C.
2001 2001 NSF Grant Progress Report: The Western Aleutians Archaeological and Paleobiological Project. National Science Foundation, Washington, D.C.
2002 2002 NSF Grant Progress Report: The Western Aleutians Archaeological and Paleobiological Project. National Science Foundation, Washington, D.C.
2003 2003 NSF Grant Progress Report: The Western Aleutians Archaeological and Paleobiological Project: Attu Island. National Science Foundation, Washington, D.C.

West, D. L., C. Lefèvre, D. Corbett
1998b Radiocarbon Dates for the Near Islands, Aleutian Islands, Alaska. *Current Research in the Pleistocene* 16(1999):83–85.

Whitmore, F. C., and L. M. Gard, Jr.
1977 Steller's Sea Cow *(Hydrodamalis gigas)* of Late Pleistocene Age from Amchitka, Aleutian Islands, Alaska. U.S. Geological Survey Professional Paper 1036.

Yesner, D. R.
1977 *Prehistoric Subsistence and Settlement in the Aleutian Islands.* Ph.D dissertation, University of Connecticut, Storrs.

4

Historical Ecology and Human Impacts on Coastal Ecosystems of the Santa Barbara Channel Region, California

Torben C. Rick, Jon M. Erlandson, Todd J. Braje, James A. Estes, Michael H. Graham, and René L. Vellanoweth

IN RECENT YEARS, a number of researchers have emphasized the importance of archaeological and historical records for elucidating contemporary environmental issues and crises (e.g., Diamond 2005; Grayson 2001; Jackson et al. 2001; Kirch 1997; Kirch and Hunt 1997; Krech 1999; Lyman and Cannon 2004; Redman 1999; Redman et al. 2004). Many of these studies have demonstrated that virtually everywhere humans have gone they have had an impact and influence on local environments. These impacts have sometimes been profound. Although still a hotly debated and contentious issue (see Grayson and Meltzer 2002, 2003; Wroe et al. 2006), some researchers argue that early humans, as they spread into new regions of the world during the Late Quaternary, actually transformed entire biotas by causing the extinctions of numerous large vertebrate species (Martin 2005). Others have speculated that the sudden loss of these large vertebrate consumers—regardless of the cause—had important ecological and evolutionary consequences (Barlow 2000; Janzen and Martin 1982). The scale and magnitude of human environmental impacts, however, varied greatly both geographically and temporally, suggesting that much remains to be learned about ancient human interactions with the environment (Redman 1999; van der Leeuw and Redman 2002). This is particularly true of coastal regions, where archaeologists stand to make important contributions to our understanding of marine ecology, the role of ancient peoples in influencing environmental change, and the best strategies for future remediation of marine habitats and fisheries (Jackson et al. 2001).

In this chapter, we examine the historical ecology of marine ecosystems of the Santa Barbara Channel region of southern California (Figure 4.1) from the terminal Pleistocene to the present, focusing on evidence for ancient human impacts on marine fisheries and habitats. The area has been occupied by Native Americans for roughly 13,000 calendar years and, in historical times, supported important sea otter, pinniped, swordfish, sardine, rockfish, abalone, urchin, lobster, and other fisheries.

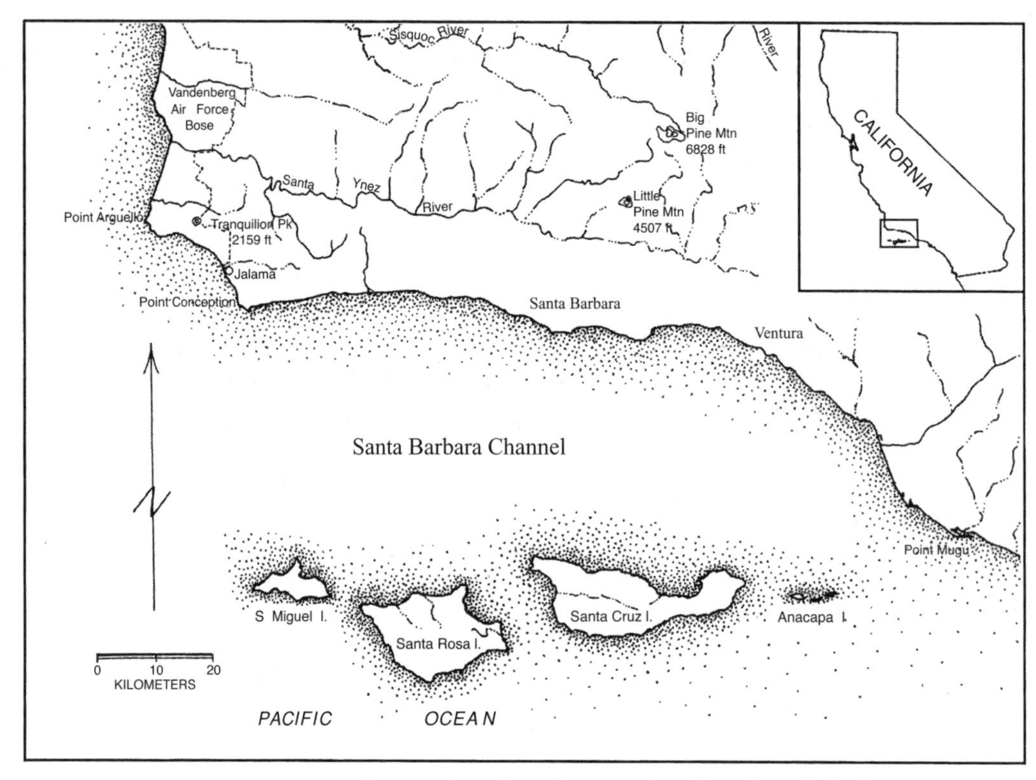

FIGURE 4.1. Location of Santa Barbara Channel region and Northern Channel Islands.

The unique combination of a long and continuous archaeological record, a wealth of historical data on the maritime Chumash, Spanish, Asian, and Euroamerican occupations of the area, and detailed ecological and paleoecological records make it an excellent laboratory for investigating the historical ecology of marine environments and bridging the divide between biology, archaeology, history, and the modern world. Our fundamental challenge is one that permeates all of modern and historical ecology—how to make strong inferences about the causes of ecological change. Although the patterns of change are often clear, the causes of these patterns are usually much less clear and thus matters of debate. The debates nearly always boil down to arguments over the relative importance of two factors: human overexploitation/predation on the one hand, and environmental (e.g., natural climate and ecological) change on the other.

Archaeological and historical records of the Santa Barbara Channel generally contain four classes of resources that speak to marine ecology: the remains of marine fishes, shellfish, mammals, and birds. The habitats each of these organisms occupied and their behaviors provide proxies for the type of habitats present in the past and their exploitation by ancient peoples. We provide a general overview of the structure of each of these faunal communities in the Santa Barbara Channel region over the last 10,000 years, including discussions of their historic status and impacts, and then speculate on possibilities for ancient human impacts and the implications of our data for present day management and restoration efforts. This general overview serves as a baseline for more detailed investigations of individual faunal classes or species, as well as a framework in which to couch future research on the historical ecology of the region. We begin with brief environmental and cultural summaries of the area to contextualize our analysis and help distinguish natural changes in marine ecosystems from possible environmental impacts caused by humans.

ENVIRONMENTAL, CULTURAL, AND HISTORICAL CONTEXT

The Santa Barbara Channel is one of the most sheltered and protected stretches of the California Coast (Schoenherr 1992). The region is environmentally circumscribed, with a narrow mainland coastal plain flanked by the Santa Ynez Mountains to the north and by the Santa Barbara Channel and the Northern Channel Islands to the south. The geographic diversity of the area—with a variety of habitats or ecosystems stacked in close proximity to one another—has provided humans with access to a wide variety of resources for millennia. Despite this productivity and diversity, there is considerable variability in the distribution of resources, including a dearth of terrestrial resources on the Channel Islands.

The Santa Barbara Channel region has a Mediterranean climate with warm, dry summers and cool, wet winters. Average rainfall in the region is about 45 cm, although the Santa Ynez Mountains receive nearly twice this much. Santa Barbara Channel waters are characterized by relatively intense upwelling and contain a mix of southern and northern currents, promoting a diverse and distinct combination of aquatic fauna and some of the richest and most productive marine habitats in the world, including kelp forests, rocky coasts and reefs, sandy beaches, estuaries, and deeper pelagic and benthic communities. Straddling the boundary area between the Panamanian and Oregonian biotic provinces, the Santa Barbara Channel contains an unusually diverse array of marine species, high marine productivity, and relatively complex food webs (see Steneck et al. 2002).

Over the last 13,000 years environmental changes have dramatically altered the geography of the region. During the last glacial, the Northern Channel Islands were linked in a much larger landmass known as Santarosae Island, separated from the mainland by as little as 6 to 8 km at its eastern end. In the terminal Pleistocene, Santarosae separated into several smaller islands as rapidly rising postglacial sea levels transformed the geography of the area. For the Northern Channel Islands, modeling of postglacial changes in coastal and nearshore habitats suggests that kelp forests were considerably more extensive in the terminal Pleistocene and Early Holocene (Kinlan et al. 2005). Much of the mainland coast was also punctuated by estuarine embayments during the Early Holocene (Erlandson 1994), but there is limited evidence for the presence of such estuaries on the Channel Islands (Rick et al. 2005). By the Middle Holocene, most of these estuaries had disappeared, with only the largest (Goleta Slough, El Estero, Carpinteria Slough, Mugu Lagoon, etc.) persisting into historic times. In contrast, sandy beaches and nearshore habitats appear to have increased during the Middle Holocene (Graham et al. 2003). By the Late Holocene, environmental conditions were generally similar to the present, albeit with a variety of annual and decadal fluctuations (see Kennett and Kennett 2000). The Santa Barbara Channel area is susceptible to El Niño/La Niña cycles, intensive droughts, and other climatic perturbations. Analyses of long, high-resolution varved sediment sequences from the Santa Barbara Basin indicate that these environmental disturbances operated in the past (see Friddell et al. 2003; Kennett and Ingram 1995; Kennett et al. 2007) and were an important component of changes in human subsistence, settlement, demography, and cultural developments in the region (see Arnold 1992; Arnold et al. 1997; Erlandson and Rick 2002; Gamble 2005; Johnson 2000; Jones et al. 1999; Kennett 2005; Kennett and Kennett 2000; Kennett et al. 2007; Raab and Larson 1997).

At the time of European contact, the Santa Barbara Channel region and Northern Channel Islands were occupied by the coastal Chumash, who were among the most densely populated hunter-gatherers in the world, with numerous towns or villages, some with up to 1,000 residents. The Chumash had a hierarchical social organization with formal chiefs, craft specialists, and intensive exchange networks. Trade was both intensive and extensive, facilitated by a shell bead currency that was primarily produced on the Channel Islands and exchanged between island and mainland people for a variety of goods and services (Arnold 1992; King 1990). The Chumash and their predecessors lived in the Santa Barbara Channel region for roughly 13,000 calendar years. An isolated fragment of a fluted point found on the western Santa Barbara Coast suggests a Paleoindian presence in the area (see Erlandson 1994), and the Arlington Man skeleton found on Santa Rosa Island may be as old as 13,000 cal BP (Johnson et al. 2002). A basal shell midden at Daisy Cave on San Miguel Island has been radiocarbon dated to about 11,600 cal BP (Erlandson et al. 1996; Rick et al. 2001), and more than 35 shell middens from the islands and mainland date between about 10,000 and 8500 cal BP. Shell middens, lithic scatters, villages, and other site types are found throughout the Santa Barbara Channel region more or less continuously from about 10,000 years ago through the early nineteenth century. Populations grew substantially during the Holocene, with the greatest increase during the last 3,000 years when the Chumash and their ancestors established a series of large villages on the mainland coast and Channel Islands. Although the Chumash were not agriculturalists, they managed the landscape relatively intensively, including regular burning of vegetation communities to maintain their productivity (Timbrook et al. 1982). Ecologically, on both the mainland and the Channel Islands, the Chumash served as apex predators in local marine and terrestrial ecosystems. As Chumash populations grew in the circumscribed environments of the Santa Barbara Channel, it seems likely that their impacts on local ecosystems also increased.

The first Europeans to contact the Chumash were members of a Spanish maritime expedition led by Juan Rodriguez Cabrillo, who spent more than a month among the Island Chumash during the winter of AD 1542–1543 (Wagner 1929). For the next 225 years, other "protohistoric" contacts in the region were relatively brief and sporadic but may have caused some decline in Chumash populations (Erlandson and Bartoy 1995; Erlandson et al. 2001; Preston 1996). After Spanish colonization of Alta California began in AD 1769, several missions and a presidio were built in the Santa Barbara Channel vicinity, beginning with Mission San Buenaventura in AD 1782, Mission Santa Barbara in 1786, Mission La Purisima in 1787, and Mission Santa Inez in 1804 (Costello and Hornbeck 1989). By about AD 1822, the last of the Island Chumash were removed to the mainland missions, where they were incorporated into the emerging colonial economies and multiethnic communities of successive Spanish, Mexican, and American regimes.

The impacts of the missions, pueblos, and ranchos on terrestrial ecosystems of the Santa Barbara Channel area were dramatic, including massive landscape clearing and the introduction of domestic livestock (cattle, sheep, horses, etc.) and agriculture to the area, which ravaged terrestrial plant communities and watersheds. The early colonial era also saw a steep decline in Chumash populations decimated by Old World diseases, an end to native burning, the introduction of numerous exotic grasses and other plant species, and the eradication or depletion of grizzly bears and other large terrestrial predators. The impact of the Spanish and other early colonists on marine ecosystems is less well understood. Initially, the reduction of Chumash population and deterioration of their traditional maritime economy may have encouraged a rebound in some marine species (see Braje et al. 2007a; Preston 2002). During the nineteenth century, dramatic changes are evident in local ecology, however, as the fur trade driven by the emergence of a global economy caused the local extinction of sea otters and dramatic declines in the population of pinnipeds, whales, and abalones (see Scammon 1968). By AD 1822, the Mexican government assumed control of California, and the missions were secularized in AD 1834. Nonetheless, the environmental alterations of the Mission period continued to expand as the population of local pueblos (towns) grew and numerous homesteads and ranchos were established in the area. After the abandonment of the missions, many of the surviving Chumash moved into the labor force of these pueblos and ranchos. After a brief war with Mexico and the gold rush, California gained U.S. statehood in AD 1850. The area supported a sizable abalone fishery in historic times, starting with Chinese fishers in the 1850s, a "boom and bust" industry facilitated by the local eradication of sea otters (Braje et al. 2007a; Estes and Van Blaricom 1985) and the collapse of the traditional Chumash maritime economy. The next 150 years witnessed enormous population growth, landscape alteration, and unprecedented impacts on the marine and estuarine ecosystems of the southern California Coast.

Today, the Santa Barbara Channel area contains several cities (e.g., Goleta, Santa Barbara, Carpinteria, and Ventura) with a combined population of a few hundred thousand people. Commercial fishing in the region has been relatively aggressive, including fisheries for sea urchins, abalones, numerous types of near- and offshore fishes, and a variety of other resources. At various times in the past, these resources have all experienced collapses, resulting in greater regulation, temporary or more permanent moratoriums on their use, and the establishment of several marine protected areas in Channel Islands National Park. As Schroeder and Love (2002) noted, recreational fisheries in the area also have a profound impact on the structure of local fisheries. Modern ecological data suggest that although abalones, sea urchins, and other resources have experienced some rebounds, they continue to be at historic low levels and in a state of decline (UCSB Sustainable Indicators Data 2007). Given the profound alterations of marine fisheries and ecosystems of the historic era, and the need to develop reasonable ecological baselines to guide management and restoration efforts, we believe the archaeological record of 13,000 years of human intervention in marine habitats of the Santa Barbara Channel has a great deal to offer managers of local fisheries and ecosystems.

HISTORICAL ECOLOGY AND THE SANTA BARBARA CHANNEL AREA

Jackson et al. (2001) emphasized the importance of building long-term ecological records to better understand and manage present day marine ecosystems. To provide a context for long-term ecological change in the Santa Barbara Channel region, we briefly summarize the trans-Holocene archaeological record of human interaction with marine environments and resources of the Santa Barbara Channel region and the Northern Channel Islands. We synthesize archaeological data from the mainland coast and Channel Islands to trace the status of human impacts on fishes, marine mammals, birds, and shellfish over the past 10,000 years. Our goal is to provide a broad and diachronic framework that can be used as a springboard for research on various animal classes, genera, or species.

Fishes

One of the most prominent trends in the prehistory of the Santa Barbara Channel is a general increase in the amount of fishing and decrease in the amount of shellfish collecting through time (Glassow 1993a; Kennett 2005; Lambert 1993; Rick 2007; Vellanoweth et al. 2002). This shift is most apparent during the Late Holocene, when fish remains make up the vast majority of edible meat yields (>70 percent) and bone densities in most island and mainland archaeological sites. The increase in fishing is more pronounced on the Channel Islands than on the mainland due to the mainland's greater availability of terrestrial mammals and plants (Erlandson and Rick 2002). Several researchers have explored the implications of this trend, with most suggesting that it is closely linked with growing population densities, improvements in fishing technologies, and the greater biomass and availability of fishes than many other marine resources

(Erlandson and Rick 2002; Glassow 1993a; Kennett 2005). As Lambert (1993) suggested, the apparent increase in fishing on the islands was accompanied by some declines in human health, including a reduction in stature, an increase in inflammatory bone lesions, and greater evidence for interpersonal conflict and violence.

Questions remain about the implications of this dramatic increase in fishing and the possible impacts people had on local fisheries. The exponential increase in fish bones, with some island sites having more than 30,000 bones/m^3 (Kennett and Conlee 2002; Rick 2007) suggests that the impacts of Chumash fishing were increasingly significant. To date, however, the archaeological record of human impacts on island and mainland fisheries is somewhat ambiguous, with evidence for continuity in the dominant types of fishes found in some Early, Middle, and Late Holocene assemblages on San Miguel and Santa Rosa islands (Paige 2000; Rick 2007; Rick et al. 2001). One limitation is that no studies of fish size through time are available, a factor determined by the fact that surfperches and rockfish, two of the more common fishes in trans-Holocene assemblages, are generally not identified to species, making size determinations problematic. Data on temporal changes in fish size have been an important tool for investigating human impacts in other areas (see Carder et al. 2007; Newsom and Wing 2004; Reitz 2004) and would be a valuable addition for the Santa Barbara Channel area as well.

Most early fishing appears to focus on nearshore species (rockfish, surfperches, croaker, etc.), and fishes appear to have generally been supplemental to shellfish. Associated with the dramatic increases in fishing through time, especially the last 3,000 years, the types of fishes also increase significantly. Fishing technology also appears to vary through time, beginning with gorge and line in the Early Holocene and culminating with the Late Holocene appearance of single-piece shell fishhooks. As Braje (2007), Colten (2001), Noah (2005), Paige (2000), and Rick (2007) noted, however, people continued to focus largely on rockfish, surfperch, and other similar fishes (see also Bowser 1993). At some locations there are much greater numbers of deeper water fishes (Colten 2001; Pletka 2001), a factor that may be related to the variability in offshore bathymetry and the presence of submarine canyons in the area. Even at these sites, nearshore fishes still tend to be the most abundant taxa. Mainland fish assemblages are often considerably more diverse than island assemblages, a factor probably related to the presence of a wider range of habitats on the mainland (e.g., sand beaches, estuaries, etc.; Erlandson et al. 2007a; Glenn 1990; J. Johnson 1980; Rick and Glassow 1999). With some exceptions, many of the fishes identified in early mainland sites are also comparable to those found in later assemblages, although the density of fishes increases greatly.

Much has been made in recent ecological literature about the dramatic effects of commercial fishing on trophic diversity. Pauly et al. (1998) proposed "fishing down the food web" to explain the shift from high trophic level fishes to lower trophic level fishes in the wake of historical overfishing. Reitz (2004), Morales and Rosello (2004), Bourque et al. (this volume), and Kennett et al. (this volume) have begun to look for these processes in the archaeological record, with varying levels of success. Although these data are complicated and preliminary in nature, in some cases ancient peoples may have been fishing in a manner consistent with the "fishing down the food web" model. In contrast to some of these cases, preliminary data from the Santa Barbara Channel suggest the opposite pattern, where people may have been fishing up the food web (Figure 4.2). In this case, many of the higher trophic-level fishes (swordfish, tunas, sharks, barracudas, etc.) are not common in the archaeological record until after about 1,500 years ago, with some areas showing increases during the Historic period (Bernard 2004; Noah 2005; Pletka 2001). While some of these fishes may appear relatively late in the

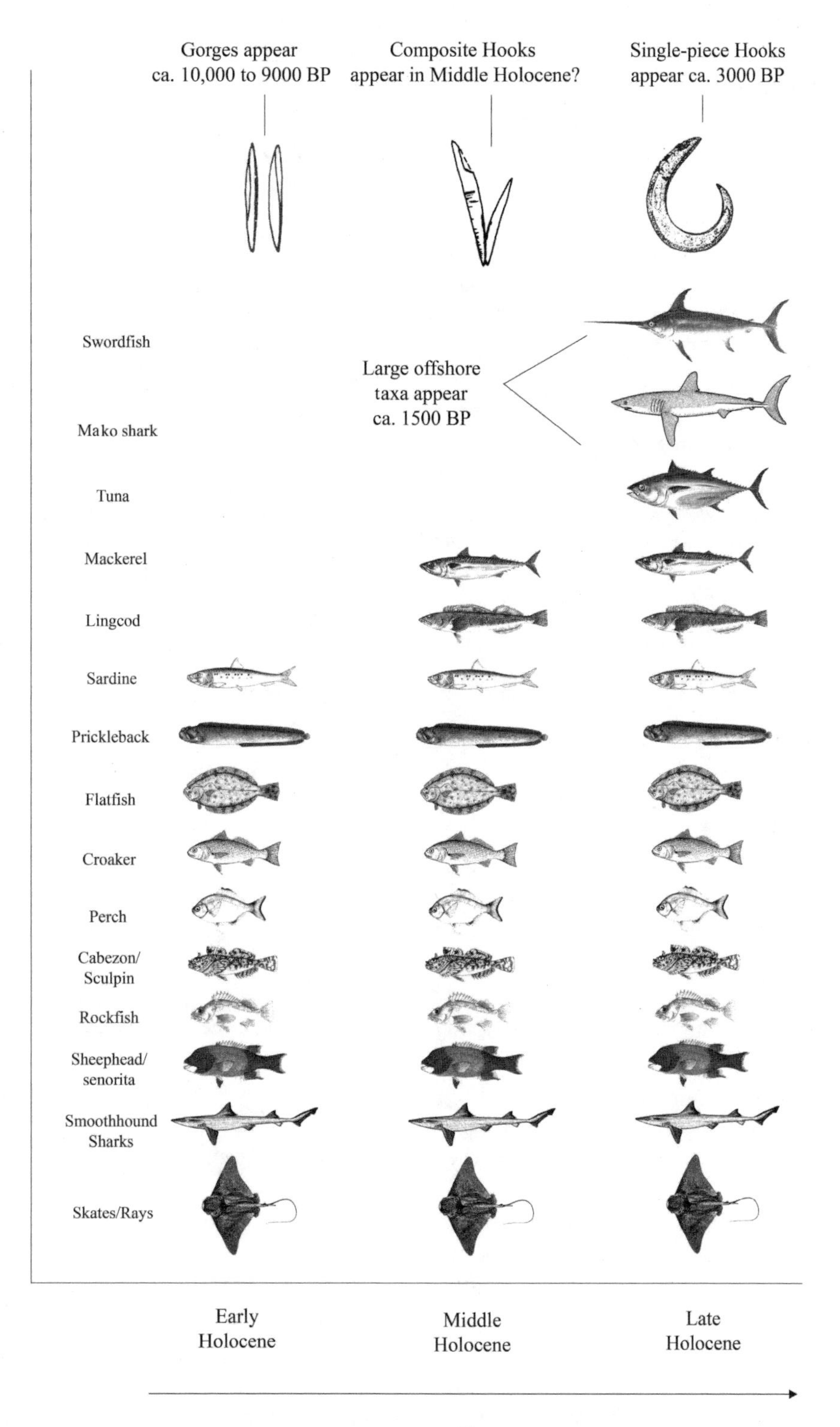

FIGURE 4.2. Diagram showing general increase in fish taxa through time, changes in fishing technology, and Late Holocene appearance of large, offshore taxa in the Santa Barbara Channel.

record due to technological improvements later in time (e.g., plank canoes, toggling harpoons), these data suggest a fishing strategy contrary to many historical patterns. The long-term trend toward fishing up the food web and incremental use of higher trophic levels is one strategy that may have proven successful in acquiring food for growing populations.

The apparent continuity and relative stability of aspects of Chumash fishing practices stand in contrast to the great declines witnessed during the historic era, a factor in part due to technological limitations and the high reproductive rates of fishes. Historically, periodic bans on the fishing of certain taxa (e.g., swordfish, sea bass, rockfish) have occurred in Santa Barbara Channel waters. The recent expansion of no fishing zones surrounding the Northern Channel Islands is an effort to curtail these historic processes, and we hope that aspects of the Chumash record will also enlighten current management strategies.

Marine Mammals

A number of archaeologists have debated the nature, significance, and consequences of sea mammal hunting along the Pacific Coast of North America (e.g., Colten and Arnold 1998; Erlandson et al. 1998; Etnier 2004; Hildebrandt and Jones 1992; Jones and Hildebrandt 1995; Jones et al. 2004; Kennett 2005; Lyman 1995; Porcasi et al. 2000; Rick 2007; Walker et al. 2002). Numerous species of marine mammals, including pinnipeds, sea otters, and cetaceans, have been identified in the regional archaeological record (Table 4.1). Although the Chumash appear to have hunted dolphins, porpoises, and pinnipeds and scavenged beached whales, there is currently no evidence that the Chumash hunted large whales. Relying on the behaviors of pinnipeds and otters, Hildebrandt and Jones (1992) argued that early Native American predation led to the abandonment of onshore rookeries, with fur seals and sea lions moving to smaller islets or more isolated localities, followed by a shift toward more intensive hunting of harbor seals and sea otters that do not use rookeries and are generally more difficult to capture. Based in part on archaeological data from the Santa Barbara Channel and Channel Islands, they proposed a prehistoric "tragedy of the commons," with human predation causing fundamental changes in the distribution and demography of Pacific Coast pinnipeds (Hildebrandt and Jones 1992; Jones and Hildebrandt 1995; Porcasi et al. 2000). Erlandson et al. (2004a) focused on the early archaeological record of the Northern Channel Islands, noting that none of these early sites contain abundant pinniped remains (see also Kennett 2005:222–223), and suggested that early foragers concentrated on shellfish supplemented by fishes, raising questions about the "tragedy of the commons" model. Evidence for early pinniped hunting is still limited, but recent excavations on San Miguel Island have identified greater evidence for early hunting technologies (see Braje 2007), suggesting that the importance of sea mammal hunting at many early island sites may be underestimated. Recent research has also documented dramatic increases in sea mammal hunting around 1,500 years ago, a date consistent with the intensification of fishing for some large pelagic fish species.

Braje (2007), Kennett (2005), Rick (2007), and Walker et al. (2002) noted significant increases in pinniped bone at three San Miguel Island sites dated between about 1500 and 1000 cal BP. At these sites pinniped bones (Guadalupe fur seals, California sea lions, elephant seals, harbor seals, Steller sea lions, and northern fur seals) are extremely abundant. By about 600 to 500 cal BP, other island sites show that pinnipeds were replaced by fishes as the most important component of the diet (Kennett 2005; Rick 2007). Colten and Arnold (1998) noted that Santa Cruz Island sites dated to this interval contain only minor amounts of pinniped remains throughout the last 1,500 years, suggesting spatial variability on the Channel Islands similar to the present day. The meaning and nuances of this trend are still being explored, but for now it appears that the

TABLE 4.1

Major Marine Mammals Identified in Santa Barbara Channel Archaeological Sites

TAXA	BEHAVIOR	ARCHAEOLOGY
Pinnipeds		
Arctocephalus townsendii (Guadalupe fur seal)	Migratory, sexually diamorphic, formerly common on Channel Islands	Common in island and some mainland sites, especially after 1500 cal BP
Callorhinus ursinus (Northern fur seal)	Migratory, sexually diamorphic, and breeds on Channel Islands	Common in island and some mainland sites, especially after 1500 cal BP
Eumetopias jubatus (Steller sea lion)	Migratory, sexually diamorphic, rare on Channel Islands	Rare in island and mainland sites
Mirounga angustirostris (elephant seal)	Migratory, sexually diamorphic, and breeds on Channel Islands	Rare in island and mainland sites; oldest known occurrence 6000 cal BP
Phoca vitulina (harbor seal)	Resident, skittish, and breeds in water	Relatively rare in mainland and island sites
Zalophus californianus (California sea lion)	Migratory, sexually diamorphic, and breeds on Channel Islands	Common in island and some mainland sites, especially after 1500 cal BP
Sea otters		
Enhydra lutris (sea otter)	Resident, skittish, and found primarily in water	Common to rare in island and mainland sites throughout last 9,000 years
Cetaceans		
Delphinus delphis (common dolphin)	Social and travels in groups in near and offshore waters	Generally rare in island and mainland sites but common in Middle Holocene context at Punta Arena, Santa Cruz Island
Globicephala macrorychus (short-finned pilot whale)	Travels in groups sometimes with other species in deep and nearshore water	Rare in island and mainland sites, hunted and/or scavenged
Grampus griseus (Risso's dolphin)	Travels in groups sometimes with other species in deep and nearshore water	Rare in island and mainland sites, hunted and/or scavenged
Lagenorynchus obliquidens (Pacific white sided dolphin)	Gregarious and travels in large groups sometimes with other species in deep and nearshore water	Generally rare in island and mainland sites but common in Middle Holocene context at Punta Arena, Santa Cruz Island
Lisodelphis borealis (northern right whale dolphin)	Travels in large groups sometimes with other species in deep and nearshore water	Generally rare in island and mainland sites but common in Middle Holocene context at Punta Arena, Santa Cruz Island
Orcinus orca (killer whale)	Travels in pods in near and offshore waters, top predator in the ocean	Rare in island and mainland sites, probably scavenged
Physeter catodon (sperm whale)	Common in deep water in pods sometimes with other species	Rare in island and mainland sites, probably scavenged
Tursiops truncatus (bottlenose dolphin)	Gregarious travels in groups in near and offshore waters	Generally rare in island and mainland sites but common in Middle Holocene context at Punta Arena, Santa Cruz Island

Chumash had a substantial impact on marine mammal populations on the islands, but that this occurred relatively late in the archaeological record with people possibly coexisting with pinnipeds for millennia. The abundance of sea mammal remains in sites dated to around 1500 cal BP seems to correspond with the appearance of the plank canoe and the bow and arrow, which may have facilitated hunting pinnipeds from the water and at offshore locales (see Kennett 2005).

Unfortunately, much remains to be learned about the structure and nature of ancient pinniped populations and human hunting on the Channel Islands. In particular, we have relatively limited data from the period between about 7000 and 1500 cal BP, leaving a substantial gap in the record. It should also be noted that pinnipeds still occur in late and historic deposits, suggesting that they were still available locally, albeit probably in reduced numbers.

In mainland archaeological sites, pinniped remains are considerably less abundant than on the Channel Islands, and many mainland faunal assemblages are also highly fragmented, making identifications to genus or species difficult. Pinnipeds were probably always less abundant along the mainland coast, where bears and other large predators limited their ability to breed or haul out onshore (Erlandson et al. 1998; Jones et al. 2004). Erlandson and Rick (2002) noted that pinnipeds make up roughly 30 to 50 percent of the dietary protein represented in several components dated between about 1,200 to 600 years ago along the Santa Barbara Coast, however, suggesting that they were also important in some mainland sites during this time period. Many of these animals may have been obtained through trade with islanders where pinnipeds were probably more abundant. Although the number of diagnostic bones was small, Erlandson et al. (2007a) noted that Late Holocene assemblages in Tecolote Canyon contained small amounts of Guadalupe fur seals, northern fur seals, harbor seals, California sea lions, and sea otters. Some of these may have been taken as sick or stranded individuals, however, as illustrated by a 2005 toxic dinoflagellate bloom that drove hundreds of incapacitated California sea lions ashore on mainland beaches. All of these species also strand occasionally today, and modern facilities have been established to care for these animals (e.g., The Marine Mammal Center, Sausalito, CA).

Modern ecological data suggest that pinnipeds in the Santa Barbara Channel area feed primarily on squid and other marine animals that reside in offshore habitats. If this pattern is true of the more distant past, human impacts on seals and sea lions may have had only limited effects on the nearshore ecological communities that would have been most visible in the archaeological record. Sea otters, in contrast, are "strong interactors" in nearshore biological communities, whose behavior and abundance has major consequences for the structure and productivity of some nearshore shellfish and fish populations, as well as broader kelp forest communities (Estes and Palmisano 1974; Simenstad et al. 1978; Steneck et al. 2002).

Until recently, few California archaeologists have recognized the importance of sea otters in structuring nearshore marine ecosystems. For the Northern Channel Islands, we explored such relationships with a preliminary record of trans-Holocene sea otter hunting. Erlandson et al. (2005a) suggested that human control of otter populations may have been crucial to the development of a productive red abalone fishery during the Middle Holocene, between about 7,500 and 3,000 years ago. Using preliminary data, we argued that people may have reduced otter densities or even eradicated otters from local village territories. In nearshore habitats, abalones and urchins compete for food and space, so intensive Native American harvesting of abalones may also have contributed to an increase in sea urchin populations. Released from intensive predation and competition pressures, sea urchin populations can alter kelp forests through overgrazing (Steneck et al. 2002), sometimes creating "urchin barrens" in which kelp and abalones cannot survive

(Graham 2004; Graham et al. 2007). Possible evidence for such perturbations in local kelp forests may come from dense accumulations of sea urchin tests identified in some San Miguel Island shell middens dating to the past 4,000 years or so, when Native populations were larger and more sedentary, and the intensity of nearshore fishing increased significantly. Nonetheless, people may have significantly altered the structure of local kelp forest communities by about 7,500 years ago by reducing or possibly even depleting local sea otter populations. A similar pattern (albeit of lesser antiquity) was proposed by Simenstad et al. (1978) for the Aleutians (see also Corbett et al., this volume).

In recent decades, marine mammal populations have rebounded significantly on the Channel Islands and are one of the major success stories of marine conservation. As of 1997, San Miguel was home to over 140,000 pinnipeds of six separate species: the California sea lion *(Zalophus californianus)*, Steller sea lion *(Eumetopias jubata)*, elephant seal *(Mirounga angustirostris)*, harbor seal *(Phoca vitulina)*, Guadalupe fur seal *(Arctocephalus townsendii)*, and northern fur seal *(Callorhinus ursinus)* (DeLong and Melin 2002). Preliminary archaeological data from the Channel Islands, however, including a late prehistoric and early historic period Chumash village in the middle of the Point Bennett rookery, suggest that these high populations are unprecedented, at least for the past several millennia. The presence of humans on the mainland and islands throughout the Holocene clearly had an impact on the demography and structure of local pinniped populations. There is no evidence for local extinctions of pinnipeds caused by Native Americans, however; and although declines are evident there appears to have been some degree of coexistence between humans and pinnipeds for 10,000 years.

Seabirds

On many Pacific Islands and other areas of the world, early human occupants have caused the extinction of numerous island bird species (Steadman 1995, 2006). Bird remains, however, are generally less common in Santa Barbara Channel archaeological sites than the remains of mammals, fishes, and shellfish. People appear to have relied on birds as a supplemental food resource, with many birds captured or scavenged for their bones and feathers rather than for food (Erlandson 1994; Guthrie 1993a; Kennett 2005). The dearth of bird remains in most sites makes it difficult to determine what impacts people had on these animals, but a few general trends deserve mention.

The best information on ancient human impacts on bird populations is from the Northern Channel Islands, especially San Miguel Island, where Guthrie (1980, 1993b, 2005) has analyzed remains from fossil localities and archaeological sites. Colten (2001) also identified the remains of at least 24 bird taxa in Santa Cruz Island sites dated to the last 1,500 years, but he noted that birds generally appear to have contributed less than 3 percent of the diet. People seem to have exploited a wide range of birds, including marine species such as cormorants, gulls, pelicans, and auklets. Many of these same birds have also been identified in mainland archaeological sites (Erlandson et al. 2007a).

Guthrie (1993b) suggested that humans caused the extinction of *Chendytes lawi*, an extinct flightless goose that lived on the islands and mainland during the late Pleistocene and much of the Holocene, coexisting with humans for several millennia (see Morejohn 1976). Although *Chendytes* bones have been found in a number of Santa Barbara Channel archaeological and paleontological sites (Table 4.2), no clearly cooked or processed *Chendytes* bones have been found on the islands (see Guthrie 1993b:413). While humans certainly would have exploited flightless birds that bred on the islands, there is currently no definitive evidence that people alone were responsible for their extinction. Along the Santa Barbara mainland coast, *Chendytes* remains have been identified in sites as young as 4,000 years old (Erlandson

TABLE 4.2

Chendytes lawi (Flightless Duck) Bones in Santa Barbara Channel and Vandenberg Archaeological Sites

SITE	NISP	AGE	REFERENCE
SBA-97	3	Early Holocene?	Erlandson et al. 1993
SBA-210	?	ca. 4500 RYBP	Glassow 1996:128
SBA-552	19	Early/Middle Holocene	Glassow 1996:128
SBA-2067	3	4860–4050 cal RYBP	Erlandson et al. 1993, 2004b
SCRI-109	17	6300–5300 cal BP	Glassow et al. 2007
SMI-1	43	Mixed Early/Middle/Late Holocene	Guthrie 1980
SMI-261	51	Early to Late Holocene?	Guthrie 1980
VEN-1	4	Early and Middle Holocene	Dallas 2004
SMI fossil sites	1,836	Pleistocene fossil localities	Guthrie 2005
West ANI fossil site	?	Pleistocene fossil localities	Miller et al. 1961

NOTE: NISP, number of identified specimens.

et al. 2004b), for instance, suggesting that they coexisted with humans in the Santa Barbara Channel for at least 8,000 years. Further research from a wider range of sites, including a more thorough chronology of *Chendytes* hunting and extinction, is needed to determine the effects of human predation on the demise of *Chendytes*.

Another potential impact is the possible exploitation of birds from offshore rocks where they breed, such as Prince Island and Castle Rock off the coast of San Miguel Island. Rick (2007) noted a dramatic increase in the density of bird bones and the dietary contribution of birds at CA-SMI-163, including a large number of auklet and cormorant bones that make up 17 percent of the dietary meat yield—one of the highest values for the Holocene. These findings suggest that sixteenth- through early nineteenth-century Chumash may have exploited local breeding colonies on Prince Island located less than a kilometer from the site. If these colonies existed in the past like they do today, the Chumash may have influenced the structure and nature of auklet and other bird breeding on the islands. This argument is hypothetical at this point, but future analysis of the age profiles of these and other bird assemblages could help evaluate this possibility.

The Island Chumash and their predecessors also had dogs since at least the Middle Holocene, and possibly for 9,000 or more years. Recently archaeologists have speculated that people may also have introduced the island fox (*Urocyon littoralis*) to the region (Kennett 2005; Vellanoweth 1998). Other than humans and dogs, the island fox was the largest "native" terrestrial mammal on the islands during the Holocene, with deer mice and a small skunk making up most of the other mammalian fauna (Schoenherr et al. 1999). The dearth of large terrestrial herbivores and carnivores on the islands suggests that the introduction of dogs and possibly foxes by humans—both of which would have hunted some birds and possibly infant sea mammals—may have had a major impact and influence on the native fauna. Similar to the situation with pinnipeds, several species of birds (e.g., gulls and cormorants) are successfully growing in population on the islands—in part because of a precipitous decline in island fox populations and the institution of a captive breeding program—while others continue to be threatened (e.g., Xantus' murrelets). If people exploited offshore breeding grounds, limited the space for birds to breed on the islands and mainland, and introduced dogs and possibly foxes to the islands, then historic populations and some modern breeding areas and behaviors for these birds may differ radically from what they were for most of the last 10,000 years.

Shellfish

Shellfish remains are extremely abundant in many Santa Barbara Channel archaeological sites. From the earliest shell middens to large historic villages, humans harvested shellfish in incredible numbers. Because shellfish were an important resource for ancient peoples in the area and they are relatively susceptible to overexploitation, they should be one of the most vulnerable faunal classes to human impact (Erlandson et al. 2004a). Indeed, a number of researchers around the world have documented significant human impacts on shellfish populations (e.g., Anderson 2001; de Boer et al. 2000; Klein et al. 2004; Mannino and Thomas 2002). People of the Santa Barbara Channel area also appear to have altered shellfish populations, which decrease in dietary importance through time, but they remain a conspicuous feature of the archaeological record. Even during the Late Holocene, when Chumash populations were relatively high, millions of individual shellfish were being exploited annually, in a fishery that appears to have been relatively sustainable in the long run (Erlandson et al. 2007b).

Along the mainland coast, dramatic environmental changes are present in the composition of shellfish remains, namely, the abundance of estuarine shellfish during the Early Holocene. The decline or disappearance of estuarine shellfish in many Middle Holocene sites, however, appears to be related primarily to environmental changes rather than human overexploitation (Erlandson 1994, 1997). Shellfish remains from other mainland sites also track environmental developments and paleogeography, including changes in rocky intertidal and sandy beach habitats (Vellanoweth and Erlandson 2004). The most dramatic alteration is a shift from shellfish contributing roughly 55 percent or more of protein yields before about 2,500 years ago, to shellfish making up generally less than 10 to 15 percent after that time. This trend illustrates the limitations of shellfish in the Santa Barbara Channel area as a viable resource for growing populations, even in mainland areas where terrestrial alternatives are abundant. Unfortunately, very few data on changes in shellfish size through time are currently available for the mainland.

Unlike the mainland with its distinct changes in shellfish and other habitats through time, there is considerable continuity on the islands in the types of shellfish found through the Holocene. Rocky intertidal taxa such as California mussels, abalones, urchins, turbans, and chitons are found throughout the last 10,000 years (Erlandson et al. 2004a). One notable exception is a small number of sites on eastern Santa Rosa Island that indicate the presence of an estuary before about 5,000 years ago, but in most of these sites rocky intertidal shells still dominate the assemblages (Rick et al. 2005). As on the mainland, shellfish appear to gradually become less important through time in Channel Island assemblages. By about 2,500 years ago, they are a relatively minor dietary contributor, although mussels and other shellfish were still harvested in huge quantities. Kennett (2005) suggested that one response to human predation on shellfish may be a general broadening of the types of shellfish exploited by Late Holocene peoples. This is a phenomenon that appears to be fairly consistent with our research, where the number of taxa being exploited generally increases through time. Drawing on foraging theory, Kennett (2005) also suggested that the Chumash may have targeted black and red abalones early on, as they were the largest and most optimal resources (see also Braje et al. 2007b). He cites the preponderance of red abalone shells in some Middle Holocene sites—a feature linked primarily with environmental change by other researchers (e.g., Glassow 1993b, 2002, 2005; Glassow et al. 1994)—and their decline in Late Holocene sites as evidence for human impact working in tandem with climate change. While humans may have had an impact on abalone populations, mussels seem to dominate the shellfish assemblages at most sites throughout the Holocene, and abalone shells may be less common in Late Holocene sites in part because they were heavily used for technological and ornamental purposes. Human predation may have been an aspect of the red abalone decline, but people probably were not the sole cause.

Several years ago, we initiated a program of measuring whole mussel, black and red abalone, and other shells from sites that span the last 10,000 years (Erlandson et al. 2004a, 2007b; Rick 2007). Other than a limited size decline about 7,500 years ago, our preliminary analysis indicated no clear pattern of changes in mussel size through time. Instead, mussel sizes appeared to oscillate randomly through time, possibly indicating human impact followed by periods of rebound. We now have nearly 10,000 measurements for mussels from San Miguel Island archaeological sites, and this larger record shows a general trend toward size reduction through time, but with a large amount of variability (Figure 4.3). Our analysis of black abalone measurements, which also includes historic data from nineteenth-century Chinese abalone camps, documents no significant change in size through the Holocene, although the average size is considerably larger in the historic Chinese occupations (see Braje et al. 2007a). The red abalone size data document a decline in size during the Middle and Late Holocene. In addition to human predation, however, this size decline may be related to environmental changes associated with a reduction of red abalone in sites around the end of the Middle Holocene (Glassow 1993b), and the extensive use of red abalone shells during the Late Holocene for technologies such as fishhooks and beads. The available data suggest that human impacts on local and regional shellfish populations may have been mediated by the periodic movement of village locations, shifting to lower-ranked shellfish taxa, and a general intensification of predation on other faunal classes (fishes, sea mammals, etc.) through time (e.g., fishing up the food web). Overall, however, the record of shellfish exploitation on the Northern Channel Islands shows that the same species of rocky intertidal shellfish were harvested by the Chumash and their ancestors for 10,000 years or more.

ACCELERATING HISTORICAL PERTURBATIONS IN SANTA BARBARA CHANNEL ECOSYSTEMS

Following the removal of most of the Chumash to the missions in the early nineteenth century, the demise of the "apex" predator from the Santa Barbara Channel waters may have allowed some key animal populations to rebound, including pinnipeds, otters, sea mussels, and fishes. If introduced diseases caused earlier human population declines during the sixteenth to eighteenth centuries, these demographic collapses may also have promoted rebounds in marine populations (see Erlandson et al. 2004a; Preston 2002). The effects of such Old World disease epidemics are heavily debated, but several researchers elsewhere in North America have speculated that Protohistoric declines in human populations promoted significant rebounds in animal populations (e.g., Broughton 1999; Butler 2000). Such regenerations of flora and fauna may have radically altered the perceptions of early Europeans about the nature and abundance of local environments (see Ames 2004; Broughton 1999; Kay and Simmons 2002; Krech 1999). In coastal California, however, such ecological rebounds would have been short-lived, as historical fisheries caused relatively rapid and dramatic changes to the environment.

Beginning in the late 1700s, the fur trade had a devastating and relatively well documented impact on sea otters and pinnipeds in the region, causing dramatic declines in their abundance and the local extinction of sea otters in the Santa Barbara Channel region (Ogden 1941; Scammon 1968). Other aspects of nineteenth-century fisheries are poorly documented in written historical records, but there is a long tradition of commercial

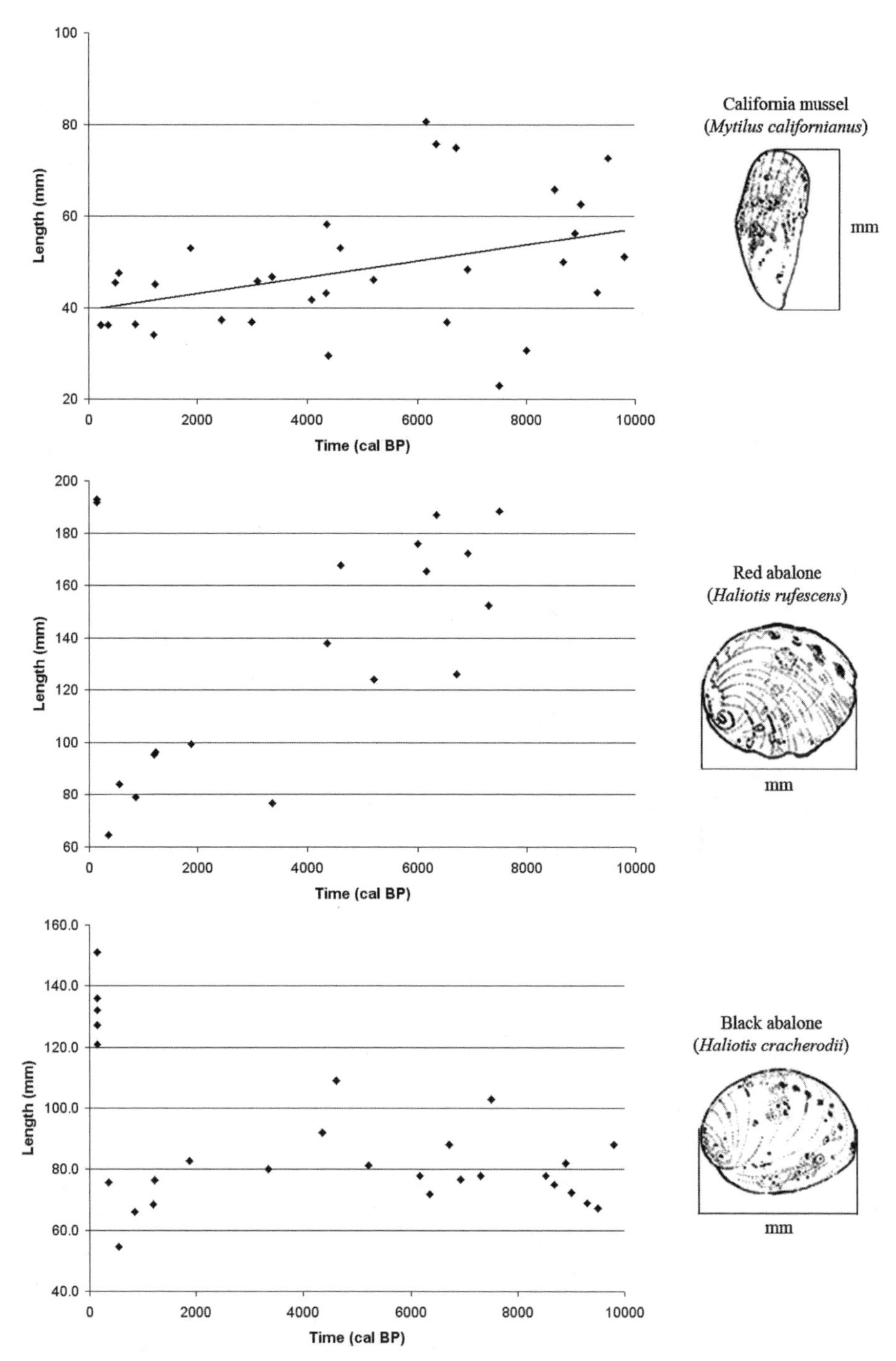

FIGURE 4.3. Average shell length (mm) for California mussel, black abalone, and red abalone from San Miguel Island, California, over the last 10,000 years.

fishing in the area. Numerous Chinese abalone camps on the Channel Islands (see Berryman 1995; Braje et al. 2007a) provide evidence of the intensive exploitation of abalones for local and overseas markets during the nineteenth and early twentieth centuries by Chinese, Japanese, and Euroamerican fishers. This historic abalone fishery was only possible because of the temporary release of abalone populations from predation by Native American foragers and sea otters that had kept them in check for millennia.

The introduction of livestock for commercial ranching operations on the Channel Islands also had devastating effects on local ecosystems, including loss of vegetation, erosion, and significant increases in the amount of sand in some nearshore environments, including the creation of a long sand spit on the east end of San Miguel (D. Johnson 1980). This human-induced increase in the littoral sand budget undoubtedly altered the structure and productivity of rocky intertidal, kelp forest, and other nearshore marine habitats. While dune building and littoral sand input also accelerated at other times in the Holocene (see Erlandson et al. 2005b), none of these approach the scale of the historic era. Grazing by livestock, agricultural activities, accelerated erosion, and land development along the mainland coast also promoted the infilling of the remaining estuaries. The Goleta Slough once sheltered several large Chumash villages and was entered historically by large ships, for instance, but is almost completely filled with sediment today, including massive grading spoils associated with the construction of the Santa Barbara Airport.

In the twentieth century, commercial and recreational fisheries have greatly intensified in the Santa Barbara Channel region. These include dramatic overfishing of key resources, including swordfish and other finfishes, sea urchins, and abalones. All too often the result has been the decline of a particular fishery, increased regulation, a moratorium on fishing activity, and periods of economic hardship. Pollution from sewage runoff, oil spills, and other modern wastes also greatly impact the productivity or health of the coastal environment. These environmental perturbations have caused great social, economic, and political tensions—conflicts that continue today between the representatives of the commercial fishing industry, recreation and tourist-related interests, conservationists, government agencies, and with resource managers often caught in the crossfire.

Along the California Coast, environmental impacts and human tensions related to marine fisheries and ecosystems are not restricted to the last few decades or centuries. As we have shown, the Chumash and their ancestors influenced the coastal ecosystems of the Santa Barbara Channel region for millennia. There appears to be a general acceleration of human impacts on marine ecosystems in the Santa Barbara Channel area through time, moreover, a process fueled by human population growth, increasing technological efficiency, and the expansion of trade and regional or global markets. With a lens spanning more than 10,000 years, however, the impacts of the Native Americans who occupied the area for millennia pale when compared to the rapid and radical ecological changes of historic times and the serial decline of fisheries associated with colonial extraction policies.

DISCUSSION AND CONCLUSIONS

Despite a long history of accelerating human impacts on marine ecosystems of the Santa Barbara Channel, there has been some significant progress in marine conservation and restoration in the area. In the past several decades, federal and state protection of marine mammals and seabirds has resulted in significant (and sometimes phenomenal) expansions in their local populations. The formation of Channel Islands National Park, a Nature Conservancy preserve covering most of Santa Cruz Island, and Channel Islands Marine Sanctuary were important steps in the protection and restoration of coastal ecosystems. The recent expansion of no fishing zones around the Channel Islands has also significantly expanded these conservation efforts. The removal of exotic animals (sheep, pigs, rats, etc.) and the reintroduction of bald eagles and sea otters native to the area are other important steps in restoring local ecosystems to some semblance of their "natural" state.

Nonetheless, more dramatic steps may be needed to rebuild the health of marine fisheries in the area, reduce the levels of various pollutants entering the system, restore the ecological complexity of local food webs, and avoid further disturbance of marine ecosystems. Until now, management efforts for local fisheries and ecosystems have been conducted primarily on a species-specific level, with only limited application of ecosystem management principles. Management efforts have also been governed almost exclusively by ecological data gathered over a few decades or less, sometimes supplemented with a few additional decades of historical fisheries data. As Pauly et al. (1998) and Jackson et al. (2001) have argued, such management approaches are often flawed because they do not account for the abundance of various species or their roles in local ecosystems prior to the effects of early fishing.

The several thousand archaeological sites of the Santa Barbara Channel region—including numerous shell middens that preserve a wide variety of shellfish, fish, sea mammal, seabird, and other animal remains—provide a remarkable opportunity to investigate human impacts on the environment with relatively high resolution and great time depth. Because European contact reaches back to the sixteenth century and includes successive Spanish, Mexican, and American occupations, the historical record is also rich. These data allow for a broad and synthetic comparison of the archaeological, historical, and modern records.

It is increasingly clear that the Santa Barbara Channel has been an anthropogenic landscape for much of the last 10,000 years, with people functioning as a distinct and increasingly dominant component of the ecosystem. The various subsistence activities of the Chumash and their ancestors—fishing, shellfish collecting, and marine mammal and bird exploitation—clearly influenced the structure and nature of coastal ecosystems (Braje et al. 2007b; Erlandson et al. 2005a; Walker et al. 2002). The introduction of dogs and possibly foxes to the Channel Islands represents another impact or transformation to island and mainland bird and other animal and plant communities. The Chumash also set fires to burn portions of the terrestrial landscape on the mainland and possibly the islands (Timbrook et al. 1982), adding to their influence on local terrestrial landscapes. As in much of California and North America, the Chumash and their ancestors actively managed and altered portions of the landscape, in some cases to increase its productivity (see Anderson 2005).

We have only scratched the surface of understanding the impacts of ancient people on the local marine environment, as well as the possibility of sustainable or alternative strategies they may have developed in light of these impacts. As several researchers have demonstrated, Native Americans had profound impacts on their environment and cases of resource depression have been documented in several areas of western North America and beyond (e.g., Broughton 1999; Butler 2000; Grayson 2001; James 2004; Martin 2005; Nagaoka 2002; Steadman et al. 2005).

Significant questions remain about the extent of these phenomena among the Chumash—although a case for the depression of certain shellfish populations seems relatively well established—as well as some apparent inconsistencies with the predictions of foraging theory and the archaeological record on the Channel Islands (Kennett 2005:220–223). The Chumash case is complex, but preliminary evidence suggests that Channel Islanders may have had an impact on the average size of intertidal California mussels as much as 7,500 years ago (Erlandson et al. 2007a) and depressed local sea otter populations through much of the Middle Holocene (Erlandson et al. 2005a). During the past 2,500 years the intensity of Chumash fishing increased significantly, with a greater emphasis on hunting and fishing at higher trophic levels (~1,500 to 250 years ago), including intensive human predation on pinnipeds on San Miguel Island about 1,500 to 1,000 years ago. Analyses of Chumash skeletal remains also suggest that human health generally declined (including increased human violence) around this time, probably due to a combination of growing populations, climatic and resource stress, possibly polluted water sources, and greater social conflict (Lambert 1993).

Despite all these potential impacts, there is a great deal of continuity in the resources people used throughout the Holocene. Whether such strategies indicate the development of conservation practices or epiphenomenal conservation strategies, or were purely accidental remains a subject that is difficult to determine with the archaeological record. For the Santa Barbara Channel and other coastal areas around the world, we need more integrative studies that look at the interrelationships of organisms in an ecosystem and how people influence those environments both positively and negatively (e.g., Reitz 2004). Complicating this issue is the fact that we often cannot distinguish the archaeological signatures of human impacts from ecological changes caused by natural climatic, geographic, or biological fluctuations (Reitz et al., this volume).

Given the fact that the subsistence efforts of the Chumash and their predecessors influenced the marine environment, one fundamental question remains: if the Chumash had ready access to the modern technologies and global markets available today, what impacts would they have had on local ecosystems? Such a question is highly political and impossible to answer with archaeological and historical records. One pattern that emerges from the Santa Barbara Channel data, however, is that the Chumash and their predecessors lived in the area for roughly 13,000 calendar years and relied on a similar suite of resources, albeit in varying degrees of abundance, throughout the Holocene (Erlandson et al. 2004a; Kennett 2005). All the while they adapted to dramatic environmental changes, including sea level rise, periodic droughts, El Niño, and other environmental perturbations. Other than the flightless goose, *Chendytes lawi*, which coexisted with people for thousands of years, the archaeological record of the area contains little evidence for human-induced extinction prior to European contact (see Walker [1980] for discussion of possible human-related rodent and bat extinctions). Declines in the local abundance of some animals are evident, but the shift in focus to different resources and relocation of villages may have promoted a fairly sustainable strategy. Of course this becomes more problematic in densely populated and circumscribed areas, such as the Goleta Slough. These are the conditions under which we would expect to see the greatest impacts.

Unlike some Pacific Islands, Greenland, and other areas of the world where human impacts were rapid and clearly discernible, the Santa Barbara Channel area is much more ambiguous. This is despite the fact that the Chumash had relatively high population densities, sophisticated maritime technologies, and extensive trade networks with both their coastal and interior neighbors. In a comparison of the history of kelp forest ecosystems in California, the Aleutians, and the Gulf of Maine, Steneck et al. (2002) noted that California kelp forests were characterized by significantly higher biodiversity and more complex food webs, traits that might provide greater resistance to human impacts and more resilience for recovery (see also Graham et al. 2007). If they are correct, the high productivity, biodiversity, and complexity of marine ecosystems in the Santa Barbara Channel area may help explain the limited evidence for large-scale ecological disruptions caused by the Chumash and their predecessors during 10,000 years of marine fishing, hunting, and gathering.

As our Santa Barbara Channel case study demonstrates, linking the archaeological, historical, and modern ecological records is difficult, as each data set has its own strengths and weaknesses. It is clear, however, that the Chumash and their predecessors, like all humans, were an integral part of their environment, functioning as one of the apex predators in coastal ecosystems for over 10,000 years. We can no longer divorce Native Americans or other people from the environments in which they hunted and foraged, as it is clear that they influenced the structure and nature of local ecosystems for millennia (Ames 2004; Grayson 2001; Krech 1999). This knowledge must become a part of the management strategies for the coastal ecosystems of the Santa Barbara Channel region, particularly when trying to establish historical baselines or targets for restoration (Dayton et al. 1998; Jackson et al. 2001; Pauly 1995). As we have shown here, archaeology can provide a lens for examining long-term ecological changes in marine habitats, reconstructing extended histories of human impacts on marine species or communities, and developing more realistic baselines for managing and restoring coastal ecosystems. We hope this chapter stimulates future research on the historical ecology of the Santa Barbara Channel region and other coastal areas around the world.

ACKNOWLEDGMENTS

Our research has been supported by the National Park Service, National Science Foundation, Foundation for Exploration and Research on Cultural Origins, Marine Conservation and Biology Institute, University of Oregon, Southern Methodist University, National Center for Ecological Analysis and Synthesis (NCEAS), U.S. Navy, Oregon Sea Grant, and Western Parks and Monuments Association. At Channel Islands National Park, we are indebted to Ann Huston, Kelly Minas, Don Morris, Ian Williams, Georganna Hawley, and Mark Senning. At NCEAS, biologists, marine ecologists, paleontologists, and archaeologists working on Jeremy Jackson's Long-Term Marine Records Group provided invaluable data that helped stimulate many of the research avenues presented here. We are especially grateful to the members of the NCEAS Kelp Bed Working Group: Bruce Bourque, Debby Corbett, Jeremy Jackson, Robert Steneck, and Mia Tegner. We also thank Bob Delong, Mike Glassow, Doug Kennett, Sharon Melin, and Phil Walker for discussions that significantly improved the chapter. We thank Steven James and David Steadman for comments that greatly improved this manuscript. Finally, we thank Blake Edgar and the editorial staff of University of Calfornia Press for help in the review and production of this manuscript. Some of the fish and shellfish images provided in Figures 4.2 and 4.3 were adapted from Eschmeyer et al. 1983 and Morris et al. 1980.

REFERENCES CITED

Ames, K. M.
2004 Political and Historical Ecologies. In *A Companion to the Anthropology of American Indians,* edited by T. Biolsi, pp. 7–23. Blackwell, Oxford.

Anderson, A. J.
2001 No Meat on that Beautiful Shore: The Prehistoric Abandonment of Subtropical Polynesian Islands. *International Journal of Osteoarchaeology* 11:14–23.

Anderson, M. K.
2005 *Tending the Wild: Native American Knowledge and the Management of California's Natural Resources.* University of California Press, Berkeley.

Arnold, J. E.
1992 Complex Hunter-Gatherer-Fishers of Prehistoric California: Chiefs, Specialists, and Maritime Adaptations of the Channel Islands. *American Antiquity* 57:60–84.

Arnold, J. E., R. Colten, and S. Pletka
1997 Contexts of Cultural Change in Insular California. *American Antiquity* 62:157–168.

Barlow, C.
2000 *The Ghosts of Evolution.* Basic Books, New York.

Bernard, J.
2004 Status and the Swordfish: The Origins of Large Species Fishing among the Chumash. In *Foundations of Chumash Complexity,* edited by J. E. Arnold, pp. 25–51. Cotsen Institute of Archaeology, University of California, Los Angeles.

Berryman, J. A.
1995 Archival Information, Abalone Shell, Broken Pots, Hearths, and Windbreaks: Clues to Identifying Nineteenth Century California Abalone Collection and Processing Sites, San Clemente Island: A Case Study. Ph.D. dissertation, University of California, Riverside. University Microfilms, Ann Arbor.

Bowser, B. J.
1993 Dead Fish Tales: Analysis of Fish Remains from Two Middle Period Sites on San Miguel Island, California. In *Archaeology of the Northern Channel Islands of California: Studies of Subsistence, Economics, and Social Organization,* edited by M. Glassow. Archives of California Prehistory 34:95–135. Coyote Press, Salinas, California.

Braje, T. J.
2007 Archaeology, Human Impacts, and Historical Ecology on San Miguel Island, California. Ph.D. dissertation, University of Oregon, Eugene.

Braje, T. J., J. M. Erlandson, and T. C. Rick
2007a An Historic Chinese Abalone Fishery on San Miguel Island, California. *Historical Archaeology* 41(4):117–128.
Braje, T. J., D. J. Kennett, J. M. Erlandson, and B. J. Culleton
2007b Human Impacts on Nearshore Shellfish Taxa: A 7000 Year Record from Santa Rosa Island, California. *American Antiquity* 72:735–756.
Broughton, J.
1999 *Resource Depression and Intensification during the Late Holocene, San Francisco Bay: Evidence from the Emeryville Shellmound Vertebrate Fauna.* University of California Anthropological Records 32. University of California Press, Berkeley.
Butler, V. L.
2000 Resource Depression on the Northwest Coast of North America. *Antiquity* 74:649–661.
Carder, N., E. J. Reitz, and J. G. Crock
2007 Fish Communities and Populations during the Post-Saladoid Period (AD 600/800–1500), Anguilla, Lesser Antilles. *Journal of Archaeological Science* 34:588–599.
Colten, R. H.
2001 Ecological and Economic Analysis of Faunal Remains from Santa Cruz Island. In *The Origins of a Pacific Coast Chiefdom: The Chumash of the Channel Islands,* edited by J. Arnold, pp. 199–219. University of Utah Press, Salt Lake City.
Colten, R. H., and J. E. Arnold
1998 Prehistoric Marine Mammal Hunting on California's Northern Channel Islands. *American Antiquity* 63:679–701.
Costello, J. G., and D. Hornbeck
1989 Alta California: An Overview. In *Columbian Consequences,* Vol. 1: *Archaeological and Historical Perspectives on the Spanish Borderlands West,* edited by D. Thomas, pp. 303–331. Smithsonian Institution, Washington, D.C.
Dallas, H.
2004 Reevaluating the Early Millingstone Complex in Coastal Southern California: VEN-1, a Case Study. *Proceedings of the Society for California Archaeology* 17:151–161.
Dayton, P. K., M. J. Tegner, P. B. Edwards, and K. L. Riser
1998 Sliding Baselines, Ghosts, and Reduced Expectations in Kelp Forest Communities. *Ecological Applications* 8(2):309–322.
deBoer, W. F., T. Pereira, and A. Guissamulo
2000 Comparing Recent and Abandoned Shell Middens to Detect the Impact of Human Exploitation on the Intertidal Ecosystem. *Aquatic Ecology* 34:287–297.
DeLong, R. L., and S. R. Melin
2002 Thirty Years of Pinniped Research at San Miguel Island. In *The Fifth California Islands Symposium,* edited by D. Brown, K. Mitchell, and H. Chaney, pp. 401–406. Santa Barbara Museum of Natural History, Santa Barbara, California.
Diamond, J.
2005 *Collapse: How Societies Choose to Fail or Succeed.* Viking, New York.
Erlandson, J. M.
1994 *Early Hunter-Gatherers of the California Coast.* Plenum, New York.
1997 The Middle Holocene on the Western Santa Barbara Coast. In *Archaeology of the California Coast during the Middle Holocene,* edited by J. Erlandson and M. Glassow, pp. 91–109. Institute of Archaeology, University of California, Los Angeles.
Erlandson, J. M., and K. Bartoy
1995 The Chumash, Cabrillo, and Old World Diseases. *Journal of California and Great Basin Anthropology* 17:153–173.
Erlandson, J. M., and T. C. Rick
2002 Late Holocene Cultural Developments along the Santa Barbara Coast. In *Catalysts to Complexity: Late Holocene Societies of the California Coast,* edited by J. Erlandson and T. Jones, pp. 166–182. Cotsen Institute of Archaeology, University of California, Los Angeles.
Erlansdon, J. M., R. Carrico, R. Dugger, L. Santoro, G. Toren, T. Cooley, and T. Hazeltine
1993 *The Archaeology of the Western Santa Barbara Coast: Results of the Chevron Point Arguello Project Cultural Resource Program.* Ogden Environmental, San Diego.
Erlandson, J. M., D. J. Kennett, B. L. Ingram, D. A. Guthrie, D. P. Morris, M. A. Tveskov, G. J. West, and P. L. Walker
1996 An Archaeological and Paleontological Chronology for Daisy Cave (CA-SMI-261), San Miguel Island, California. *Radiocarbon* 38:355–373.
Erlandson, J. M., M. A. Tveskov, and R. S. Byram
1998 The Development of Maritime Adaptations on the Southern Northwest Coast of North America. *Arctic Anthropology* 35:6–22.
Erlandson, J. M., T. C. Rick, D. J. Kennett, and P. L. Walker
2001 Dates, Demography, and Disease: Cultural Contacts and Possible Evidence for Old World Epidemics among the Island Chumash. *Pacific Coast Archaeological Society Quarterly* 37(3):11–26.
Erlandson, J. M., T. C. Rick, and R. L. Vellanoweth
2004a Human Impacts on Ancient Environments: A Case Study from California's Northern

Channel Islands. In *Voyages of Discovery: The Archaeology of Islands,* edited by S.M. Fitzpatrick, pp. 51–83. Praeger Publishers, Westport Connecticut.

2004b CA-SBA-2067: A Buried Middle Holocene Site at Gaviota, Santa Barbara County, California. *Proceedings of the Society for California Archaeology* 14:47–54.

2008 *A Canyon through Time: The Archaeology, History, and Ecology of the Tecolote Canyon Area, Santa Barbara County, California.* University of Utah Press, Salt Lake City, in press.

Erlandson, J.M., T.C. Rick, J.A. Estes, M.H. Graham, T.J. Braje, and R.L. Vellanoweth

2005a Sea Otters, Shellfish, and Humans: 10,000 Years of Ecological Interaction on San Miguel Island, California. In *Proceedings of the Sixth California Islands Symposium,* edited by D. Garcelon and C. Schwemm, pp. 9–21. National Park Service Technical Publication CHIS-05-01. Institute for Wildlife Studies, Arcata, California.

Erlandson, J.M., T.C. Rick, and C. Peterson

2005b A Geoarchaeological Chronology for Holocene Dune Building on San Miguel Island, California. *The Holocene* 15:1227–135.

Erlandson, J.M., T.C. Rick, T.J. Braje, A. Steinberg, and R.L.Vellanoweth

2007b Human Impacts on Ancient Shellfish: A 10,000 Year Record from San Miguel Island, California. *Journal of Archaeological Science,* in press.

Eschmeyer, W.N., E.S. Herald, and H. Hammann.

1983 *A Field Guide to Pacific Coast Fishes of North America.* Boston: Houghton Mifflin.

Estes, J.A., and J.F. Palmisano

1974 Sea Otters: Their Role in Structuring Nearshore Communities. *Science* 185:1058–1060.

Estes, J.A., and G.R. VanBlaricom

1985 Sea Otters and Shellfisheries. In *Conflicts between Marine Mammals and Fisheries,* edited by R.H. Beverton, D. Lavigne, and J. Beddington, pp. 187–235. Allen and Unwin, London.

Etnier, M.A.

2004 The Potential of Zooarchaeological Data to Guide Pinniped Management Decisions in the Eastern North Pacific. In *Zooarchaeology and Conservation Biology,* edited by R.L. Lyman and K. Cannon, pp. 88–102. University of Utah Press, Salt Lake City.

Friddell, J.E., R.C. Thunell, T.P. Guilderson, and M. Kashgarian

2003 Increased Northeast Pacific Climatic Variability during the Warm Middle Holocene. *Geophysical Research Letters* 30(11):14-1 to 14-4.

Gamble, L.H.

2005 Culture and Climate: Reconsidering the Effect of Paleoclimatic Variability among Southern California Hunter-Gatherer Societies. *World Archaeology* 37:92–108.

Glassow, M.A.

1993a Changes in Subsistence on Marine Resources through 7,000 years of Prehistory on Santa Cruz Island. In *Archaeology of the Northern Channel Islands of California: Studies of Subsistence, Economics, and Social Organization,* edited by M. Glassow. Archives of California Prehistory 34:75–94. Coyote Press, Salinas, California.

1993b The Occurrence of Red Abalone Shells in Northern Channel Island Archaeological Middens. In *Third California Islands Symposium: Recent Advances in Research on the California Islands,* edited by F.G. Hochberg, pp. 567–576. Santa Barbara Museum of Natural History, Santa Barbara, California.

1996 *Purismeño Chumash Prehistory: Maritime Adaptations along the Southern California Coast.* Harcourt Brace College Publishers, Fort Worth, Texas.

2002 Prehistoric Chronology and Environmental Change at the Punta Arena Site, Santa Cruz Island, California. In *Proceedings of the Fifth California Islands Symposium,* edited by D. Browne, K. Mitchell, and H. Chaney, pp. 555–562. Santa Barbara Museum of Natural History, Santa Barbara, California.

2005 Variation in Marine Fauna Utilization by Middle Holocene Occupants of Santa Cruz Island. In *Proceedings of the Sixth California Islands Symposium,* edited by D. Garcelon and C. Schwemm, pp. 23–34. National Park Service Technical Publication CHIS-05-01. Institute for Wildlife Studies, Arcata, California.

Glassow, M.A., D.J. Kennett, J.P. Kennett, and L.R. Wilcoxon

1994 Confirmation of Middle Holocene Ocean Cooling Inferred from Stable Isotopic Analysis of Prehistoric Shells from Santa Cruz Island, California. In *The Fourth California Islands Symposium: Update on the Status of Resources,* edited by W. Halvorson and G. Maender, pp. 223–232. Santa Barbara Museum of Natural History, Santa Barbara, California.

Glassow, M.A., P. Paige, and J. Perry

2007 *The Punta Arena Site and Early and Middle Holocene Cultural Development on Santa Cruz Island, California.* SBMNH Contributions in Anthropology. Santa Barbara Museum of Natural History, Santa Barbara, California, in press.

Glenn, B. K.

1990 Fish Exploitation: Analysis of Vertebrae and Otoliths. In *Archaeological Investigations at Helo' on Mescalitan Island,* edited by L. H. Gamble, pp. 17–1 through 17–34. Report on file, Central Coast Information Center, Department of Anthropology, University of California, Santa Barbara.

Graham, M. H.

2004 Effects of Local Deforestation on the Diversity and Structure of Southern California Giant Kelp Forest Food Webs. *Ecosystems* 7:341–357.

Graham, M. H., P. K. Dayton, and J. M. Erlandson

2003 Ice Ages and Ecological Transitions on Temperate Coasts. *Trends in Ecology and Evolution* 18(1):33–40.

Graham, M., B. Halpern, and M. Carr

2007 Diversity and Dynamics of Californian Subtidal Kelp Forests. In *Food Webs and the Dynamics of Marine Reefs,* edited by T. R. McClanahan and G. M. Branch. Oxford University Press, New York, in press.

Grayson, D. K.

2001 The Archaeological Record of Human Impact on Animal Populations. *Journal of World Prehistory* 15:1–68.

Grayson, D. K., and D. J. Meltzer

2002 Clovis Hunting and Large Mammal Extinction: A Critical Review of the Evidence. *Journal of World Prehistory* 16:313–359.

2003 Requiem for North American Overkill. *Journal of Archaeological Science* 30:585–593.

Guthrie, D. A.

1980 Analysis of Avifaunal and Bat Remains from Midden Sites on San Miguel Island. In *The California Channel Islands: Proceedings of a Multidisciplinary Symposium,* edited by D. Power, pp. 689–702. Santa Barbara Museum of Natural History, Santa Barbara, California.

1993a Listen to the Birds? The Use of Avian Remains in Channel Islands Archaeology. In *Archaeology on the Northern Channel Islands, California,* edited by M. Glassow, Archives of California Prehistory 34:153–167. Coyote Press, Salinas, California.

1993b New Information on the Prehistoric Fauna of San Miguel Island, California. In *Third California Islands Symposium: Recent Advances in Research on the California Islands,* edited by F. G. Hochberg, pp. 405–416. Santa Barbara Museum of Natural History, Santa Barbara, California.

2005 Distribution and Provenance of Fossil Avifauna on San Miguel Island. In *Proceedings of the Sixth California Islands Symposium,* edited by D. Garcelon and C. Schwemm, pp. 34–39. National Park Service Technical Publication CHIS-05-01. Institute for Wildlife Studies, Arcata, California.

Hildebrandt, W. R., and T. Jones

1992 Evolution of Marine Mammal Hunting: A View from the California and Oregon Coasts. *Journal of Anthropological Archaeology* 11: 360–401.

Jackson, J. B. C., M. Kirby, W. Berger, K. Bjorndal, L. Botsford, B. Bourque, R. Bradbury, R. Cooke, J. Erlandson, A. James, J. Estes, T. Hughes, S. Kidwell, C. Lange, H. Lenihan, J. Pandolfi, C. Peterson, R. Steneck, M. Tegner, and R. Warner

2001 Historical Overfishing and the Recent Collapse of Coastal Ecosystems. *Science* 293: 629–637.

James, S. R.

2004 Hunting, Fishing, and Resource Depression in Prehistoric Southwest North America. In *The Archaeology of Global Environmental Change: The Impact of Humans on Their Environment,* edited by C. L. Redman, S. R. James, P. R. Fish, and J. D. Rogers, pp. 28–62. Smithsonian Institution, Washington, D.C.

Janzen, D., and P. S. Martin

1982 Neotropical Anachronisms: The Fruits the Gomphotheres Ate. *Science* 215:19–27.

Johnson, D. L.

1980 Episodic Vegetation Stripping, Soil Erosion, and Landscape Modification in Prehistoric and Recent Historic Time, San Miguel Island, California. In *The California Channel Islands: Proceedings of a Multidisciplinary Symposium,* edited by D. Power, pp. 103–121. Santa Barbara Museum of Natural History, Santa Barbara, California.

Johnson, J. R.

1980 Archaeological Analysis of Fish Remains from SBA-1 Rincon Point. In *Cultural Resources Technical Report: Rincon Tract No. 12,932,* assembled by M. Kornfeld, pp. 11–1 to 11–18. Report on file, Central Coast Information Center, Department of Anthropology, University of California, Santa Barbara.

2000 Social Responses to Climate Change among the Chumash Indians of South-Central California. In *The Way the Wind Blows: Climate, History, and Human Action,* edited by R. McIntosh, J. Tainter, and S. McIntosh, pp. 301–327. Columbia University Press, New York.

Johnson, J. R., T. W. Stafford, H. O. Ajie, and D. P. Morris

2002 Arlington Springs Revisited. In *Proceedings of the Fifth California Islands Symposium,* edited by D. Browne, K. Mitchell, and H. Chaney, pp. 541–545. Santa Barbara Museum of Natural History, Santa Barbara, California.

Jones, T. L., and W. Hildebrandt
1995 Reasserting a Prehistoric Tragedy of the Commons: Reply to Lyman. *Journal of Anthropological Archaeology* 14:78–98.

Jones, T. L., G. M. Brown, L. M. Raab, J. L. McVickar, W. G. Spaulding, D. J. Kennett, A. York, and P. L. Walker
1999 Environmental Imperatives Reconsidered: Demographic Crises in Western North America during the Medieval Climatic Anomaly. *Current Anthropology* 40(2):137–170.

Jones, T. L., W. R. Hildebrandt, D. J. Kennett, and J. F. Porcasi
2004 Prehistoric Marine Mammal Overkill. *Journal of California and Great Basin Anthropology* 24:69–80.

Kay, C. E., and R. T. Simmons (editors)
2002 *Wilderness and Political Ecology: Aboriginal Influences and the Original State of Nature.* University of Utah Press, Salt Lake City.

Kennett, D. J.
2005 *The Island Chumash: Behavioral Ecology of a Maritime Society.* University of California Press, Berkeley.

Kennett, D. J., and C. A. Conlee
2002 Emergence of Late Holocene Sociopolitical Complexity on Santa Rosa and San Miguel Islands. In *Catalysts to Complexity: Late Holocene Societies of the California Coast,* edited by J. M. Erlandson and T. L. Jones, 147–165. Cotsen Institute of Archaeology, University of California, Los Angeles.

Kennett, D. J., and J. P. Kennett
2000 Competitive and Cooperative Responses to Climatic Instability in Coastal Southern California. *American Antiquity* 65:379–396.

Kennett, D. J., J. P. Kennett, J. M. Erlandson, and K. G. Cannariato
2007 Human Responses to Middle Holocene Climate Change on California's Channel Islands. *Quaternary Science Reviews* 26:351–367.

Kennett, J. P., and B. L. Ingram
1995 A 20,000 Year Record of Ocean Circulation and Climatic Change from the Santa Barbara Basin. *Nature* 377:510–514.

King, C. D.
1990 *Evolution of Chumash Society: A Comparative Study of Artifacts Used for Social System Maintenance in the Santa Barbara Channel Region before AD 1804.* Garland Publishing, New York.

Kinlan, B. P., M. H. Graham, and J. M. Erlandson
2005 Late Quaternary Changes in the Size and Shape of the California Channel Islands: Implications for Marine Subsidies to Terrestrial Communities. In *Proceedings of the Sixth California Islands Symposium,* edited by D. Garcelon and C. Schwemm, pp. 131–142. National Park Service Technical Publication CHIS-05-01. Institute for Wildlife Studies, Arcata, California.

Kirch, P. V.
1997 Microcosmic Histories: Island Perspectives on "Global" Change. *American Anthropologist* 99:30–42.

Kirch, P. V., and T. L. Hunt (editors)
1997 *Historical Ecology in the Pacific Islands: Prehistoric Environmental and Landscape Change.* Yale University Press, New Haven.

Klein, R. G., G. Avery, K. Cruz-Uribe, D. Halkett, J. E. Parkington, T. Steele, T. P. Volman, and R. Yates
2004 The Yserfontein 1 Middle Stone Age Site, South Africa, and Early Human Exploitation of Coastal Resources. *Proceedings of the National Academy of Sciences* 101:5708–5715.

Krech, S.
1999 *The Ecological Indian: Myth and History.* W. W. Norton, New York.

Lambert, P. M.
1993 Health in Prehistoric Populations of the Santa Barbara Channel Islands. *American Antiquity* 58:509–522.

Lyman, R. L.
1995 On the Evolution of Marine Mammal Hunting on the West Coast of North America. *Journal of Anthropological Archaeology* 14:45–77.

Lyman, R. L., and K. Cannon (editors)
2004 *Zooarchaeology and Conservation Biology.* University of Utah Press, Salt Lake City.

Mannino, M. A., and K. D. Thomas
2002 Depletion of a Resource? The Impact of Prehistoric Human Foraging on Intertidal Mollusk Communities and Its Significance for Human Settlement, Mobility, and Dispersal. *World Archaeology* 33:452–474.

Martin, P. S.
2005 *Twilight of the Mammoths: Ice Age Extinctions and the Rewilding of the Americas.* University of California Press, Berkeley.

Miller, L., E. Mitchell, and J. Lipps
1961 New Light on the Flightless Goose *Chendytes lawi. Contributions in Science* 43:3–11.

Morales, A., and E. Rosello
2004 Fishing down the Food Web in Iberian Prehistory? A New Look at the Fishes from Cueva de Nerja (Malaga, Spain). In *Pettis Animaux Et Societes Humaines Du Complement Alimentaire Aux Ressources Utilitaires XXIV,* edited by J. Brugal and J. Desse, pp. 111–123. APDCA, Antibes.

Morejohn, G. V.
1976 Evidence of the Survival to Recent Times of the Extinct Flightless Duck *Chendytes lawi* Miller. In *Collected Papers in Avian Paleontology Honoring the 90th Birthday of Alexander Wetmore,* edited by S. L. Olson, pp. 207–211. Smithsonian Contributions to Paleobiology 27. U.S. Government Printing Office, Washington, D.C.

Morris, R. H., D. P. Abbott, and E. C. Haderlie
1980 *Intertidal Invertebrates of California.* Stanford University Press, Stanford, California.

Nagaoka, L.
2002 The Effects of Resource Depression on Foraging Efficiency, Diet Breadth, and Patch Use in Southern New Zealand. *Journal of Anthropological Archaeology* 21:419–442.

Newsom, L. A., and E. Wing
2004 *On Land and Sea: Native American Uses of Biological Resources in the West Indies.* University of Alabama Press, Tuscaloosa.

Noah, A. C.
2005 Household Economies: The Role of Animals in a Historic Period Chiefdom on the California Coast. Ph.D. dissertation, University of California, Los Angeles. University Microfilms, Ann Arbor.

Ogden, A.
1941 *The California Sea Otter Trade 1784–1848.* University of California Publications in History 26. University of California Press, Berkeley.

Paige, P.
2000 Archaeological Fish Remains as Indicators of Paleoenvironmental Change: An Ichthyofaunal Analysis of Two Multi-component Sites on Western Santa Rosa Island. Master's paper, Department of Anthropology, University of California, Santa Barbara.

Pauly, D.
1995 Anecdotes and the Shifting Baseline Syndrome of Fisheries. *Trends in Ecology and Evolution* 10:430.

Pauly, D., V. Christensen, J. Dalsgaard, F. Rainer, and F. Torres
1998 Fishing down Marine Food Webs. *Science* 279:860–863.

Pletka, S. M.
2001 The Economics of Island Chumash Fishing Practices. In *The Origins of a Pacific Coast Chiefdom: The Chumash of the Channel Islands,* edited by J. Arnold, pp. 221–244. University of Utah Press, Salt Lake City.

Porcasi, J. F., T. L. Jones, and L. Mark Raab
2000 Trans-Holocene Marine Mammal Exploitation on San Clemente Island: A Tragedy of the Commons Revisited. *Journal of Anthropological Archaeology* 19:200–220.

Preston, W. L.
1996 Serpent in Eden: Dispersal of Foreign Diseases into Pre-Mission California. *Journal of California and Great Basin Anthropology* 18:2–37.
2002 Post-Columbian Wildlife Irruptions in California: Implications for Cultural and Environmental Understanding. In *Wilderness and Political Ecology: Aboriginal Influences and the Original State of Nature,* edited by C. E. Kay and R. T. Simmons, pp. 111–140. University of Utah Press, Salt Lake City.

Raab, L. M., and D. O. Larson
1997 Medieval Climatic Anomaly and Punctuated Cultural Evolution in Coastal Southern California. *American Antiquity* 62:319–336.

Redman, C. L.
1999 *Human Impact on Ancient Environments.* University of Arizona Press, Tucson.

Redman, C. L., S. R. James, P. R. Fish, and J. D. Rogers (editors)
2004 *The Archaeology of Global Change: The Impact of Humans on their Environment.* Smithsonian Institution Press, Washington, D.C.

Reitz, E. J.
2004 "Fishing down the Food Web": A Case Study from St. Augustine, Florida, U.S.A. *American Antiquity* 69:63–83.

Rick, T. C.
2007 *The Archaeology and Historical Ecology of Late Holocene San Miguel Island.* Cotsen Institute of Archaeology, University of California, Los Angeles.

Rick, T. C., and M. A. Glassow
1999 Middle Holocene Fisheries of the Central Santa Barbara Channel Region, California: Investigations at CA-SBA-53. *Journal of California and Great Basin Anthropology* 21:236–256.

Rick, T. C., J. M. Erlandson, and R. L. Vellanoweth
2001 Paleocoastal Marine Fishing on the Pacific Coast of the Americas: Perspectives from Daisy Cave, California. *American Antiquity* 66: 595–613.

Rick, T. C., D. J. Kennett, and J. M. Erlandson
2005 Preliminary Report on the Archaeology and Paleoecology of the Abalone Rocks Estuary, Santa Rosa Island, California. In *Proceedings of the Sixth California Islands Symposium,* edited by D. Garcelon and C. Schwemm, pp. 55–63. National Park Service Technical Publication CHIS-05-01. Institute for Wildlife Studies, Arcata, California.

Scammon, C. M.
1968 *The Marine Mammals of the Northwestern Coast of North America.* Originally published in 1874. Dover, New York.

Schoenherr, A.
1992 *A Natural History of California*. University of California Press, Berkeley.
Schoenherr, A., C. R. Feldmath, and M. Emerson
1999 *Natural History of the Islands of California*. University of California Press, Berkeley.
Schroeder, D. M., and M. S. Love
2002 Recreational Fishing and Marine Fish Populations in California. *CalCOFI Report* 43:182–190.
Simenstad, C. A., J. A. Estes, and K. W. Kenyon
1978 Aleuts, Sea Otters, and Alternate Stable-State Communities. *Science* 200:403–411.
Steadman, D. W.
1995 Prehistoric Extinctions of Pacific Islands Birds: Biodiversity Meets Zooarchaeology. *Science* 267:1123–1131.
2006 *Extinction and Biogeography of Tropical Pacific Birds*. University of Chicago Press, Chicago.
Steadman, D. W., P. S. Martin, R. D. E. MacPhee, A. J. T. Jull, H. G. McDonald, C. A. Woods, M. Iturralde-Vinent, and G. W. L. Hodgins
2005 Asynchronous Extinction of Late Quaternary Sloths on Continents and Island. *Proceedings of the National Academy of Sciences* 102: 11763–11768.
Steneck, R., M. Graham, B. Bourque, D. Corbett, J. M. Erlandson, and J. Estes
2002 Kelp Forest Ecosystems: Biodiversity, Stability, Resilience, and Their Future. *Environmental Conservation* 29:436–459.
Timbrook, J., J. Johnson, and D. Earle
1982 Vegetation Burning by the Chumash. *Journal of California and Great Basin Anthropology* 4:163–186.
UCSB Sustainable Indicators
2007 Sustainable Environmental Indicators. Electronic document, www.es.ucsb.edu/proj/135Bindicators2/santabarbara/environmental/, accessed July 2007.
van der Leeuw, S., and C. L. Redman
2002 Placing Archaeology at the Center of Socionatural Studies. *American Antiquity* 67: 597–605.
Vellanoweth, R. L.
1998 Earliest Island Fox Remains on the Southern Channel Islands: Evidence from San Nicolas Island, California. *Journal of California and Great Basin Anthropology* 20:100–108.
Vellanoweth, R. L., and J. M. Erlandson
2004 Coastal Paleogeography and Human Land Use at Tecolote Canyon, Southern California, USA. *Geoarchaeology* 19:141–165.
Vellanoweth, R. L., T. C. Rick, and J. M. Erlandson
2002 Middle and Late Holocene Maritime Adaptations on Northeastern San Miguel Island, California. In *Proceedings of the Fifth California Islands Symposium*, edited by D. Browne, K. Mitchell, and H. Chaney, pp. 607–614. Santa Barbara Museum of Natural History, Santa Barbara, California.
Wagner, H. R.
1929 *Spanish Voyages to the Northwest Coast of North America in the Sixteenth Century*. California Historical Society, San Francisco.
Walker, P. L.
1980 Archaeological Evidence for the Recent Extinction of Three Terrestrial Mammals on San Miguel Island. In *The California Islands: Proceedings of a Multidisciplinary Symposium*, edited by D. M. Power, pp. 703–717. Santa Barbara Museum of Natural History, Santa Barbara.
Walker, P. L., D. J. Kennett, T. Jones, and R. Delong
2002 Archaeological Investigations of the Point Bennett Pinniped Rookery on San Miguel Island. In *The Fifth California Islands Symposium*, edited by D. Brown, K. Mitchell, and H. Chaney, pp. 628–632. Santa Barbara Museum of Natural History, Santa Barbara, California.
Wroe, S., J. Field, and D. K. Grayson
2006 Megafaunal Extinction: Climate, Humans, and Assumptions. *Trends in Ecology and Evolution* 21:61–62.

5

Long-Term Effects of Human Predation on Marine Ecosystems in Guerrero, Mexico

Douglas J. Kennett, Barbara Voorhies, Thomas A. Wake, and Natalia Martínez

COASTAL AND MARINE ecosystems have long played a central role in the economies of people inhabiting Mexico, where today they are of paramount importance in the modern economy. Twenty-nine percent of the country's 107 million people live in coastal settings, with annual capture rates of fish rising from ~.4 to 1.2 million metric tons since the 1970s (Earthtrends 2006). Increasing populations in coastal areas coupled with technological advancements and the expansion of Mexico's fishing fleet contribute to concerns regarding long-term effects on these ecosystems. A new government agency, the Secretaría de Agricultura, Ganadería, Desarrollo Rural, Pesca y Alimentación (SAGARPA[1]), was created in 2000 to establish better linkage between fishery production and environmental issues, with the idea of creating a more sustainable fishery. These concerns in Mexico occur within the broader context of a global fisheries "crisis" signaled by industrial fishing declines caused by overfishing (Pauly and Christensen 1995; Pauly et al. 1998, 2000).

In this chapter, we explore the long history of human use of coastal and marine ecosystems in the Mexican state of Guerrero (Figure 5.1) using the archaeological and historical records available for the region. The primary aim of this work is to (1) begin creating a historical and environmental framework to improve our understanding of the complex relationship between people and marine ecosystems; (2) provide an ecological baseline for future remediation of coastal and marine habitats in this region; and (3) establish a general approach that incorporates archaeological and historical data to put the modern exploitation of fisheries into perspective, an approach that can be applied elsewhere in Mesoamerica—the area encompassing the modern countries of Mexico, Guatemala, Belize, Honduras, El Salvador, and parts of Nicaragua.

Historical use and transformation of marine ecosystems in Mexico cannot be separated from the demographic, economic, and environmental effects of developing agricultural systems during the last 10,000 years. At the time of European contact, much of the Mesoamerican landscape was altered and transformed by these long-standing agrarian

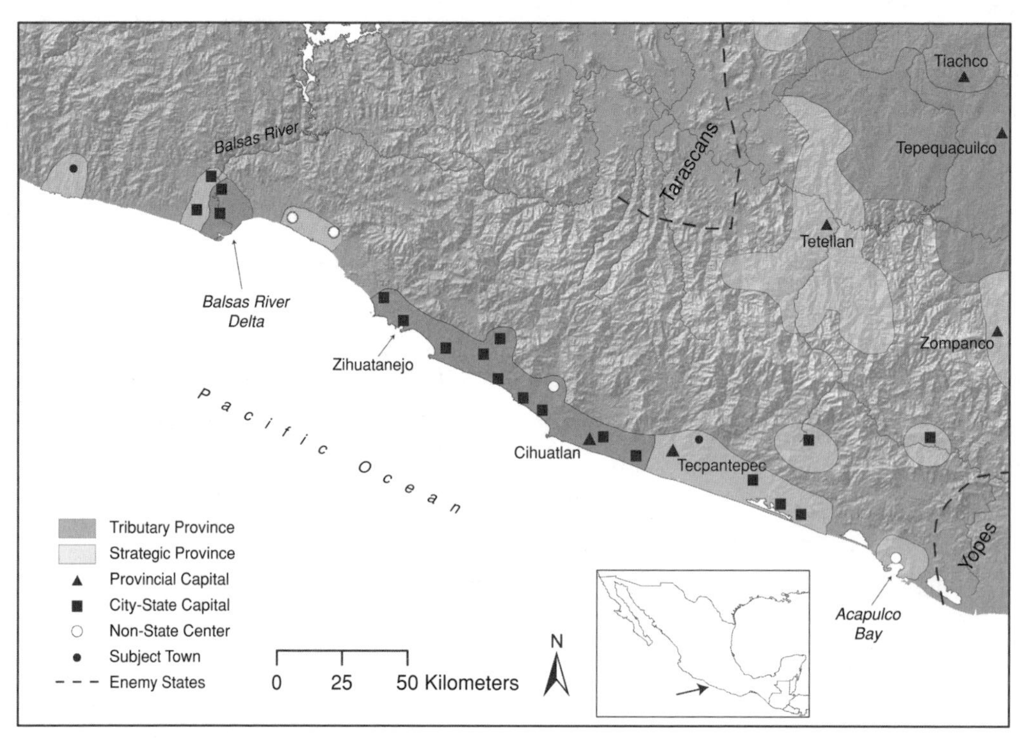

FIGURE 5.1. Map showing the Guerrero Coast and Aztec strategic and tributary provinces.

economies that sustained human populations between 22.8 and 25 million people (Crawford 1998; Denevan 1992). The cultivation of maize, beans, and squash formed the basis of these agricultural systems, cultigens domesticated by foragers during the Early Holocene (Piperno and Pearsall 1998). Plant domestication occurred within the context of major changes in the distribution of plant and animal communities associated with new Holocene climatic regimes, and significant changes in coastal ecosystems due to rapid postglacial sea level rise, and in the wake of well-known and widespread megafaunal extinctions that occurred throughout the Americas at the end of the Pleistocene (Grayson 2001; Grayson and Meltzer 2002; Hodell et al. 2000; Piperno 2006; Piperno and Pearsall 1998). Biogeographical, genetic, and archaeological studies indicate that the initial use and experimentation leading to the domestication of these key cultigens occurred in central Mexico. This was followed by wide geographic dispersal and a long period of local experimentation by Mesoamerican peoples, who practiced a range of mixed foraging and farming strategies during much of the Early and Middle Holocene (see Kennett et al. 2006; Smith 2001).

Mesoamerican communities expanded and contracted during the Early and Middle Holocene, with gradual population increases in many locations. This resulted in (1) the loss of forested habitat, (2) the expansion of grasslands along with human commensal species (e.g., bottle gourd), and (3) the reduction of animal populations due to direct hunting or loss of habitat from forest burning and maize horticulture (Kennett et al. 2006). Maize-based food production spread rapidly throughout much of Mesoamerica after about 4000 cal BP (Late Archaic to Formative period), and this favored large-scale expansion of populations, more aggregated settlements, and the associated emergence of hierarchically organized ranked societies (Clark and Blake 1994; Kennett et al.

2006). This formed the foundation for the earliest large and highly integrated political systems marked by administrative hierarchies and rulers with significant power and authority—so called state-level societies (Feinman and Marcus 1998). Intensive forms of agriculture involving terracing, irrigation (Doolittle 1990), and other more sophisticated systems (e.g., raised fields; Fedick 1996) fueled the development of socially stratified, politically centralized, and technologically innovative state-level societies in Mesoamerica but also underwrote exponential population growth, urbanization, and environmental destruction.

The environmental impacts of expanding agricultural systems (e.g., deforestation, erosion, soil depletion) are well known in Mesoamerica (Lentz 2000; Webster 2002), and the potential effects of these impacts for the disintegration of societies has received wide public exposure in the case of the Maya (Diamond 2005). Less is known about the effects expanding populations and highly extractive state-level societies had on marine ecosystems in Mesoamerica. The slowing of postglacial sea level rise about 6,000 years ago and the stabilization of aquatic habitats along the Pacific, Gulf, and Caribbean coasts favored increased population densities, group formation, community stability, enhanced maritime trade, and the emergence of social hierarchies. Marine ecosystems also provided important sources of protein for agricultural peoples whose diet was rich in carbohydrates. The natural diversity of resources in coastal/aquatic habitats, in combination with newly domesticated plants and animals, provided the economic foundation for these developing communities, as they did elsewhere during the Early and Middle Holocene (Binford 1968; Clark and Blake 1994; Moseley 1975). In this chapter we explore the history of marine resource use during the Middle and Late Holocene (~5500 cal BP to present) on the Pacific Coast of Guerrero, Mexico, focusing on the impact of expanding populations and intensive marine resource use at Puerto Marqués, a coastal community that was occupied from the Archaic through Classic periods (~5500–1200 cal BP).

ENVIRONMENTAL AND HISTORICAL CONTEXT

Acapulco, a city of close to one million people world-renowned for its beach resorts and sportfishing, forms the boundary between the Costa Grande (north to Zacatala) and the Costa Chica (south to the Oaxacan border). Both Acapulco and Zijuatanejo, a smaller tourist destination to the northwest, were early Spanish fishing villages established on bays that have served as important harbors for maritime exploration and trade since the sixteenth century. The crescent-shaped Acapulco Bay, linked to Mexico City and ultimately to Spain by the Camino de Asia, served as the principle harbor for trade between Manila (Philippines) and New Spain (Mexico) between AD 1571 and 1814 (Meyer and Beezley 2000:131–132). One or more ships per year generally arrived from the Orient laden with silks, jade, ivory, and perfumes, and Acapulco burgeoned as a trade and ship-building center (Meyer and Sherman 1991:180; Miller 1985:107–108). The settlement's strategic importance was marked by the Fuerte de San Diego, a fort built in AD 1616 to help protect Manila galleons from Dutch and English buccaneers. The establishment of Acapulco as an important economic center signals the emergence of one of the first truly global markets based on maritime trade and exploitation.

Outside of Acapulco and Zijuatenejo, smaller fishing and farming communities today dot the coastal plain, a narrow sliver of arable land wedged between the Pacific Ocean and the Sierra Madre del Sur, the mountain range that rises precipitously into the central Mexican Highlands and the Valley of Mexico—the cultural epicenter of Mexico in the late prehistoric past and present. The complex tectonic history of subsidence and uplift (Ramírez-Herrera and Urrutia-Fueugachi 1999; Sedor 2005) combined with postglacial sea-level rise (Berger 1983; Curray et al.

1969; Fairbanks 1989), and perhaps anthropogenic geomorphological changes (Goman et al. 2005), has resulted in a sinuous coastline composed of rocky shores, offshore reefs, and sandy beaches. The Río Balsas, one of Mexico's largest river systems, flows out of the central Mexican highlands to the Pacific Coast just north of the Costa Grande. Sediments from this large system and smaller drainages along the coast are swept south by longshore currents to create a series of barrier beaches and associated lagoons that began to form with the stabilization of sea level between 7,000 and 5,000 years ago. Artisanal fishing communities ply these sheltered waters in search of schooling fish and shrimp, as they do elsewhere along the Pacific Coast of Mexico (McGoodwin 1990, 1994, 2001). The stabilization of sea level is coincident with evidence for the first communities along the Pacific Coast of Mesoamerica (Voorhies 2004), but evidence for earlier occupations may have been submerged by postglacial sea-level rise or lie deeply buried under alluvium.

The fishery along the Pacific Coast of Mexico is highly productive, providing 79 percent (1,237,693 tons) of the 1,564,966 tons produced throughout Mexico in 2003. The Instituto Nacional de Pesca (INP), a government institution charged with exploring new fisheries technology and creating a link between fish production and ecological sustainability, regulates fishing in coastal Guerrero and elsewhere. In 2003, there were 276 fishing cooperatives and 12,505 fishermen registered with this agency along the coast of Guerrero. There were also numerous companies and commercial outfits participating in the larger fishing industry. Daily hauls are recorded by registered fisherman, and these data are compiled annually (see Anuario Estadístico de Pesca 2003; Carta Nacional Pesquera 2006); but unregistered fishing is common and apparently has a major impact on the overall productivity and health of the fishery. The primary species taken in 2003 included mojarra (*Gerreidae*, 2,020 tons), snapper (*Lutjanus*, 459 tons), jack (*Caranx*, 357 tons), bandera (*Ariidae*, 276 tons), mullet (*Mugil*, 127 tons), and oyster (*Crassostrea iridiscens*, 238 tons).

Data collected over the last decade suggest that Guerrero´s fishery production has declined steadily. Starting in 1993, the total annual volume of fish production was 30,558 tons. This decreased to 8,885 tons in 1998 and kept decreasing to 3,962 tons in 2000, with a slight rebound in 2003 (6,153 tons). Some species also disappeared from the record. For instance, fishing for catfish (*Ariopsis* sp.) was economically viable in 1994 (222 tons), but no longer appears in the record after 1998. Red lobster *(Panulirus interruptus)* was a viable part of the fishery in 1997 (46 tons), but the quantities captured dropped significantly the following year (1 ton), and it disappears from the record in 2001. The lagoon and mangrove systems of Guerrero are particularly sensitive, and the INP has identified a number of environmental issues related to the overexploitation of these habitats. Jackson et al. (2001) have suggested, however, that the management of marine fisheries can be improved with the deeper historical perspectives that archaeological and historical data can provide on the nature of human impacts on marine ecosystems before monitoring of fishery yields and modern ecological data began.

In Mexico, interest in Pacific Coast marine resources by highly centralized state-level societies extends back into the prehistoric past and is well documented historically during the Aztec period. The Aztecs established control in the Basin of Mexico (~AD 1440) by allying themselves with the political elite of Texcoco *(Acolhua)* and Tlacopan *(Tepaneca)* to form the Triple Alliance. The main interest of the alliance was to expand the limits of the empire throughout Mesoamerica by means of war and suppression. In so doing they established tributary and strategic outer provinces, the former created for the regularly scheduled payment of tribute in the form of goods or labor, and the latter providing soldiers and corvée labor strategically located in frontier lands. With this expansion they obtained tribute, which included a great variety of goods not available in the Basin of Mexico. Within this framework of Aztec expansion and domination, the Gulf of Mexico and the Pacific Ocean played important roles in strategies for the acquisition of goods (Manzanilla López 2000:36). This started with the fifth Aztec emperor, Ahuítzotl (AD 1486–1502), who became obsessed with the idea of reaching the Pacific Ocean. He started with the Balsas River, subduing the Tepuztecos, establishing a mountain corridor by the Río de las Truchas to later reach the Costa Grande in Xolochiyuhyan. From there he expanded to Coyuca and Acapulco, conquering the 16 towns that comprise the Cihuatlán province (Figure 5.1; Barlow 1990:93, 130, 137; Smith and Berdan 1996:300). After conquering several towns along the Pacific Coast, the Aztecs formed a tributary province centered on the town of Cihuatlán, today known as San Luis la Loma (Manzanilla López 2000:36).

Cihuatlán province was composed of 12 to 16 towns that served as regional capitals of small states (Smith and Berdan 1996:277). The largest and most dominant of these states was Zacotallan, located near the mouth of the Balsas River northwest of Acapulco and Zijuatanejo. According to the Relación de Zacatula (1945), each of these states was governed by a *Capitán* who served principally as a war chief. Several ethnic and linguistic groups existed (e.g., Cuitlatecos, Tepuztecos; Acuña 1987), and considerable interpolity warfare occurred in the region prior to integration into the Aztec Empire (Smith and Berdan 1996). The Cuitlatecas, Tepuztecas, and Yopes resisted the Aztecs for many years before Ahuítzotl conquered the coastal regions on either side of Yopi territory, a mountainous area along the coast southeast of Acapulco (Barlow 1990:71). Yopitzinco remained an independent state with its northwestern boundary maintained by Tecpantepec, an Aztec strategic province composed of 12 towns that stretched along the Costa Grande, buffering the Cihuatlán tributary province on both its northwestern and southeastern ends from Tarascan and Yopi territories, respectively (Smith 1996:141; Smith and Berdan 1996:277).

Information on tribute from the Cihuatlán province comes from three historic documents, the Matrícula de Tributos, the Codex Mendoza, and the Relación de Citlaltomahua (Berdan 1996; Litvak 1971). Twelve towns are listed in the Matrícula de Tributos and the Codex Mendoza: Cihuatlán (Zihuatanejo), Coliman, Panotlan (Pantla), Nochcoc (Nuxco), Iztapan (Ixtapa), Xolochiuhyan (Julucha), Petlatlan, Xihuacan, Apancalecan, Cocohuipilecan, Coyucac (Coyuquilla), and Cacatulan (Zacatula) (Mohar 2002:507; Smith and Berdan 1996:277). These tributaries provided cotton blankets, cotton, flowers *(coyichcatl)*, cacao, and marine shells to the Aztecs. Mollusk shells are the only significant marine resources identified in these documents, but fish were certainly potential tribute items and may have circulated as important interregional exchange items outside of the tribute system (Smith and Berdan 1996:315). The Cihuatlán province provided a range of marine shells in tribute to the Aztecs. Every 80 days the province was required to send 800 nacre shells *(tapachtli)*, and every six months to send 800 colored shells (*Pinctada mazatlanica;* Barlow 1990:21). Red shells (likely *Spondylus princeps*) are also listed in Codex Mendoza, and according to Sahagún and Diego de Landa, they were used as a medium of exchange or as offerings (Mohar 2002:511–512). Marine shells were also common offerings in burials within the Templo Mayor (Velázquez Castro 1999, 2000), one of the primary ceremonial complexes in the Aztec capital of Tenochtitlán. These include shells of *Anadara multicostata, Noetia reversa, Pinctada mazatlanica, Spondylus calcifer, S. princeps, Pitar lupanaria, Megapitaria aurantiaca, Dosinia ponderosa, Mactrellona alata, Astrea olivacea, A. unguis, Nerita scabricosta, Polinices uber, Cassis centiquadrata, Morum tuberculosum, Cymatium lignarium, Murex recurvirostris, Hexaplex erythrostomus, Cantharus sanguinolentus, Columbella fuscata, Opeatostoma pseudodon,* and *Harpa crenata.* These are not listed in the tributary texts but are believed to come from Guerrero because they are of Pacific Coast origin and were associated with jade figurines and other objects from artistic traditions known only in Guerrero (Mezcala, Chontal, and Sultepec).

Shrimp procurement in the towns of Xolochiuhcan, Xihuacán, and Cihua is mentioned in the Relación de Citlaltomahua (Litvak 1971:83), but we can only speculate about the significance of dried shrimp within the regional exchange system. According to Fabiola Guzmán and Polanco (2000:154), the earliest ethnohistoric record of marine fish arriving as tribute to the Aztec imperial capital was in AD 1440, at the time of the coronation of emperor Motecuhzoma Ilhuicamina (AD 1440–1469). This coincides with the political expansion of the Aztec empire that embraced both coasts, but the sources do not specify exactly where this tribute originated. A well-studied offering (Offering 23) at the Templo Mayor de Tenochtitlan, which is believed to have been deposited during the reign of that emperor, contains fish and mollusk remains from both the Atlantic and Pacific coasts. The best ethnohistoric evidence concerning types of marine fish and their use in the Mexico heartland is during the reign of Motechuzoma Xocoyotzin (AD 1502–1520), about 50 years after the deposition of Offering 23 (Fabiola Guzmán and Polanco 2000:155) and just prior to the Spanish conquest. The sources include a single explicit reference that the inhabitants of Ayutla, a settlement near the port of Acapulco, sent marine fish to Motecuhzoma Xocoyotzin about twice a year (Fabiola Guzmán and Polanco 2000:155). However, marine fish appear not to have been regular tribute items since they are not itemized in either the Matrícula de Tributos or the Codex Mendoza. None of the ethnohistorical sources indicate whether these fish were brought to the Valley of Mexico alive or dead, whole or filleted, dried, salted, smoked, or fresh, and so forth (Fabiola Guzmán and Polanco 2000:155). Although the exact impact of the Aztec tribute system on marine ecosystems in Guerrero is unknown, we can surmise that the selective extraction of certain species for nonlocal use had a significant ecological effect.

PUERTO MARQUÉS: ENVIRONMENTAL, DEMOGRAPHIC, AND HISTORICAL CONTEXT

Our investigation of prehistoric human impacts on marine ecosystems in Guerrero is centered on Puerto Marqués, where a well-stratified archaeological sequence contains material dating from the Late Archaic through Classic periods (~5500–1200 cal BP; Figure 5.2). The small bay on which this coastal community was established is located on the northern end of the Costa Chica and, during the Aztec conquest period, was incorporated into the Tecpantepec strategic province, buffering the Cihuatlán tributary province from Yopi territory. The bay is formed by two promontories: one separating Acapulco Bay from Puerto Marqués, and the other by a rocky point that forms the southern side of the bay (Punta Diamante). A flat sandy beach (Playa Majahua) connects these two promontories and separates the bay from a mangrove swamp that extends away from the coast and was likely connected to the Tres Palos Lagoon during the Early and Middle Holocene (Kennett et al. 2004).

Early investigations at Puerto Marqués defined a substantial Classic period (1800–1200 cal BP) coastal community with a long prehistory that was positioned on a series of terraces overlooking the Playa Majahua and the sheltered bay (Brush 1969). Prehistoric materials are concentrated in the southern portion of this bay at the edge of a small drainage that flows from the northern flanks of Punta Diamante. A low-lying mound on the south side of this small drainage contains a deeply buried (6.6 m below the surface) shell midden (~1 m thick) dating to the Late Archaic period (~5500 and 4000 cal BP; Brush 1969; Kennett et al. 2004; Manzanilla López et al. 1991). A very limited range of tools was also identified in these early deposits, with much of the assemblage consisting of stone flakes and chipping debris fashioned from a variety of local (quartz, chert, chalcedony) and exotic (obsidian) materials. Occupation at Puerto Marqués continued into

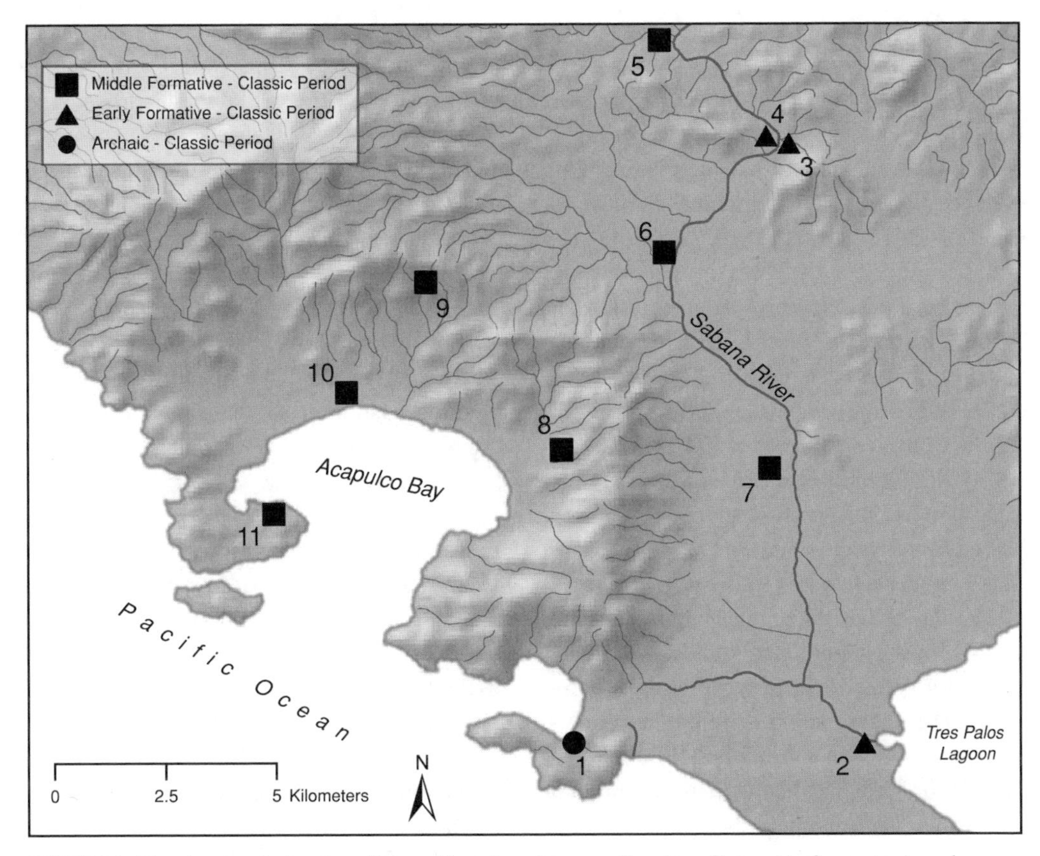

FIGURE 5.2. Map showing the location of Puerto Marqués and surrounding sites of interest in the greater Acapulco area. Symbols represent chronological estimates based on ceramic assemblages and radiocarbon dates. 1, Puerto Marqués; 2, La Zanja; 3, Arroyo Seco; 4, El Recreativo; 5, Barrio Nuevo; 6, La Sabana; 7, Infonavit; 8, La Picuda; 9, Palma Sola; 10, Hornos; 11, Tambuco.

the Early Formative period (~3800–2000 cal BP) when pottery appeared in the archaeological record for the first time (~3800 cal BP). Stone tool assemblages diversified to include a number of formal or specialized tools (microdrills, macrodrills) and a bipolar flaking technology that is well known in other parts of Mesoamerica (Clark 1981).

Early Formative period deposits (3100 cal BP) are also present at the site of La Zanja, an earthen mound site located 5 km east of Puerto Marqués and ~3 km away from the coast at the mouth of the Río de la Sabana where it enters the Tres Palos Lagoon. Similar ceramic and stone tool deposits are evident at this location. The appearance of obsidian prismatic blades during the Early to Middle Formative period transition is coincident with a diversification in the ceramic assemblage. Formative period settlement is also evident at the Tambuco site located on the western periphery of Acapulco Bay (Ekholm 1948).

The expansion of settlements in the Acapulco region during the Formative period coincides with clear evidence for settled village life and an increasing commitment to maize-based food production within the region. Biogeographical and genetic studies suggest that maize was domesticated in the Balsas River Valley or the Valley of Oaxaca some time before 7500 cal BP (Doebley 1990; Matsuoaka et al. 2002). Microbotanical data (phytoliths and pollen) from a range of lowland tropical environments extending as far south as Panama suggest a far-reaching dispersal of maize through the lowland tropics between 7500 and

7000 cal BP (Piperno 2006). The initial use of early maize varied from place to place, but the early low-level use of this cultigen by foragers was followed by gradual increases in consumption rates. At this point there is no compelling evidence that people living along the coast of Guerrero cultivated maize in a significant way until late in the Formative period. Pollen identified in a sediment core from the Tetitlán Lagoon, just to the north, indicates that *Zea mays* was being cultivated in that area by the end of the Early Formative period (3170 ± 280 RYBP; González-Quintero and Mora-Echeverriá 1978). This date coincides with the first expansion of populations away from the coast into stillwater wetland habitats along the Costa Chica, as represented by La Zanja.

Significant population expansion along the coast is most evident during the Classic period (1800–1200 cal BP), when settlements along the coast and the bordering mountain slopes became more concentrated and organized. These more-urbanized settlements were composed of housing compounds formed by patios surrounded by three or four houses. The best known sites with habitation and agricultural terraces are Playa Hornos, Zanja, Puerto Marqués, Ciudad Perdida, Coyuca, San Jerónimo, Las Peñas, Atoyac, Nuxco, San Luis la Loma, Soledad de Maciel, La Corea, El Cabrito, El Zopilote, Tierras Prietas, Victorino Rodríguez, V3, V38, and V42 (Manzanilla López 2000:196). Stone-lined agricultural terraces that were carved into more mountainous slopes indicate that intensive maize-based food production was well established in the region (Manzanilla López 2000:195). Evidence of exchange activities and cultural interaction with other groups in west Mexico increases during this time, and trade connections were established with Teotihuacán and Monte Albán—two expansionistic states from the central Mexican highlands that were interested in marine shells, bird feathers, cotton, and cacao in this region. Greater articulation with Teotihuacán and Monte Albán parallels the establishment of a ceremonial complex (e.g., mounds, platforms, stone monuments, and altars) in settlements along the Costa Grande and Costa Chica (Manzanilla López 2000:199), and the use of highland Mexican religious symbols by aspiring elites (e.g., depictions of elites with Tlaloc [the Teotihuacán rain and war god] or Zapotec-style year glyphs as at Vella Rotaria; Manzanilla López 2000:204–205). During the Classic period evidence for the intensive extraction of aquatic habitats appears for the first time, with the best evidence for these activities coming from three large shell mounds positioned on the landward side of Laguna Coyuca, north of Acapulco Bay (Kennett et al. 2004).

Settlement persisted at both Puerto Marqués and La Zanja during the Classic period, the former a thriving coastal community composed of a series of terraces with substantial domestic architecture extending up the slopes of Punta Diamante (Manzanilla López 2000). Larger settlements were well established at several locations along the Río de la Sabana by this time (Figure 5.2), and many of these were first occupied during the Middle to Late Formative. The sites of El Recreativo, La Sabana, Infonavit, Arroyo Seco, and Palma Sola are all larger than Puerto Marqués and La Zanja and are composed of terraces and large platforms excavated into the mountainous terrain with substantial stone structures, pyramids, and sunken plazas evident in many instances (Cabrera Guerrero 1990). These data are consistent with evidence for substantial settlements along the shores of Acapulco Bay and to the north along the Costa Grande (Brush 1969; Ekholm 1948).

The persistence of settlement at Puerto Marqués over a 2,000-year period as populations expanded in the region, coupled with well-preserved faunal materials, provides an excellent opportunity to track changes in marine resource use through time and to explore the potential impacts of expanding populations in the region on marine ecosystems. We now turn to the detailed faunal records spanning the Archaic and Formative periods to explore the impacts that expanding human populations had at this location.

TABLE 5.1

Stratigraphic, Chronological, and Basic Assemblage Information from Puerto Marqués

PERIOD/PHASE	DEPTH (M)	^{14}C AGE	ERROR	CAL BP (1 SIG.)	CHARACTER
Late Formative	0.60–1.20	2460	40	2710–2360	Diverse pottery assemblage
		2490	35	2713–2470	Red wares
		3090	75	2745–2602	Obsidian prismatic blades
Middle Formative	1.20–3.00	2790	40	2947–2847	Diverse pottery assemblage
		2890	65	3159–2893	Red and white wares
					Obsidian prismatic blades
Early Formative	3.00–4.00	3980	50	3814–3637	Small quantities of pottery "pox"
		4020	45	3840–3690	Diverse stone tool assemblage
Late Archaic	4.00–5.40	4120	45	3973–3830	Aceramic shell midden
		4180	70	4088–3879	Limited stone tool assemblage
		4560	40	5313–5085	Flakes/debitage
		4800	40	5592–5484	

NOTE: Volumes excavated are 1.6 m^3 (Late Archaic), 2.4 m^3 (Early Formative), 3.6 m^3 (Middle Formative), and 0.39 m^3 (Late Formative). The Archaic period mollusk data is corrected based on a smaller volume analyzed (0.013 m^3) due to the large quantities of shell involved.

Fish and Other Aquatic Fauna

The vertebrate assemblage from the Puerto Marqués site consists of 5,447 bones representing all major vertebrate classes. Fish dominate the assemblage (92.6 percent; 5,048 specimens), followed by reptiles (6 percent), mammals (1.3 percent) amphibians (<.1 percent), and birds (<.1 percent). Fish are the most varied vertebrate class present in the assemblage with 35 genera and 26 species identified. Virtually all of the identified fish are ray finned (Actinopterygii) with the most common fish families being the Scrombridae (tunas and mackerels), the Carangidae (jacks), and the Lutjanidae (snappers). Cartilaginous fish (sharks and rays; Elasmobranchii) are rare. Two sea turtle species (green sea turtle, *Chelonia agasizii*; and hawksbill, *Eretmochelys imbricata*), both endangered today, were identified and were the only other major marine vertebrates represented in the assemblage. The large Central American crocodile *(Crocodilus acutus)*, common in Mexico's Pacific Coast estuaries (Alvarez del Toro 1983), was also identified, along with a variety of smaller animals and fish from brackish and freshwater habitats. The mollusk assemblage from Puerto Marqués consists of over 17,000 individual specimens (number of identified specimens [NISP]) representing 50 different taxa. These taxa come from a range of habitats including stillwater estuaries, sandy beaches, rocky shoreline, and mangrove swamps. Mollusk shells are abundant in these deposits, and it is clear that shellfish played an important dietary role for the inhabitants of this community, but the density of shellfish remains was highest in the Late Archaic period deposits and decreased through time.

To explore changes in the faunal assemblage through time we divided the collection into four temporal categories—Late Archaic, Early Formative, Middle Formative, and Late Formative—based on stratigraphic, artifactual, and chronometric data (Table 5.1). Late Archaic people living at Puerto Marqués targeted fish from a range of habitats while focusing on open-water (epipelagic) species such as tunas, jacks, and roosterfish (70 percent; Table 5.2). Fish common in rocky reef (23 percent; e.g., snappers, wrasses surgeonfish, triggerfish, porcupinefish, grunts), nearshore beachfront habitats (e.g., mojarras, croakers, mullet), and stillwater estuarine habitats (2 percent; snook) were also identified. Sea turtle (Cheloniidae) bones and carapace fragments were the most common reptile

TABLE 5.2

Identified Fish Remains from Puerto Marqués

SCIENTIFIC NAME	COMMON NAME	TROPHIC LEVEL	HABITAT	ARCHAIC		FORMATIVE Early		FORMATIVE Middle		FORMATIVE Late		TOTAL	
				NISP	*MNI*	*NISP*	*MNI*	*NISP*	*MNI*	*NISP*	*MNI*	*NISP*	*MNI*
Arius sp.	Sea catfish	Mid	Estuarine							9	2	9	2
Centropomus pectinatus	Black snook	High	Estuarine							1	1	1	1
Centropomus sp.	Snook	High	Estuarine	15	2	1	1	6	1	1	1	21	1
Strongylura exilis	California needlefish	High	Estuarine/inshore	7	1					1	1	8	1
Myliobatidae	Bat rays	Mid	Inshore	1	1							1	1
Epinephelus sp.	Grouper	High	Inshore	9	1							9	1
Diapterus sp.	Mojarra	Mid	Inshore			1	1					1	1
Eugerres sp.	Mojarra	Mid	Inshore			1	1					1	1
Gerres cinereus	Yellowfin Mojarra	Mid	Inshore	3	1			7	1			10	2
Haemulon sp.	Grunt	Mid	Inshore	2	1							2	1
Haemulopsis sp.	Grunt	Mid	Inshore	1	1			1	1			2	2
Pomadasys sp.	Grunt	Mid	Inshore	1	1							1	1
Cynoscion sp.	Weakfish	High	Inshore	4	1			2	1			6	2
Micropogonias altipinnis	Tallfin croaker	Mid	Inshore			1	1					1	1
Micropogonias sp.	Croaker	Mid	Inshore	1	1							1	1
Umbrina sp.	Drum	Mid	Inshore	3	2							3	2
Mugil sp.	Mullet	Low	Inshore	1	1			2	1	2	1	5	3
Serranidae	Sea basses	High	Inshore			1	1					1	1
Gerreidae	Mojarras	Mid	Inshore	8	1			1	1	4	1	13	3
Haemulidae	Grunts	Mid	Inshore	10	1			2	1	1	1	13	3
Sciaenidae	Corvinas	Mid	Inshore			1	1					1	1
Caulolatilus sp.	Whitefish	High	Open water	1	1							1	1
Caranx caninus	Crevalle jack	High	Open water	4	4			3	1			7	5
Caranx sp.	Jack	High	Open water	36	3	7	1	9	1	2	1	54	6
Hemicaranx sp.	Scad	Mid	Open water	137	7	11	1	8	1	4	1	160	13
Seriola sp.	Amberjack	High	Open water					1	1			1	1
Trachinotus sp.	Pompano	Mid	Open water	8	1			2	1			10	2
Nematistius pectoralis	Roosterfish	High	Open water	4	1			4	1			8	2

Kyphosus elegans	Cortez sea chub	High	Open water	1	1							1	1
Sphyraena ensis	Barracuda (Mex.)	High	Open water	5	2	5	3	14	1	3	1	27	7
Sphyraeana sp.	Barracuda	High	Open water			1	1					1	1
Euthynnus lineatus	Black skipjack	High	Open water	343	9	131	6	136	3	4	1	614	19
Katsuwonis sp.	Skipjack	High	Open water	1	1							1	1
Thunnus alalanga	Albacore	High	Open water	1	1							1	1
Thunnus sp.	Tuna	High	Open water	44	1	3	1	1	1			48	3
Carangidae	Jacks	High	Open water	36	1	5	1	2	1	3	1	46	4
Thunninae	Tunas	High	Open water	32	1	13	1	5	1			50	3
Scombridae	Tunas	High	Open water	11	1	6	1	4	1	2	1	23	4
Carcharhinidae	Requiem sharks	High	Open water/inshore					4	1			4	1
Lutjanus argentiventris	Yellow snapper	Mid	Reef	2	1			2	2			4	3
Lutjanus colorado	Red snapper	Mid	Reef					7	2			7	2
Lutjanus guttatus	Spotted rose snapper	Mid	Reef					1	1			1	1
Lutjanus jordani	Jordan's snapper	Mid	Reef	2	1			2	1			4	2
Lutjanus novemfasciatus	Dog snapper	High	Reef	2	1	1	1	13	2			16	4
Lutjanus sp.	Snapper	Mid	Reef	105	4	13	2	75	2	4	1	197	9
Bodianus diplotaenea	Mexican hogfish	Mid	Reef	8	2	1	1					9	3
Halichoeres nicholsi	Spinster wrasse	Mid	Reef	3	3							3	3
Scarus ghobban	Blue-barred parrotfish	Low	Reef					2	1			2	1
Scarus perrico	Bumphead parrotfish	Low	Reef	1	1							1	1
Scarus sp.	Parrotfish	Low	Reef	4	2			1	1			5	3
Acanthurus sp.	Surgeonfish	Mid	Reef	2	1							2	1
Acanthurus xanthopterus	Yellow surgeonfish	Mid	Reef					1	1			1	1
Prionurus punctatus	Yellowtail surgeonfish	Mid	Reef	6	1	1	1	3	1			10	3
Balistes polylepis	Finescale triggerfish	Mid	Reef	7	4			4	1			11	5
Sufflamen verres	Orangeside triggerfish	Mid	Reef					1	1			1	1
cf. *Xanthihthys mento*	Redtail triggerfish	Mid	Reef					1	1			1	1
Diodon hystrix	Porcupinefish	Low	Reef	46	4	1	1	5	2			52	7
Diodon sp.	Porcupinefish	Low	Reef	28	1	1	1					29	2
Acanthuridae	Surgeonfish	Mid	Reef					1	1			1	1
Balistidae	Triggerfish	Mid	Reef			1	1					1	1

NOTE: The relative abundance of vertebrates is reported in MNI (minimum number of individuals) and NISP (number of identified specimens). All excavated soil was screened with 6-mm mesh, with a smaller sample from each level screened using 3-mm mesh (see Kennett et al. [2004] for details).

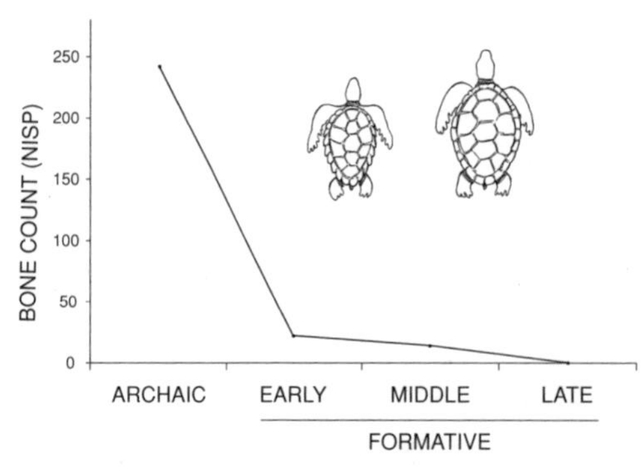

FIGURE 5.3. Frequency of sea turtle bones in deposits. Numbers are not volumetrically corrected and simply show a general decreasing trend in the abundance of sea turtle bones in these deposits. Volumetrically corrected values are 151, 9, 4, and 0. NISP, number of identified specimens. See Table 5.1 for volumetric data.

remains in these early deposits (Figure 5.3). The presence of crocodile bones (n = 6) suggests that this species was hunted in nearby estuarine habitats. Three freshwater turtle species *(Kinosternon integrum, Trachemys scripta, Rhinoclemmys pulcherrima)* and one amphibian (toad, *Bufo* sp.) were also identified in these deposits. The bones occur within a matrix of mollusk shells dominated by oyster (*Ostrea palmula*, Table 5.3), a species that adheres to mangrove roots or on reefs exposed to surf (Keen 1971:84). The strong presence of schooling fish from open-ocean habitats (Black skipjack tuna, *Euthynnus* cf. *lineatus*, n = 343; Yellowfin jack, *Hemicaranx* sp., n = 137) suggests that people living at this location during the Archaic period employed boats and nets to capture these elusive offshore species (see below).

The Early Formative period deposits at Puerto Marqués were even more strongly dominated by fish from open-water habitats (89 percent, tunas, jacks and roosterfish) with the black skipjack tuna (n = 343) topping the list of most important prey species. A smaller percentage of fish from reef and nearshore soft-bottom beachfront habitats constitute the remainder of the Early Formative period assemblage, with species from estuarine and freshwater environments virtually absent. Sea turtles, most likely green sea turtle, were the next most common vertebrates in the assemblage. However, sea turtle remains were less common when compared to the Late Archaic period deposits. No crocodilians were identified in this subassemblage. Changes visible in the vertebrate assemblage parallel an overall reduction in the quantity of mollusk shell along with an expansion in the variety of species targeted. This included a sharp reduction in the use of oysters. Two large clam species, *Chione californiensis* and *Megapitaria aurantiaca*, dominate the Early Formative period assemblage, and both of these species are found in sandy beach habitats like those found near the Puerto Marqués site today. These two large clams are no longer available in the region for commercial exploitation (personal communication from local informant to B. Voorhies).

The ratio of open-water to reef fish species in the Middle Formative period deposits is more balanced compared with the Late Archaic and Early Formative period assemblages. Fish from open-water habitats continue to dominate in the Middle Formative period faunal assemblage (56 percent), but inshore reef oriented species are more common (37 percent). Tunas continue to dominate the assemblage, but snappers (n = 100), a species that prefers rocky or coral reef habitats, are much more common compared with previous periods. Few fish species commonly found in estuaries are identified. Sea turtle remains, mainly small fragments of carapace, are present but less common compared with previous periods. The two fragments of the Central American crocodile were identified. A large variety of mollusk species continued to be collected with continued preference for *Chione* and *Megapitaria*, but with a broadening reliance upon a wider range of smaller and less accessible taxa, as well as species most likely used for decorative purposes (e.g., *Oliva uncrassata*).

The Late Formative faunal collection is the smallest and most different at Puerto Marqués. The small size of this assemblage is due to the small volume of Late Formative deposits penetrated in the 2005 excavations, but the fish represented come from a more diverse array of

TABLE 5.3
Identified Mollusk Remains from Puerto Marqués

SPECIES	LATE ARCHAIC		FORMATIVE					
			Early		Middle		Late	
	NISP	per m^3	NISP	per m^3	NISP	per m^3	NISP	per m^3
Chione californiensis	18	1384.6	1230	512.5	1650	457.3	927	2376.9
Megapitaria aurantiaca	23	1769.2	403	167.9	1195	331.2	46	117.9
Ostrea palmula	69	5307.7	83	34.6	15	4.2	65	166.7
Polymesoda			4	1.7	22	6.1	217	556.4
Anadara multicostata	1	76.9	46	19.2	95	26.3	40	102.6
Trachycardium consors	2	153.8	15	6.3	85	23.6	19	48.7
Hexaplex erythrostomus	4	307.7	39	16.3	51	14.1	45	115.4
Glycymeris gigantea			12	5.0	36	10.0	39	100.0
Chama mexicana			25	10.4	5	1.4	19	48.7
Pododesmus macrochisma			5	2.1	7	1.9	13	33.3
Stramonita biserialis			39	16.3	17	4.7	7	17.9
Undulostrea megodea			22	9.2	6	1.7	6	15.4
Plicopupura pansa			1	0.4	18	5.0	38	97.4
Melongena patula			24	10.0	13	3.6	5	12.8
Dosinia ponderosa			4	1.7	19	5.3	1	2.6
Theodoxus luteofasiastus			3	1.3	2	0.6	40	102.6
Oliva uncrassata			2	0.8	24	6.7	12	30.8
Stombus granulatus			3	1.3	24	6.7	21	53.8
Pinnidae			1	0.4	5	1.4	11	28.2
Cymatlum wiegmanni			1	0.4	11	3.0	5	12.8
Crucibulum scutellatum			3	1.3	2	0.6	8	20.5
Iphigenia altior			1	0.4	3	0.8	4	10.3
Strombus galeatus			3	1.3	4	1.1	4	10.3
Cerithidea mazatlantica			1	0.4			10	25.6
Turritella leucostoma			1	0.4	3	0.8	1	2.6
Lyropecten subnodosus			1	0.4			2	5.1
Donax puctatostriatus			3	1.3	1	0.3	1	2.6
Fissurella virescens					4	1.1	1	2.6
Chione subrugosa					1	0.3	1	2.6
Cardita megastropha					1	0.3	1	2.6
Tucetona multicostata			1	0.4	2	0.6		
Spondylus calcifer					2	0.6	1	2.6
Spondylus spp.			1	0.4	2	0.6		
Unknown gastropod					1	0.3	1	2.6
Anadara formosa			1	0.4	1	0.3		
Pleuroploca princeps					2	0.6		
Unknown univalve					1	0.3		
Semicassis centriquadrata					1	0.3		
Tagelus affinis							1	2.6
Cypraea arabicula							1	2.6
Ancistromesus mexicana							1	2.6
Crepidula lessonii							1	2.6
Columbella fuscata							1	2.6
Littorina modesta							1	2.6
Mytella strigata	1	76.9						
Ancistromesus mexicana			1	0.4				

NOTE: NISP, number of identified specimens.

species and habitats. Fish continue to dominate the vertebrate assemblage, including fish from open-water (44 percent), estuarine (29 percent), inshore (17 percent), and reef (10 percent) habitats. Tunas dominate the open-water fishes, and snappers are more common than in the Formative period strata. Carangid fish are less common compared to previous phases and sea catfish (*Arius* sp.), and snook are the dominant fish from estuarine habitats. Sea turtle and crocodile are both absent from the Late Formative period assemblage. Continued decreases in the number of mollusks harvested parallels further increases in species diversity. *Chione californiensis* continued to be targeted, but the abundance of *Megapiteria aurantiaca* is reduced significantly. Increased importance of the marsh clam species *Polymesoda,* in conjunction with increases in the number of estuarine fish, suggests the expanding importance of this more distant environmental zone.

The strong presence of schooling fish from open-ocean habitats (black skipjack tuna, n = 343; yellowfin jack, n = 137) suggests that people living at this location employed boats and nets to capture fast-moving offshore species starting as early as the Late Archaic period. Many of these species will strike a baited hook or gorge, but drop line hooks, gorges, or trolling rigs, perhaps similar to those used in Polynesia to capture blue water scombrids and carangids, are absent from the tool assemblage at Puerto Marqués. Suitable conditions for epipelagic fish exist just offshore and within the confines of the sheltered bay where this prehistoric community was established due to a deep submarine canyon offshore that funnels nutrient-rich deep waters into these sheltered nearshore environments. Regardless, it is most likely that gill nets were used to capture these types of fish, rather than beach seines, because the latter would produce a fish fauna heavier in croakers (Scieanidae), mullet (Mugilidae), and other beachfront species. Suitable watercraft was necessary to deliver gill nets to nearshore locations. Within the sheltered confines of this bay these might be simple flotation devices made from reeds, but we have not ruled out the use of open-hulled canoes or balsas, given the specialized nature of the fishery. Other types of technology (e.g., fish traps) may also have been used to capture nearshore soft-bottom and reef fishes.

Richness and Diversity of Fish and Shellfish Taxa

To further explore changes in the local subsistence ecology through time and possible impacts to food webs due to human harvesting strategies, we employ summary measures of fish and shellfish assemblage richness and diversity (Figure 5.3). Species richness is simply the number of fish or shellfish taxa represented in each subassemblage (Reitz and Wing 1999:102). It is a complex product of (1) the overall diversity and character of an ecological community or communities, (2) prehistoric human subsistence and processing decisions, and (3) variations in sample size and bone or shell preservation. Species diversity is a measure that combines species richness with abundance in each category and thus is a measure of the heterogeneity of an assemblage. We employed the Shannon-Weaver (1949) index to estimate the overall diversity (H′) of each subassemblage, calculated as

$$\overset{s}{\underset{i=1}{H'}} = \sum (p_i)(\log p_i)$$

where p_i is the number of ith taxon within the sample, log p_i is the natural log of p_i, and s equals the total number of taxa represented. In this view of diversity, equally rich faunal assemblages that contain an even abundance of taxa are more diverse than those that have larger quantities of certain species. This accounts for the overall dietary contribution of each species, rather than its simple presence or absence (see Reitz and Wing 1999). In this study we use volumetrically corrected NISP values for both fish and shellfish taxa due to different sized soil samples processed for each period of time.

Based on these calculations, there are significant changes in the richness and diversity of fish and shellfish assemblages through time

(Figure 5.4; Table 5.4). The highest richness of fish taxa is evident during the Late Archaic period ($H' = 41$), and assemblage richness declines through the Formative period to a low of $H' = 14$, with a rebound ($H' = 36$) during the Middle Formative. Fish assemblage richness and diversity are both high during the Late Archaic period, reflecting the overall intensity of marine resource use prior to increasing dependence on maize-based food production in the region. Fish assemblage diversity plummets in the Early Formative with increased exploitation of large schooling fishes (e.g., tuna) followed by reduced capture of these high trophic level species and expanding use of lower trophic level species from a range of marine and estuarine habitats. Reductions in the abundance of large gregarious fish species coupled with expanding use of behaviorally dissimilar small and medium size fish from a range of habitats are consistent with the idea that human predation influenced the availability of highly valued species and that this triggered dietary expansion to smaller, less-desirable species through time (e.g., Pauly et al. 1998, 2000). This occurred during the Early Formative period when maize-based food production was becoming established on this coastal plain and populations were expanding into the interior.

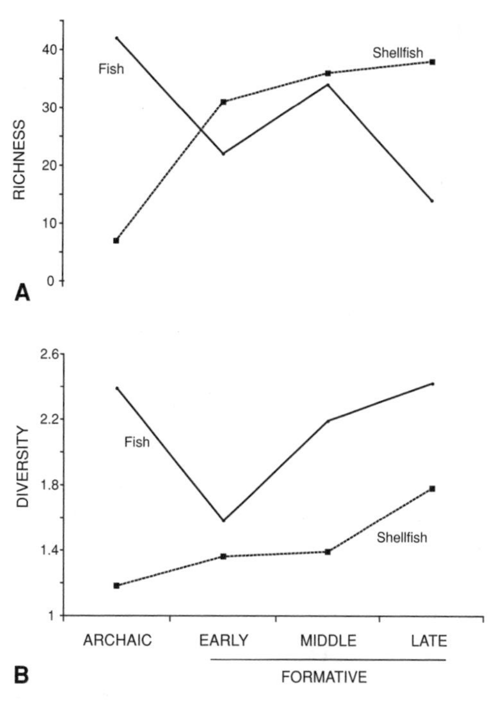

FIGURE 5.4. Relationship between richness (A) and diversity (B) in fish and mollusk assemblages from the Late Archaic through Late Formative periods. Numbers (NISP [number of identified specimens]) from Tables 5.2 and 5.3 are volumetrically corrected (per m^3). See Table 5.1 for volumetric data.

Increases in the diversity of fish at Puerto Marqués also parallel increases in the richness of the shellfish assemblage through time from a low of $H' = 7$ during the Late Archaic period to a high of $H' = 38$ in the Late Formative. Parallel increases are evident in the shellfish diversity index. Late Archaic period foragers targeted a small number of species with a primary emphasis on oysters, a species that was locally abundant at this time. Two large beach clams, *Chione californiensis* and *Megapitaria aurantiaca*, also feature prominently in the Late Archaic assemblage and are the dominant species collected through the Formative period—after significant reductions in the abundance of oyster occurred. Oysters adhere to mangrove roots, and their reduction, coupled with the increased importance of sandy beach species, suggests that the small mangrove swamp near Puerto Marqués today was more extensive during the Late Archaic period and that increased sediment loading, perhaps associated with expanding agricultural economies and sea-level stabilization, expanded the beach and reduced the mangrove oyster habitat. Given the high density of oyster shell found in the Late Archaic midden, however, direct human predation on oysters may also have contributed to the shift in shellfish exploitation patterns through time. One way or the other, reductions in the abundance of the three most desirable molluskan taxa *(Ostrea, Chione,* and *Megapitaria)* associated with parallel increases in smaller and less accessible species seem consistent with the idea that faunal assemblage changes are at least partly anthropogenic in origin.

In sum, we argue that an overall decrease in fish assemblage richness coupled with increasing

TABLE 5.4

Richness and Diversity Estimates of the Puerto Marqués Fish and Mollusk Assemblage

	ARCHAIC	FORMATIVE		
		Early	*Middle*	*Late*
Fish richness	42.00	22.00	34.00	14.00
Fish diversity	2.39	1.58	2.19	2.42
Shellfish richness	7.00	31.00	36.00	38.00
Shellfish diversity	1.18	1.36	1.39	1.78

NOTE: See Table 5.1 for volumetric details.

diversity through the Formative period resulted from the depletion of the local fishery due largely to human overexploitation in the context of regional population expansion. Increased exploitation for expanding trade networks may have also played a role in depressing locally available fish stocks. Reductions in the most highly valued shellfish species through time and increases in the variety of species harvested are also consistent with this hypothesis. These interpretations will need to be carefully tested with future paleoenvironmental work, but preliminary oxygen and carbon isotopic work on marine mollusks from these deposits indicates no major changes in ocean circulation or productivity.

Fishing down Food Webs

Pauly et al. (1998, 2000) have documented a general pattern of declining trophic levels in recent world fisheries, in which humans focused on larger fish from higher trophic levels, "fish down food webs," and reduce the diversity and complexity of marine ecosystems. Archaeologists have just begun to explore deeper historical records for evidence of similar patterns among ancient fishing societies, with mixed results (see Bourque et al., this volume; Erlandson and Rick, this volume; Morales and Rosello 2004; Reitz 2004; Wing 2001). Shifts in the composition of fish assemblages from high to low trophic level species have been used elsewhere to gauge overfishing or the depletion of fish stocks (Wing 2001), but these trends are often complex and dependent upon cultural and environmental context (Reitz 2004). Larger open-water predatory fish, like those dominating the Puerto Marqués assemblage, eat other fish, whereas small reef herbivores or schooling baitfish like anchovies are on a lower trophic level. Here we explore whether the patterns identified by Pauly et al. (1998) can be extrapolated backward in time and test the hypothesis that people have a tendency to target high trophic level fish species and that lower trophic level species are targeted more frequently as higher trophic level fish stocks are depleted.

To explore potential trophic cascade effects in the context of expanding populations in coastal Guerrero we calculated the mean trophic level of the four intervals represented at Puerto Marqués (Late Archaic to Late Formative) using methods described by Reitz (2004:70). Mean trophic level (TL_i) for each time period was calculated using the equation

$$TL_i = \sum_i (TL_{ij})(NISP_{ij}) / \sum_i NISP_i$$

where the trophic level for each taxon per time period (TL_{ij}) is summed together and multiplied by the NISP for the same interval. The product is divided by the total sum of NISP for the same period of interest. Trophic levels for each fish taxon, or the closest taxonomic category, are assigned using Fishbase 2006 (www.fishbase.org). These values range from 2 and 4.5, with the lowest values representing small fish that mainly eat plant and detritus, and the highest representing large fish that eat other

TABLE 5.5

Percent Mean Trophic Level of Fish at Puerto Marqués through Time

TROPHIC LEVELS	ARCHAIC	FORMATIVE		
		Early	*Middle*	*Late*
2.0–2.99	2.33	0.335	3.99	5.128
3.0–3.99	30.97	8.027	9.42	58.97
4–4.5	66.7	91.63	86.62	35.89
	100	100	100	100

fish. We substitute NISP for biomass in this formula due to the lack of allometric data available for fish species in coastal Guerrero and the great diversity of the assemblage at Puerto Marqués.

Mean trophic level from the Late Archaic through Late Formative periods ranges from 4.23 to 3.71 and indicates that people in this early coastal community generally targeted high-ranked fish species (Figure 5.5). The mean trophic level for the Late Archaic period is relatively high (4.06), followed by a peak during the Early Formative (4.23), and a decline during the Middle (4.15) and Late (3.71) Formative periods. Late Archaic period populations took 67 percent of fish biomass from high trophic level sources, primarily large schooling fish (e.g., tuna and jacks), and much of the remainder (31 percent) from a variety of fish at middle range levels (Table 5.5). The overall high diversity of the Late Archaic period fish assemblage, coupled with the intensive use of mollusks, is not surprising for a coastal foraging population. More intensive use of the high trophic level fish taxa (92 percent) during the Early Formative period occurred with a greater commitment to maize-based food production regionally. Gradual decreases in the percentage of high trophic level fish taxa occur after this time as the importance of species from medium and low trophic levels increased. The largest shift occurs in the Late Formative period, and this is unlikely a product of sample size because a large diversity of small, medium, and large fishes were recovered from these levels. It is also unlikely related to the dietary changes associated with increases in the dependence of maize-based food production because this occurred much earlier during the Late Archaic to Early Formative period transition with no visible effect. Reductions in high trophic level species could be the product of overfishing during the Formative period. However, this reduction could have more to do with dietary expansion and more localized reductions in nearshore prey species that stimulated schooling pelagic fishes to forage elsewhere along the coast. Therefore, although these data are consistent with the idea that people fish down food webs (Pauly et al. 1998), the trends visible in the data could also result from the cascading effect of local human predation on the prey of these high trophic level species.

CONCLUSIONS

The archaeological record at Puerto Marqués testifies to a once rich and diverse nearshore marine ecosystem. Qualitative assessment suggests that the richness and diversity of these same habitats today are greatly reduced, with an estimated one million people living in the greater Acapulco region. Quantitative data from the Guerrero fishery also indicate that productivity has declined significantly during the last 10 years and that certain species (e.g., red lobster and possibly large clams) have been extirpated or reduced to such low levels that they are no longer economically viable. The reduction in fisheries productivity due to overfishing, habitat destruction, and pollution is not unique to Guerrero but is a problem facing all of Mexico's coastal populations. The economic importance of this fishery for Mexico has stimulated new

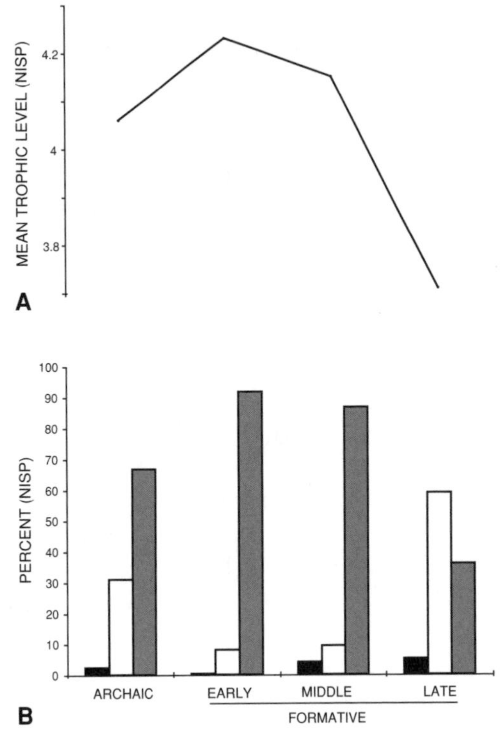

FIGURE 5.5. Relationship between mean trophic level of fish (A) and the percentage of fish represented from low (2.00–2.99), medium (3.0–3.99), and high (4.0–4.5) trophic levels (B). NISP, number of identified specimens. Values are volumetrically corrected as in Figure 5.4. Trophic levels: ■ 2.0–2.99; □ 3.0–3.99; ▩ 4–4.5.

legislation and the formation of a government institution to develop more sustainable fisheries that balance the health of marine ecosystems and coastal environments with rates of production.

The severe human impacts evident in the last 10 years occur within a broader historical context that likely begins during the terminal Pleistocene or Holocene, but with the first visible manifestations during the Middle Holocene at ~5500 cal BP. This is followed by clear demographic expansion in the region during the Formative, Classic, and Postclassic periods. Settlements increased in size, people developed a greater reliance upon maize-based food production, and small and medium-sized settlements were incorporated into increasingly larger political systems ruled by an elite class. By the Classic period these small polities were heavily influenced by the large, state-level societies that had developed in the central highlands of Mexico—Teotihuacán and Monte Albán—and sociopolitical integration culminated with the incorporation of these polities into the Aztec Empire. Trade flourished with increasing societal sophistication and integration, and the use of marine resources expanded from local subsistence use to items sought after in increasingly larger systems of trade and tribute. The best evidence for the imperial extraction of marine resources from coastal Guerrero comes from the remaining Aztec tribute codices. Given the limited amount of archaeological work done to address this question in coastal Guerrero, our best glimpse comes from the archaeological record from Puerto Marqués dating from the Archaic and Formative periods. We offer this work as a first step toward creating a historical and environmental framework to improve our understanding of the impacts of human fishing on marine ecosystems in this region. The rich and diverse character of the faunal assemblages at Puerto Marqués most immediately provides a sobering view of the accumulated impacts that expanding human populations have had in this region.

At the local level, six broad trends are visible in the faunal assemblage at Puerto Marqués from the Late Archaic to the Late Formative. First, there is a continuous dominance of open-water fish species, overwhelmingly skipjack tuna, until the Late Formative period, when there is a significant decrease of these high trophic level species. Second, sea turtles are best represented in the earliest levels, decrease in frequency through time, and drop out of the record in the Late Formative period. Third, there is a general shift away from the dominance of high trophic level open-water fish starting in the Middle Formative, with expanding use of lower trophic level species from a range of habitats into the Late Formative. Fourth, a parallel increase in the frequency of fish most commonly found in estuarine habitats suggests an increased reliance on a greater variety of habitats from a larger catchment area and an expanding subsistence base. Fifth, decreases in the overall density of shellfish parallel increases in mollusk diversity. Sixth, the variety of mollusk species collected increases through time, with the largest difference occurring between the aceramic Late Archaic period levels and the overlying ceramic-bearing Formative period levels. We argue that the changes evident in the faunal assemblage during this time are largely the product of human predation, but further work is needed to explore the role of environmental change. Regardless, given the small size of the population at Puerto Marqués during the Late Archaic and Formative periods (~100–500 people), our work shows the highly sensitive nature of at least some marine ecosystems to the cascading effects of human predation.

ACKNOWLEDGMENTS

We thank Arqlgo. Cuauhtémoc Reyes Alvarez (Central Regional Guerrero del Instituto Nacional de Antroplogía e Historia) for sharing his knowledge about the region and facilitating our research, Dr. Rubén Manzanilla López for generously sharing his unpublished dissertation and insight about the prehistory of Puerto Marqués, and Martha Eugenia Cabrera Guerrero for sharing her environmental and archaeological knowledge of the region. Ing. Paul Rangel Merkley kindly provided access to the Puerto Marqués site, whereas Ing. Víctor Hugo Martínez and Ing. Francisco Rodríguez from Desarrollo Integral de Ingenería SAdeCV (DEIN) aided us in innumerable ways. The vertebrate assemblage from Puerto Marqués was analyzed at the Zooarchaeology Laboratory at the Cotsen Institute of Archaeology (CIOA) at UCLA. Identifications were confirmed using comparative vertebrate materials housed in the CIOA, the UCLA Department of Biology, and the Los Angeles County Museum of Natural History. Finally, special thanks to our field crew: Cassy Albush, Jorge Morales, Amparo Robles, and Nathan Wilson. This work was funded by the National Science Foundation (BCS-0211215).

–1. SAGARPA establishes and enforces regulations in the "Ley de Pesca," its "Reglamento de la Ley de Pesca," and the "Normas oficales Mexicanas" (see http://www.cddhcu.gob.mx).

REFERENCES CITED

Acuña, R.
1987 Relación de la Villa de Zacatula, Relaciones Geográficas del Siglo XVI: México, Tomo Noveno, pp. 437–462. Instituto de Investigaciones Antropológicas, UNAM, México City.

Alvarez del Toro, M.
1983 *Los Reptiles de Chiapas.* Instituto de Historia Natural, Tuxtla Gutiérrez, México.

Anuario Estadístico de Pesca
2003 www.sagarpa.com.mx/conapesca/planeacion/anuario/anuario2003.pdf, accessed August 2006.

Barlow, R.
1990 *Los Mexicas y la Triple Alianza,* Vol. III, edited by J. Monjarás-Ruiz, E. Limón, and M. De la Cruz. Instituto Nacional de Antropología and Universidad de las Américas, Puebla, Mexico.

Berdan, F. F.
1996 The Tributary Provinces. In *Aztec Imperial Strategies,* edited by F. F. Berdan, R. E. Blanton, E. H. Boone, M. G., Hodge, M. E., Smith, and E. Umberger, pp. 115–135. Dumbarton Oaks, Washington, D.C.

Berger, R.
1983 Sea Levels and Tree-Ring Calibrated Dating. In *Quaternary Coastlines and Marine Archaeology: Toward the Prehistory of Land Bridges and Continental Shelves,* edited by P. M. Masters and N. C. Flemming, pp. 51–61. Academic Press, New York.

Binford, L. R.
1968 Post-Pleistocene Adaptations. In *New Perspectives in Archaeology,* edited by S. R. Binford and L. R. Binford, pp. 313–341. Aldine, Chicago.

Brush, C. F.
1969 A Contribution to the Archaeology of Coastal Guerrero, Mexico. Unpublished Ph.D. dissertation, Department of Anthropology, Columbia University.

Cabrera Guerrero, M. E.
1990 Los Pobladores prehispánicos de Acapulco. Colección Cientifica, Serie Arqueología. Instituto Nacional de Antropología e Historía, Mexico City.

Carta Nacional Pesquera
2006 www.sagarpa.gob.mx/conapesca/ordenamiento/carta nacional pesquera/cnp.htm, accessed August, 2006.

Clark, J. E.
1981 The Early Preclassic Obsidian Industry of Paso de la Amada, Chiapas, Mexico. *Estudios de Cultura Maya* 13:265–284.

Clark, J. E., and M. Blake
1994 Power and Prestige: Competitive Generosity and the Emergence of Rank in Lowland Mesoamerica. In *Factional Competition and*

Political Development in the New World., edited by E. M. Brumfiel and J. W. Fox. pp. 17–30. Cambridge University Press, Cambridge.

Crawford, M. H.
1998 *The Origins of Native Americans: Evidence from Anthropological Genetics.* Cambridge University Press, Cambridge.

Curray, J. R., F. J. Emmel, and P. J. Crampton
1969 Holocene History of a Strand Plain Lagoonal Coast, Nayarit, Mexico. In *Lagunas Costeras: Un Simposio,* edited by A. Ayala Castañares and F. Phleger, pp. 63–100. Universidad Autónoma de México, México, D. F.

Denevan, W. M.
1992 *The Native Populations of the Americas in 1492.* University of Wisconsin Press, Madison.

Diamond, J.
2005 *Collapse: How Societies Choose to Fail or Succeed.* Viking, New York.

Doebley, J.
1990 Molecular Evidence and the Evolution of Maize. *Economic Botany* 44(3 Supplement):6–27.

Doolittle, W. E.
1990 *Canal Irrigation in Prehistoric Mexico: The Sequence of Technological Change.* University of Texas Press, Austin.

Earthtrends
2006 Earthtrends, http://earthtrends.wri.org, accessed July 2007.

Ekholm, G. F.
1948 Ceramic Stratigraphy at Acapulco, Guerrero. IV *Reunión de Mesa Redonda: Sociedad Méxicana de Antropología,* México City. El Occidente de México, pp. 95–104.

Fairbanks, R. G.
1989 A 17,000-year Glacio-eustatic Sea Level Record: Influence of Glacial Melting Rates on the Younger Dryas Event and Deep Ocean Circulation. *Nature* 342:637–641.

Fedick, S. (editor)
1996 *The Managed Mosaic: Ancient Maya Agriculture and Resource Use.* University of Utah Press, Salt Lake City.

Feinman, G. M., and J. Marcus (editors)
1998 *Archaic States.* School of American Research, Santa Fe, New Mexico.

Goman, M., A. Joyce, and R. Mueller
2005 Stratigraphic Evidence for Anthropogenically Induced Coastal Environmental Change from Oaxaca, Mexico. *Quaternary Research* 63:250–260.

González-Quintero, L., and J. Mora-Echeverría
1978 Estudio arqueológico-ecologico de un caso de explotación de recursos litorales en el Pacífico Mexicano. In *Arqueobotánica (Métodos y Aplicaciones),* edited by F. Sanchez Martínez, pp. 51–66, Colección Científica núm. 63, INAH, México City.

Grayson, D. K.
2001 The Archaeological Record of Human Impacts on Animal Populations. *Journal of World Prehistory* 15:1–68.

Grayson, D. K., and D. J. Meltzer
2002 Clovis Hunting and Large Mammal Extinction: A Critical Review of the Evidence. *Journal of World Prehistory* 16:313–359.

Guzmán F., A., and O. J. Polanco
2000 *Los Peces Arqueológicos de la Ofrenda 23 el Templo Mayor de Tenochtitlan.* Serie Arqueologiía. Instituto Nacional de Antropología e Historia, México City.

Hodell, D. A., M. Brenner, and J. H. Curtis
2000 Climate Change in the Northern American Tropics since the Last Ice Age: Implications for Environment and Culture. In *Imperfect Balance: Landscape Transformations in the Precolumbian Americas,* edited by D. L. Lentz, pp. 13–38. Columbia University Press, New York.

Jackson, J. B. C., M. X. Kirby, W. H. Berger, K. A. Bjorndal, L. W. Botsford, B. J. Bourque, R. H. Bradbury, R. Cooke, J. Erlandson, J. A. Estes, T. P. Hughes, S. Kidwell, C. B. Lange, H. S. Lenihan, J. M. Pandolfi, C. H. Peterson, R. S. Steneck, M. J. Tegner, and R. R. Warner
2001 Historical Overfishing and the Recent Collapse of Coastal Ecosystems. *Science* 293: 629–638.

Keen, A. M.
1971 *Seashells of Tropical West America: Marine Mollusks from Lower California to Peru.* Stanford University Press, Stanford, California.

Kennett, D. J., B. Voorhies, J. Iriate, J. G. Jones, D. Piperno, M. T. Ramírez Herrera, and T. A. Wake
2004 *Avances en el Proyecto Arcaico-Formativo: Costa de Guerrero.* Technical report submitted to the Instituto Nacional de Antropología e Historia, Mexico.

Kennett, D. J., B. Voorhies, and D. Martorana
2006 An Ecological Model for the Origins of Maize-Based Food Production on the Pacific Coast of Southern Mexico. In *Behavioral Ecology and the Transition to Agriculture,* edited by D. J. Kennett and B. Winterhalder, pp. 103–136. University of California Press, Berkeley.

Lentz, D. L.
2000 Anthropocentric Food Webs in the Precolumbian Americas. In *Imperfect Balance: Landscape Transformations in the Precolumbian Americas,* edited by D. L. Lentz, pp. 89–119. Columbia University Press, New York.

Litvak, J.

1971 *Cihuatlán y Tepecoacuilco, Provincias Tributarias de México en el Siglo XVI.* Instituto de Investigaciones Históricas, Serie Antropológica 12. Universidad Nacional Autónoma de México, México City.

McGoodwin, J. R.

1990 *Crisis in the World's Fisheries: People, Problems, and Policies.* Stanford University Press, Stanford, California.

1994 Nowadays, Nobody Has Any Respect: The Demise of Folk Management in a Rural Mexican Fishery. In *Folk Management in the World's Fisheries: Lessons for Modern Fisheries Management,* edited by C. L. Dyer and J. R. McGoodwin, pp. 43–54. University Press of Colorado, Niwot.

2001 *Understanding the Cultures of Fishing Communities: A Key to Fisheries Management and Food Security.* Food and Agriculture Organization of the United Nations, Rome.

Manzanilla López, R. A.

2000 La región arqueológica de la costa grande de guerrero: su definición a través de la organización social y territorialidad prehispánicas. Tesis doctorado en antropología, Escuela Nacional de Antropología e Historia, Mexico City.

Manzanilla López, R., A. Talavera González, and E. Rodriquez Sanchez

1991 *Informe técnico de campo de la primera etapa del Proyecto de Investigación y Salvamento Aqueológico en Puerto Marquéz, Estado Guerrero.* Entregrado al INAH Mexicanos y Centroamericanos, Gobierno del Estado de Guerrero and Instituto Nacional de Antropología e Historia, México City.

Matsuoka, Y., Y. Vigouroux, M. M. Goodman, J. Sanchez, E. Buckler, and J. Doebley

2002 A Single Domestication for Maize Shown by Multilocus Microsatellite Genotyping. *Proceedings of the National Academy of Sciences* 99: 6080–6084.

Meyer, M. C., and W. H. Beezley

2000 *The Oxford History of Mexico.* Oxford University Press, New York.

Meyer, M. C., and W. L. Sherman

1991 *The Course of Mexican History.* Oxford University Press, New York.

Miller, R. R.

1985 *Mexico: A History.* University of Oklahoma Press, Norman.

Mohar, L. M.

2002 Tributos Guerrerenses a los Señores de Tenochtitlan. In *El Pasado Arqueológico de Guerrero,* edited by C. Niederberger and R. Reyna, pp. 505–531. Centro Francés de Estudios Mexicanos y Centroamericanos, Gobierno del Estado de Guerrero and Instituto Nacional de Antropología e Historia, México City.

Morales, A., and E. Rosello

2004 Fishing down the Food Web in Iberian Prehistory? A New Look at the Fishes from Cueva de Nerja (Malaga, Spain). In *Pettis Animaux Et Societes Humaines Du Complement Alimentaire Aux Ressources Utilitaires* 24, edited by J. Brugal and J. Desse, pp. 111–123. APDCA, Antibes.

Moseley, M.

1975 *The Maritime Foundations of Andean Civilization.* Cummings, Menlo Park, California.

Pauly, D., and V. Christensen

1995 Primary Production Required to Sustain Global Fisheries. *Nature* 374:255–257.

Pauly, D., V. Christensen, J. Dalsgaard, R. Froese, and F. Torres, Jr.

1998 Fishing down Marine Food Webs. *Science* 279:860–863.

Pauly, D., V. Christensen, R. Froese, and M. L. Palomares

2000 Fishing down Aquatic Food Webs. *American Scientist* 88:46–51.

Piperno, D. R.

2006 The Origins of Plant Cultivation and Domestication in the Neotropics: A Behavioral Ecological Perspective. In *Behavioral Ecology and the Transtion to Agriculture,* edited by D. J. Kennett and B. Winterhalder, pp. 137–166. University of California Press, Berkeley.

Piperno, D. R., and D. M. Pearsall

1998 *The Origins of Agriculture in the Lowland Neotropics.* Academic Press, San Diego.

Ramírez-Herrera M. T., and J. Urrutia-Fucugauchi

1999 Morphotectonic Zones along the Coast of the Pacific Continental Margin, Southern Mexico. *Geomorphology* 28:237–250.

Reitz, E. J.

2004 "Fishing down the Food Web": A Case Study from St. Augustine, Florida, USA. *American Antiquity* 69:63–83.

Reitz, E. J., and E. S. Wing

1999 *Zooarchaeology.* Cambridge University Press, Cambridge.

Relación de Zacatula

1945 Relación de Zacatula, 1580. *Tlalocan* 2: 258–268.

Sedor, M. D.

2005 Interpretations of Long Term Tectonic Deformation in and around the Guerrero Seismic Gap, Mexico. Master's thesis, Department of Geological Sciences, California State University, Long Beach.

Shannon, C. E., and W. Weaver

1949 *The Mathematical Theory of Communication.* University of Illinois Press, Urbana.

Smith, B. D.

2001 Low-Level Food Production. *Journal of Archaeological Research* 9:1–43.

Smith, M. E.
1996 The Strategic Provinces. In *Aztec Imperial Strategies*, edited by F. F. Berdan, R. E. Blanton, E. H. Boone, M. G. Hodge, M. E. Smith, and E. Umberger, pp. 115–135. Dumbarton Oaks, Washington, D.C.

Smith, M. E., and F. F. Berdan
1996 Appendix 4: Province Descriptions. In *Aztec Imperial Strategies*, edited by F. F. Berdan, R. E. Blanton, E. H. Boone, M. G., Hodge, M. E., Smith, and E. Umberger, pp. 115–135. Dumbarton Oaks, Washington, D.C.

Velázquez Castro, A.
1999 *Tipología de los objetos de concha del Templo Mayor de Tenochtitlan*. Serie Historia. Instituto Nacional de Antropología e Historia, Mexico City.
2000 *El Simbolismo de los Objetos de Concha Encontrados en las Ofrendas del Templo Mayor de Tenochtitlan*. Serie Arqueología. Instituto Nacional de Antropología e Historia, Mexico City.

Voorhies, B.
2004 *Coastal Collectors in the Holocene: The Chantuto People of Southwest Mexico*. University Press of Florida, Gainesville.

Webster, D.
2002 *The Fall of the Ancient Maya: Solving the Mystery of the Maya Collapse*. Thames and Hudson, New York.

Wing, E. S.
2001 The Sustainability of Resources Used by Native Americans on Four Caribbean Islands. *International Journal of Osteoarchaeology* 11:112–126.

6

Ancient Fisheries and Marine Ecology of Coastal Peru

Elizabeth J. Reitz, C. Fred T. Andrus, and Daniel H. Sandweiss

RECENT RESEARCH, EXEMPLIFIED by chapters in this volume, documents the profound impact of people on marine fisheries and ecology in many areas (e.g., Lauwerier and Plug 2004; Pauly 1995; Pauly and Christensen 1995; Pauly et al. 1998, 2000). Although we do not disagree with evidence for the human role in precipitating some environmental changes, the Peruvian case suggests that in some instances evidence for the impact of ancient fisheries on marine ecology is subtle and should be evaluated in the context of other environmental forces. Climate models, zooarchaeology, and stable isotopes highlight the antiquity of marine resource use in coastal Peru, the importance of the sea as the primary source of animal protein in the human diet throughout the Early and Middle Holocene, and the role of oceanic and atmospheric dynamics in the resilience of this ecosystem in the face of human fishing pressure.

Peru has one of the longest archaeological fishery records in the Americas. Exploitation of marine invertebrates and vertebrates began during the Terminal Pleistocene (e.g., Sandweiss et al. 1998) and continues today. Such a record is also found in Chile and Ecuador (e.g., Jerardino et al. 1992; Llagostera 1979, 1992; though see Núñez et al. 1994; Reitz and Masucci 2004; True 1975). In contrast with regions in which human fishing strategies altered marine ecosystems, multiple lines of evidence suggest that the biotic effects of natural variability in oceanic and atmospheric forces moderated anthropogenic effects in Peru. The region is characterized by changes in the frequency of El Niño–Southern Oscillation (ENSO) which influence marine and terrestrial productivity in complex but significant ways. ENSO itself operates in the presence of tremendous productivity supported by the Peru Current upwelling system. This productivity does not appear to have been severely affected by fishing pressure until the twentieth century. Even now, some argue that natural, ENSO-created variability damages the fishery as much as does industrial fishing methods, perhaps more so (e.g., Barber and Chavez 1977, 1983, 1986; Chavez et al. 2003). Thus, the impact of ancient fisheries on the marine ecosystems of Peru must be considered in the context of oceanic and atmospheric forcing, as well as human behavior. In this chapter, the environmental background of Peru's marine ecosystem, the antiquity and economic significance of the fishing tradition, human impact on marine ecosystems, and the biotic effects of variability in ENSO and the Peru Current are reviewed, with emphasis on archaeological evidence from the Early Preceramic (13,000–11,000 cal BP) through the Initial (4200–2800 cal BP) periods.

THE PERU CURRENT AND ENSO

The Peru Current and ENSO dominate both terrestrial and marine ecosystems of the Peruvian Coast today. To assess the impact of anthropogenic and nonanthropogenic forces on the marine ecosystems of Peru, the impact of ancient fisheries as well as of ENSO and the oceanic conditions within which ENSO operates must be considered. Though both the Peru Current and ENSO vary through time, evidence suggests that their combined influence has deep roots. These forces act on multiple time scales (Chavez et al. 2003).

The coastal upwelling ecosystem supported by the strong flow of the Peru Current is one of the most productive in the world (Bakun 1996; Briggs 1974:137, 1995:253; Idyll 1973; Santander 1980; Schweigger 1964). During most years, the cool Antarctic Peru Current flows north along the coast of Chile and Peru, turning westward from the coast near the Peru-Ecuador border. Typically Peruvian waters are temperate, with cool, nutrient-rich waters upwelling near shore in response to the Peru Current and steady trade winds. One of the world's richest fisheries is supported by nutrients brought to the surface by cold benthic waters rising from the ocean floor in response to prevailing winds and currents.

In contrast to the biological abundance of the sea, terrestrial resources are limited. The combination of the Andean rain shadow and the Peru Current contributes to one of the world's driest coastal deserts, with a relatively narrow seasonal cycle of sea and terrestrial temperatures. The only significant, quasi-permanent freshwater sources are braided streams draining high-altitude glaciers in the Andes.

Thus, the primary characteristic of the Peruvian Coast is the exceptionally rich marine ecosystem in which the upwelled nutrients support an abundant quantity of zooplankton and phytoplankton. A characteristic complex of invertebrates and vertebrates feed on these microorganisms (Chirichigno 1982; Hildebrand 1945; Reitz 1988a, 1988b; Reitz and Sandweiss 2001; Schweigger 1964). Typical Peru Current invertebrates include echinoderms (Echinodermata), chitons (Polyplacophora), scallops and mussels (e.g., *Argopecten purpuratus, Aulacomya* spp., *Choromytilus chorus, Perumytilus purpuratus, Semimytilus algosus*), clams (e.g., *Mesodesma donacium, Protothaca thaca, Donax obesulus*), gastropods (e.g., *Fissurella* spp., *Acmaea* spp., *Thais* spp.), and crustaceans (e.g., *Balanus* spp.). Typical Peru Current vertebrates include whales and dolphins (Cetaceans), sea lions and seals (Pinnipedia), pelicans (*Pelecanus* spp.), cormorants (*Phalacrocorax* spp.), boobies (*Sula* spp.), herrings (Clupeidae), and anchovies (Engraulidae). In addition, some members of other fish families are common along the Peruvian Coast, particularly jacks (Carangidae), grunts (Haemulidae), drums (Sciaenidae), and mackerels (Scombridae). Members of these groups are found in both inshore and offshore waters; most are typical of nearshore waters associated with rocky headlands, sandy beaches, open littoral, and the lower reaches of the few intermittent or perennial streams that occur along the coast. The precise complex of organisms accessible from each archaeological site depends on the latitude of the site, local topography and geomorphology, coastline orientation, the width of the continental shelf, and the tectonic history of the specific locality (e.g., Navarrete et al. 2005). The resources available at each site also vary through time in response to long-term variations in ENSO cycles (Reitz and Sandweiss 2001; Sandweiss 2003; Sandweiss et al. 2004).

The dry, temperate climate is punctuated at irregular intervals by complex oceanic and atmospheric events referred to jointly under the term "ENSO." Often, "El Niño" refers to the oceanic component and "Southern Oscillation" to the atmospheric one, with "ENSO" encompassing the complementary El Niño (warm events) and La Niña (cool events) states in this natural mode of oscillation (Fedorov and Philander 2000). At irregular intervals, El Niño events depress the thermocline in the eastern equatorial Pacific, resulting in coastal warming in Peru and northern Chile and a notable decrease in nearshore biological productivity. The trade winds slacken or reverse during the warming phase of the ENSO cycle, and rain falls along the desert coast and western slopes of the Andes. Today, warming events recur at 2- to 15-year intervals and persist from several months to over a year. During this time, tropical surface water displaces cool, upwelled water upon which local ecosystems rely. At such times, sea surface temperatures (SST) can reach 28 to 29°C, exceeding "normal" temperatures by 5°C or more at some locations (Chavez et al. 1999). Mobile temperate organisms die or migrate to cooler waters, immobile organisms die or decline dramatically, and floods inundate coastal deserts. El Niño conditions contrast with those of La Niña, in which local coastal waters are cooler than average; these may drop as much as 8°C (Chavez et al. 1999). The phases of ENSO cycles are associated with striking changes in biological productivity (Chavez et al. 1999; Chavez et al. 2003). At the same time, other large-scale changes occur on different time scales.

Though chronologies vary, many proxies indicate changes in frequency and intensity of the ENSO cycle from the Terminal Pleistocene to the present and its impact on marine ecosystems throughout the Pacific Basin (e.g., Trenberth and Otto-Bliesner 2003). These include, among others, lake records from the Ecuadorian highlands (Moy et al. 2002; Rodbell et al. 1999), central Chile (Jenny et al. 2002; Maldonado and Villagrán 2002), the Galapagos (Riedinger et al. 2002), Australia (McGlone et al. 1993), and New Zealand (Shulmeister and Lees 1995); Andean ice cores (Thompson et al. 1985, 1995); corals from Australia (Gagan et al. 1998) and New Guinea (Tudhope et al. 2001); eastern equatorial Pacific deep-sea cores (Loubere et al. 2003); and Peruvian coastal flood records (Fontugne et al. 1999; Keefer et al. 2003). The frequency and impact of ENSO also can be tracked for the most recent 13,000 years through analysis of marine organisms in archaeological sites (e.g., Andrus et al. 2002a; Rollins et al. 1986a, 1986b; Sandweiss 2003; Sandweiss et al. 1996, 1999, 2001a, 2004). These proxy data may be summarized as follows: (1) prior to 9000 cal BP, ENSO operated, but information on the frequency of that variation is not available; (2) from 9000 to 5800 cal BP, ENSO cycled at very long intervals, and coastal waters in northern Peru were generally warm compared to today; (3) from 5800 to 3000 cal BP, ENSO cycled less frequently than it does today, while the sea off northern Peru was cool; and (4) since 3000 cal BP, ENSO has varied within the range of historically known frequencies, and coastal waters generally are cool.

For evidence of environmental change unrelated to human behavior one need look no further than the devastating effects of ENSO in 1982–1983 and 1997–1998 on primary productivity and on invertebrate and vertebrate populations. These well-studied nonanthropogenic impacts are orders of magnitude greater than any known human impact. In fact, it took the 1972 ENSO combined with a decade or more of intensive industrial fishing to depress the anchoveta *(Engraulis ringens)* fishery, which rebounded reasonably well between the 1982–1983 and 1997–1998 events.

ENSO is a powerful and overwhelming factor in the Peru-Chilean Province and in Peruvian history. Although it undoubtedly is true that predator-prey relationships such as those between sea otters and sea urchins in the California kelp forest exist in the zoogeographic Peru-Chilean Province defined by the Peru Current, the nutrient upwelling associated with the current is the major ecological driver at the base of the food chain (Schweigger

1964). This source of nutrients supports all higher organisms, including brown (Phaeophyceae) and red (Rhodophyceae) algae (Schweigger 1964:195–198) and seabirds such as cormorants (*Phalacrocorax gaimardi*; Zavalaga et al. 2002). Species requiring reliable cool waters are vulnerable to direct and indirect impact, and their populations are likely limited by warm El Niño water temperatures. For example, the intertidal kelp *(Lessonia nigrescens)* experienced massive mortality as a result of the ENSO event of 1982–1983 in northern Chile, and recovery of this population was very slow (Martínez et al. 2003). Species that prefer continental shelf habitats are further limited by the size and availability of such habitats in Peru (Richardson 1981). With the possible exception of recent human behavior, people could not disrupt the ENSO cycle or directly access the nutrients borne by the upwelling. However, when the upwelling is disrupted by the ENSO cycle, disaster strikes every organism that relies upon it. Thus the driver in this province is more clearly physical and chemical than biological. If one were unaware of the impact of long-term variability in ENSO, it would be simple to conclude that the changes in the faunal record seen in these data were due to biological dynamics within the ecosystem, human impact on the Peruvian ecosystem, or to changes in cultural patterns unrelated to environmental factors. Yet, as will be shown below, the most important changes in the Peruvian fishery can be linked to climatic change. This complicates interpretations of the impact of people on marine fisheries and ecology in this area.

THE ANTIQUITY OF THE FISHING TRADITION

As on other maritime coasts, the archaeological record of western South America is biased against coastal sites prior to sea level stabilization around 6,000 calendar years ago (Richardson 1981). Surviving evidence of maritime-oriented sites is most likely to be associated with a narrow, steep continental shelf where the rising sea caused relatively little horizontal displacement of the shoreline. In Peru, the shelf is narrowest near Talara and from the Paracas Peninsula south into northern Chile. Field work in these two regions has located many maritime-adapted sites predating sea level stabilization. Four of these sites were occupied between 13,000 and 11,000 cal BP and yield clear evidence for the antiquity and extent of the Peruvian fishing tradition (Table 6.1).

Peru's early cultural history is divided into Preceramic and Ceramic stages. The Preceramic stage is subdivided into Early (13,000–9000 cal BP), Middle (9000–5800 cal BP), and Late (5800–4200 cal BP) periods based upon a variety of criteria (see Benfer 1984; Keefer et al. 1998; Pozorski and Pozorski 1987; Quilter 1991; Sandweiss 1996b; Sandweiss et al. 1989, 1998). Early and Middle Preceramic coastal sites have neither cotton nor ceramics. Although by definition Late Preceramic sites lack pottery, many have evidence of intensive agriculture that focused on industrial crops such as gourds and cotton, associated with monumental architecture (e.g., Béarez and Miranda 2000; Bonavia 1982; Haas et al. 2004; Pozorski 1983; Pozorski and Pozorski 1979a; Quilter 1991; Quilter et al. 1991; Sandweiss et al. 2001b; Shady Solís et al. 2001). The Late Preceramic Period is also known as the Cotton Preceramic (Engel 1957:138–142; Moseley 1975:21–22). The subsequent Initial Period (4200–2800 cal BP; e.g., Pozorski and Pozorski 1979b, 1987, 1992) is characterized by ceramics, agriculture, monumental structures, and many other aspects of a complex cultural life.

Reliable quantitative estimates of demographic trends during this time are not available for the Peruvian Coast (for one attempt, see Rick 1987). The general trend is for population to grow over the millennia from what must have been small groups of semisedentary fishers to much larger communities by the fifteenth-century Spanish conquest, when individual coastal valleys had populations often numbering in the tens of thousands (Cook 1981).

TABLE 6.1

Percentage of Vertebrate Minimum Number of Individuals (MNI) from Marine and Terrestrial Habitats

SITE (LATITUDE)	% MARINE	% TERRESTRIAL	MNI	CAL BP
Southern Peru-Chilean Province (17°–12° S)				
Quebrada Tacahuay	91.7	8.3	72	12,750–11,950
Ring Site	99.1	0.9	436	11,400–5850
Quebrada Jaguay	83.9	16.1	248	13,050–8250
Paloma Probability Samples	97.7	2.3	132	8500–5450
Northern Peru-Chilean Province (9°–8° S)				
Cardal 1/4-inch Samples	85.3	14.7	34	3250–2800
Almejas 1/4-inch Samples	93.0	7.0	114	7500
Pampa de las Llamas 1/4-inch Samples	87.0	13.0	77	4500–3500
Ostra Base Camp	100.0	–	144	7100–6200
Alto Salaverry	100.0	–	64	4950–4050
Paiján Complex	16.2	83.8	761	12,250–9150
Panamanian Province (4°–2° S)				
Sitio Siches				
Honda Phase	99.6	0.4	512	5850–5150
Siches Phase	99.8	0.2	2195	7900–6750
Amotape Phase	92.0	8.0	50	10,650–10,050
Real Alto	94.9	5.1	236	Middle Valdivia
Las Vegas	42.8	57.2	145	11,400–7,450

NOTE: Dates from Benfer (1984), Chauchat (1988, 1992), Sandweiss et al. (1983, 1989, 1996), and Stothert (1988), calibrated via Calib 5.0 (Hughen et al. 2004; McCormac et al. 2004; Reimer et al. 2004; Stuiver and Reimer 1993). Age ranges from means of oldest and most recent ^{14}C dates where available, discounting outliers rejected by the excavators. No ^{14}C dates are available for Alto Salaverry, dated by association with similar sites and artifacts. Faunal data, excluding human and invertebrate MNI, from Byrd (1976, 1996), Chase (1988), deFrance et al. (2001), Pozorski and Pozorski (1979a), Reitz (1987, 1988a, 1988b, 1995, 1999, 2003, 2005), Reitz and Cannzaroni (2004), Reitz and Sandweiss (2001), Sandweiss et al. (1989), and Wing (1986). Data for Las Vegas omits earlier, Pre-Vegas deposits.

Likewise, patterns of residential mobility and sedentism remain unresolved. Perhaps during the Terminal Pleistocene small groups of coastal fishing communities were seasonally migratory, at least within the coastal sector. Although there are indications that early coastal populations migrated between the coastal setting and the highlands (Sandweiss et al. 1998), for most of Peruvian history coastal populations were sedentary. Even under highland influence or domination by the Wari and Inca (ca. AD 600–1000 and AD 1470–1532 (or 1350–950 and 480–418 BP), most coast-dwellers were the traditional peoples of the shore with their specialized knowledge of coastal resource extraction and traditional claims to fishing grounds (see, for instance, Rostworowski 1981; Sandweiss 1992). The complexity of making a living safely and reliably from the sea calls upon tools, skills, and knowledge that are incompatible with part-time fishing. Evidence of extensive trade networks linking various parts of the Andean region appears very early in the Peruvian sequence, suggesting that exchange rather than residential mobility was the preferred way to obtain resources not locally available (e.g., deFrance 2005; Pozorski 1983; Pozorski and Pozorski 1979b, 1987; Reitz 1988a; Sandweiss 1996a; Shady Solís et al. 2001; Wing 1972:332, 1986, 1992), and isotopic data reviewed below likewise suggest year-round occupation at Siches and Ostra.

A fishing tradition combined with the use of a few, mainly small, terrestrial animals is the oldest verifiable protein-acquisition strategy in coastal Peru (Table 6.1). Although some continue to argue on theoretical grounds that these were hunting economies focused on large terrestrial animals, or at least fishing economies that grew out of earlier coastal (or interior) hunting traditions, the antiquity and intensity of fishing traditions is evident at all of the oldest coastal sites for which vertebrate data are available. These sites include Quebrada Tacahuay (deFrance et al. 2001; Keefer et al. 1998), the Ring Site (Sandweiss et al. 1989), Quebrada Jaguay (Reitz 2005; Sandweiss et al. 1998), and the complex of sites known as Paiján (Chauchat 1988, 1992; Wing 1986). The early levels of these sites date at least between 13,000 and 11,400 cal BP, if not earlier, and they extend from the southernmost boundary of Peru to the northernmost (Figure 6.1). Theories relying upon a migratory highland-coastal terrestrial hunting tradition lack zooarchaeological support. Although other indicators point to coastal-highland connections at some sites, we do not know if these are evidence of migrations or of exchange networks.

Except for the Paiján example, echinoderms, mollusks, crustaceans, fishes, seabirds, and some marine mammals were used to the exclusion of almost all terrestrial animals. The few terrestrial animals in collections from these sites are primarily lizards, small birds, and mice. Although terrestrial plant remains are limited at these sites, they are present at least at Quebrada Tacahuay and Quebrada Jaguay (deFrance et al. 2001; Sandweiss et al. 1998). It is presumed also that marine plants, including algae, were used throughout the archaeological sequence as they were during Inca times as well as today (Sandweiss 1992:125–126; Schweigger 1964:195–197; Zavalaga et al. 2002). Direct archaeological evidence for this use will necessarily be limited.

The Paiján sites are an interesting exception in that they document the extent to which people used marine resources even when the coast was relatively distant. Although most early sites lie within 3 km of the modern shore, the Paiján sites are 14 to 36 km from the present-day coastline. Because the continental shelf is relatively wide in the Paiján region, these sites were even further from the shore when they were occupied. Despite the distance, marine animals comprise 16 percent of the MNI (minimum number of individuals), testifying to the presence of the maritime tradition at locations not directly on the coast. Among the few terrestrial animals identified in the Paiján collections, lizards comprise 70 percent of the individuals, while deer make up less than 1 percent (Wing 1986).

Data from other sites demonstrate that this tradition extended throughout the Holocene (e.g., Reitz 2004). Faunal data from Middle Preceramic sites such as Quebrada de los Burros (Béarez 2000; P. Béarez, personal communication 2005), Paloma (Reitz 1988a, 1988b, 2003), Ostra Base Camp (Reitz 2001; Reitz and Sandweiss 2001), and Sitio Siches (Reitz and Cannarozzi 2004) document intense use of marine resources. The Quebrada de los Burros study focused on fishes, but most of the other vertebrates are marine birds rather than terrestrial animals (P. Béarez, personal communication 2005). The Paloma, Ostra, and Siches studies include all vertebrate classes to the extent that they were present in the samples studied. Essentially 100 percent of the vertebrate individuals are from the sea (Table 6.1). Quebrada de los Burros, Paloma, and Siches all contain evidence of domesticated plants, but the emphasis on marine animals continued unabated (Lavallée et al. 1999; D. Piperno, personal communication 2004; Piperno and Pearsall 1998:271; see Piperno and Stothert [2003] for examples from Ecuador).

Over 80 percent of the vertebrate individuals are marine even at Late Preceramic and Initial Period sites with more abundant evidence for domesticated plants, such as Alto Salaverry (Pozorski and Pozorski 1979a), Los Gavilanes (Bonavia 1982; Wing and Reitz 1982), Cardal (Burger and Salazar-Burger 1991; Reitz 1987; Umlauf 1993), El Paraíso (Quilter and Stocker

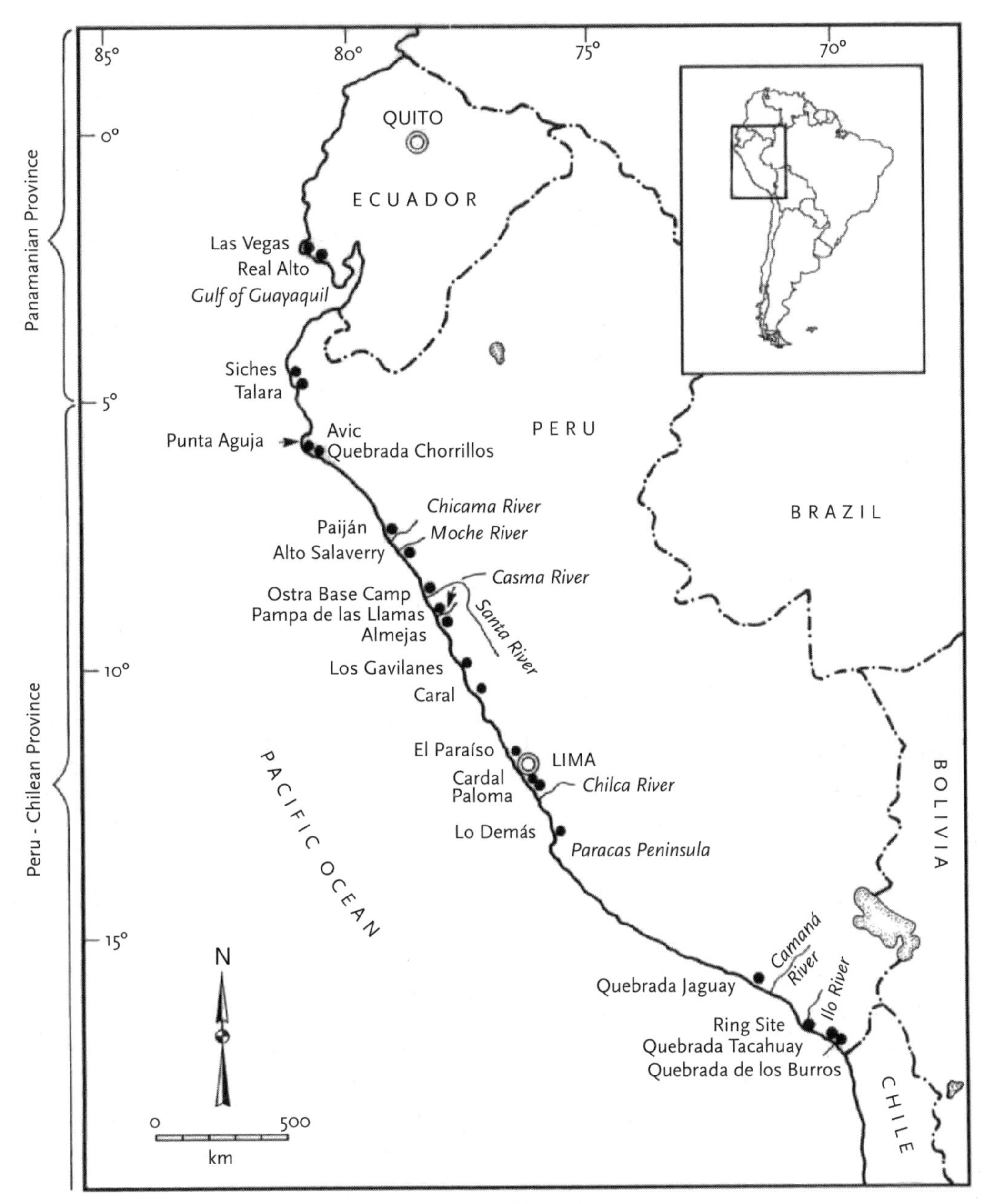

FIGURE 6.1. Map of the study area.

1983; Quilter et al. 1991), and Pampa de las Llamas (Pozorski and Pozorski 1987; Reitz 1999). Although data for classes other than fishes are not presently available from Caral, a similar pattern is likely at that site (P. Béarez, personal communication 2005; Béarez and Miranda 2000; Shady Solís et al. 2001; also known as Chupacigarro Grande [East]). This is the case at sites located within a few kilometers of the coast as well as at sites 23 to 50 km from the sea. In many cases, the crops consist primarily of bottle gourd *(Lagenaria siceraria)* and cotton *(Gossypium barbadense)*, industrial plants that enhanced the fishing effort but were not themselves new sources of nutrients (e.g., Grobman et al. 1982:149; Piperno and Pearsall 1998:267–280).

The fishing tradition continued as an important source of protein beyond the Initial Period (e.g., Chavez et al. 2003; Marcus et al. 1999; Roselló et al. 2001; Rostworowski 1977; Sandweiss 1992; Sandweiss et al. 2004). Although marine resource use continued past the Initial Period, our focus is on the early part of the Peruvian cultural sequence for several reasons. Firstly, the issue of the primacy of fishing in early economies remains unresolved, and the overwhelming evidence in favor of a long tradition is underappreciated by many researchers. Exploring the impact of people on marine fisheries requires establishing the antiquity of the fishing tradition in Peru, as well as identifying the anthropogenic and nonanthropogenic forces that could have produced changes in the Peruvian fishery as seen through the filter of the zooarchaeological record. Our focus on the early part of the sequence is mandated for another reason as well. Ironically, zooarchaeological evidence for fishing after the Initial Period is extremely limited, particularly considering that geographical location is an important variable. Data from the few collections that have been published suggest that during at least a portion of this later period the archaeological record reflects ENSO's impact on sardine and anchoveta populations (Sandweiss et al. 2004), a phenomenon that needs to be studied further with improved data from throughout the coastal strand.

Thus, until the end of the Initial Period most animal-based nutrients were obtained from a fishing tradition that was fully developed as early as 13,000 cal BP, and most plant nutrients were obtained from nondomestic sources until around 3500 cal BP. In view of the regional dominance of this tradition, it seems likely that any disruption in marine productivity would profoundly influence human life there.

ECONOMIC SIGNIFICANCE OF ANCIENT FISHERIES

Archaeologists have long debated the economic importance of marine resources relative to terrestrial animals and/or plants (see Erlandson [2001] and Yesner [1980], among others). In 1975, Michael Moseley proposed what is known as the Maritime Foundations of Andean Civilization (MFAC) hypothesis to explain the appearance of large sites with complex architecture before large-scale food production from irrigation agriculture on the central Andean Coast (Moseley 1975, 1992). Although the MFAC hypothesis is important in discussions of Andean technology, settlement patterns, and social organization, we will not enter into it here in the interest of focusing on the ancient fisheries themselves.

We emphasize, however, that the dichotomy of terrestrial versus marine resources as a basis for the development of Andean architectural and social traditions is a false one. Typically this argument degenerates into a terrestrial plant versus marine animal argument, that terrestrial plants played a more important role in cultural florescence than did marine animals or vice versa. Humans are biologically broad-spectrum omnivores; they require fats, proteins, some minerals, and fat-soluble vitamins from animal sources, and they require carbohydrates, other minerals, and water-soluble vitamins from plant sources. There are few energetically efficient ways around this nutritional imperative. Fundamentally, people on the coast of Peru always used nutrients from both the plant and the animal kingdoms. Due to the low diversity of terrestrial plants and animals in the Peruvian coastal desert, some plant (e.g., Piperno and Pearsall 1998) and most animal (Reitz 2001) nutrients, but by no means all, likely were obtained from the sea until such time as production of domestic food plants became part of the subsistence system. None of the data presently available suggest that large terrestrial animals played a dominant role in the diet of early coastal Peruvians.

Thus, from a nutritional perspective, it is unlikely that people ever concentrated all of their economic efforts exclusively upon either marine animals or terrestrial plants. The earliest sites tend to lack well-preserved macrobotanical remains except under unusual circumstances, thus the range of wild plants included in these maritime economies is poorly documented, which does not mean they were not used. From the limited evidence available from early sites, it is clear that fishing did not preclude using terrestrial plants (e.g., Benfer 1990, 1999; A. Cano, personal communication 2001; deFrance et al. 2001; Piperno and Pearsall 1998; Quilter 1991; Quilter et al. 1991; Sandweiss et al. 1989; Weir and Dering 1986; Weir et al. 1988). Equally clearly, using plants did not interrupt the use of marine resources; the earliest domestic plants (gourds and cotton) actually facilitated the fishing tradition. There is no reason that either activity would preclude the other; in this region there are many reasons why they should not.

Even after domestic sources of plant-based nutrients were available, the need for energy, high-quality protein, other nutrients, and raw materials from animal sources remained. Ancient fisheries supplied those needs, a tradition that continued even after domestic animals were available to meet the animal-based nutritional needs. This does not mean that terrestrial resources were excluded from the human diet before domestic plants became available. Biologically that seems extremely unlikely.

The debate on the development of Peruvian culture history should be recast using an energetically and nutritionally realistic model in which the contribution of both plants and animals to the development of Andean cultures is acknowledged. Human and nonhuman biological evidence indicates that ancient fisheries were combined with nondomestic plant foods to support early cultural florescence. Nutrients from domestic plant sources grew in importance over the centuries, with domestic animal nutrients eventually added to the mix. Cultural florescence would not have occurred without critical proteins, fats, vitamins, and minerals from the sea.

It seems likely that one of the problems that had to be resolved as people moved inland to grow crops was how to retain access to their traditional protein base. Improvements in technology offered by cotton netting and reliable exchange networks between coastal and coastal plain communities were the solution that developed in the Peruvian case. One might even wonder if settlements initially moved to better-watered locations away from the sea in order to grow more cotton so more fish could be caught. Eventually these inland peoples found themselves committed to life away from the sea, but still dependent on it for protein. Much of subsequent Peruvian history may reflect solutions to the problem of maintaining, or controlling, access to the sea while living some distance away from it.

EVIDENCE FOR HUMAN IMPACT ON MARINE ECOSYSTEMS

Given the length and the intensity of the Peruvian fishing tradition, evidence of human impact on the marine ecology, including resource depletion, would be expected. Indeed, archaeological evidence suggests that fishing specialization grew at the community level and that the types of marine organisms used did change (e.g., Reitz 2001, 2003, 2004; Reitz and Cannarozzi 2004; Reitz and Sandweiss 2001; Sandweiss et al. 2004; see also Table 6.1). Such changes could be evidence of a response to anthropogenic damage to the system, or they could have been culturally generated and subsequently led to such damage. Changes in many other cultural practices accompany these technological and biological shifts (e.g., Moore 1991; Sandweiss 1996a, 1996b; Sandweiss et al. 2001a). The parsimonious explanation is that these changes in fishing outcomes, as well as in the cultural patterns associated with them, are related to ENSO and Peru Current fluctuations rather than to overfishing (e.g., Sandweiss et al. 2001a, 2004). Nonetheless, the possibility that this is evidence for human responses to a marine ecosystem altered by overfishing, other aspects of resource depletion, or changes in cultural behavior unrelated to fishing must be considered. We focus here on two aspects of human behavior that could have affected the marine ecosystem: technology and seasonal fishing schedules.

Technology is an important ingredient in fishing traditions. Changes in technology could be a response to changes in the resource base, cause a change in the resource base, or, at the very least, result in an altered zooarchaeological assemblage. In any case, the archaeological record would contain a different suite of marine organisms. One need only look at the changes in marine ecosystems during the twentieth century to appreciate the impact that a new technology can have. Unfortunately, most of the evidence currently available only supports the conclusion that more sophisticated research is needed. In particular, distinctions between nearshore and offshore fishing need to be examined more closely. Reasons and outcomes for deep diving and for using tools such as nets, leisters, traps, boats, and hooks should also be considered. A third aspect that should be studied more closely are differences along a continuum between small-scale fishing by family groups to serve local needs, and large-scale, specialized fishing producing a surplus for trade under state control.

Cotton netting is one of the most important technological innovations that correlates with changes in the Peruvian archaeological record. If overfishing or other forms of resource depletion were important in Peruvian history, widespread deployment of cotton nets could be related to increased yields of small, low trophic level fishes such as anchovetas and small herrings. The strength of this as a causal relationship is weakened by the presence of cotton at sites that have few small fishes (Pozorski and Pozorski 1979a) and the presence of small fishes at sites without cotton (Chauchat 1988, 1992; Pozorski and Pozorski 1984; Reitz 1995, 2003; Reitz and Cannarozzi 2004; Reitz and Sandweiss 2001; Wing 1992). Prior to the introduction of cotton netting, small fishes may have been captured with devices made of noncotton fibers such as the twined and knotted cordage found at Quebrada Jaguay (Sandweiss et al. 1998). The important points are that cotton is not essential to the fishing tradition or to the capture of low trophic level fishes, and that cotton netting is not directly linked to fishing outcomes.

Another important animal-related technological change is the appearance of domestic animals on the coast during the Initial Period (e.g., Bonavia 1996:130–135, 141–145, 152–157). Domestic animals consisted of guinea pig *(Cavia porcellus)* and two members of the family Camelidae, the llama *(Lama glama)* and alpaca *(Vicugna pacos)*. The incorporation of domesticated animals into coastal economies was part of a broader diversification strategy. None of these animals were domesticated on the coast, instead they were production alternatives received from elsewhere, as were most of the crops. In spite of this domestic meat source, fishing remained an important source of animal protein (e.g., Moore 1991), in part, perhaps, because natural pasture is scarce on the coast and most human-derived pasturage must be irrigated, diverting water away from crops of more direct economic and nutritional value to people.

Another cultural change that might lead to overfishing and other forms of resource depletion is year-round use of marine resources. A common problem in distinguishing between anthropogenic and nonanthropogenic factors is that much of the evidence for changes in seasonal fishing schedules could instead be evidence for climate change. The same markers that indicate seasonal periodicity in resource use also may indicate a change over time in temperature and other climate variables. We can discriminate between evidence for climate change and seasonal periodicity by combining biogeography, geochemistry, and growth habits of environmentally sensitive animals to obtain multiyear records of water temperatures. Such data can then be compared to climate models to determine if the animal died during a cooler or warmer part of the year, and whether the seasonal temperature variations are similar or dissimilar to modern conditions. For instance, the sea catfish *(Galeichthys peruvianus)* from Sitio Siches in Figure 6.2 died when water temperatures were cool.

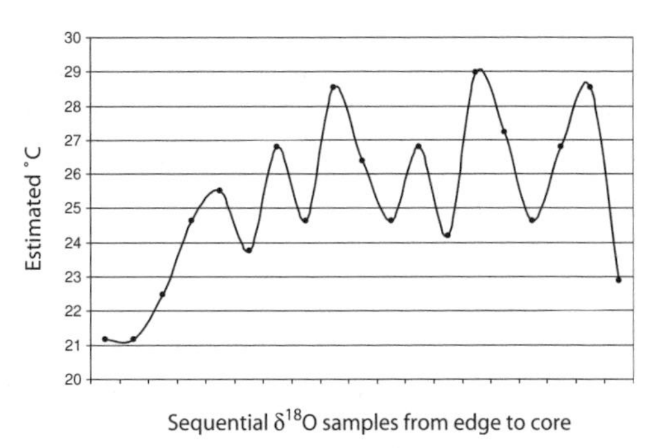

FIGURE 6.2. Season of capture estimated using $\delta^{18}O$-based seasonal temperature oscillations in a sea catfish (*Galeichthys peruvianus*) otolith from Sitio Siches. Temperatures are estimated from Grossman and Ku (1986) equation B, assuming a constant .5‰ $\delta^{18}O$ water value. Temperatures on *y* axis are listed for relative comparative value and should not necessarily be interpreted as accurate absolute values. The *x* axis represents sequential samples micromilled through ontogeny. $\delta^{18}O$ precision is better than .1‰ (parts per mil).

Oxygen isotope measurements in fish otoliths and mollusk valves also can define season of capture of individual organisms, and, by extension, seasonal fishing schedules. Using incremental growth rings to estimate season of capture for Peruvian examples is problematic in that ENSO fluctuations result in a variable seasonal cycle. One way to circumvent this variability is to analyze isotopes sequentially throughout the life of individuals (Figure 6.2). As temperature oscillated seasonally during the life of this individual, so too did the oxygen isotopes precipitated in the otolith aragonite as they grew. Season of capture is denoted by the $\delta^{18}O$ values in the last isotopic measurement compared to earlier seasonal oscillations survived by the organism, with the $\delta^{18}O$ values having a negative relationship with ambient temperature. This assessment offers a minimum season of occupation. The $\delta^{18}O$ values in the terminal growth band in sea catfish otoliths and cockle (*Trachycardium procerum*) valves from Ostra and Sitio Siches indicate that fishing occurred during both warm and cool parts of the year, but particularly during the cool season (Table 6.2).

Isotope data were generated via laser and conventional micromill extraction as described by Andrus et al. (2002b). Data generated via conventional extraction are more precise and of a finer spatial resolution than those generated via laser. As a result of these and other differences in the methods, conventional micromill data should be considered more robust. Both methods indicated that Ostra specimens were captured in both warm and cool seasons, thus lending strength to the assessment of seasonality at that site. The two micromilled samples from Siches indicated a cool season occupation, while the two laser samples suggest a warm season occupation.

TABLE 6.2

Combined Sea Catfish Otolith (Galeichthys peruvianus) *and Cockle Valve* (Trachycardium procerum) *Season of Capture Estimates from Ostra Base Camp and Sitio Siches Based on* $\delta^{18}O$

SITE/SEASON	WARM SEASON	COOL SEASON
Ostra Base Camp	2	5
Sitio Siches	2	2

When isotopic and biogeographic data are combined, we are led to the conclusion that Paloma, Ostra, and Sitio Siches are multiseasonal occupations, if not permanent ones (Reitz 2003; Reitz and Cannarozzi 2004; Reitz and Sandweiss 2001). Such data strengthen the argument that the predominance of warm-water fishes and mollusks at these sites and not at more recent ones reflects climatic conditions rather than changes in targeted species or fishing schedules. If fishing was mainly a cool-season activity then the assemblage should be dominated by cool-water animals. Instead, warm-water animals currently more typical of northern locales are found in these assemblages (Reitz 2001).

If it is true that these sites were occupied during multiple seasons (if not throughout the year) then marine organisms at these sites experienced sustained, intense, multiseasonal fishing for centuries. Though such data do not distinguish among periodic seasonal use of these sites, frequent but not continuous occupation, and essentially permanent occupation, it seems likely that many of the early sites were used during several seasons, if not permanently. Fishermen, of course, guard their fishing grounds jealously, and it is unlikely that people left any productive traditional fishing claim, or gear too heavy to transport inland, unattended for long (e.g., Acheson 1981; Yesner 1980). Additionally, these seasonality data demonstrate that the apparent full-time dependence on marine resources in this region occurred very early, with no measurable negative impact on the fishery. Much more work needs to be done to (1) associate changes in fishes represented in archaeological deposits with fishing technology and other cultural patterns, and (2) assess the possibility that some of these changes were driven by overfishing rather than by nonanthropogenic changes. For the time being, the evidence for fisheries change due to technological innovations or to overfishing is inconclusive at best.

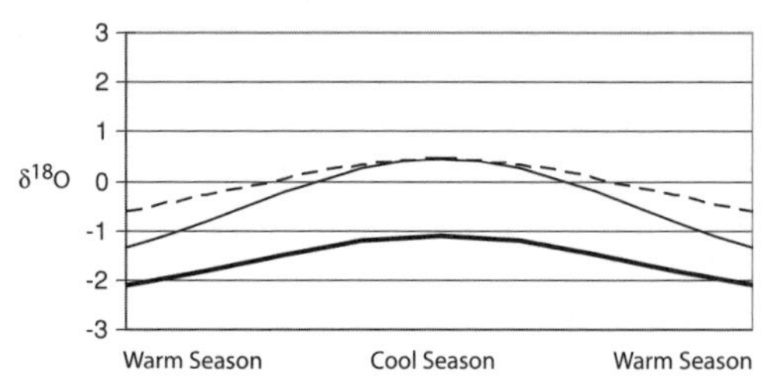

FIGURE 6.3. Sea surface temperatures based on $\delta^{18}O$ in modern and archaeological sea catfish (*Galeichthys peruvianus*) otoliths. Modern otoliths (dashed line) are from north-central Peru, and archaeological otoliths (fine line) are from Ostra Base Camp and Sitio Siches (bold line) (modified from Andrus et al. 2002a, 2002b).

EVIDENCE FOR BIOTIC EFFECTS OF VARIABILITY IN ENSO AND THE PERU CURRENT

Whereas the evidence for anthropogenic drivers for changes in fishing is limited or problematic, evidence for a nonanthropogenic source of variability is compelling. Stable oxygen isotopes and faunal biogeography suggest that the combination of change-resistant species adapted to the natural variability of ENSO and Peru Current productivity buffered the ecosystem from adverse impacts of human predation until very recently. The changes found in the archaeological record appear to reflect that natural variability.

The oxygen isotope profiles of archaeological and modern sea catfish otoliths from Ostra and Siches suggest that at one time seasonal SST variability was greater than it is today (Andrus et al. 2002a, 2003; see Carré et al. [2005] for Early Holocene data from southern Peru for the wedge clam *[Mesodesma donacium]*). The oxygen isotope profile in Ostra sea catfish otoliths indicates that the annual temperature at that site was similar to that experienced today, but that the seasonal range was greater (Figure 6.3; Andrus et al. 2002a, 2002b, 2003). We cannot associate months of the year with these warm and cold seasons because our hypothesized climate change might alter the months in these seasons in some unknown way. Elsewhere it is reported that associating seasonal patterns with specific months is problematic (e.g., Claassen 1998:150–152). The temperature range at Ostra was nearly 7 to 8°C compared to an approximately 5°C range in modern otoliths. Most of the above-modern temperatures evidenced in the Ostra otoliths occurred in the summer months; temperatures averaged almost 3°C warmer during the Ostra occupation compared to temperatures today. In contrast, at Sitio Siches between 7900 and 5150 cal BP, the amplitude of the seasonal cycle was similar to modern, but offset by about +3°C.

The Ostra otolith data are consistent with those derived from cockle *(T. procerum)* valves. The mean $\delta^{18}O$ range of valves that grew during the powerful 1982–1983 El Niño was 1.3‰ (Andrus et al. 2003; Rollins et al. 1987). In contrast, the mean $\delta^{18}O$ range in shells from Ostra is approximately 1.5‰ (Andrus et al. 2003; Perrier et al. 1994). Thus shells from Ostra, none of which contain evidence of El Niño in their growth structure, suggest a wider seasonal SST range than experienced by modern cockles during the 1982–1983 El Niño event. Unfortunately, absolute temperatures cannot be calculated because the former bay at Ostra was an evaporative environment with different oxygen isotope chemistry (U. Brand, personal communication and unpublished data 2002; Perrier et al. 1994).

These data support the hypothesis that ENSO events were less frequent or absent for some millennia prior to about 5800 cal BP and the seasonal range in water temperatures was greater in north-central, coastal Peru than it is today. The isotope profiles themselves are too short to infer the frequency and duration of ENSO cycling, but the seasonal temperature range they suggest explains the warm-water fauna found at Ostra and other sites north of 10° S between ca. 9000 and 5800 cal BP (Reitz 2001, 2004; Reitz and Sandweiss 2001; Sandweiss et al. 1996). Temperature changes of only a few degrees can fundamentally alter a marine ecosystem, impacting the productivity of coastal waters, the reproductive and growth habits of traditional prey species, and the distribution of these animals (Attrill and Power 2002; Perry et al. 2005; Richardson and Schoeman 2004; Sielfeld et al. 2002; Ware and Thomson 2005).

Thus, some changes in the archaeological record may result from natural variability influencing marine biogeography, resource availability, and, hence, fishing strategies (Navarrete et al. 2005; Sandweiss et al. 2001a, 2004). For example, a shift in the frequency of ENSO may be responsible for the disappearance of two bivalves *(Choromytilus chorus* and *Mesodesma donacium)* from north coast archaeological sites around 3000 cal BP (Sandweiss et al. 2001a). The modern ranges of these two mollusks suggest that they would be present in northern Peru only under cool-water conditions with very infrequent ENSO warming events.

In another example, twentieth-century records show that the cycles of fisheries dominated by either sardines (Clupeidae) or anchoveta (Engraulidae) correlate with multidecadal climate variability (Chavez et al. 2003). The "sardine regime" is associated with warmer conditions and higher-frequency ENSO events, and the "anchovy regime" is characterized by cooler conditions and lower-frequency ENSO cycles. Fish remains at Lo Demás, occupied between approximately AD 1480 and 1540, document a similar shift from a fishery dominated by anchoveta to one dominated by sardines by about AD 1500 (Sandweiss 1992; Sandweiss et al. 2004). This shift correlates with records for increasing ENSO frequency at the same time. Earlier sites from the Middle and Late Holocene also have fish assemblages that suggest similar regime changes.

ANCIENT FISHERIES AND MARINE ECOLOGY OF COASTAL PERU

Environmental variation is a key factor in the antiquity and role of the fishing tradition in Peru. That variation is associated with changes in the types of marine resources used and also with cultural changes (e.g., Quilter 1991, 1992; Quilter and Stocker 1983; Quilter et al. 1991; Sandweiss 1996b; Sandweiss et al. 2001a). In spite of millennia in which people focused on a suite of marine animals for much, if not all, of their animal-derived nutrients, in at least some cases throughout the year, we see no strong evidence for human-induced changes in the marine ecosystem; at least none that cannot be ascribed to nonanthropogenic factors.

As of this writing, no one has examined the later archaeological record for human impacts on shellfish, seabird, sea mammals, or fish populations along the Peruvian Coast. Our examination of the types of fishes in the archaeological record finds that changes in these materials are more likely related to ENSO variability on fish populations than to human predation, and that changes in the types of fishes taken throughout the early sequence are a result of ENSO variability more than of human fishing pressure (see also Sandweiss et al. 2004). It remains possible that some changes, such as an increase in offshore or benthic species, might be technologically driven later in the sequence (e.g., Marcus et al. 1999).

The most parsimonious interpretation of this evidence is that changes in fishing outcomes, as well as in the cultural patterns associated with them, are related to ENSO and Peru Current fluctuations rather than to anthropogenic impact. At the moment, neither fishing technologies nor fishing schedules are strong candidates as the cause or the response to anthropogenic impact on the Peruvian fishery. We expect that applications of isotopic and related approaches will soon provide new data to further distinguish between anthropogenic and nonanthropogenic signatures in the Peruvian zooarchaeological record. Such analysis will improve our understanding of resource depletion and other consequences of ancient fisheries for the marine ecology of coastal Peru unrelated to ENSO.

Levels of predation pressure high enough to trigger rapid change seem not to have been reached during prehispanic times. Clearly, the recent human impact on the marine ecosystems of Peru may be attributed to the advent of several new conditions that altered the entire ecosystem. These include a commercial fishing technology that extracts huge numbers of fishes and other marine organisms, serious exploitation of a resource area not previously exploited, and intentional or unintentional extraction of organisms from the entire food chain. This is quite different from the characteristics of earlier fishing traditions (Jackson et al. 2001). If we add to this the trend for increasingly warmer waters during the twentieth century, then even a rich fishery, one that sustained human life for millennia through many variations in ENSO frequency, duration, and strength as well as variations in the Peru Current, at last could do so no longer. This most recent assault clearly has an anthropogenic component.

Modern fisheries managers need to consider this story carefully. Ancient fishing was not a simple, inflexible strategy (Andrus et al. 2002a; Reitz 2001, 2004; Sandweiss et al. 2004). Some of the changes in fishing strategy were probably responses to nonanthropogenic changes in the resource base associated with changes in ENSO and other with geological, atmospheric, and oceanic phenomena. The primary role we ascribe to nonanthropogenic factors for changes in the ancient Peruvian fishery may or may not be supported by additional research. The important point is that the fishing strategy, and probably the structure of the fishery itself, changed markedly in the twentieth century and that the consequences of those changes have severely affected an ecosystem that otherwise appears extremely resilient.

If resource managers and conservation biologists use archaeological data to manage modern fisheries, they must recognize that the answers will not be simple dichotomies of human versus nonhuman causality and anthropogenic resource depletion. The historical record presented by archaeological data should be viewed as a complex web in which environmental and cultural variables are woven together into a fabric rich with variety and surprises. The Holocene environment itself was not a uniform, stable stage upon which people could depend. Flexibility is evident in both the cultural and the noncultural record. It remains to be seen if the need to be flexible was entirely due to nonanthropogenic environmental changes, or if people share some responsibility for changes in the Peruvian resource base. Not only are many recent historical baselines derived from depleted or collapsed fisheries (Jackson et al. 2001), their use in management plans presumes static ecological conditions that the Peruvian archaeological record suggests was not characteristic of this region.

CONCLUSIONS

After approximately 13,000 years of a major focus on the sea for protein and what must at times have been intense fishing pressure, only in the last 100 or so years has the human impact on the resource base exceeded the influence of natural variability. This is counterintuitive but can be attributed to two factors unique to this region. The most obvious is the productivity of the coastal upwelling system. With so much nearshore productivity, it may be hard to disrupt the system. Related to this is ENSO, which passed through many phases in the Pleistocene and Holocene, alternating among stages where it is "on" or "off," slowed or accelerated. It may be that organisms in these systems are supported by the Peru Current and adapted to the massive disruptions caused by ENSO to such an extent that the resource base was able to sustain intense mortality caused by fishing until very recently. In other words, what may be interesting in the Peruvian case is that it took the combined pressure of a series of intense El Niños in the later half of the twentieth century coupled with international, industrial-scale fishing to affect this region in any significant way. More robust assessment of this hypothesis requires stronger evidence. Regrettably, many of the limitations identified by Yesner (1980) continue to characterize studies of maritime peoples. We look forward to continued research into the relationship between humans and marine ecosystems in Peru with more sophisticated techniques.

ACKNOWLEDGMENTS

We are grateful to Philippe Béarez, Uwe Brand, Asunción Cano, and Dolores Piperno for their kind permission to cite their unpublished work. An earlier version of this chapter was presented at the 70th Annual Meeting of the Society for American Archaeology, Salt Lake City, Utah, April 2, 2005, in the symposium Archaeology, Historical Ecology, and Human Impacts on Marine Environments organized by Torben Rick and Jon Erlandson.

REFERENCES CITED

Acheson, J. M.
1981 Anthropology of Fishing. *Annual Review of Anthropology* 10:275–316.

Andrus, C. F. T., D. E. Crowe, D. H. Sandweiss, E. J. Reitz, and C. S. Romanek
2002a Otolith $\delta^{18}O$ Record of Mid-Holocene Sea Surface Temperatures in Peru. *Science* 295:1508–1511.

Andrus, C. F. T., D. E. Crowe, and C. S. Romanek
2002b Oxygen Isotope Record of the 1997–1998 El Niño in Peruvian sea catfish *(Galeichthys peruvianus)* Otoliths. *Paleoceanography* 17:1053–1060.

Andrus, C. F. T., D. E. Crowe, D. H. Sandweiss, E. J. Reitz, C. S. Romanek, and K. A. Maasch
2003 Response to Comment on "Otolith $\delta^{18}O$ Record of Mid-Holocene Sea Surface Temperatures in Peru." *Science* 209:203b.

Attrill, M. J., and M. Power
2002 Climate Influence on a Marine Fish Assemblage. *Nature* 417:275–278.

Bakun, A.
1996 Patterns in the Ocean: Ocean Processes and Marine Population Dynamics. California Sea Grant College System Report T-037. California Sea Grant College System, National Oceanic and Atmospheric Administration, La Jolla, California.

Barber, R. T., and F. P. Chavez
1977 Biological Consequences of the 1975 El Niño. *Science* 195:285–287.
1983 Biological Consequences of El Niño. *Science* 222:1203–1210.
1986 Ocean Variability in Relation to Living Resources during the 1982/83 El Niño. *Nature* 319:279–285.

Béarez, P.
2000 Archaic Fishing at Quebrada de los Burros, Southern Coast of Peru. Reconstruction of Fish Size Using Otoliths. *Archaeofauna* 9:29–34.

Béarez, P., and L. Miranda
2000 Análisis Arqueo-ictiológico del Sector Residencial del Sitio Arqueológico de Caral-Supe, Costa Central del Perú. *Arqueología y Sociedad* 13:67–77. Museo de Arqueología y Antropología, Universidad Nacional Mayor de San Marcos, Lima, Perú.

Benfer, R. A.
1984 The Challenges and Rewards of Sedentism: The Preceramic Village of Paloma, Peru. In *Paleopathology at the Origins of Agriculture*, edited by M. N. Cohen and G. J. Armelagos, pp. 531–558. Academic Press, New York.
1990 The Preceramic Period Site of Paloma, Peru: Bioindications of Improving Adaptation to Sedentism. *Latin American Antiquity* 1:284–318.
1999 Proyecto de Excavaciones en Paloma, Valle de Chilca, Peru. *Boletín de Arqueologia PUCP* 3:213–237.

Bonavia, D.
1982 *Precerámico Peruano: Los Gavilanes, Mar, Desierto y Oásis en la Historia del Hombre.* Editorial Ausonia-Talleres Gráficos, Lima, Peru.
1996 *Los Camélidos Sudamericanos: Una Introducción a su Estudio.* Instituto Francés de Estudios Andinos–UPCH–Conservation International, Lluvia Editores, Lima, Peru.

Briggs, J. C.
1974 *Marine Zoogeography.* McGraw-Hill, New York.
1995 *Global Biogeography.* Elsevier, Amsterdam.

Burger, R. L., and L. Salazar-Burger
1991 The Second Season of Investigations at the Initial Period Center of Cardal, Peru. *Journal of Field Archaeology* 18:275–296.

Byrd, K. M.
1976 Changing Animal Utilization Patterns and Their Implications: Southwest Ecuador (6500 BC–AD 1400). Ph.D. dissertation, University of Florida, Gainesville. University Microfilms, Ann Arbor.
1996 Subsistence Strategies in Coastal Ecuador. In *Case Studies in Environmental Archaeology*, edited by E. J. Reitz, L. A. Newsom, and S. J. Scudder, pp. 87–101. Plenum Press, New York.
Carré, M., I. Bentaleb, M. Fontugne, and D. Lavallée
2005 Strong El Niño Events during the Early Holocene: Stable Isotope Evidence from Peruvian Sea Shells. *The Holocene* 15:42–47.
Chase, T.
1988 Restos Fáunicos. In *La Prehistoria Temprana de la Península de Santa Elena, Ecuador: Cultura Las Vegas*, by K. E. Stothert, pp. 171–178. Miscelánea Antropológica Ecuatoriana Serie Monográfica 10. Museos del Banco Central del Ecuador, Guayaquil.
Chauchat, C.
1988 Early Hunter-Gatherers on the Peruvian Coast. In *Peruvian Prehistory*, edited by R. W. Keatinge, pp. 41–66. University of Cambridge Press, Cambridge.
1992 *Préhistoire de la Côte Nord du Pérou Le Paijanien de Cupisnique.* Cahiers du Quaternaire 18. CNRS-Éditions, Paris.
Chavez, F. P., J. Ryan, S. E. Lluch-Cota, and M. Ñiquen
2003 From Anchovies to Sardines and Back: Multidecadal Change in the Pacific Ocean. *Science* 299:217–221.
Chavez, F. P., P. G. Strutton, G. E. Friederich, R. A. Feely, G. C. Feldman, D. G. Foley, and M. J. McPhaden
1999 Biological and Chemical Response of the Equatorial Pacific Ocean to the 1997–98 El Niño. *Science* 286:2126–2131.
Chirichigno F.
1982 *Catálogo de Especies Marinas de Interés Económico Actual o Potencial para America Latina.* Parte II: Pacífico Centro y Suroriental. FAO, Rome.
Claassen, C.
1998 *Shells.* Cambridge University Press, Cambridge.
Cook, N. D.
1981 *Demographic Collapse, Indian Peru, 1520–1620.* Cambridge University Press, New York.
deFrance, S. D.
2005 Late Pleistocene Marine Birds from Southern Peru: Distinguishing Human Capture from El Niño-Induced Windfall. *Journal of Archaeological Science*: 32:1131–1146.
deFrance, S. D., D. K. Keefer, J. B. Richardson, III, and A. Umire Alvarez
2001 Late Paleo-Indian Coastal Foragers: Specialized Extractive Behavior at Quebrada Tacahuay, Peru. *Latin American Antiquity* 12:413–426.
Engel, F.
1957 Sites et établissements sans céramique de la côte péruvienne. *Journal de la Société des Américanistes Nouvelle Serie* 46:67–153.
Erlandson, J. M.
2001 The Archaeology of Aquatic Adaptations: Paradigms for a New Millennium. *Journal of Archaeological Research* 9:287–350.
Fedorov, A. V., and S. G. Philander
2000 Is El Niño Changing? *Science* 288:1997–2002.
Fontugne, M., P. Usselmann, D. Lavallée, M. Julien, and C. Hatté
1999 El Niño Variability in the Coastal Desert of Southern Peru during the Mid-Holocene. *Quaternary Research* 52:171–179.
Gagan, M. K., L. K. Ayliffe, D. Hopley, J. A. Cali, G. E. Mortimer, J. Chappell, M. T. McCulloch, and M. J. Head
1998 Temperature and Surface-Ocean Water Balance of the Mid-Holocene Tropical Western Pacific. *Science* 279:1014–1018.
Grobman, A., L. Kaplan, C. A. Maran V., V. Popper, and S. G. Stephens
1982 Restos Botánicos. In *Precerámico Peruano: Los Gavilanes, Mar, Desierto y Oásis en la Historia del Hombre,* edited by D. Bonavia, pp. 147–182. Editorial Ausonia-Talleres Gráficos, Lima, Peru.
Grossman, E. L., and T. L. Ku
1986 Oxygen and Carbon Isotope Fractionation in Biogenic Aragonite: Temperature Effects. *Chemical Geology* 59:59–74.
Haas, J., W. Creamer, and A. Ruiz
2004 Dating the Late Archaic Occupation of the Norte Chico Region in Peru. *Nature* 432:1020–1023.
Hildebrand, S. F.
1945 *A Descriptive Catalog of the Shore Fishes of Peru.* USNM Bulletin 189. U.S. National Museum, Washington, D.C.
Hughen, K. A., M. G. L. Baillie, E. Bard, J. W. Beck, C. J. H. Bertrand, P. G. Blackwell, C. E. Buck, G. S. Burr, K. B. Cutler, P. E. Damon, R. L. Edwards, R. G. Fairbanks, M. Friedrich, T. P. Guilderson, B. Kromer, G. McCormac, S. Manning, C. Bronk Ramsey, P. J. Reimer, R. W. Reimer, S. Remmele, J. R. Southon, M. Stuiver, S. Talamo, F. W. Taylor, J. van der Plicht, and C. E. Weyhenmeyer
2004 Marine04 Marine Radiocarbon Age Calibration, 0–26 Cal Kyr BP. *Radiocarbon* 46: 1059–1086.

Idyll, C. P.
1973 The Anchovy Crisis. *Scientific American* 228(6):22–29.

Jackson, J. B. C., M. X. Kirby, W. H. Berger, K. A. Bjorndal, L. W. Botsford, B. J. Bourque, R. H. Bradbury, R. Cooke, J. Erlandson, J. A. Estes, T. P. Hughes, S. Kidwell, C. B. Lange, H. S. Lenihan, J. M. Pandolfi, C. H. Peterson, R. S. Steneck, M. J. Tegner, and R. R. Warner
2001 Historical Overfishing and the Recent Collapse of Coastal Ecosystems. *Science* 293: 629–638.

Jenny, B., B. L. Valero-Garcés, R. Villa-Martínez, R. Urrutia, M. Geyh, and H. Veit
2002 Early to Mid-Holocene Aridity in Central Chile and the Southern Westerlies: The Laguna Aculeo Record (34°S). *Quaternary Research* 58:160–170.

Jerardino, A., J. C. Castilla, J. Miguel Ramírez, and N. Hermosilla
1992 Early Coastal Subsistence Patterns in Central Chile: A Systematic Study of the Marine-Invertebrate Fauna from the Site of Curaumilla-1. *Latin American Antiquity* 3(1):43–62.

Keefer, D. K., S. D. deFrance, M. E. Moseley, J. B. Richardson, III, D. R. Satterlee, and A. Day-Lewis
1998 Early Maritime Economy and El Niño Events at Quebrada Tacahuay, Peru. *Science* 281: 1833–1835.

Keefer, D. K., M. E. Moseley, and S. D. deFrance
2003 A 38,000-year Record of Floods and Debris Flows in the Ilo Region of Southern Peru and Its Relation to El Niño Events and Great Earthquakes. *Palaeogeography, Palaeoclimatology, Palaeoecology* 194:41–77.

Lauwerier, R. C. G. M., and I. Plug (editors)
2004 *The Future from the Past: Archaeozoology in Wildlife Conservation and Heritage Management.* Oxbow Books, Oxford.

Lavallée, D., P. Béarez, A. Chevalier, M. Julien, P. Usselmann, and M. Fontugne
1999 Paleoambiente y Ocupación Prehistórica del Litoral Extremo-sur del Perú: Las Ocupaciones del Arcaico en la Quebrada de los Burros y Alrededores (Tacna, Perú). *Boletín de Arqueología PUCP* 3:393–416.

Llagostera Martinez, A.
1979 9,700 Years of Maritime Subsistence on the Pacific: An Analysis by Means of Bioindicators in the North of Chile. *American Antiquity* 44:309–323.
1992 Early Occupations and the Emergence of Fishermen on the Pacific Coast of South America. *Andean Past* 3:87–109.

Loubere, P., M. Richaud, Z. Liu, and F. Mekik
2003 Oceanic Conditions in the Eastern Equatorial Pacific During the Onset of ENSO in the Holocene. *Quaternary Research* 60:142–148.

McCormac, F. G., A. G. Hogg, P. G. Blackwell, C. E. Buck, T. F. G. Higham, and P. J. Reimer
2004 ShCal04 Southern Hemisphere Calibration, 0–11.0 Cal Kyr BP. *Radiocarbon* 46:1087–1092.

McGlone, M. S., A. P. Kershaw, and V. Markgraf
1993 El Niño/Southern Oscillation Climatic Variability in Australasian and South American Paleoenvironmental Records. In *El Niño Historical and Paleoclimatic Aspects of the Southern Oscillation*, edited by H. F. Diaz and V. Markgraf, pp. 435–462. Cambridge University Press, Cambridge.

Maldonado, A., and C. Villagrán
2002 Paleoenvironmental Changes in the Semiarid Coast of Chile (~32°S) during the Last 6200 cal Years Inferred from a Swamp-Forest Pollen Record. *Quaternary Research* 58:130–138.

Marcus, J., J. D. Sommer, and C. P. Glew
1999 Fish and Mammals in the Economy of an Ancient Peruvian Kingdom. *Proceedings of the National Academy of Science* 96:6564–6570.

Martínez, E. A., L. Cárdenas, and R. Pinto
2003 Recovery and Genetic Diversity of the Intertidal Kelp *Lessonia nigrescens* (Phaeophyceae) 20 Years after El Niño 1982/83. *Journal of Phycology* 39:504–508.

Moore, J. D.
1991 Cultural Responses to Environmental Catastrophes: Post-El Niño Subsistence on the Prehistoric North Coast of Peru. *Latin American Antiquity* 2(1):27–47.

Moseley, M. E.
1975 *The Maritime Foundations of Andean Civilization.* Cummings, Menlo Park, California.
1992 Maritime Foundations and Multilinear Evolution: Retrospect and Prospect. *Andean Past* 3:5–42.

Moy, C. M., G. O. Seltzer, D. T. Rodbell, and D. M. Anderson
2002 Variability of El Niño/Southern Oscillation at Millennial Timescales during the Holocene Epoch. *Nature* 420:162–165.

Navarrete, S. A., E. A. Wieters, B. R. Broitman, and J. C. Castilla
2005 Scales of Benthic-Pelagic Coupling and the Intensity of Species Interactions: From Recruitment Limitation to Top-down Control. *Proceedings of the National Academy of Sciences* 102: 18046–18051.

Núñez, L., J. Varela, R. Casamiquela, and C. Villagrán
1994 Reconstrucción Multidisciplinaria de la Ocupación Prehistórica de Quereo, Centro de Chile. *Latin American Antiquity* 5:99–118.

Pauly, D.
1995 Anecdotes and the Shifting Baselines Syndrome of Fisheries. *Trends in Ecology and Evolution* 10:430.

Pauly, D., and V. Christensen
1995 Primary Production Required to Sustain Global Fisheries. *Nature* 374:255–257.

Pauly, D., V. Christensen, J. Dalsgaard, R. Froese, and F. Torres, Jr.
1998 Fishing down Marine Food Webs. *Science* 279:860–863.

Pauly, D., V. Christensen, R. Froese, and M. L. Palomares
2000 Fishing down Aquatic Food Webs. *American Scientist* 88(1):46–51.

Perrier, C., C. Hillaire-Marcel, and L. Ortlieb
1994 Paléogéographie Littorale et Enregistrement Isotopique ($^{13}C^{18}O$) D'événements de Type El Niño par les Mollusques Holocènes et Récents du Nord-ouest Péruvien. *Géographie Physique et Quaternaire* 48:23–38.

Perry, A. L., P. J. Low, J. R. Ellis, J. D. Reynolds
2005 Climate Change and Distribution Shifts in Marine Fishes. *Science* 308:1912–1915.

Piperno, D. R., and D. M. Pearsall
1998 *The Origins of Agriculture in the Lowland Neotropics.* Academic Press, New York.

Piperno, D. R., and K. E. Stothert
2003 Phytolith Evidence for Early Holocene *Cucurbita* Domestication in Southwest Ecuador. *Science* 299:1054–1057.

Pozorski, S.
1983 Changing Subsistence Priorities and Early Settlement Patterns on the North Coast of Peru. *Journal of Ethnobiology* 3:15–38.

Pozorski, S., and T. Pozorski
1979a Alto Salaverry: A Peruvian Coastal Preceramic Site. *Annals of Carnegie Museum* 48: 337–375.
1979b An Early Subsistence Exchange System in the Moche Valley, Peru. *Journal of Field Archaeology* 6:413–432.
1987 *Early Settlement and Subsistence in the Casma Valley, Peru.* University of Iowa Press, Iowa City.
1992 Early Civilization in the Casma Valley, Peru. *Antiquity* 66:845–870.

Pozorski, T., and S. Pozorski
1984 Almejas: Early Exploitation of Warm-water Shellfish on the North Central Coast of Peru. Final Report to the National Science Foundation, Grant BNS 82–03452. University of Texas, Pan-American, Edinburg, Texas.

Quilter, J.
1991 Late Preceramic Peru. *Journal of World Prehistory* 5:387–438.
1992 To Fish in the Afternoon: Beyond Subsistence Economies in the Study of Early Andean Civilization. *Andean Past* 3:111–125.

Quilter, J., and T. Stocker
1983 Subsistence Economies and the Origins of Andean Complex Societies. *American Anthropologist* 85:545–562.

Quilter, J., B. Ojeda, D. M. Pearsall, D. H. Sandweiss, J. G. Jones, and E. S. Wing
1991 Subsistence Economy of El Paraíso, An Early Peruvian Site. *Science* 251:277–283.

Reimer, P. J., M. G. L. Baillie, E. Bard, A. Bayliss, J. W. Beck, C. Bertrand, P. G. B., C. E. Buck, G. Burr, K. B. Cutler, P. E. Damon, R. L. Edwards, R. G. Fairbanks, M. Friedrich, T. P. Guilderson, A. G. Hogg, K. A. Hughen, B. Kromer, G. McCormac, S. Manning, C. B. Ramsey, R. W. Reimer, S. Remmele, J. R. Southon, M. Stuiver, S. Talamo, F. W. Taylor, J. van der Plicht, and C. E. Weyhenmeyer
2004 IntCal04 Terrestrial Radiocarbon Age Calibration, 0–26 Cal Kyr BP. *Radiocarbon* 46: 1029–1058.

Reitz, E. J.
1987 Preliminary Faunal Study for Cardal, Peru. On file, Zooarchaeology Laboratory, Georgia Museum of Natural History, University of Georgia, Athens.
1988a Faunal Remains from Paloma, An Archaic Site in Peru. *American Anthropologist* 90: 310–322.
1988b Preceramic Animal Use on the Central Coast. In *Economic Prehistory of the Central Andes,* edited by E. S. Wing and J. C. Wheeler, pp. 31–55. BAR International Series 427. British Archaeological Reports, Oxford.
1995 Environmental Change at Almejas, Peru. On file, Zooarchaeology Laboratory, Georgia Museum of Natural History, University of Georgia, Athens.
1999 Vertebrate Fauna from Pampa de Las Llamas-Moxeke. On file, Zooarchaeology Laboratory, Georgia Museum of Natural History, University of Georgia, Athens.
2001 Fishing in Peru between 10,000 and 3750 BP. *International Journal of Osteoarchaeology* 11:163–171.
2003 Resource Use through Time at Paloma, Peru. *Bulletin of the Florida Museum of Natural History* 44(1):65–80.
2004 The Use of Archaeofaunal Data in Fish Management. In *The Future from the Past,* edited by R. C. G. M. Lauwerier and I. Plug, pp. 19–33. Oxbow Books, Oxford.
2005 Vertebrate Remains from Terminal Pleistocene Deposits (Sector II), Quebrada Jaguay

(QJ 280 1999), Peru. On file, Zooarchaeology Laboratory, Georgia Museum of Natural History, University of Georgia, Athens.

Reitz, E. J., and N. R. Cannarozzi
2004 Vertebrate Remains from Honda, Siches, and Amotape Phase Occupations at Sitio Siches (PV 7-19), Peru. On file, Zooarchaeology Laboratory, Georgia Museum of Natural History, University of Georgia, Athens.

Reitz, E. J., and M. A. Masucci
2004 *Guangala Fishers and Farmers: A Case Study of Animal Use at El Azúcar, Southwestern Ecuador.* University of Pittsburgh Latin American Archaeology Publications 14, Pittsburgh, Pennsylvania.

Reitz, E. J., and D. H. Sandweiss
2001 Environmental Change at Ostra Base Camp, A Peruvian Pre-ceramic Site. *Journal of Archaeological Science* 28:1085–1100.

Richardson, A. J., and D. S. Schoeman
2004 Climate Impact on Plankton Ecosystems in the Northeast Atlantic. *Science* 305:1609–1612.

Richardson, J. B., III
1981 Modeling the Development of Sedentary Maritime Economies on the Coast of Peru: A Preliminary Statement. *Annals of Carnegie Museum* 50(5):139–150.

Rick, J. W.
1987 Dates as Data: An Examination of the Peruvian Preceramic Radiocarbon Record. *American Antiquity* 52:55–73.

Riedinger, M. A., M. Steinitz-Kannan, W. M. Last, and M. Brenner
2002 A 6100 ^{14}C yr record of El Niño Activity from the Galápagos Islands. *Journal of Paleolimnology* 27:1–7.

Rodbell, D. T., G. O. Seltzer, D. M. Anderson, M. B. Abbott, D. B. Enfield, and J. H. Newman
1999 An ~15,000-Year Record of El Niño-Driven Alluviation in Southwestern Ecuador. *Science* 283:516–520.

Rollins, H. B., J. B. Richardson, III, and D. H. Sandweiss
1986a The Birth of El Niño: Geoarchaeological Evidence and Implications. *Geoarchaeology* 1:3–16.

Rollins, H. B., D. H. Sandweiss, and J. C. Rollins
1986b Effect of the 1982–1983 El Niño on Bivalve Mollusks. *National Geographic Research* 2(1): 106–112.

Rollins, H. B., D. H. Sandweiss, U. Brand, and J. C. Rollins
1987 Growth Increment and Stable Isotope Analysis of Marine Bivalves: Implications for the Geoarchaeological Record of El Niño. *Geoarchaeology* 2:181–197.

Roselló, E, V. Vásquez, A. Morales, and T. Rosales
2001 Marine Resources from an Urban Moche (470–600 AD) Area in the "Huacas-del-Sol-y-de-la-Luna" Archaeological Complex (Trujillo, Peru). *International Journal of Osteoarchaeology* 11:72–87.

Rostworowski de Diez Canseco, M.
1977 Coastal Fishermen, Merchants, and Artisans in Prehispanic Peru. In *The Sea in the Pre-Columbian World,* edited by E. P. Benson, pp. 167–188. Dumbarton Oaks, Washington, D.C.
1981 *Recursos Naturales Renovables y Pesca, Siglos XVI y XVII.* Instituto de Estudios Peruanos, Lima, Peru.

Sandweiss, D. H.
1992 *The Archaeology of Chincha Fishermen: Specialization and Status in Inka Peru.* Bulletin of Carnegie Museum 29. Carnegie Museum of Natural History, Pittsburgh, Pennsylvania.
1996a Mid-Holocene Cultural Interaction on the North Coast of Peru and Ecuador. *Latin American Antiquity* 7:41–50.
1996b The Development of Fishing Specialization on the Central Andean Coast. In *Prehistoric Hunter-Gatherer Fishing Strategies,* edited by M. G. Plew, pp. 41–63. Department of Anthropology, Boise State University, Boise, Idaho.
2003 Terminal Pleistocene through Mid-Holocene Archaeological Sites as Paleoclimatic Archives for the Peruvian Coast. *Palaeogeography, Palaeoclimatology, Palaeoecology* 194:23–40.

Sandweiss, D. H., H. B. Rollins, and J. B. Richardson, III
1983 Landscape Alteration and Prehistoric Human Occupation on the North Coast of Peru. *Annals of Carnegie Museum* 52:277–298.

Sandweiss, D. H., J. B. Richardson, III, E. J. Reitz, J. T. Hsu, and R. A. Feldman
1989 Early Maritime Adaptations in the Andes: Preliminary Studies at the Ring Site, Peru. In *Ecology, Settlement and History in the Osmore Drainage, Peru,* edited by D. Rice, C. Stanish, and P. R. Scarr, pp. 35–84. BAR International Series 545i. British Archaeological Reports, Oxford.

Sandweiss, D. H., J. B. Richardson, III, E. J. Reitz, H. B. Rollins, and K. A. Maasch
1996 Geoarchaeological Evidence from Peru for a 5000 Years BP Onset of El Niño. *Science* 273:1531–1533.

Sandweiss, D. H., H. McInnis, R. L. Burger, A. Cano, B. Ojeda, R. Paredes, M. del Carmen Sandweiss, and M. D. Glascock
1998 Quebrada Jaguay: Early Maritime Adaptations in South America. *Science* 281:1830–1832.

Sandweiss, D. H., K. A. Maasch, and D. G. Anderson
1999 Climate and Culture: Transitions in the Mid-Holocene. *Science* 283:499–500.

Sandweiss, D. H., K. A. Maasch, R. L. Burger, J. B. Richardson, III, H. B. Rollins, and A. Clement
2001a Variation in Holocene El Niño Frequencies: Climate Records and Cultural Consequences in Ancient Peru. *Geology* 29:603–606.

Sandweiss, D. H., M. E. Moseley, J. Haas, and W. Creamer
2001b Amplifying Importance of New Research in Peru. *Science* 294:1651–1653.

Sandweiss, D. H., K. A. Maasch, F. Chai, C. F. T. Andrus, and E. J. Reitz
2004 Geoarchaeological Evidence for Multidecadal Natural Climatic Variability and Ancient Peruvian Fisheries. *Quaternary Research* 61:330–334.

Santander, H.
1980 The Peru Current System. 2: Biological Aspects. In *Proceedings of the Workshop on the Phenomenon Known as "El Niño,"* pp. 217–227. UNESCO, Paris.

Schulmeister, J., and B. G. Lees
1995 Pollen Evidence from Tropical Australia for the Onset of an ENSO-Dominated Climate at c. 4000 BP. *The Holocene* 5:10–18.

Schweigger, E.
1964 *El Litoral Peruano*. Universidad Nacional Federico Villarreal, Lima, Peru.

Sielfeld, W., M. Vargas, V. Berrios, and G. Aguirre
2002 Warm ENSO Events and Their Effects on the Coastal Fish Fauna of Northern Chile. *Investigaciones Marinas* 30(1, Supplement):122–124.

Shady Solís, R., J. Haas, and W. Creamer
2001 Dating Caral, a Preceramic Site in the Supe Valley on the Central Coast of Peru. *Science* 292:723–726.

Stothert, K. E.
1988 *La Prehistoria Temprana de la Península de Santa Elena, Ecuador: Cultura Las Vegas*. Miscelánea Antropológica Ecuatoriana Serie Monográfica 10. Museos del Banco Central del Ecuador, Guayaquil.

Stuiver, M., and P. J. Reimer
1993 Extended ^{14}C data base and revised CALIB 3.0 ^{14}C Age calibration program. *Radiocarbon* 35:215–230.

Thompson, L. G., E. Mosley-Thompson, J. F. Bolzan, and B. R. Koci
1985 A 1500 Year Record of Tropical Precipitation Recorded in Ice Cores from the Quelccaya Ice Cap, Peru. *Science* 229:971–973.

Thompson, L. G., E. Mosley-Thompson, R. Mulvaney, M. E. Davis, P.-N. Lin, K. A. Henderson, J. Cole-Dai, J. F. Bolzan, and K.-B. Liu
1995 Late Glacial Stage and Holocene Tropical Ice Core Records from Huascarán, Peru, *Science* 269:46–50.

Trenberth, K. E., and B. L. Otto-Bliesner
2003 Toward Integrated Reconstruction of Past Climates. *Science* 300:589–591.

True, D. L.
1975 Early Maritime Cultural Orientations in Prehistoric Chile. In *Maritime Adaptations of the Pacific*, edited by R. W. Casteel and G. I. Quimby, pp. 89–143. Mouton, The Hague.

Tudhope, A. W., C. P. Chilcott, M. T. McCulloch, E. R. Cook, J. Chappell, R. M. Ellam, D. W. Lea, J. M. Lough, and G. B. Shimmiel
2001 Variability in the El Niño-Southern Oscillation through a Glacial-Interglacial Cycle. *Science* 291:1511–1517.

Umlauf, M.
1993 Phytolith Evidence for Initial Period Maize at Cardal, Central Coast of Peru. In *Current Research in Phytolith Analysis: Applications in Archaeology and Paleoecology*, edited by D. M. Pearsall and D. R. Piperno. MASCA Research Papers in Science and Archaeology 10:125–129. The University Museum of Archaeology and Anthropology, Philadelphia.

Ware, D. M., and R. E. Thomson
2005 Bottom-up Ecosystem Trophic Dynamics Determine Fish Production in the Northeast Pacific. *Science* 308:1280–1284.

Weir, G., and P. Dering
1986 The Lomas of Paloma: Human-Environment Relations in a Central Peruvian Fog Oasis-Archaeobotany and Palynology. In *Andean Archaeology*, edited by R. Matos M., S. A. Turpin, and H. H. Eling, pp. 18–44. Monographs in Archaeology No. 27. UCLA Institute of Archaeology, Los Angeles.

Weir, G.. H., R. A. Benfer, and J. G. Jones
1988 Preceramic to Early Formative Subsistence on the Central Coast. Preceramic Animal Use on the Central Coast. In *Economic Prehistory of the Central Andes*, edited by E. S. Wing and J. C. Wheeler, pp. 56–94. BAR International Series 427. British Archaeological Reports, Oxford.

Wing, E. S.
1972 Utilization of Animal Resources in the Peruvian Andes. In *Andes 4: Excavations at Kotosh, Peru*. edited by S. Izumi and K. Terada, Appendix IV. University of Tokyo Press, Tokyo.
1986 Methods Employed in the Identification and Analysis of the Vertebrate Remains Associated with Sites of the Paiján Culture. On file, Florida Museum of Natural History, University of Florida, Gainesville.

1992 Los Restes de Vertébrés. In *Préhistoire de la Côte Nord du Pérou Le Paijanien de Cupisnique*, edited by C. Chauchat, pp. 42–47. Cahiers du Quaternaire 18. CNRS-Éditions, Paris.

Wing, E. S., and E. J. Reitz

1982 Pisces, Reptilia, Aves, Mammalia. In *Precerámico Peruano: Los Gavilanes, Mar, Desierto y Oásis en la Historia del Hombre*, edited by D. Bonavia, pp. 191–200. Editorial Ausonia-Talleres Gráficos, Lima, Peru.

Yesner, D. R.

1980 Maritime Hunter-Gatherers: Ecology and Prehistory. *Current Anthropology* 21:727–750.

Zavalaga, C. B., E. Frere, and P. Gandini

2002 Status of Red-legged Cormorant in Peru: What Factors Affect Distribution and Numbers? *Waterbirds* 25(1):8–15.

7

Human Impacts on Marine Environments in the West Indies during the Middle to Late Holocene

Scott M. Fitzpatrick, William F. Keegan, and Kathleen Sullivan Sealey

We badly need an historical ecology of sea monsters to determine the pristine abundances and sizes of megafauna before they were fished, and to provide the basic data for modeling their former ecological interactions with other, smaller species and their effects on biological habitats so that we can figure out what we have lost and decide what to do about it if we want to.

JACKSON AND SALA 2001:279

THE RECENT COLLAPSE of fisheries around the world (Jackson et al. 2001) from the overharvesting of resources, expanding coastal development, and dumping of various industrial and domestic waste products, has confirmed what many researchers in both the natural and social sciences had already suspected: humans are drastically influencing marine ecosystems to the point that many may never fully recover. The islands of the Caribbean are not immune to these impacts and are playing an increasingly important role in helping to determine the degree to which marine taxa have been altered by human activities over time. As continuing research in archaeology, history, and ecology shows, islands in the Caribbean have experienced a variety of impacts to marine environments by different human groups over the course of at least seven millennia. What effects did humans have on these once pristine environments, and how did this differ between populations?

An increasing amount of evidence has shown that a number of extinctions of both terrestrial and marine vertebrates have occurred, much of it related to human occupation (e.g., Adam 2004; James 2004; MacPhee and Marx 1997; MacPhee et al. 1989; Morgan and Woods 1986; Steadman and Jones 2006; Steadman and Stokes 2002; Steadman et al. 1984). In fact, within the past 4,500 years (well within the range of human settlement), at least 37 mammalian species have gone extinct (Morgan and Woods 1986). The extinction rate for mammals during the Late Holocene is 122 years per species, quite high when compared with an extinction rate of 299 years per species during the late Pleistocene (Morgan and Woods 1986). Fitzpatrick and Keegan (2007) have taken an historical ecology approach to analyzing the impacts of humans on insular biota in the Caribbean, stressing the need for a far deeper historical perspective to more effectively understand the magnitude of recent collapses and help restore marine ecosystems.

Here we focus specifically on illustrating how marine resources were utilized by prehistoric Amerindians, historic European colonists, and modern populations, and how these changes can be tracked archaeologically, historically, and biologically. By synthesizing the known data from prehistoric sites ranging from the earliest period of colonization dating to ca. 4000 BC to more modern times, a picture emerges that intimately ties the level of impact to population size and technological expertise. Current research suggests that smaller and fairly mobile Archaic hunter-gatherer-forager groups during the Middle Holocene had an impact on marine resources, but to a lesser degree than terrestrial ones. Subsequent ceramic-making horticulturalists who migrated from South America beginning around 500–200 BC, however, focused heavily on marine resources and the cultivation of manioc *(Manihot esculenta)*. Each of these activities in some fashion affected sea turtles, finfish, and mollusk populations. After European contact, larger vertebrates, especially turtles, were heavily targeted and are now largely ecologically extinct in the Caribbean. In conjunction with overfishing, development, pollution, and a host of other anthropogenic activities and natural processes, there is an ecological shift to habitats devoid of larger vertebrate taxa which has had a profound impact on coral reef structures and fisheries in the region. In addition, it is probably the smallest islands in the Caribbean such as the Bahamas, Turks and Caicos, and Grenadines that have, and will continue to be, the most susceptible to human impacts due to their high levels of biodiversity, extensive coral reef systems, and attractive marine resources.

ENVIRONMENTAL BACKGROUND

The Caribbean is the world's second largest sea and seventh largest body of water. It encompasses an area of 2,754,000 km^2 and stretches 1,700 km north-south from Florida to Panama and 2,300 km east-west from the Antillean chain of islands to the Yucatán (Figure 7.1). The West Indies are typically subdivided into three main island groups: the Greater Antilles, the Lesser Antilles, and the Bahamian archipelago. Other groups such as the Cayman Islands, Virgin Islands, Turks and Caicos, Trinidad and Tobago, and those adjacent to the South American mainland, including Curaçao, Bonaire, and Aruba, do not readily fit into these categories, yet are still important for examining the region's history and biodiversity.

The Greater Antilles are composed largely of limestone, while the Lesser Antilles arc is characterized by mostly smaller volcanic islands formed as a result of tectonic interaction. The Bahamas and numerous other islands are low-lying carbonate formations. The Caribbean is extremely diverse ecologically, home to 2.3 percent of the world's endemic plant species and 2.9 percent of endemic vertebrate species. These are exceptionally significant percentages considering that the Caribbean contributes only .15 percent of the Earth's surface. To date, over 1,500 species of fish, 25 coral genera, more than 630 mollusk species, and numerous echinoderms, crustaceans, sea mammals, sponges, birds, and reptiles have been recorded in aquatic and terrestrial environments. This has prompted Conservation International to designate the Caribbean as fourth on the list of the world's 25 "biodiversity hotspots" (see www.biodiversityhotspots.org). These are regions that are relatively small, but contain high percentages of endemic species.

Today, the availability of marine resources in each of these island groups can be characterized by three factors: (1) the overall area of shallow water habitat, (2) the extent of island perimeters (e.g., kilometers of shoreline), and (3) rainfall patterns. The production of marine life in shallows is very high on the island-bank

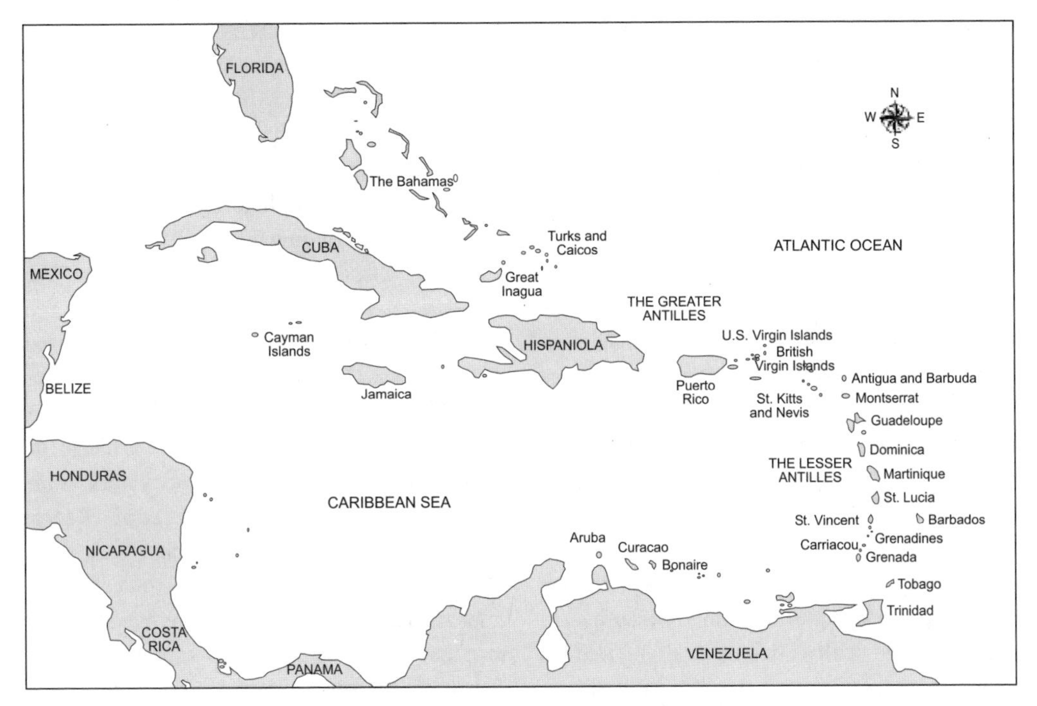

FIGURE 7.1. Map of the Caribbean.

systems around Cuba and Hispaniola, for example, because these islands have complex shorelines with extensive offshore cays and embayments, as well as surface water resources and sufficient rainfall, critical components for creating extensive coastal wetland and estuarine environments. This spatial extent and complexity of habitats supports a wider diversity and abundance of both benthic invertebrates and finfish, but rainfall is perhaps the most important factor. Columbus described Guantanamo Bay as one of the most impressive harbors he had seen, but the area around the mouth of the bay lacked permanent human settlement, a likely result of extremely low rainfall (Sara and Keegan 2004).

Of the three main regions, the Lesser Antilles probably has the lowest island-bank productivity simply because the associated bank systems of each island are quite small. There are few submerged bank systems (e.g., Saba Banks) to support extensive shallow-water habitats such as seagrass beds and coral reefs. A major exception is the shallow banks of the Grenadines chain between Grenada and St. Vincent. One of the islands here, Carriacou, takes its name from an ancient Carib word meaning "land of many reefs," a testament to the importance of these ecosystems to peoples prehistorically.

The Bahamian archipelago is a group of carbonate bank-island systems with less (and more variable) rainfall than the other island groups. The climate here typically varies from wetter and cooler in the extreme north to hot and dry in the south. In this immense archipelago that stretches across six degrees of latitude, there are five distinct types of bank-island system, each with its own unique assemblages of marine and terrestrial organisms. The sheer size of the bank systems (Keegan et al. 2007) would support larger populations of marine species such as the queen conch *(Strombus gigas)* that are no doubt highly aggregated on spatial scales but nevertheless have greater reproductive potential than smaller bank populations.

Large sheltered bank systems such as Caicos Bank or Long Island provide the spatial extent and diversity of habitats to support high abundances of marine life. Although high concentrations of marine resources (e.g., shellfish or turtles) can be highly patchy both locally and regionally (similar to terrestrial resources), and concentrated in unique geographic features (e.g., mangrove creek systems), they are often more easily exploitable. Studies by Gilliam and Sullivan (1993) of southern stingray feeding in the central Bahamas illustrate this point. Benthic surveys of sandy bank areas adjacent to the Exuma Cays with grabs, cores, and benthic sleds indicate a relatively low density of infaunal invertebrates such as mollusks, polychaetes, and crustaceans. However, the stomach contents of foraging stingrays caught in the same areas held about twice the number of species and large numbers of individuals. The foraging techniques of an 80-kg stingray allowed the animal to identify patches of high-density food items.

Mangrove creeks and coastal wetlands are also associated with concentrations of marine resources accessible and close to land. However, there is a great deal of variability between creek systems based on their spatial extent and the size and proximity of associated seagrass beds (Nero and Sullivan Sealey 2005). The size, species composition, and seasonal changes in fish assemblages are all related to basic habitat features of mangrove creeks.

In general, there is a great diversity of marine habitats in the West Indies, with smaller islands often having the most productive marine habitats (Keegan et al. 2007). Prehistoric and historic populations were drawn to these areas, and it is quite probable that their settlements will tell us more about patterns of exploitation than do settlements on the larger islands as archaeological research continues (see Keegan et al. 2007).

HUMAN SETTLEMENT OF THE CARIBBEAN

The earliest human occupation of islands in the Caribbean comes from two distinct and geographically separated groups that date between about 6000 and 3000 BC (Table 7.1). The first, and possibly earlier, of the two migrations originated from South America into Trinidad, although sea levels were lower and would not have involved long-distance ocean voyaging (see Keegan 1994; Steadman and Stokes 2002). This makes the settlement of Trinidad and Tobago different from that of other major settlements of hunter-gatherer groups in the region. The second was a Lithic/Archaic group that probably migrated from somewhere in Mesoamerica to Cuba and Hispaniola (also known as Casimiroid) (Keegan 1994; Rouse 1992; Wilson et al. 1998) by around 4000–3000 BC, although there is a lack of good chronologies for these sites. A later movement of Archaic peoples took place between 2000 and 500 BC into the Lesser Antilles from South America (e.g., Davis 2000; Hofman and Hoogland 2003; Hofman et al. 2006; Keegan 1994; Richard 1994; Watters et al. 1992). Based on a limited, but now growing, set of evidence it appears that Archaic peoples were fisher-foragers and that each site's occupants subsisted on a different set of primary foods (Keegan 1994:270) consisting of fish, turtles, and invertebrates from reef and nearshore habitats (Davis 1988; Narganes Storde 1991), as well as some seasonally available foods (e.g., Hofman et al. 2006).

During the Ceramic age (ca. 500 BC–AD 1400) there was an expansion of Amerindian groups from South America into virtually every island in the Caribbean except for the Caymans (Stokes and Keegan 1996). These ceramic-making horticulturalists introduced several new species of plants and animals. Based on archaeological evidence from numerous island groups, marine resources were probably the prime attraction to these colonists (Keegan 2000), the most important of which were sea turtles (Cheloniidae), parrotfish (Scaridae), jacks (Carangidae), snappers (Lutjanidae), grunts (Haemulidae), groupers (Serranidae), the queen conch, and West Indian topshell *(Citarrium pica)* (Newsom and Wing 2004:67–68). Technologies for fishing included hook and line, spears, traps, poison, and nets, to name a few (see Keegan 1986). Turtles could be easily captured on beaches during egg-laying cycles, and many mollusks could simply be collected from sandy beaches and rocky shores without any special tools.

TABLE 7.1

Major Islands in the West Indies (Listed North to South) with Size of Islands (See Keegan 1994:258) and Earliest Acceptable Prehistoric Occupation

	ISLAND	AREA (KM^2)	EARLIEST DATE
Bahamas	San Salvador	149	AD 600
Turks and Caicos	Grand Turk	20	AD 700
Caymans	Grand Cayman	197	X
	Little Cayman	26	X
	Cayman Brac	33	X
Greater Antilles	Cuba	110,922	2900–2000 BC
	Hispaniola	76,484	3520–3020 BC
	Jamaica	11,424	AD 720–1180
	Puerto Rico	8,897	1870–1520 BC
Northern Lesser Antilles	U.S. Virgin Islands	344	1270–900 BC
	British Virgin Islands	174	AD 400?
	Anegada	39	AD 1010–1300
	Anguilla	188	1670–1430 BC
	St. Martin	34	1870–1660 BC
	Saba	13	1880–1640 BC
	St. Eustatius	21	AD 240–380
	St. Kitts	176	2800–2500 BC?
	Nevis	130	750 BC?
	Barbuda	161	1900–1800 BC
	Antigua	280	2300–500 BC
	Montserrat	84	800–400 BC
	Guadeloupe	1,702	AD 30–210
	Dominica	790	X
Southern Lesser Antilles	Martinique	1,090	900–400 BC?
	St. Lucia	603	AD 640–1020
	St. Vincent	389	AD 0–400?
	Barbados	440	AD 260–660
	Carriacou	32	AD 400–600
	Grenada	345	40 BC–AD 240
Adjacent to South America	Tobago	300	2950–2770 BC
	Trinidad	4,828	6220–5840 BC
	Margarita	1,150	2290–1530 BC
	Bonaire	288	X
	Aruba	193	450 BC–AD 240
	Curaçao	443	3370–2920 BC

NOTE: Earliest acceptable prehistoric occupation based on an assessment of radiocarbon dates (see Fitzpatrick 2006). X, no acceptable ^{14}C chronologies to satisfactorily establish the presence and/or extent of prehistoric occupation.

The last major migration of peoples into the Caribbean began in the fifteenth century when Columbus "discovered" the New World. This was followed by three more expeditions by the well-known explorer and numerous other Europeans. This led to an increasing amount of Old World plants, animals, insects, and diseases being transported across the Atlantic to help support the new colonizing populations. As this "Columbian exchange" (Crosby 1972; Salvaggio 1992) continued, native peoples were decimated by a variety of illnesses to which they had no immunity, as well as slavery and indiscriminate killing. Later, insular landscapes and seascapes were transformed with the establishment of plantation agriculture, land clearance for food production and animal grazing, the arrival of tens of thousands of African slaves, and the harvesting of local resources. It was shortly after European arrival that marine resources, especially turtles, sea mammals, and reef fisheries, began playing a major role in sustaining these fledgling populations, which then began to ultimately support burgeoning trade networks developing across the Atlantic triangle.

EXPLOITATION OF MARINE RESOURCES THROUGH TIME

Evidence of Native American Impacts

Archaeological research suggests that subsistence strategies during the Archaic period involved capturing around 10 or so major mollusk species from shallow marine environments such as mangroves and both sandy and rocky shores, although there are a few other shellfish taxa, fish, turtle, and manatee found at archaeological sites dating to this time period (Table 7.2). Site inhabitants at Jolly Beach on Antigua appear to have relied more heavily on terrestrial species including lizards and birds (Davis 2000), but this is much different than other early Archaic sites where subsistence has been investigated in detail (e.g., Crock et al. 1995; Figueredo 1974; Hofman and Hoogland 2003; Hofman et al. 2006; Lundberg 1989).

Although there have been relatively few Archaic period sites found, archaeological studies demonstrate that there was heavy exploitation of marine fish, particularly parrotfish, but also grouper, snapper, and grunts. Terrestrial vertebrates in general were uncommon in these sites, and there is no evidence of introduced mammals (Newsom and Wing 2004:129). However, new dates suggest that Archaic peoples were responsible for the extirpation of ground sloths (Steadman et al. 2005). Data from Archaic sites in the Caribbean so far suggest that peoples were fairly mobile, exploited foods that were locally available, and did this by occupying sites seasonally to take advantage of certain resource concentrations (Hofman and Hoogland 2003; Hofman et al. 2006).

During the Ceramic age beginning with the Saladoid period (ca. 400 BC–AD 600), people subsisted on both introduced (e.g., hutia, agouti) and indigenous terrestrial (e.g., land crab, iguana, tortoise, birds) and marine taxa. Over time, however, there is a significant decrease in terrestrial foods such as rice rats and land crabs and an increase in *Donax* shells at some sites, suggesting a move from more terrestrially based fauna to marine foods. The average weight of the major fish and land crabs decreases through time at the Nevis sites, Hope Estate on St. Martin, and Tutu on St. Thomas, for example (Newsom and Wing 2004:102–104, 139), indicating that certain taxa were being overharvested. The effect of increased land clearance for manioc production, along with a spike in the number of archaeological sites found during the Ceramic age (Keegan 2000), seems to have placed greater pressure on the marine environment.

There is good evidence that marine fish, particularly those from reef environments, were being overharvested throughout the Caribbean during this time (Wing and Wing 2001). It is important to note that recent research of post-Saladoid (AD 600/800–1500) settlements on Anguilla suggests this may not have been the case everywhere (Carder et al. 2007). The general trend, however, seems to be a decrease in numbers of certain fish taken, accompanied by a decline in the estimated reef fish biomass and the mean trophic level. The average size of the specimens within each family gradually decreases, and there is a transition to more herbivorous and omnivorous species with a corresponding decrease in aggressive carnivorous species (see Newsom and Wing 2004:111; Tables 6.7 and 6.8). These robust data sets, which include vertebrate fauna from archaeological sites on Puerto Rico, St. Thomas, St. Martin, Saba, and Nevis, demonstrate that

TABLE 7.2

List of the Most Common Marine Taxa Exploited over Time by Humans in the Caribbean Based on Current Data

	COMMON NAME	SCIENTIFIC NAME	HARVESTED PREHISTORICALLY?	HARVESTED HISTORICALLY?	CURRENTLY FISHED LEGALLY?	CURRENT GENERAL STATUS (YEAR LISTED ON ENDANGERED LIST)	REFERENCE
Mollusks	Queen conch	*Strombus gigas*	X	X	X	Overfished; threatened	Keegan et al. 2003; Newsom and Wing 2004; Theile 2001
	West Indian topsnail	*Cittarium pica*	X	X	X	Unknown	Newsom and Wing 2004
Finfish	Parrotfish	Scaridae	X	X	X	Locally harvested	Carlson and Keegan 2004; Newsom and Wing 2004; NOAA 2004
	Grouper	Serranidae	X	X	X	Overfished (e.g., The Bahamas)	Newsom and Wing 2004; NOAA 2004
	Snapper	Lutjanidae	X	X	X	Many species overfished	Newsom and Wing 2004; NOAA 2004
	Grunts	Haemulidae	X	X	X	Unknown	Newsom and Wing 2004; NOAA 2004
	Surgeonfish	Acanthuridae	X	X	X	Unknown	Newsom and Wing 2004; NOAA 2004
	Jacks	Carangidae	X	X	X	Protection proposed for spawning	Newsom and Wing 2004; NOAA 2004
Sea turtles	Hawksbill	*Eretmochelys imbricata*	X	X	N	Endangered (1970)	Newsom and Wing 2004; NOAA 2004
	Loggerhead	*Caretta caretta*	X	X	N	Endangered (1978)	Carlson and Keegan 2004; Newsom and Wing 2004

(continued)

TABLE 7.2 (*continued*)

	COMMON NAME	SCIENTIFIC NAME	HARVESTED PREHISTORICALLY?	HARVESTED HISTORICALLY?	CURRENTLY FISHED LEGALLY?	CURRENT GENERAL STATUS (YEAR LISTED ON ENDANGERED LIST)	REFERENCE
	Green	*Chelonia mydas*	X	X	N	Endangered (1978)	Carlson and Keegan 2004; Newsom and Wing 2004
	Leatherback	*Dermochelys coriacea*	X	X	N	Endangered (1970)	NOAA Office of Sustainable Fisheries
	Kemp's ridley	*Lepidochelys kempii*	O	X	N	Endangered (1970)	NOAA Office of Sustainable Fisheries
	Olive ridley	*Lepidochelys olivacea*	O	X	N	Endangered (1978)	NOAA Office of Sustainable Fisheries
Marine mammals	West Indian manatee	*Trichechus manatus*	O	X	N	Endangered (1973)	McKillop 1985; Mignucci-Giannoni et al. 2000
	Caribbean monk seal	*Monachus tropicalis*	N	X	–	Extinct (1967)	NOAA Office of Sustainable Fisheries
	Humpback whale	*Megaptera novaeangliae*	N	X	N	Endangered (1970)	NOAA Office of Sustainable Fisheries

NOTE: The intensity of exploitation of certain species such as queen conch has varied geographically, and some populations may be under less threat than others depending on legislation and restrictions implemented by host countries (see Theile 2001:51–54). X, intensively; O, occasionally; N, never.

FIGURE 7.2. *Strombus gigas* was one of the most important resources in the Caribbean for producing tools such as adzes, and for food. This species is a common constituent of Ceramic age sites, as can be seen from this photo from the Grand Bay site on Carriacou (note the archaeologically recovered shells in the foreground and those in situ that remain in the excavation unit).

these changes are characteristic of overfishing currently taking place (see Roberts 1995; Russ 1991; Wing and Wing 2001).

These trends are particularly evident at the sites of Hichmans and Indian Castle (Nevis), where the abundance of carnivorous reef fish (e.g., groupers) in earlier deposits are twice that of herbivorous fish (e.g., parrotfish, surgeonfish), which then decline to only 10 percent of the total (Wing 2001). Along with changes in reef fish assemblages is an increase in pelagic fish such as tuna, jacks, and flying fish (Newsom and Wing 2004:111–112). The move from nearshore to offshore fisheries would have been a riskier endeavor and required different technologies such as hook and line fishing. This may have resulted from the overpredation of reef taxa, leading to a heavier focus on important food crops such as manioc to compensate for marine species that were more widely available but less and easily captured. This is a classic "top-down" approach whereby humans hit higher trophic levels first (or most intensively), a phenomenon that also matches modern patterns. With populations growing, land clearance for producing fuel and arable land would have also increased and with it, erosion and infilling of local embayments and the expansion of mangrove habitats. The resulting sedimentation is also lethal to coral reefs and likely destroyed some of these habitats.

The overexploitation of shellfish species such as queen conch has also been documented during the Late Ceramic periods in Jamaica (Keegan et al. 2003; Figure 7.2). From Ostionan to Meillacan times in Jamaica (ca. AD 800–1400), there are shifts in the use of mollusks that appear to be a result of mangrove and muddy substrates replacing seagrass habitats, and overfishing of strombid species, including *S. gigas*, that were continually reduced as environments changed. Part of this depletion may also be related to the need for additional molluskan biomass as sea turtles, larger fish, and conchs were reduced in size and numbers (Keegan et al. 2003:1615). The increase of West Indian topsnail collected from rocky intertidal zones during the early Ceramic age, possibly as a response to land crab overharvesting, tends to decrease in size over time at the Tutu site on St. Thomas (Newsom and Wing 2004:140–41).

Evidence shows that mollusks such as the queen conch, a ubiquitous component of Caribbean archaeological assemblages, were being overharvested prehistorically; they have since declined from overfishing in modern times throughout the circum-Caribbean (Brownell and Steveley 1981; Stager and Chen 1996; Torres 2003; Weil and Laughlin 1984). It should be noted, however, that although fishing pressure prehistorically may have been selecting for faster maturation rates of *S. gigas* (Stager and Chen 1996:18) at different points in time, this could also be related to the alteration of suitable habitats for planktonic veligers into nurseries, leading to a decline in nearshore populations (Keegan et al. 2003:1614). It is critical to note that in these cases, when attempting to model changes in the quantity and size of certain animal species through time, it is necessary to determine the availability of resources, environmental conditions, and cultural factors (Torres 2003), which are not always easy to discern in the archaeological and paleoecological records.

The depletion of other faunal resources, many of them marine, is also evident throughout the northern Caribbean in Jamaica, the Bahamas, and Haiti (Carlson and Keegan 2004). Sea turtles, birds, iguanas, and land mammals were typically targeted first when available. In many cases, later inhabitants could no longer exploit them. On Grand Turk, both marine and terrestrial resources were overexploited but not completely eliminated (Carlson and Keegan 2004:102). Turtles appear to have been heavily exploited on the island of Carriacou as well (S. M. Fitzpatrick, personal observation 2005) and were probably always taken when available. The Cayman Islands may be the one exception whose turtle populations lay untouched by human hands until after European contact because it was not settled prehistorically. It is important to note that the deepest part of the Caribbean is adjacent to the Cayman Islands, extending nearly 7,700 m in depth. This allows for only shallow fringing reef to develop, not the extensive reef systems that are seen in the Bahamas, Turks and Caicos, and Grenadines. Thus, the Cayman Islands represent somewhat of an enigma in the Caribbean in terms of prehistoric settlement and are much different in regards to island resource activity.

Postcontact European Patterns

After European contact, other industries such as whaling and capturing finfish and turtles became increasingly important to the livelihoods of people in the Caribbean (Jackson 1997). The most commercially harvested were humpback whales (Romero et al. 2002), sea turtles (Jackson 1997), and the now extinct Caribbean monk seal, which appears to have been hunted by Amerindians only infrequently. This may have been a result of its wider distribution in the western Gulf of Mexico (Adam 2004; Timm et al. 1997). Efforts to capitalize on the local availability of these and other large sea mammals and vertebrates have been intense. Needless to say, the capturing of these animals was typically not managed with conservation in mind and historically has led to significant reductions in populations over a comparatively short period of time either from direct capture or human-related impacts to particular environments where manatees congregate.

Within the past few decades, it has not been the direct targeting of marine species that has probably had the most profound impact on coral reef and benthic ecosystems, but widespread deforestation, increasingly sophisticated forms of agriculture, eutrophication, oil pollution, and industrialization that are causing major declines in coral reef structures and associated marine communities over vast swaths of the West Indies (Cortes and Risk 1985; Jackson 1997; Roberts 1995). In Barbados, for example, aerial photographs taken from 1950 to 1991, along with direct examination of coral reef health, shows extensive damage to these ecosystems (Lewis 2002). Both natural and cultural factors are probably to blame (Lewis 2002:54), and storm damage could have also played a role. Sugarcane production on Barbados and many other islands, one of the prime causes behind deforestation, has led to a drastic reduction and almost complete annihilation of the original island flora that was present at contact.

Historically, the effects of new European-based technologies for farming, extensive land clearing for agriculture and settlement, the introduction of Old World plants and animals such as cattle, horses, pigs, and sheep as part of the "Columbian exchange" (Crosby 1972), and overfishing and hunting have drastically altered Caribbean island ecosystems. One of the greatest impacts to marine fauna since the arrival of Columbus is the decline of turtle populations from overhunting (Jackson 1997). Other large marine vertebrates such as manatees, stingrays, and whales, which are also extremely important ecologically, are much rarer today than historical records suggest at contact, but none of these are known to have been widely targeted by Amerindians. Today there are very few groups of West Indian manatees *(Trichechus manatus)* left in the Caribbean due to accidental deaths (often associated with boat propellers as the manatees congregate near shallow areas to feed), poaching, coastal habitat contamination from industrial discharge, strandings, and incidental capture (Mignucci-Giannoni et al. 2000; National Oceanic and Atmospheric Administration [NOAA] 2004; Romero et al. 2002; Timm et al. 1997). Since the late 1800s, there has been a general increase in the number of manatee deaths, which has dramatically increased in the last two decades, comprising about 80 percent of known manatee deaths in the last century or so (Mignucci-Giannoni et al. 2000:192). Today, less than 100 individuals have been documented in Puerto Rico through aerial surveys (Mignucci-Giannoni 2000:189) and even less in Trinidad (Romero et al. 2002). In southeast Florida, manatee deaths have increased annually to the point that extinction for the species is likely in the next 1,000 years given a 10 percent adult mortality rate, unless proactive conservation policies are put into effect in the near future (Marmontel et al. 1997).

Whaling has also depleted local populations of humpbacks *(Megaptera novaeangliae)* in the West Indies (Reeves et al. 2001; Romero et al. 2002; Smith and Reeves 2003). The hunting of whales was never a focus of prehistoric subsistence, and the occasional bone found in archaeological sites is probably the result of the scavenging of individuals who beached themselves or washed up on shore. Commercial whaling that began in the 1820s, particularly in the southern Caribbean (in contrast to present-day wintering areas for the species, which occur primarily around Hispaniola; Reeves et al. 2001), has led to depletion of the species within the region. Although passive acoustics and visual observations used by Swartz et al. (2003) found that humpbacks still occupy areas where they were actively hunted historically, these numbers are considerably lower than during the nineteenth century. High numbers of humpbacks taken by locals for modern subsistence hunting on the island of Bequia in the northern Grenadines, for example, was the result of coming into contact with more efficient commercial whaling tactics such as guns and explosive harpoons (Baker and Clapham 2002). Populations continually declined, and by the end of the 1800s it was no longer economically feasible to hunt whales in the area.

After the British captured Jamaica in 1655, locals began harvesting green turtles on Grand Cayman, which provided a majority of meat consumed up until the 1730s (Long 1774; Sloane 1707–1725 [from Jackson 1997:S26]). This was critical for island communities here because agriculture had not yet been satisfactorily established. Turtle populations declined rapidly in the Caymans, and turtlers were then forced to travel farther afield to "Far Tortuga" along the coast of Central America.

Based on calculations using historical sources, Jackson (1997:S26–27) estimated that there may have been between 33 and 39 million total adult turtles in the Caribbean. This seems to support accounts from chroniclers such as Andres Bernaldez on Columbus's second voyage in 1494 who said that "in those twenty leagues . . . the sea was thick with them, and they were of the very largest, so numerous that it seemed that the ships would run aground on them and were as if bathing in them" (from Jackson 1997:S27). Although these numbers may be exaggerated, it is clear that sea turtle numbers today, based on chronicler accounts and more detailed surveys (e.g., Carrillo et al. 1999), are nowhere near the numbers that were present before European contact, suggesting that native Amerindians, although exploiting turtles, were probably not impacting them in a manner that was unsustainable.

Sea turtles have been heavily exploited by humans for centuries, but not at the scale of what occurred in the 1700s and 1800s when hawksbill *(Eretmochelys imbricata)* shells became highly prized in Europe, and the eggs and meat were harvested for consumption. Even in the late 1800s, people were still migrating to Cuba to establish turtle fisheries (Carrillo et al. 1999:266), and the heavy exploitation of turtles here continued throughout the 1900s. It was not even until the early 1990s that the Cuban government moved away from hunting turtles—not under pressure from the Convention on International Trade in Endangered Species, but so they could focus on more profitable fishery exports. Carrillo et al. (1999:264) report that in Cuba between 1935 and 1994, approximately 170,000 hawksbill turtles were harvested, equaling 8,600 metric tons live body weight. Although it is difficult to quantify the numbers of turtles taken prior to detailed record keeping and whether historical captures were actually sustainable, research does suggest that Cuba did not necessarily overexploit hawksbill populations. This may be a result of most nesting areas being located on uninhabited offshore islands (Carrillo et al. 1999:278). Needless to say, the six main species of sea turtles in the Caribbean today—the hawksbill, loggerhead *(Caretta caretta)*, green *(Chelonia mydas)*, leatherback *(Dermochelys coriacea)*, Kemp's ridley *(Lepidochelys kempii)*, and olive ridley *(Lepidochelys olivacea)*—are all in danger of extinction from human overexploitation, loss of nesting beaches, disease, incidental capturing by fishers, and habitat destruction (Fish et al. 2005:483). These problems will only be exacerbated as sea levels continue to rise from changes in global climatic patterns and reduce the size and number of nesting grounds (Fish et al. 2005).

DISCUSSION AND CONCLUSIONS

The Caribbean Islands have a long history of human settlement dating back at least 7,000 years, much of it focused along coastlines. Archaeological research provides a great deal of support that prehistoric human populations in the region exploited a wide range of native flora and fauna (both terrestrial and marine) and brought with them nonnative species that were critical components to the prehistoric diet. However, as research suggests, people were primarily maritime oriented, with an emphasis on marine resources for subsistence and tool production.

Until recently, it was thought that prehistoric peoples who occupied the West Indies had little impact on marine environments. Archaeology and historical ecology is changing this perception. Archaic groups were thought to have a very limited impact due to their small populations, and although we do not know a great deal about their marine impacts, they were probably responsible for the extinction of ground sloths on Hispaniola (Steadman et al. 2005). It has also been assumed by marine ecologists that Ceramic age groups had a limited impact on island ecosystems and that human impacts in general were minimal during all of prehistory with regard to nearshore habitats. Certain marine taxa (e.g., sea turtle) could be harvested in sustainable numbers (K. Bjorndal, personal communication 2001). Archaeological evidence from the Caribbean is quickly changing this view, especially now that we are gaining a better understanding of when humans got to these islands, the plants (e.g., manioc) and animals (e.g., hutia, agouti, opossum, guinea pig, dog) they brought with them, and how these new taxa and people may have influenced these insular environments. On Grand Turk, prehistoric peoples visited the island seasonally beginning around AD 750 and drove iguanas, large fish, sea turtles, numerous bird species (including red-footed and masked boobies), and an indigenous tortoise to extirpation (Carlson 1999; Carlson and Keegan 2004; Keegan et al. 2007). In southwestern Jamaica around AD 800, land clearance by the first human settlers likely contributed to greater sediment loads that transformed Bluefields Bay from a free-circulating seagrass habitat to a muddy mangrove habitat (Keegan et al. 2003).

Analyses of faunal assemblages in the Caribbean demonstrate that human groups were overexploiting a number of marine taxa. This is especially evident after Saladoid peoples entered the Antilles from South America around 500–200 BC. A number of finfish and shellfish species on St. Martin, St. Thomas, Nevis, and Saba decline in numbers and size, probably as a result of overexploitation. The decline in reef fish also led to an increase in the capture of offshore fish. On other islands, the prehistoric record shows a reliance on *S. gigas* in the one- to two-year age class (Carlson and Keegan 2004). Only recently have biologists begun to investigate the local impacts on mollusk populations (Torres 2003).

Although prehistoric peoples in the West Indies were heavily exploiting marine resources, some to the point of being overharvested, one aspect of prehistoric marine exploitation that needs further investigation is the role that smaller islands played in subsistence strategies and settlement patterns. As Keegan et al. (2007) suggest, archaeological investigations on Grand Turk, Carriacou, and Middle Caicos demonstrate that marine and other resources (e.g., salt) on smaller islands were much more abundant and deliberately sought after by Amerindians, even by those who lived on larger islands such as Hispaniola. Some of the attractants include terrestrial species such as birds, giant iguanas *(C. carinata)*, and an extinct tortoise, but also large marine vertebrates, including enormous loggerhead turtles. There also appears to be an unusually large number of turtle remains and other marine foods on Carriacou in the southern Grenadines (S. M. Fitzpatrick, personal observation 2005), which were also heavily targeted by Europeans shortly after contact (Richardson 1975). Extensive seagrass flats and patch reefs on Middle Caicos (along with salt deposits for preserving meats), seem to have made these and other smaller islands extremely important for prehistoric peoples (Keegan et al. 2007).

On smaller islands around the Caribbean, humans have probably always sought out turtle rookeries and fisheries. Jamaican settlers targeted the turtle populations during the late 1600s and early 1700s prior to the advent of an established agricultural system. The loss of turtles to feed people on Jamaica eventually led to heavy overfishing of other species. Unfortunately, fisheries in Jamaica have never recovered, a situation that began centuries earlier (Keegan et al. 2003) and is not unique to the island, but seen regionwide.

The examples we have presented here suggest that marine resources provided the main protein for groups living on the Caribbean Islands. On some of these islands, turtles were one reason why smaller islands played such a prolific role in human subsistence activities. As we have argued elsewhere (Keegan et al. 2007), smaller islands, because of their extremely diverse and plentiful resource base and abundant evidence for exploitation of marine taxa prehistorically, are critical to understanding human adaptation in archipelagoes.

During prehistoric times, native Amerindians exploited marine resources, including large marine vertebrates, nearshore finfish, and numerous invertebrate taxa (e.g., Carlson and Keegan 2004; Fitzpatrick and Keegan 2007; Keegan et al. 2003; Newsom and Wing 2004; Wing 2001; Wing and Wing 2001). Although these impacts may have been limited in comparison to European impacts, the local extirpation of "keystone" species (e.g., *S. gigas*, sea turtles, and carnivorous fish) has had a profound effect on the local ecological structures.

After European contact, the trend of exploiting these resources continued, although the impacts to fisheries have been far greater than any other point in time. It is now known that large herbivores such as turtles (Carrillo et al. 1999), manatees (see McKillop 1985; Mignucci-Giannoni et al. 2000), and numerous carnivores are ecologically extinct on Caribbean seagrass beds and coral reefs, with food chains now dominated by invertebrates and small fish (Jackson 1997:S28). The West Indian or Caribbean monk seal *(Monachus tropicalis)*, the only seal native to the Gulf of Mexico, was first reported by Columbus in 1494 but was last seen in 1952 on the banks between Jamaica and Honduras. *Monachus tropicalis* remains have been found in Caribbean archaeological sites on Puerto Rico, the Virgin Islands, St. Eustatius, Nevis, and Curaçao (Adam 2004:3), but they never appear to have been an important food source despite the fact that they are large packages of meat and inhabited islands and banks throughout the circum-Caribbean (with the exception, perhaps, of the southern Lesser Antilles [Adam 2004:2; Timm et al. 1997]). It is unclear why prehistoric peoples chose not to hunt them more intensively (or at all), but it is widely known that they lacked escape responses and were easily captured by hunters and scientists (primarily by the former for its oil), which led to this species' early demise; it was officially declared extinct in 1996.

Both terrestrial and marine ecosystems were further and more severely degraded after European contact, long before ecologists began intensively studying them in the late 1950s (Goreau 1959; Jackson 1997; Randall 1965). There is great concern for coral reef health because anthropogenic activities such as "increased predation pressure, hyper- and hypothermic stress, reduced water quality associated with terrestrial runoff and poor watershed management, overgrowth by macroalgae, boat groundings and anchor damage, and disease" (Precht 2002:42). These have all led to long-term or irreversible impacts to these environments. Gardner et al. (2003) demonstrated that in the last 30 years, coral cover in the Caribbean has been reduced from about 50 to 10 percent. As Hughes et al. (2003:929) note, many reefs worldwide have experienced reduced stocks of herbivorous fish and added nutrients from land-based activities that have "caused ecological shifts, from the original dominance by corals to a preponderance of fleshy seaweed." Greenhouse gases are also causing global climate changes and are thought to weaken coral skeletons and prevent further accretion. Tropical storms such as hurricanes, which appear to be occurring in greater frequency, can shorten the time for recovery (Hughes et al. 2003). For now, it appears that the coral reef crisis is relegated to shallow waters and not the deeper reef systems (Bak et al. 2005:1), but as these influences continue, this will surely change (Bellwood et al. 2004).

The overfishing of herbivorous species, particularly parrotfish and surgeonfish, not only affects the size of harvestable stocks, but completely alters the dynamics of a reef system (Jackson and Sala 2001; Jackson et al. 2001). Although natural processes such as disease, storms, and temperature stress must be factored into equations that attempt to explain why reef systems are changing, direct (e.g., overfishing, habitat destruction) and indirect (sedimentation, eutrophication) anthropogenic factors are also to blame for a variety of reef stressors. The widespread mass mortality of the urchin *Diadema antillarum* in the early 1980s due to disease led to an overgrowth of macroalgae that decimated coral reefs (see Gardner et al. 2003:960; Jackson et al. 2001:631; Lessios 1988). This phenomenon appears related to overfishing of larger herbivorous species, which then increased the density of species in lower trophic levels, creating an environment for diseases to transmit more easily (Jackson et al. 2001:635).

Although the transport of nonindigenous plants and animals from the Old World (e.g., grasses, palms, vegetables, sugarcane, breadfruit, cattle, horses, chickens, pigs) has drastically altered Caribbean island environments to the point where many no longer have even a small percentage of their endemic flora and fauna, the decrease or extermination of several indigenous species of fish, crabs, rats, and birds, and an increased reliance on terrestrial and/or horticulturally important foods that required the clearance of forests, has severely affected marine taxa through time. As humans have continued to transform insular Caribbean landscapes historically and in more recent times, these problems have only been exacerbated (Jackson et al. 2001). As Jackson and Sala (2001 p. 279) note:

> The size of animals [worldwide] is falling precipitously, not just in the loss of megafauna, but also in the slow, unrelenting decrease in average size and trophic level of the species that remain. Vertebrates are disappearing rapidly and being replaced by smaller and smaller invertebrates and superabundant microbes.

Similar to many other regions worldwide, we still lack sufficient archaeological and ecological data that would give us better resolution on how marine fisheries have changed over time. As we continue to conduct further archaeological research on islands throughout the West Indies, it is becoming more apparent that conservation biologists have sorely underestimated the impact that native peoples in the West Indies have had, especially on nearshore marine resources. Archaeological evidence is now showing that the baselines set by such studies and fisheries in general, whether they involve sea mammals, sea turtles, finfish, or mollusks, have been far more depleted than originally thought. We sincerely hope that as archaeologists, ecologists, conservationists, environmental historians, and others in the Caribbean begin to better recognize the need and urgency for conducting synthetic collaborative projects, that we can establish measures to protect and preserve the dwindling ecological diversity of not only West Indian islands, but other marine environments where human agency has already caused extensive, and in some cases, irreversible damage (Hughes et al. 2003).

ACKNOWLEDGMENTS

We thank Torben Rick and Jon Erlandson for inviting us to contribute a chapter to this important collection. We greatly appreciate the contributions of Betsy Carlson, Michelle LeFebvre, and Liz Wing to the analysis of faunal remains that are "baselines" for our work. We also appreciate the comments provided by the editors, David Steadman, Steven James, and outside readers that helped to improve this chapter.

REFERENCES CITED

Adam, P. J.
2004 *Monachus tropicalis. Mammalian Species* 747:1–9.

Bak, R. P. M., G. Nieuwland, and E. Meesters
2005 Coral Reef Crisis in Deep and Shallow Reefs: 30 Years of Constancy and Change in Reefs of Curaçao and Bonaire. *Coral Reefs* 24:475–479.

Baker, C. S., and P. J. Clapham
2002 Marine Mammal Exploitation: Whales and Whaling. In *Encyclopedia of Global Environmental Change,* Vol. 3: *Causes and Consequences of Global Environmental Change,* edited by I. Douglas, pp. 446–450. John Wiley and Sons, Chichester, U.K.

Bellwood, D. R., T. P. Hughes, C. Folke, and M. Nyström
2004 Confronting the Coral Reef Crisis. *Nature* 429:827–833.

Brownell, W. N., and J. M. Stevely
1981 The Biology, Fisheries, and Management of the Queen Conch, *Strombus gigas. Marine Fisheries Review* 43:1–12.

Carder, N., E. Reitz, and J. Crock
2007 Fish Communities and Populations during the Post-Saladoid Period (AD 600/800–1500), Anguilla, Lesser Antilles. *Journal of Archaeological Science* 34:588–599.

Carlson, L. A.
1999 First Contact: the Coralie Site, Grand Turk, Turks and Caicos Islands. Unpublished Ph.D. dissertation, Department of Anthropology, University of Florida, Gainesville.

Carlson, L. A., and W. F. Keegan
2004 Resource Depletion in the Prehistoric Northern West Indies. In *Voyages of Discovery: The Archaeology of Islands,* edited by S. M. Fitzpatrick, pp. 85–107. Praeger, Westport, Connecticut.

Carrillo, E., G. J. W. Webb, and S. C. Manolis
1999 Hawksbill Turtles *(Eretmochelys imbricate)* in Cuba: An Assessment of the Historical Harvest and Its Impacts. *Chelonian Conservation and Biology* 3:264–280.

Cortes, J., and M. J. Risk
1985 A Reef under Siltation Stress: Cahuita, Costa Rica. *Bulletin of Marine Science* 36:339–356.
Crock, J. G., J. B. Petersen, and N. Douglas
1995 Preceramic Anguilla: A View from the Whitehead's Bluff Site. In *Proceedings of the XVth International Congress of Caribbean Archaeology* (Puerto Rico), edited by R. E. Alegría and M. Rodríguez, pp. 283–294. Centro de Estudios Avanzados de Puerto Rico y el Caribe, San Juan, Puerto Rico.
Crosby, A. E.
1972 *The Columbian Exchange: Biological and Cultural Consequences of 1492*. Greenwood Press, Westport, Connecticut.
Davis, D. D.
1988 Coastal Biogeography and Human Subsistence: Examples from the West Indies. *Archaeology of Eastern North America* 16:177–185.
2000 *Jolly Beach and the Preceramic Occupation of Antigua, West Indies*. Yale University Publications in Anthropology 84. Yale University Press, New Haven, Connecticut.
Figueredo, A. E.
1974 The Archaic Period of St. Thomas, Virgin Islands: New Evidence and Interpretations. Paper presented at the 39th Society for American Archaeology Meetings (April), Washington, D.C.
Fish, M., R., I. M. Côté, J. A. Gill, A. P. Jones, S. Renshoff, and A. R. Watkinson
2005 Predicting the Impact of Sea-Level Rise on Caribbean Sea Turtle Nesting Habitat. *Conservation Biology* 19:482–491.
Fitzpatrick, S. M.
2006 A Critical Approach to ^{14}C Dating in the Caribbean: Using Chronometric Hygiene to Evaluate Chronological Control and Prehistoric Settlement. *Latin American Antiquity* 17:389–418.
Fitzpatrick, S. M., and W. F. Keegan
2007 Human Impacts and Adaptation in the Caribbean Islands: An Historical Ecology Approach. *Earth and Environmental Science: Transactions of the Royal Society of Edinburgh* 98:29–45.
Gardner, T. A., I. M. Cote, J. A. Gill, A. Gant, and A. R. Watkinson
2003 Long-Term Region-Wide Declines in Caribbean Corals. *Science* 301:958–960.
Gilliam, D. S., and K. M. Sullivan
1993 Diet and Feeding Habits of the Southern Stingray *Dasyatis americana* in the Central Bahamas. *Bulletin of Marine Science* 52:1007–1013.
Goreau, T. F.
1959 The Ecology of Jamaican Coral Reefs. I. Species and Composition and Zonation. *Ecology* 40:67–90.
Hofman, C. L., and M. L. P. Hoogland
2003 Plum Piece: Evidence for Archaic Seasonal Occupation on Saba, Northern Lesser Antilles around 3300 BP. *Journal of Caribbean Archaeology* 4:1–16.
Hofman, C., A. J. Bright, and M. L. P. Hoogland
2006 Archipelagic Resource Procurement and Mobility in the Northern Lesser Antilles: The View from a 3000-year-old Tropical Forest Campsite on Saba. *Journal of Island and Coastal Archaeology* 1:145–164.
Hughes, T. P., A. H. Baird, D. R. Bellwood, M. Card, S. R. Connolly, C. Folke, R. Grosberg, O. Hoegh-Guldberg, J. B. C. Jackson, J. Kleypas, J. M. Lough, P. Marshall, M. Nystrom, S. R. Palumbi, J. M. Pandolfi, B. Rosen, and J. Roughgarden
2003 Climate Change, Human Impacts, and the Resilience of Coral Reefs. *Science* 301:929–933.
Jackson, J. B. C.
1997 Reefs Since Columbus. *Coral Reefs* 16:S23–S32.
Jackson, J. B. C., and E. Sala
2001 Unnatural Oceans. *Scientia Marina* 65:273–281.
Jackson, J. B. C., M. X. Kirby, W. H. Berger, K. A. Bjorndal, L. W. Botsford, B. J. Bourque, R. H. Bradbury, R. Cooke, J. Erlandson, J. A. Estes, T. P. Hughes, S. Kidwell, C. B. Lange, H. S. Lenihan, J. M. Pandolfi, C. H. Peterson, R. S. Steneck, M. J. Tegner, and R. R. Warner
2001 Historical Overfishing and the Recent Collapse of Coastal Ecosystems. *Science* 293:629–638.
James, S. R.
2004 Hunting, Fishing, and Resource Depression: Prehistoric Cultural Impacts on Animals in Southwest North America. In *The Archaeology of Global Change: The Impact of Humans on Their Environment*, edited by C. L. Redman, S. R. James, P. R. Fish, and J. D. Rogers, pp. 28–62. Smithsonian Books, Washington, D.C.
Keegan, W. F.
1986 The Ecology of Lucayan Arawak Fishing Practices. *American Antiquity* 51:816–825.
1994 West Indian Archaeology. 1. Overview and Foragers. *Journal of Archaeological Research* 2:255–284.
2000West Indian Archaeology. 3. Ceramic Age. *Journal of Archaeological Research* 8:135–167.
Keegan, W. F., R. W. Portell, and J. Slapcinsky
2003 Changes in Invertebrate Taxa at Two Pre-Columbian Sites in Southwestern Jamaica, AD 800–1500. *Journal of Archaeological Science* 30:1607–1617.
Keegan, W. F., S. M. Fitzpatrick, K. Sullivan Sealey, M. LeFebvre, and P. T. Sinelli
2007 The Role of Small Islands in Marine Subsistence Strategies: Case Studies from the Caribbean. In review.

Lessios, H. A.
1988 Mass Mortality of *Diadema Antillarum* in the Caribbean: What Have We Learned? *Annual Review of Ecology and Systematics* 19:371–393.

Lewis, J. B.
2002 Evidence from Aerial Photography of Structural Loss of Coral Reefs at Barbados, West Indies. *Coral Reefs* 21:49–56.

Long, E.
1774 *The History of Jamaica, or General Survey of the Ancient and Modern State of that Island: With Reflections on Its Situations, Settlements, Inhabitants, Climate, Products, Commerce, Laws, and Government.* Reprinted 1970. Frank Cass, London.

Lundberg, E.
1989 Preceramic Procurement Patterns at Krum Bay, Virgin Islands. Ph.D. dissertation, University of Illinois, Urbana-Champagne. University Microfilms, Ann Arbor.

McKillop, H. I.
1985 Prehistoric Exploitation of the Manatee in the Maya and Circum-Caribbean Areas. *World Archaeology* 16:337–353.

MacPhee, R. D. E., and P. A. Marx
1997 The 40,000-year Plague: Humans, Hyperdisease and First-Contact Extinctions. In *Natural Change and Human Impact in Madagascar,* edited by S. M. Goodman and B. D. Patterson, pp. 169–217. Smithsonian Institution Press, Washington, D.C.

MacPhee, Ross D. E., Derek C. Ford, and Donald A. McFarlane
1989 Pre-Wisconsinan Mammals from Jamaica and Models of Late Quaternary Extinction in the Greater Antilles. *Quaternary Research* 31:94–106.

Marmontel, M., S. R. Humphrey, and T. J. O'Shea
1997 Population Viability Analysis of the Florida Manatee *(Trichechus manatus latirostris),* 1976–1991. *Conservation Biology* 11:467–481.

Mignucci-Giannoni, A. A., R. A. Montoya-Ospina, N. M. Jiminéz-Marrero, M. A. Rodríguez-López, E. H. Williams, Jr., and R. K. Bonde
2000 Manatee Mortality in Puerto Rico. *Environmental Management* 25:189–198.

Morgan, G. S., and C. A. Woods
1986 Extinction and the Zoogeography of West Indian Land Mammals. *Biological Journal of the Linnean Society* 28:167–203.

Narganes Storde, Y. M.
1991 Los Restos Faunísticos del Sitio de Puerto Ferro Vieques, Puerto Rico. In *Proceedings of the Thirteenth International Congress for Caribbean Archaeology,* edited by A. Cummins and P. King, pp. 94–114. Barbados Museum and Historical Society, Bridgetown.

National Oceanic and Atmospheric Administration
2004 Office of Sustainable Fisheries Stock Assessment Report. U.S. Fish and Wildlife Service, Jacksonville, FL.

Nero, V. L., and K. Sullivan Sealey
2005 Characterization of Tropical Near Shore Fish Communities by Coastal Habitat Status on Spatially Complex Island Systems. *Environmental Biology of Fishes* 73:437–444

Newsom, L. A., and E. S. Wing
2004 *On Land and Sea: Native American Uses of Biological Resources in the West Indies.* The University of Alabama Press, Tuscaloosa.

Precht, W. F.
2002 Endangered Acroporid Corals of the Caribbean. *Coral Reefs* 21:41–42.

Randall, J. E.
1965 Grazing Effect on Sea Grasses by Herbivorous Reef Fishes in the West Indies. *Ecology* 46:255–260.

Reeves, R. R., S. L. Swartz, S. E. Wetmore, and P. J. Clapham
2001 Historical Occurrence and Distribution of Humpback Whales in the Eastern and Southern Caribbean Sea, Based on Data from American Whaling Logbooks. *Journal of Cetacean Research and Management* 3:117–129.

Richard, G.
1994 Premier Indice d'une Occupation Précéramique en Guadeloupe Continentale. *Journal de la Société des Américanistes* 80:241–242.

Richardson, B., C.
1975 The Overdevelopment of Carriacou. *Geographical Review* 65:390–399.

Roberts, C. M.
1995 Rapid Build-up of Fish Biomass in a Caribbean Marine Reserve. *Conservation Biology* 9:815–826.

Romero, A., R. Baker, J. E. Cresswell., A. Singh, A. McKie, and M. Manna
2002 Environmental History of Marine Mammal Exploitation in Trinidad and Tobago, W.I., and Its Ecological Impact. *Environment and History* 8:255–274.

Rouse, I.
1992 *The Taínos: Rise and Decline of the People who Greeted Columbus.* Yale University Press, New Haven, Connecticut.

Russ, G. R.
1991 Coral Reef Fisheries: Effects and Yields. In *The Ecology of Fishes on Coral Reefs,* edited by P. F. Sale, pp. 601–635. Academic Press, San Diego.

Salvaggio J. E.
1992 Fauna, Flora, Fowl, and Fruit: Effects of the Columbian Exchange on the Allergic Response

of New and Old World Inhabitants. *Allergy Proceedings* 13:335–344.

Sara, T. R., and W. F. Keegan (editors)
2004 Archaeological Survey and Paleoenvironmental Investigations of Portions of U.S. Naval Station, Guantanamo Bay, Cuba. Geo-Marine, Inc., Newport News, Virginia.

Sloan, H.
1707–1725 *A Voyage to the Islands Madera, Barbados, Nieves, St. Christopher's, and Jamaica; with a Natural History of the Herbs and Trees, Four-footed Beasts, Fishes, Birds, Insects, Reptiles, &c. of the Last of those Islands.* In two volumes. Printed by the author, London.

Smith, T. D., and R. R. Reeves
2003 Estimating American Nineteenth-Century Catches of Humpback Whales in the West Indies and Cape Verde Islands. *Caribbean Journal of Science* 39:286–297.

Stager, J. C., and W. Chen
1996 Fossil Evidence of Shell Length Decline in Queen Conch (*Strombus gigas* L.) at Middleton Cay, Turks and Caicos Islands, British West Indies. *Caribbean Journal of Science* 32:14–20.

Steadman, D. W., and S. Jones
2006 Long-Term Trends in Prehistoric Fishing and Hunting on Tobago, West Indies. *Latin American Antiquity* 17:316–334.

Steadman, D. W., and A. V. Stokes
2002 Changing Exploitation of Terrestrial Vertebrates during the Past 3000 Years on Tobago, West Indies. *Human Ecology* 30:339–367.

Steadman, D. W., G. K. Pregill, and S. L. Olson
1984 Fossil Vertebrates from Antigua, Lesser Antilles: Evidence for Late Holocene Human-Caused Extinctions in the West Indies. *Proceedings of the National Academy of Sciences* 81:448–4451.

Steadman, D. W., P. S. Martin, R. D. E. MacPhee, A. J. T. Jull, H. G. McDonald, C. A. Woods, M. Iturralde-Vinent, and G. W. L. Hodgins
2005 Asynchronous Extinction of Late Quaternary Sloths on Continents and Islands. *Proceedings of the National Academy of Sciences* 102:11763–11768.

Stokes, A. V., and W. F. Keegan
1996 A Reconnaissance for Prehistoric Archaeological Sites on Grand Cayman. *Caribbean Journal of Science* 32:425–430.

Swartz, S. L., T. Cole, M. A. McDonald, J. A. Hildebrand, E. M. Oleson, A. Martinez, P. J. Clapham, J. Barlows, and M. L. Jones
2003 Acoustic and Visual Survey of Humpback Whale *(Megaptera novaeangliae)* Distribution in the Eastern and Southeastern Caribbean Sea. *Caribbean Journal of Science* 39:195–208.

Theile, S.
2001 Queen Conch Fisheries and their Management in the Caribbean. Technical Report to the CITES Secretariat. TRAFFIC Europe, Cambridge.

Timm, R. M., R. M. Salazar, and A. T. Peterson
1997 Historical Distribution of the Extinct Tropical Seal, *Monachus tropicalis* (Carnivora: Phocidae). *Conservation Biology* 11:549–551.

Torres, R. E.
2003 Ecological Consequences and Population Changes from Intensive Fishing of a Shallow-Water Mollusk, *Strombus gigas*: A Study of Parque Nacional del Este, Dominican Republic. Unpublished Ph.D. dissertation, University of Miami, Coral Gables.

Watters, D. R., J. Donahue, and R. Struckenrath
1992 Paleoshorelines and the Prehistory of Barbuda, West Indies. In *Paleoshorelines and Prehistory: An Investigation of Method*, edited by L. L. Johnson, pp. 15–52. CRC Press, Boca Raton, Florida.

Weil, E. M., and R. Laughlin
1984 Biology, Population Dynamics, and Reproduction of the Queen Conch, *Strombus gigas* Linné in the Archipelago De Los Roques National Park. *Journal of Shellfish Research* 4:48–53.

Wilson, S., H. B. Iceland, and T. R. Hester
1998 Preceramic Connections between Yucatan and the Caribbean. *Latin American Antiquity* 9:342–352.

Wing, E. S.
2001 The Sustainability of Resources Used by Native Americans on Four Caribbean islands. *International Journal of Osteoarchaeology* 11:112–126.

Wing, S. R., and E. S. Wing
2001 Prehistoric Fisheries in the Caribbean. *Coral Reefs* 20:1–8.

8

Possible Prehistoric Fishing Effects on Coastal Marine Food Webs in the Gulf of Maine

Bruce J. Bourque, Beverly J. Johnson, and Robert S. Steneck

THE GULF OF MAINE is one of the world's most productive marine ecosystems. Its coastal codfish stocks attracted European colonists, including the Pilgrims at Plymouth, Massachusetts, on the shores of the then appropriately named Cape Cod. Today, however, Atlantic cod *(Gadus morhua)* and virtually all large-bodied fishes are rare and "ecologically extinct" (sensu Estes et al. 1989) from coastal zones of the Gulf of Maine. The decline of cod and other groundfishes is widely believed to be the result of overfishing (Jackson et al. 2001). Further, because cod were the dominant predator in Gulf of Maine waters (Steneck and Carlton 2001) their functional absence affects the entire ecosystem. Declines in apex predators release prey species at lower trophic levels, so mesopredators and herbivores often increase in abundance, becoming new fisheries targets (Jackson et al. 2001; Steneck and Sala 2005; Steneck et al. 2004). This process is known as "fishing down marine food webs" (Pauly et al. 1998).

Conventional wisdom assumes that fish stocks remained "pristine" until targeted by commercial fishing fleets to supply foreign markets, thus initiating a historical process of fishing down the marine food web (e.g., Jackson et al. 2001; Lotze et al. 2006). Thus, in the western North Atlantic a pristine state—one unaffected by human activity—is assumed to have persisted until European contact (Jackson et al. 2001; Lotze and Milewski 2004). Archaeological studies in the Aleutians (Simenstad et al. 1978), California (Erlandson et al. 2004, 2005), and the Caribbean (Wing and Wing 2001), however, have presented compelling evidence for prehistoric declines of apex predators due to fishing. If such prehistoric effects were widespread, then perhaps we will have to reset our timeline for when coastal ecosystems first departed from their pristine baseline (Erlandson and Fitzpatrick 2006; see also Erlandson and Rick, this volume).

In this chapter, we present archaeological and isotope data from a coastal site in Maine that suggests localized fishing down of nearshore coastal food webs may have begun thousands of years before European colonization. Specifically, we report on changes in the relative abundance of faunal remains in the well-preserved and best-studied midden in coastal Maine, the Turner Farm site in Penobscot Bay. We also use stable nitrogen and carbon isotope analysis of prehistoric and modern bone collagen of cod, sculpin, flounder, and humans to estimate relative trophic positions of each species and the degree to which coastal, kelp-derived organic matter supported the food web. We do this to provide long-term data on the magnitude and scale (spatial and temporal) of environmental change in nearshore coastal marine settings in the western North Atlantic to better understand when the region departed from pristine conditions, so managers can set realistic goals for ecosystem restoration.

GULF OF MAINE KELP FOREST ECOSYSTEM AND FOOD WEBS PAST AND PRESENT

Although the Gulf of Maine is highly productive, it has never been highly diverse (Steneck et al. 2002; Witman et al. 2004). This naturally low diversity results from three factors. First, the North Atlantic is relatively young and only recently inoculated with a subset of higher taxa from the eastern North Pacific (Vermeij 1991). Second, of the possible species relatively few could withstand the western North Atlantic's shallow water temperature extremes (Adey and Steneck 2001). Finally, North American glaciers expanded as recently as 18,000 BP (uncorrected radiocarbon years BP) to cover most of North America's rocky shores, resulting in local extinctions and biogeographic zone compression. As a result, the comparatively few species that lived in the Gulf of Maine were found in high abundance and provided key ecological services in the ecosystem.

The Gulf of Maine's coastal ecosystem comprises four dominant trophic levels: apex and mesopredators, herbivores, and algal primary producers (Figure 8.1). The algae include large structure-producing kelp such as *Laminaria* spp. (Steneck et al. 2002) and other algal forms. Kelp forests dominate nearshore rocky habitats from zero to 20–45 m deep, depending on water clarity (Vadas and Steneck 1988). Algae contribute significantly to the energy flow in nearshore food webs (Duggins et al. 1989; Mann 1973).

Most kelp forest ecosystems are sensitive to changes in herbivore populations, particularly the sea urchin *Strongylocentrotus droebachiensis* (Figure 8.1; Steneck et al. 2002). Strong interactions between adjacent trophic levels can create "trophic cascades" (sensu Paine 1980) in which predator declines release limits on herbivorous sea urchin population growth, resulting in urchin population explosions and widespread kelp deforestation (Steneck et al. 2002). Such sea urchin–induced algal deforestations are common globally and often result from the overfishing of predators (reviewed by Steneck et al. [2002]). However, in species-depauperate ecosystems such as the Gulf of Maine, there are so few species at each trophic level in the food web that population declines in a few key species can trigger significant changes throughout the system (Steneck et al. 2002).

Steneck et al. (2004) examined long-term ecological change in the Gulf of Maine and proposed that its coastal kelp forest ecosystem had remained relatively stable, dominated by apex predators and kelp for over 4,000 years (Figure 8.1). Evidence for this phase came from pooled archaeological records, mainly from the Penobscot Bay area of Maine, indicating a long-term abundance of large predatory fish, especially the demersal cod in nearshore waters beginning 4,200 to 4,000 years ago and ending early in the last century. Steneck et al. (2004) inferred the persistence of the three-trophic-level system of apex predators (primarily Atlantic cod), herbivores (the green sea urchin), and algae (mostly kelp) throughout that phase (Figure 8.1) with no hint of major changes in food webs. Mesopredators such as lobsters and crabs are generally not considered strong interactors. Then, in the last century, two significant ecosystem phase changes occurred. The first was a rapid, fisheries-induced decline in predatory finfish that began around AD 1930 with large-scale mechanized coastal trawling (Steneck and Carlton 2001). Affected species included cod and most other large-bodied, commercially important species such as haddock, hake, and halibut. Their decline allowed the

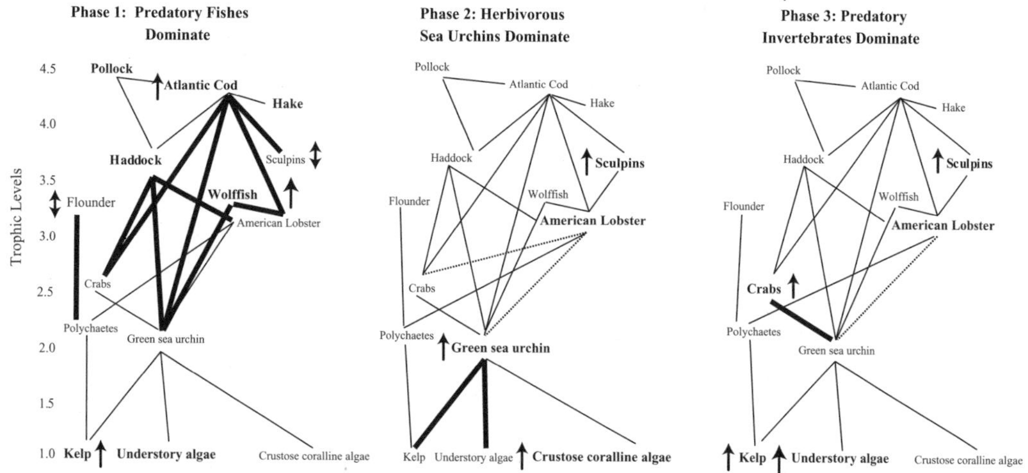

FIGURE 8.1. Three phases in marine food webs in coastal Maine over the past 5,000 years (modified from Steneck et al. 2004). All species determined to have been abundant at one time were plotted according to their assigned trophic level (Table 8.1). Only fully marine organisms are included; birds and terrestrial mammals (e.g., sea mink) were excluded. Abundant species are identified by larger font and boldface; rare or low-abundance species are shown in smaller regular type. Most trophic linkages (lines connecting species) have been demonstrated with ecological studies (see Steneck et al. 2004). Apex predators were all fish with a fractional above trophic level (TL) $\geq$ 4.They feed on mesopredators (TL >2, <4) and herbivores (TL ~ 2). Algae are primary producers in this system (TL 1). All TL values and scientific binomials are in Table 8.1. Interaction strengths correspond to the width of trophic linkage lines. Note that some species are weak interactors in this system. Lobsters' trophic linkages are weak despite their abundance in recent years because they feed primarily on lobster bait. Functionally dominant taxa at each trophic level are illustrated with an arrow indicating abundance. Double-headed arrows (pointing up and down) indicate taxa that fluctuate in importance during the identified phase.

expansion of prey species such as crabs, lobsters, and eventually sea urchins, which in turn caused a decrease in kelp forests, amounting to a functional phase shift in the system from three to two trophic levels (Figure 8.1b). The second sudden change was triggered by the unprecedented fishing of sea urchins, which began in AD 1987, peaked in AD 1993, and then quickly culminated in widespread stock collapse and another phase shift from two to one trophic level that persists to the present. Here kelp forests have expanded significantly, and crabs and lobster are the top predators (Figure 8.1c; Steneck et al. 2004).

These two recent phase shifts are examples of "trophic level dysfunction," where the abundance of organisms at a highly interactive trophic level declined to the point that they no longer limited the abundance of their prey, and their functional role was lost. The inability of the Gulf of Maine's coastal ecosystem to resist these phase shifts stems from its low biodiversity, which fails to provide ecologically equivalent species to buffer against trophic level dysfunction (Steneck at al. 2004). To date, no study has considered that such phase shifts may have occurred prehistorically. Such a consideration requires that we examine the archaeological record and cultural history of the region.

THE PENOBSCOT BAY ARCHAEOLOGICAL RECORD AND CULTURAL HISTORY

Penobscot Bay is centrally located on the Maine Coast within the Gulf of Maine. Most important archaeological sites in the bay are located on its many islands, including the Turner Farm site on North Haven Island (Figure 8.2). These sites have been the focus of archaeological research since 1970, and the bay now ranks among the better-studied archaeological regions of the western Atlantic Coast of North America. The number of excavated sites currently stands at over 40, the number of catalogued artifacts exceeds 10,000, and the number of faunal

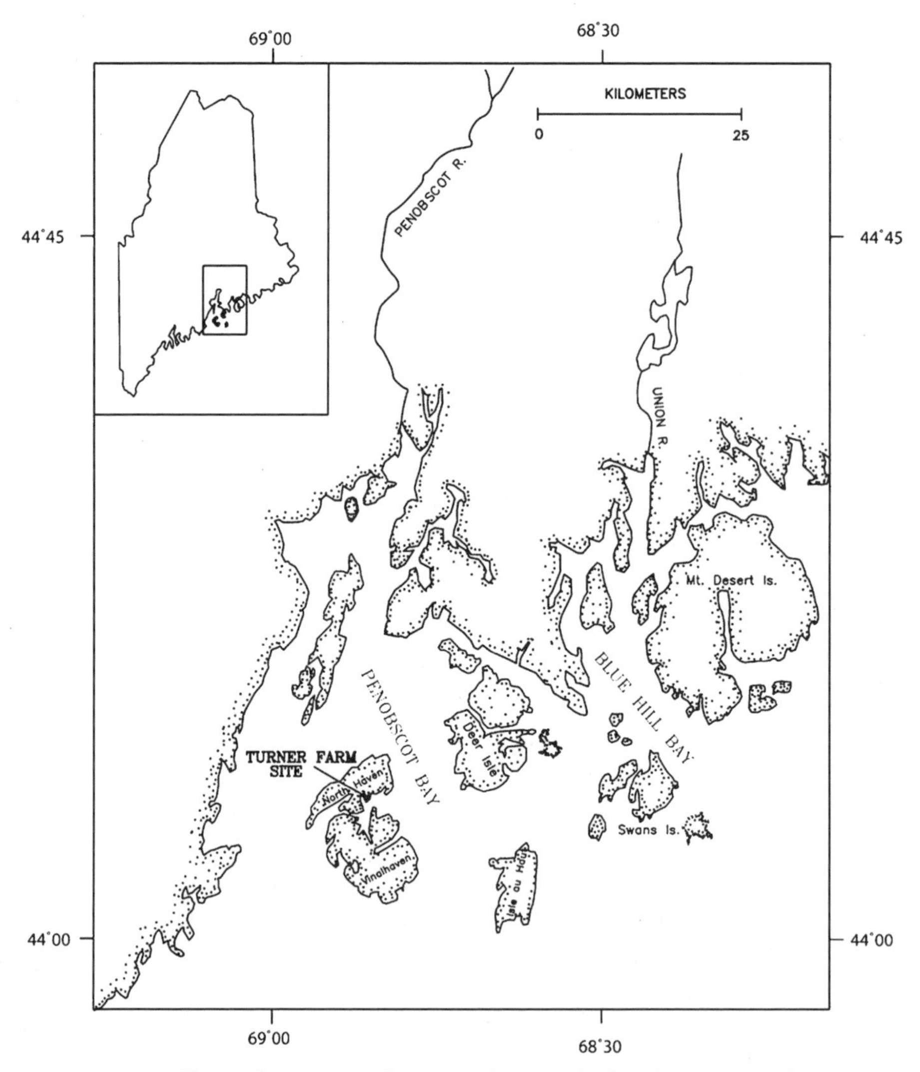

FIGURE 8.2. Map of the Penobscot Bay area showing North Haven Island on the outer coast of Maine, where most midden sites, including the Turner Farm site, were located.

specimens is exponentially larger. For purposes of this analysis, however, we focus on faunal samples from the Turner Farm site, a large, extensively analyzed shell midden on North Haven Island (see Bourque 1995; Spiess and Lewis 2001). The faunal samples from this site span, more or less continuously, the period from about 5000 to 400 BP. We grouped the samples into five occupation periods, although the record for the earliest and latest present problems. The faunal sample from the earliest occupation is very small, but we include it because it provides a glimpse of what early people ate in coastal Maine. The latest sample dates to around the time of first sustained European contact (~400 BP) but came from a stratum disturbed by historic agricultural plowing, Bourque's (1995) "plow zone" (Table 8.1). We include it in the present analysis despite the likelihood that it is to some extent contaminated by late prehistoric material because it exhibits trends we regard as significant. We use published chronostratigraphic dates (i.e., Bourque 1995) to show the overall trends, although the dates will likely be somewhat revised in the near future.

TABLE 8.1

Percent Bone Fragments from Turner Farm Site

COMMON NAME	SCIENTIFIC NAME	FRACTIONAL TROPHIC LEVEL	OCC. 1 4350 BP	OCC. 2 4100 BP	OCC. 3 3550 BP	OCC. 4 1600 BP	PLOW ZONE 400 BP
Atlantic cod	*Gadus morhua*	4.4	74.36	22.32	16.05	2.17	2.15
Flounder total	*Pleuronectes* sp.	3.2	12.82	12.02	16.36	37.68	31.79
Sea mink	*Mustela macrodon*	4.0	0.00	12.02	23.77	9.86	11.06
Fish unidentified	Class Actinopterygii	3.8	12.82	5.02	5.56	20.29	11.82
Flounder, winter	*Pleuronectes americanus*	3.2	0.00	6.61	9.10	14.76	17.62
Bivalve, softshell clams	*Mya arenaria*	2.1	0.00	12.95	15.12	1.50	10.74
Swordfish	*Xiphias gladius*	4.5	0.00	16.78	3.70	0.42	0.11
Sculpin	*Myoxocephalus* spp.	3.6	0.00	1.85	2.93	7.60	7.41
Atlantic sturgeon	*Acipenser oxyrhynchus*	3.4	0.00	1.59	1.39	1.16	2.04
Flounder, yellowtail	*Limanda ferruginea*	3.2	0.00	0.40	1.23	1.87	1.29
Atlantic tomcod	*Microgadus tomcod*	3.3	0.00	4.23	0.15	0.02	0.21
Seal, harbor	*Phoca vitulina*	4.0	0.00	0.79	0.93	0.39	0.64
Seal (unidentified)	*Phoca* sp.	4.0	0.00	0.13	1.70	0.74	0.00
Seal, gray	*Halichoerus grypus*	4.0	0.00	0.53	0.46	0.20	0.54
Flounder, American dab	*Hippoglossoides platessoides*	3.7	0.00	0.00	0.31	0.47	0.75
Spiny dogfish	*Squalus acanthias*	4.3	0.00	0.13	0.00	0.23	0.97
Cunner	*Tautogolabrus adsperus*	3.5	0.00	0.00	0.62	0.06	0.00
Atlantic herring	*Clupea harengus*	3.2	0.00	0.66	0.00	0.02	0.00
Blue mussel	*Mytilus edulis*	2.1	0.00	0.26	0.31	0.02	0.00
Haddock	*Melanogrammus aeglefinus*	3.6	0.00	0.40	0.00	0.05	0.11
Atlantic halibut	*Hippoglossus hippoglossus*	4.6	0.00	0.00	0.00	0.29	0.21

TABLE 8.1 (*continued*)

COMMON NAME	SCIENTIFIC NAME	FRACTIONAL TROPHIC LEVEL	OCC. 1 4350 BP	OCC. 2 4100 BP	OCC. 3 3550 BP	OCC. 4 1600 BP	PLOW ZONE 400 BP
Pollock	*Pollachius virens*	4.5	0.00	0.40	0.00	0.02	0.00
Atlantic wolffish	*Anarhichas lupus*	3.2	0.00	0.00	0.00	0.00	0.32
Flounder, windowpane	*Scophthalmus aquosus*	3.2	0.00	0.00	0.15	0.02	0.11
Atlantic salmon	*Salmo salar*	4.5	0.00	0.26	0.00	0.00	0.00
American eel	*Angulla rostrata*	3.7	0.00	0.00	0.15	0.02	0.00
Sea urchin	*Strongylocentrotus droebachiensis*	2.0	0.00	0.13	0.00	0.02	0.00
Seal, harp	*Phoca groenlandica*	4.0	0.00	0.13	0.00	0.00	0.00
Alewife	*Alosa pseudoharengus*	3.5	0.00	0.13	0.00	0.00	0.00
Gastropod, waved whelk	*Buccinum undatum*	2.6	0.00	0.13	0.00	0.00	0.00
Bivalve, ocean quahog	*Arctica islandica*	2.1	0.00	0.13	0.00	0.00	0.00
White shark	*Carcharodon carcharias*	4.6	0.00	0.00	0.00	0.00	0.11
Harbor porpoise	*Phocoena phocoena*	4.2	0.00	0.00	0.00	0.06	0.00
Cusk	*Brosme brosme*	4.0	0.00	0.00	0.00	0.05	0.00
Right whale	*Eubalaena glacialis*	4.2	0.00	0.00	0.00	0.03	0.00
Atlantic mackerel	*Scomber scombrus*	3.5	0.00	0.00	0.00	0.03	0.00
Bluefish	*Pomatomus saltatrix*	4.5	0.00	0.00	0.00	0.02	0.00

NOTE: From Spiess and Lewis (2001).Species ordered from highest to lowest average percent bone fragments pooled across the four distinct occupations ("Occ.") with median ages of occupation (from Bourque 1995). Fractional trophic levels for each species are from Froese and Pauly (2002) and Steneck et al. (2004).

The earliest clear evidence of human occupation in the Penobscot Bay area comprises a scattering of Early and Middle Archaic projectile points in styles found over large portions of northeastern North America that date between about 8500 and 6000 BP (Bourque 2001). The earliest intact archaeological components, however, are manifestations of the Small Stemmed point tradition, found primarily from southern New York to the mid-Maine Coast and dating between about 5000 and 4500 BP (Bourque 1995). The Turner Farm site produced the only significant faunal sample from this period.

The next clear cultural manifestation in the region, known as the Moorehead Phase, seems to be a regional descendant of the Small Stemmed point tradition and dates between about 4500 and 3800 BP (Bourque 1995). This was a population located between the Kennebec and St. John's rivers and devoted to a maritime lifestyle that included extensive fishing for cod and swordfish, as well as terrestrial hunting (Figures 8.3 and 8.4). Also characteristic of the Moorehead Phase is a complex pattern of mortuary behavior that included multiple, large cemeteries with ocher-filled graves furnished with beautifully crafted locally made artifacts and exotics from other regions. Some of the exotic artifacts come from sources (e.g., Ramah Bay, Labrador) as far as 1,000 miles away (Bourque 1995).

All traces of the Moorehead Phase disappeared around 3800 BP. Immediately thereafter a very different cultural manifestation appeared on the scene. Known as the Susquehanna Tradition, it took over territory not only of Moorehead Phase people but also apparently of all contemporaneous Northeastern cultures as far north as the St. Lawrence River (Bourque 1995:244–254). The earliest reliable dates for the Susquehanna Tradition in Maine do not exceed 3700 BP. There is now an archaeological consensus that the Susquehanna Tradition represents a complete break with the Moorehead Phase, making the arrival of a new population the likely explanation for its appearance. In southern New England, where the Susquehanna Tradition also looks like a population replacement, the pattern remained in place and changed over time for nearly 1,000 years. In Maine, however, its tenure was apparently brief, with the latest reliable radiocarbon dates falling no later than 3500 BP, and the artifact styles that typify the later phases of the Susquehanna Tradition to the south are rare or absent there.

The archaeological record becomes weak and ambiguous after the disappearance of the Susquehanna Tradition and does not revive until around 2800 BP when the earliest ceramic pottery appears and midden accumulation resumes. Thereafter, the rate of midden accumulation at the Turner Farm site and elsewhere in Penobscot Bay appears to increase throughout the remainder of the prehistoric period (Bourque 1995:169–222).

Beginning in the mid-sixteenth century, the presence of Europeans in the Gulf of St. Lawrence began to impact populations in the Gulf of Maine and, by AD 1600 (~400 BP), Europeans began to appear there on a regular basis. By that time, European demand for beaver pelts had caused indigenous economies to shift toward terrestrial beaver hunting at the expense of pursuing marine foods (Bourque 1995; Bourque and Whitehead 1994). Thereafter, native occupation of the coastal zone decreased as epidemics drastically reduced populations, and as new population amalgamations established villages in the interior.

FAUNAL TRENDS AT THE TURNER FARM SITE

To search for trends in faunal assemblages and fishing activity over time, we quantified the abundance and trophic level of all dominant species found in the Turner Farm midden (Table 8.1). Specifically, we assigned each species a fractional trophic level (TL) value (Froese and Pauly 2002) based on the trophic level of prey it consumed. Thus, primary producers, herbi-

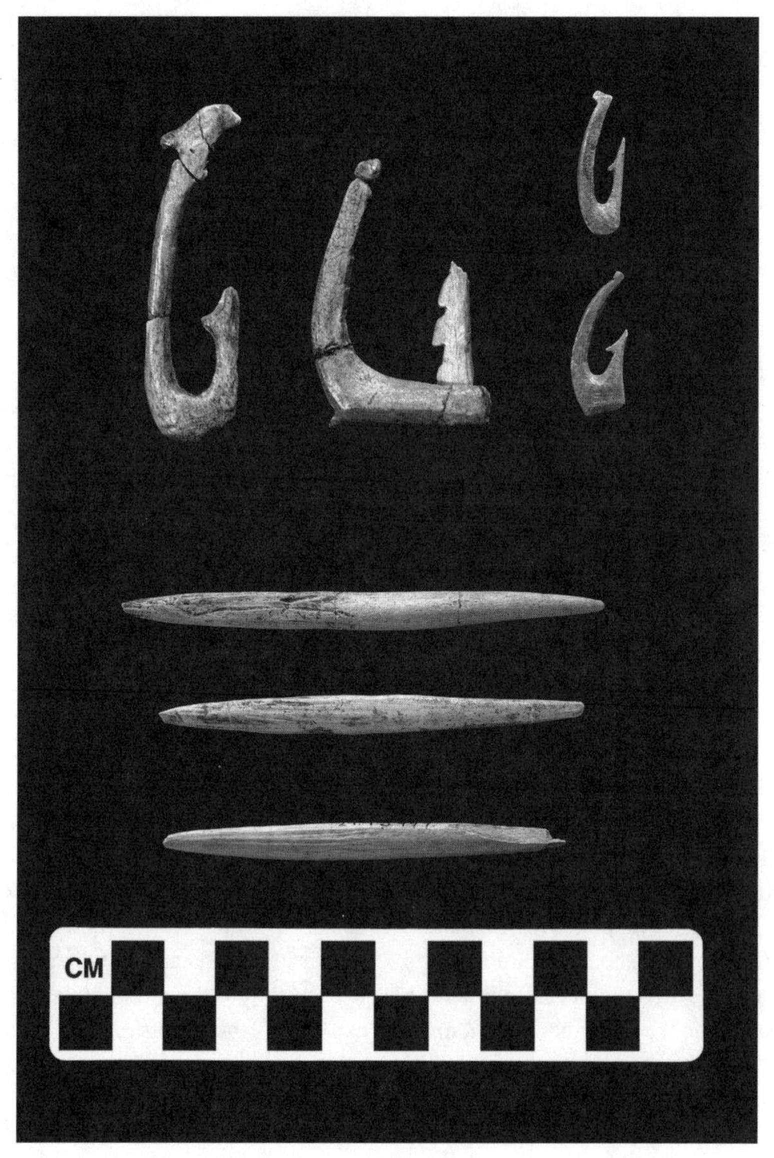

FIGURE 8.3. Bone artifacts from the Turner Farm site. The large hooks in the upper row are from Occupation 2 (~4350 BP) and probably were used to catch cod. The small hooks are from late prehistoric strata and are appropriately sized to catch flounder, the bones of which were extremely abundant in these strata. The bone points below the hooks represent an artifact type commonly found in late prehistoric contexts where flounder bone is abundant, and may have served as the central piercing element in leisters, such as the example shown in Figure 8.4.

vores, mesopredators, and apex predators were assigned TL numbers ranging from 1 to 4.6 (Table 8.1). All taxa were compiled at the level of species or at the TL possible (some bones could only be identified as seal or flounder, for example). Operationally we defined apex predators as species with a fractional trophic level of four or more. Mesopredators were assigned TLs between 3.0 and 4.0. We assigned lower fractional TLs to invertebrate mesopredators such as crabs (Figure 8.1), although they were absent from the midden (Table 8.1). Species with

FIGURE 8.4. Leisters were used historically throughout the Maine–Maritime Provinces region to spear fish, including flounder. The central piercing element of this nineteenth-century Penobscot specimen is of steel and may represent a modern modification of the bone point used prehistorically. The shaft of this example has been sawed off.

TLs < 3.0, including suspension feeders and herbivores, were relatively rare and thus were pooled for this analysis.

Apex predators declined and mesopredators increased proportionally over the five-occupation sequence (Figure 8.5). The most important trends were the decline in Atlantic cod and increase in flatfish (four species of flounder including American dab). Another group of mesopredators, the sculpins (probably several species), also increased. Today, sculpins are the most abundant fish in Maine's cod-depleted nearshore benthic communities (Figure 8.1b and c; Steneck 1997). There was no clear trend among the suspension feeders or herbivores.

Two groups of apex predators, seals and dogfish, increased slightly in abundance during the prehistoric period (Figure 8.5, Table 8.1), possi-

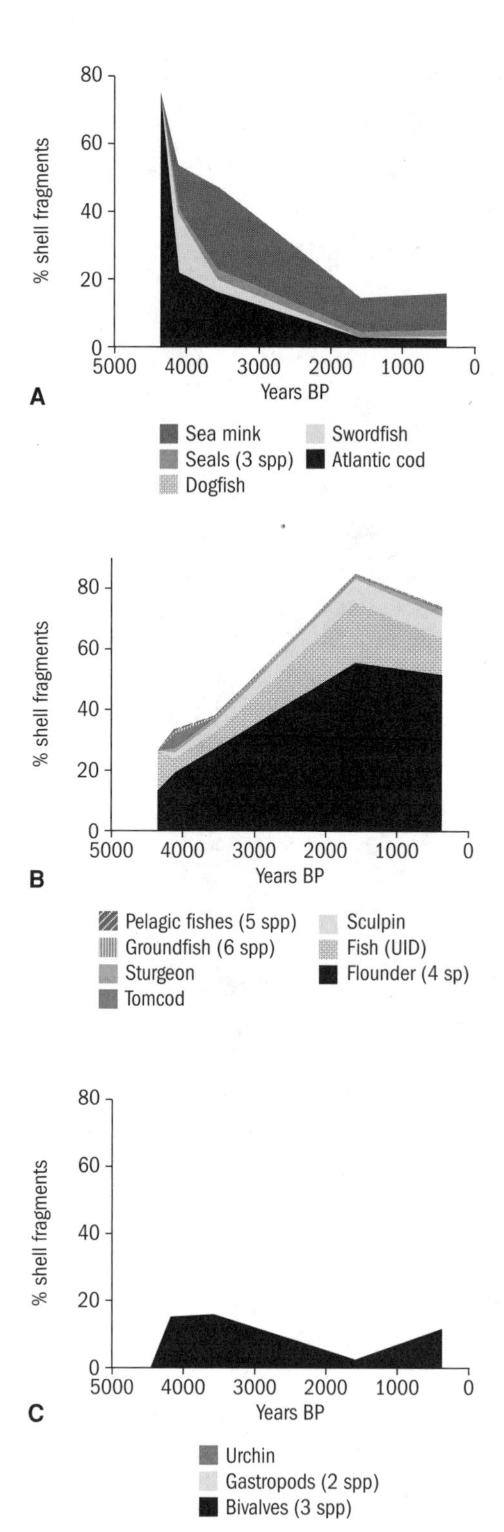

FIGURE 8.5. The proportion of bone and shell fragments in middens with median dates ranging from 4350 to 400 BP separated by trophic level groups of (A) apex predators, TL ≥ 40; (B) mesopredators TL ≥ 3.0, <4.0; and (C) suspension feeders and herbivores TL < 3.0 (data in Table 8.1).

bly because their populations were being released from competitive suppression as cod decreased in abundance and size over time. Recent declines in cod stocks in the 1980s to 1990s were followed by such an increase in dogfish (Fogarty and Murawski 1998) and seals (Trzcinski et al. 2006). The upward trend in seals is particularly interesting because they swallow their prey (predominantly fish) whole and would have been at a competitive disadvantage against large cod, which eat a wider size range of fish, including all species consumed by seals. The large cod found in the Turner Farm middens (i.e., averaging 1 m long) may well have reduced the prey available for seals, thereby putting them at a competitive advantage over the seals.

We suspect that the changes we have observed at the Turner Farm site were caused by a human-induced reduction of apex predators brought about by fishing practices (e.g., via enhanced fishing pressures, altered fishing methods, technology, or traditions, such as the abrupt culture change evident during the brief occupation by people of the Susquehanna Tradition). We regard climate change as an unlikely cause because climate conditions for cod in coastal zones should have improved during the cooling of the Little Ice Age of the thirteenth to nineteenth centuries AD (Grove 2001; Planque and Fredou 1999), when their abundance at Turner Farm reached a low point. Further, by the 1600s AD, large and abundant cod were reported at numerous other nearby coastal areas, such as in Rosier's AD 1605 account of fishing at nearby Pemaquid (Quinn and Quinn 1983:25–311).

STABLE ISOTOPES AND MARINE FOOD WEB MODELS

Faunal remains from the Turner Farm site described briefly above and elsewhere (Spiess and Lewis 2001) indicate that prehistoric people consumed many species of marine fishes for at least 5,000 years before European colonization (Table 8.1). However, as Spiess and

Lewis (2001:86) pointed out, "it is difficult to estimate the relative contribution of fish versus mammals [primarily deer] to the diet of the site's inhabitants" based only on faunal identification. Nor can faunal identification alone reveal where an organism was captured. To shed light on these issues, we turn to stable isotope analyses. We then consider if the several millennia of prehistoric harvesting could have affected relative abundances and even altered marine food webs.

Where isotopically distinct primary producers occur at the base of the food web, it is possible to use stable carbon and nitrogen isotopes of animal tissue (e.g., bone collagen, muscle, lipids) to reconstruct animal diets and energy flow within the ecosystem, as well as animal foraging behavior (Michener and Schell 1994; Michener and Kaufman 2007). In most temperate marine settings, for example, kelp and sea grasses incorporate more ^{13}C-enriched dissolved inorganic carbon during photosynthesis than most species of phytoplankton (Fry and Sherr 1984). Consequently, kelp and sea grasses have more enriched, or more positive, carbon isotope values (expressed as $\delta^{13}C$) than phytoplankton. The $\delta^{13}C$ value of the food sources is passed on, with some modification (~1–2‰ enrichment with each trophic level [Fry and Sherr 1984]), to the tissues of the consumer (Figure 8.6). Similarly, the $\delta^{15}N$ value of the food sources is passed on, with some modification (~3‰ enrichment for each trophic level [Ambrose and DeNiro 1986; Fry 1988; Minagawa and Wada 1984]) to the consumer such that nitrogen isotopes can be used to discern the relative trophic positions of organisms living within an ecosystem (Wada et al. 1991; Figure 8.6).

In Penobscot Bay, the dominant marine primary producers are benthic macroalgae (e.g., kelp), sea grass, and phytoplankton. For various species of macroalgae, carbon and nitrogen isotope values range between −12 and −27‰, and between 5 and 8‰, respectively (Fry 1988; McMahon et al. 2005). Sea grass isotope values range between −3 and −15‰ for carbon (McMillan et al. 1980) and 4 and 6‰ for nitrogen (B. Johnson, unpublished data). For Gulf of Maine phytoplankton, carbon isotope values range between −18 and −27‰, with nitrogen isotope values between 5 and 9‰ (Fry 1988; McMahon et al. 2005). Detritus and dissolved organic matter derived from these photosynthesizing organisms is consumed by passive suspension feeders and deposit feeders such as polychaetes, amphipods, isopods, mollusks, and sea urchin (Josefson et al. 2002; Duggins and Eckman 1994), then passed up the food web, with the appropriate isotope fractionations occurring at each trophic level (e.g., Lesage et al. 2001). In certain settings, it is possible to use isotopes to determine the degree to which marine species depend on various primary producers (Bustamante and Branch 1996; Stephensen et al. 1986; Duggins et al. 1989).

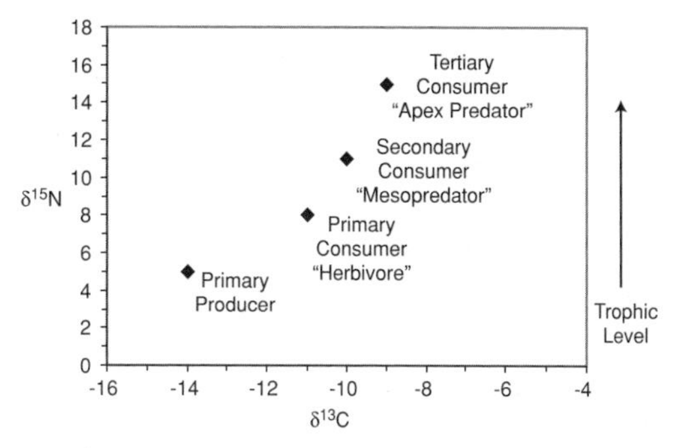

FIGURE 8.6. Simplified models of carbon and nitrogen isotope fractionation in a marine food web where kelp is the only primary producer, and each consumer is subsisting solely on organisms from the next lower trophic level. The largest carbon fractionation occurs at the lowest trophic level energy transfer, when the carbohydrate-bound carbon (i.e., primary producers) is converted to protein-bound carbon (i.e., muscle tissue of consumers). Names used for consumers in food web interactions are included in quotes (see text).

The dominant primary producers in Penobscot Bay generally occupy different ecological settings. Kelp and other benthic macroalgae grow on rocks in relatively shallow nearshore environments, whereas sea grasses (primarily *Zostera marina*) are the dominant producers in more protected, shallow, sediment-dominated environments. Rocky shores dominate most outer exposed habitats near the islands of North Haven and nearby Vinalhaven, as well as the corresponding outer coastal region (Figure 8.2), so the nearshore setting would have been kelp and benthic algal dominated. Sea grass was probably locally abundant primarily in the Fox Island Thoroughfare adjacent to the Turner Farm site.

The food web contribution of kelp and other algae relative to pelagic phytoplankton declines with distance from the shore (Mann 1973; Steneck and Dethier 1994), and phytoplankton become the dominant primary producer in pelagic, offshore systems. Within the nearshore kelp forest ecosystem, sea urchins are dominant herbivores, but they appear in low abundance in the middens at Turner Farm (Spiess and Lewis 2001), implying that they probably would have had little impact on the standing kelp biomass. Sea grass is slower than kelp to break down and is consumed by fewer organisms (Harrison 1989). Epiphitic algae in grass beds, however, is taken up by primary consumers and has isotopic values similar to that of the sea grass (Hoshiko et al. 2006). These trophic pathways presented in Figure 8.1 are localized to these areas of production.

The isotopic composition of marine animal tissues can be used to determine the degree to which these animals forage in nearshore versus offshore settings (Aurioles et al. 2006; Burton et al. 2001; Kaehler et al. 2000; Lesage et al. 2001). Tissue from animals that forage on kelp or sea grass in nearshore waters are enriched in ^{13}C, whereas those that forage in more pelagic, offshore waters are depleted of ^{13}C. When more than two sources of primary carbon are present (e.g., kelp, sea grass, phytoplankton, epiphytic algae), the use of multiple chemical tracers (e.g., nitrogen, carbon, and sulfur isotope compositions and C/N values) and isotope mixing models can elucidate more specific information on energy transfer and food web structure (e.g., Phillips and Koch 2002).

For this study, we analyzed the isotopic composition of well-preserved Middle and Late Holocene deer, bear, cod, sculpin and flounder bone collagen from the Turner Farm site to determine if the major species found in middens are trophically linked to the prehistoric people. We evaluated the degree to which members of the marine community show isotope signatures indicative of nearshore kelp and sea grass communities. We also analyzed modern muscle tissue from cod, flounder, and sculpin collected from the Gulf of Maine to compare to the Holocene record of fish diets and evaluate the degree to which fish diets may have changed over the last 4,000 years.

Cod are trophic generalists that feed on small and large crustaceans and, as they grow and mature, on other fishes (discussed further below). We predict that if cod and other coastal fish were abundant in coastal zones near the Turner Farm site, they would be enriched in ^{13}C, reflecting the presence of kelp and/or sea grass beds. Isotopically depleted cod would reflect a shift to a more phytoplankton-based, offshore food web. Flounder and to a lesser extent, sculpins commonly live in shallow sediment-dominated habitats colonized by the eelgrass *Zostera marina*. Thus, we predict that the isotopic composition of flounder and sculpin will be more enriched in ^{13}C than the cod, reflecting coastal kelp and perhaps some sea grass-derived organics at the base of the food web.

ISOTOPIC TRENDS IN PENOBSCOT BAY

Our study includes stable isotope analysis of one to four samples of deer, bear, cod, sculpin, and flounder bones picked from four different strata in the Turner Farm midden. Two samples of modern cod were analyzed, one caught by a local lobsterman in July 2005 not far from the Turner Farm site, and the other collected by the state of Maine's Department of Marine Resources (DMR) inshore trawl surveys in May 2006. Three modern flounder and sculpin were also collected by the DMR inshore surveys in May 2006. The archaeological bone collagen was extracted and prepared after Harrison and Katzenberg (2003). White muscle tissue was lipid-extracted from the modern fish.

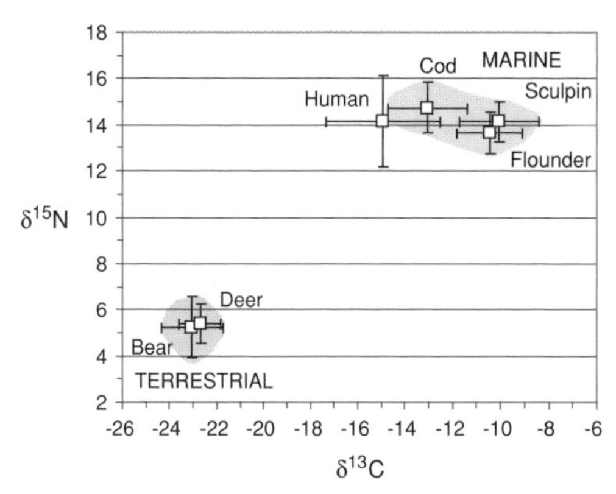

FIGURE 8.7. The average stable isotope composition (±1 SD) of Holocene (4000–400 BP) bone collagen from the dominant species in the Turner Farm midden. This figure illustrates the vastly different isotopic fields occupied by marine verses terrestrial animals (shaded regions). The human isotope data (from a 3500 BP cemetery at the Turner Farm site; Bourque and Krueger 1994) plot very close to the marine field, implying that human diets were heavily influenced by marine fish.

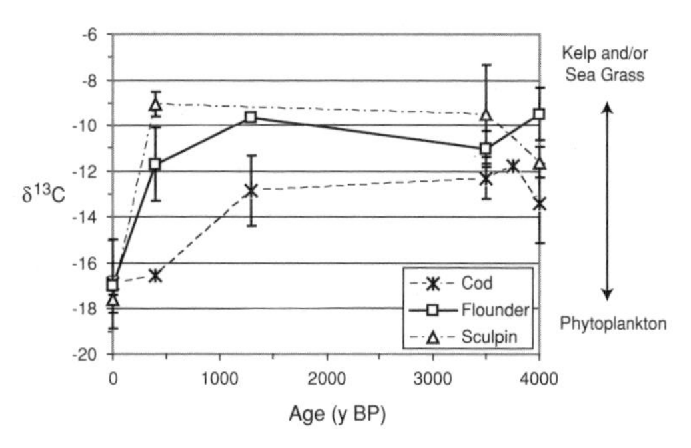

FIGURE 8.8. The average carbon isotope composition (±1 SD) of archaeological cod, flounder, and sculpin bone collagen and modern muscle tissue, where age is plotted in thousands of years (ka) BP. Dramatic shifts in cod diets occurred sometime between 1300 and 400 BP.

All samples were run in the Environmental Geochemistry Laboratory at Bates College using a ThermoFinnigan Delta Plus Advantage stable isotope ratio mass spectrometer interfaced to a Costech elemental analyzer via the combustion interface. Atomic C/N ratios between 2.9 and 4 and the presence of collagen "ghosts" were used to verify the presence of intact collagen in the archaeological samples (after Tuross et al. 1988).

The isotope data for the Holocene deer and bear bones were not statistically different from each other (Figure 8.7), suggesting that they ate similar diets comprised almost exclusively of C_3 vegetation (e.g., leaves, shoots, berries, etc.). In contrast, the Holocene cod, flounder, and sculpin samples were significantly more enriched in ^{15}N and ^{13}C relative to the bear and deer samples (Figure 8.7). The marine fish incorporated more isotopically enriched and variable carbon at the base of the food web (e.g., phytoplankton, kelp, sea grasses) and occupied a higher trophic level relative to the terrestrial animals analyzed.

The human isotope data were also enriched relative to the bear and deer data, implying that marine resources were an important component to the human diet (in agreement with Bourque and Krueger [1994]). Thus, it appears that the Turner Farm humans were eating a diet dominated by marine fish, as well as some terrestrial animals. At this stage, it is impossible to be more precise about the degree to which these different animals were consumed, and the importance of shellfish to the diet.

Temporal changes in the carbon isotope composition of the flounder, sculpin, and cod provide insight into the types of primary consumers at the base of the food web through the Middle to Late Holocene (Figure 8.8). Between 4000 and 1300 BP, the isotope data for cod, flounder, and sculpin were relatively consistent. In general, more ^{13}C-depleted values were measured in cod and more enriched values were measured in sculpin and flounder through the time series. This suggests that flounder and sculpin fed on nearshore kelp and/or sea grass–associated organisms compared to cod that had a much larger foraging range often taking them far from kelp-associated organisms in the coastal zone (Figure 8.8).

The most significant prehistoric decline in carbon isotope values appears to have occurred in cod at 400 BP (Figure 8.6). This may indicate that large cod were extirpated from nearshore environments, while sculpin and flounder persisted and were hunted there. It is possible that cod's high food value (especially its preservability) stimulated offshore and distant fishing following coastal extirpation. While cod may have been overfished at some point before 400 BP, there is no evidence that sea urchin populations expanded as a result. The carbon isotope signature of nearshore sculpins and flounders remained enriched indicating that kelp and/or sea grass dominated throughout the prehistoric period. Importantly, sea urchin remains did not increase during the later occupations (Figure 8.5). In contrast, other researchers have reported an increase in sea urchin fragments and other herbivore remains in archaeological sites where predators were extirpated, including Simenstad et al. (1978) in the Aleutians and Erlandson et al. (2005) in California.

In general, the biomass of benthic macroalgae is currently high in coastal Maine (post-1995) due to overfishing of sea urchins in the early 1990s (Steneck and Sala 2005). As fishery trawl surveys generally avoid shallow rocky areas where kelp is most abundant, the relatively depleted ^{13}C signal of modern fish samples may reflect the more offshore, phytoplankton-based nature of the sampling locations.

The convergence of isotopic values in the modern fish may represent evidence for a loss of biodiversity accompanied by overfishing of the marine food web (e.g., Steneck et al. 2004). Alternatively, the isotope data may merely reflect homogeneity in the food sources available at the collection sites. Sampling at finer spatial scale resolution will be necessary to determine the extent and significance of the recent carbon isotope depletion and the loss of biodiversity.

DISCUSSION

Cod were ecologically important as the largest abundant apex predator in Gulf of Maine coastal ecosystems (Steneck and Carlton 2001). Individuals more than 180 cm long and weighing over 95 kg were recorded as recently as the 1800s (Collette and Klein-MacPhee 2002). Significantly, Turner Farm site faunal remains indicate average body lengths of about a meter (Jackson et al. 2001; Steneck and Carlton 2001).

Atlantic cod were targeted by the earliest inhabitants of the Turner Farm site and dominated midden deposits until around 3500 BP. The carbon and nitrogen isotope signatures from both cod and human bone indicate that the marine coastal ecosystem supported populations of people and cod for at least several thousand years. The prominence of cod in this and other middens throughout the region from Maine's Boothbay Harbor (Carlson 1986) to Canada's Bay of Fundy (Lotze and Milewski 2004) is easy to understand. Cod have high food value, are easy to catch because they do not resist capture, and can be preserved with simple techniques available to prehistoric people (Kurlansky 1997).

Because of its large size and high food value, the decline of Atlantic cod through time at the Turner Farm site probably reflects a real decline in cod stocks/populations rather than a change in climate or fishing preference. Further, while most of the fish bone isotope signatures were consistent with those associated with kelp forest (or benthic algal-dominated) ecosystems, the carbon isotope values from 400 BP suggest a more pelagic, phytoplankton-based food web (Figure 8.6). If cod had become locally rare in the nearshore kelp forests surrounding North Haven and Vinalhaven, prehistoric cod fishers may have been forced to travel offshore.

We suspect that large and old cod living close to shore were locally overfished. Large cod may live for 30 to 50 years, but their rate of growth slows with age (Collette and Klein-MacPhee 2002; Scott and Scott 1988). The archaeological record suggests that the prehistoric human population of North Haven grew over time, and fishing pressure could have eventually extirpated stocks locally. The decline of this large apex predator is important because it could have relaxed population controls at lower trophic levels, resulting in a mesopredator release. In a recent case study from the Canadian Maritimes, a similar release occurred when collapsed cod stocks seem to have resulted in increases in shrimp and crab populations (Worm and Myers 2003).

The increase of flounders, sculpins, and dogfish over time in the Turner Farm samples is also in accord with declining cod stocks. Importantly, cod diets change as they grow. When they first reach the benthos, they feed on small crustaceans; later in their juvenile life they feed on decapods such as shrimp, crabs, and lobsters (Langton and Bowman 1980). As they grow to larger sizes, adult cod become increasingly piscivorous, feeding on larger fish. Since cod longer than a meter are known to feed on several species of flounder (Langton and Bowman 1980), it is possible that the increase of flounders in the Turner Farm middens reflects a mesopredator release from the loss of large cod in the areas. Similarly, flounder, sculpins, and dogfish all increased in abundance following the recent collapse of Atlantic cod in the southern Gulf of St. Lawrence (Hanson and Lanteigne 2000) and in the Bay of Fundy (Lotze and Milewski 2004). It is important to note that such a mesopredator release could have occurred if only the largest cod in the population were extirpated, since smaller cod function differently as predators only of invertebrates. Thus, an absence of large cod does not necessarily indicate that the species was absent from the region—the situation in modern times that Steneck et al. (2004) identified as trophic level dysfunction in the Gulf of Maine.

At the Turner Farm site, seal abundance varies inversely with cod, increasing dramatically with time. Given their high value for food and pelts, seals would likely have been hunted by early coastal peoples as they were in the North Pacific (Hildebrandt and Jones 2002). The relative scarcity of seal bone in prehistoric Gulf of Maine middens reported by Lotze and Milewski (2004) and Spiess and Lewis (2001; Table 8.1) suggests that they may have been less abundant in the region's prehistoric coastal ecosystems where cod were limiting their prey. According to Frank et al. (2005), recent increases in seal abundances may likewise reflect the collapse of cod populations because of a mesopredator release in small forage fish that seals can swallow, although marine mammal protection legislation has also likely had a strong influence on the dramatic increase in seal populations in the United States and Canada (Baraff and Loughlin 2000; Trzcinski et al. 2006).

Lobsters and crabs are strikingly absent from the Turner Farm site faunal samples (Table 8.1; Steneck et al. 2004). In fact, none have been found at archaeological sites anywhere in the Gulf of Maine (Lotze and Milewski 2004). This is surprising given their current hyperabundance in Maine's coastal zone, where lobster population densities can exceed two per meter square in boulder habitats (Butler et al. 2006; Steneck 2006; Steneck and Wilson 2001). Their abundance in early historic times is suggested by reports that Europeans captured very large ones with boat hooks (Quinn and Quinn 1983:283, 307). One possible explanation for this absence is a prehistoric scarcity of lobsters and crabs resulting from cod predation that allowed them to survive only in low numbers and/or hidden in sheltered refugia were they would have been particularly difficult to catch. A prehistoric depletion of cod in nearshore coastal zones could have had significant demographic consequences for lobsters. Lobsters begin life on the benthos, in shallow-water nearshore coastal zones (i.e., <20 m; Butler et al. 2006). Thus, a loss of predators could have contributed to local increases of lobsters by the time of early European colonial fisheries. Supporting such an inference is the fact that lobsters and crabs were commonly eaten by cod in Maine's coastal zone during the 1800s (Herrick 1911; Smith 1879), suggesting that these predators did indeed suppress lobster and crab population densities in coastal zones during early prehistoric times (Butler et al. 2006). Another possibility, however, is that lobster and crab shells may not preserve well in shell middens. Their exoskeletons are well calcified and so, like sea urchins, might be expected to preserve well in the alkaline environment of shell middens. Indeed, both decapods can preserve well in the fossil record (Bishop 1986). However their shells are made of chitin, a polysaccharide embedded in a hardened proteinaceous matrix that may succumb to biological degradation in the well-oxygenated context of a shell midden. Furthermore, if they were at all abundant prehistorically, their shells would, as today, likely be commonly found along the shore. If so, an occasional specimen would likely have been collected along with small gastropods that are often found in the lenses of beach gravel used to cover prehistoric house floors. In any case, the high abundances of today may well be another manifestation of mesopredator release (Steneck and Carlton 2001).

Despite our evidence for declines in apex predators and increases in mesopredators, we found no evidence of change in sea urchin abundance and, by inference, kelp forests (Steneck et al. 2002). In other coastal zones, predator declines were followed by increases of sea urchins in middens (e.g., Erlandson et al. 2004, 2005; Simenstad et al. 1978). Thus, the possible changes to coastal food webs by prehistoric cultures in Maine were relatively subtle by being confined to apex and mesopredators, compared to the larger changes that affected herbivory recently (Steneck et al. 2002, 2004).

That prehistoric peoples negatively affected ecosystems is no longer novel, but there have been few reported marine examples, and we suspect the effects we describe here were very localized. Many past studies have focused on agricultural or large-scale societies affecting terrestrial ecosystems (Redman 1999). We suggest that prehistoric hunter-gatherers in Penobscot Bay negatively affected a highly productive marine coastal ecosystem but did so very locally. As an example of how a simple fishing technology might have fished down a coastal Maine marine food web, we offer two examples of local cod stock depletion from the seventeenth century.

In AD 1614, Captain John Smith described Monhegan Island (about 20 km from the Turner Farm site) as a marvelous fishpond. He reported 15 to 18 fishermen using small boats to catch cod at a rate of 60,000 fish per month. This developed into a fishing station where as many as 80 fishing boats were based in the island's harbor between AD 1616 and 1622. Yet the station closed a few years later in AD 1626 (McLane 1992). About a decade later, a very small English fishing station was established at a very rich fishing location on Richmond Island near Casco Bay in Maine (about 100 km southwest of the Turner Farm site). This station, initially operating just three small vessels, produced 2,000 quintals (100 tons) of salt cod in AD 1639. Thereafter, however, catches rapidly declined to only 257 quintals by AD 1641 despite a threefold increase in fishing effort (Baxter 1884:155, 163–169, 215, 283, 312, 335).

Importantly, these very early fishing stations operating close to shore at a spatial scale comparable to prehistoric fishing efforts in Penobscot Bay appear to have rapidly depleted local cod stocks. The history of the early English fishing stations suggests that localized nearshore fisheries suffered nonsynchronous booms and busts. This asynchrony is important because it argues against cod stock declines having been climate driven.

The evolution toward larger boats with greater range continued following local coastal extirpations. By AD 1840, a "Report of the Joint Select Committee on the Fisheries, Maine legislature, 1841" (reported in O'Leary 1996) stated: "The coast of Maine, is in some parts sterile." From that period forward a greater proportion of Maine's landings came from more distant locations. By the mid 1800s, the fishing fleet from Massachusetts was fishing primarily on Canada's Scotian Shelf (Rosenberg et al. 2006). Although coastal cod of that era may not have been economical to harvest using the hook and line methods of the day, they were not absent. For example, scientists studying crabs collected them from the stomachs of cod in Casco Bay (Smith 1879). However, as discussed earlier, decapods are food for juvenile cod, so it is unlikely that abundant and large inshore cod such as the meter-long cod in Maine's prehistoric middens survived in Maine's coastal zones into the mid-nineteenth century.

Both the Turner Farm site faunal sample and early historic records of fishing in Maine suggest that depletions can occur at small spatial and temporal scales. Recent research on cod population structure has revealed why this is true. Rather than vast schools of interbreeding fish, as cod stocks have traditionally been characterized, they are actually composed of mosaics of loosely connected metapopulations in which some substocks are very localized (Bentzen et al. 1996). For example, a small stock was identified in the mouth of the Sheepscot River in Maine, and tagging studies showed a high proportion were recaptured in the same area over a six-year period (Perkins et al. 1997). The local nature of cod stocks helps explain why the chronology of extirpation can be so asynchronous. Even at larger scales, evidence suggests that coastal stocks in Maine collapsed in the 1930s (Steneck 1997), whereas offshore Canada and U.S. stocks declined decades later (Myers et al. 1997). Thus, localized depletions of cod due to low-tech fishing may have happened repeatedly beginning in prehistoric times.

CONCLUSIONS

The Gulf of Maine is one of the world's most productive and species-depauperate marine ecosystems, leaving it relatively susceptible to changes in trophic structure and function. A prime driver of ecosystem change is overfishing that first depletes key predators in upper trophic levels, causing former prey species to increase and become the new targets for fishing effort. Such cases of fishing down food webs have significantly affected other nearshore ecosystems on historical (and archaeological) timescales. While it is tempting to assume Maine's coastal ecosystem was pristine at the time of European contact, we have presented archaeological and isotope evidence from the Turner Farm site in Penobscot Bay suggesting that localized fishing down of nearshore coastal food webs may have begun thousands to hundreds of years before European fishers first arrived. Apex predators in this ecosystem were the first targeted and dominate the midden bone mass in the earliest strata (~4350 BP). However, cod were also among the first to decline in relative abundance. Thus by ~3550 BP, cod no longer were the dominant species represented in the middens. By ~1600 and 400 BP cod were a minor midden constituent. Coincident with this cod decline, mesopredators such as flounder and sculpins increased over the next 3,500 years. Although the nearshore fauna appears to have changed locally due to prehistoric fishing pressures, we found minimal indication in the stable carbon isotope composition of cod, sculpin, or flounder bone collagen between 4350 and 1200 BP that sea urchin populations expanded enough to induce kelp deforestation. Cod did show an isotope change before European contact, however, suggesting that fished individuals were no longer coming from a kelp forest ecosystem. By about 400 BP, it is possible that cod had been extirpated from the nearshore kelp-dominated coastal zone, forcing the site's occupants to travel farther to catch them. The ease in capture and preservation of cod may have made them sufficiently valuable to prehistoric people that they expended extra effort to pursue this species. In sum, prehistoric changes to food web structure and functioning at the Turner Farm site suggest that significant human impacts in Gulf of Maine coastal ecosystems—although probably localized in nature—may have started earlier and been of greater magnitude than previously thought. As a result, the fragility of this coastal ecosystem may be underestimated.

ACKNOWLEDGMENTS

This research was an outgrowth of the Long-Term Ecological Records of Marine Environments, Populations, and Communities Working Group supported by the National Center for Ecological Analysis and Synthesis (funded by NSF grant DEB-0072909, the University of California, and the University of California, Santa Barbara). Additional support for RSS came from the National Undersea Research Program (Avery Point, Connecticut) and the University of Maine's Sea Grant Program. We thank Charlotte Lehmann, Kim Rodgers, and Carl Noblitt for sample preparation and analysis. Bob Lewis assisted in figure preparation. This research was funded in part by Bates College and the Maine Marine Research Fund, established by the 120th Session of the Maine Legislature, administered by the Maine Technology Institute, and funded through a bond issue approved by Maine voters.

REFERENCES CITED

Adey, W. H., and R. S. Steneck
2001 Thermogeography over Time Creates Biogeographic Regions: A Temperature/Space/Time-Integrated Model and an Abundance-Weighted Test for Benthic Marine Algae. *Journal of Phycology* 37:677–698.

Ambrose, S. H., and M. J. DeNiro
1986 The Isotope Ecology of East African Mammals. *Oecologia* 69:395–406.

Aurioles, D., P. L. Koch, and B. J. Le Boeuf
2006 Differences in Foraging Location of Mexican and California Elephant Seals: Evidence from Stable Isotopes in Pups. *Marine Mammal Science* 22:326–338.

Baraff, L. S., and T. R. Loughlin
2000 Trends and Potential Interactions between Pinnipeds and Fisheries of New England and the U.S. West Coast. *Marine Fisheries Review* 62:1–39.

Baxter, J. P.
1884 *Documentary History of the State of Maine, Vol. 3: The Trelawny Papers*. Hoyt, Fogg, and Donham, Portland, Maine.

Bentzen, P., C. T. Taggart, D. E. Ruzzante, and D. Cook
1996 Microsatellite Polymorphism and the Population Structure of Atlantic Cod *(Gadus morhua)* in the Northwest Atlantic. *Canadian Journal of Fisheries and Aquatic Science* 53:2706–2721.

Bishop, G.
1986 Taphonomy of the North American Decapods. *Journal of Crustacean Biology* 6:326–355.

Bourque, B. J.
1995 *Diversity and Complexity in Prehistoric Maritime Societies: A Gulf of Maine Perspective.* Plenum Press, New York.
2001 *Twelve Thousand Years: American Indians in Maine.* University of Nebraska Press, Lincoln.

Bourque, B. J., and H. W. Krueger
1994 Dietary Reconstruction from Human Bone Isotopes for Five New England Coastal Populations. In *Paleonutrition: The Diet and Health of Prehistoric Americans*, edited by K. D. Sobolik, pp. 195–209. Southern Illinois University Center for Archaeological Investigations, Carbondale.

Bourque, B. J., and R. H. Whitehead
1994 Trade and Alliances in the Contact Period. In *American Beginnings: Exploration, Culture, and Cartography in the Land of Norumbega*, edited by E. W. Baker, E. A. Churchill, R. S. D'Abate, K. Jones, V. A. Konrad, and H. E. Prins, pp. 131–147. University of Nebraska Press, Lincoln. Reprint of Tarrentines and the Introduction of European Trade Goods in the Gulf of Maine. *Ethnohistory* 32:327–341.

Burton, R. K., J. J. Snodgrass, D. Gifford-Gonzalez, T. Guilderson, T. Brown, and P. L. Koch
2001 Holocene Changes in the Ecology of Northern Fur Seals: Insights from Stable Isotopes and Archaeofauna. *Oecologia* 128:107–115.

Bustamante, R. H., and G. M. Branch
1996 The Dependence of Intertidal Consumers of Kelp-derived Organic Matter on the West Coast of South Africa. *Journal of Experimental Marine Biology and Ecology* 196:1–28.

Butler, M., R. S. Steneck, and W. Herrnkind
2006 Ecology of Juvenile and Adult Lobsters. In *Lobsters: The Biology, Management, Aquaculture and Fisheries*, edited by R. Phillips, pp. 263–309. Blackwell, Oxford.

Carlson, C. C.
1986 Maritime Catchment Areas: An Analysis of Prehistoric Fishing Strategies in the Boothbay Region of Maine. Unpublished Master's thesis, University of Maine, Orono.

Collette, B. B., and G. Klein-MacPhee
2002 *Bigelow and Schroeder's Fishes of the Gulf of Maine.* Smithsonian Institution Press, Washington, D. C.

Duggins, D. O., and J. E. Eckman
1994 The Role of Kelp Detritus in the Growth of Benthic Suspension Feeders in an Understory Kelp Forest. *Journal of Experimental Marine Biology and Ecology* 176:53–68.

Duggins, D., C. S. Simenstad, and J. A. Estes
1989 Magnification of Secondary Production by Kelp Detritus in Coastal Marine Ecosystems. *Science* 245:101–232.

Erlandson, J. M., and S. M. Fitzpatrick
2006 Oceans, Islands, and Coasts: Current Perspectives on the Role of the Sea in Human Prehistory. *Journal of Island and Coastal Archaeology* 1(1):5–33.

Erlandson, J. M., T. C. Rick, and R. L. Vellanoweth
2004 Human Impacts on Ancient Environments: A Case Study from California's Northern Channel Islands. In *Voyages of Discovery: The Archaeology of Islands*, edited by S. M. Fitzpatrick, pp. 51–83. Praeger, Westport, Connecticut.

Erlandson, J. M., T. C. Rick, M. Graham, J. Estes, T. Braje, and R. Vellanoweth
2005 Sea Otters, Shellfish, and Humans: 10,000 Years of Ecological Interaction on San Miguel Island, California. In *Proceedings of the Sixth California Islands Symposium*, Ventura, California, edited by D. K. Garcelon and C. A. Schwemm, pp. 58–69. Institute for Wildlife Studies and National Park Service, Arcata, California.

Estes, J. A., D. O. Duggins, and G. B. Rathbun
1989 The Ecology of Extinctions in Kelp Forest Communities. *Conservation Biology* 3:252–264.

Fogarty, M. J., and S. A. Murawski
1998 Large-Scale Disturbance and the Structure of Marine Systems: Fisheries Impacts on Georges Bank. *Ecological Applications* 8:S6–S22.

Frank, K. T., B. Petrie, J. S. Choi, and W. C. Leggett
2005 Trophic Cascades in a Formerly Cod-Dominated Ecosystem. *Science* 308:1621–1623.

Froese, R., and D. Pauly (editors)
2002 FishBase. Electronic document, www.fishbase.org, accessed July 2007.

Fry, B.
1988 Food Web Structure on Georges Bank from Stable C, N, and S Isotopic Compositions. *Limnology and Oceanography* 33:1182–1190.

Fry, B., and E. B. Sherr
1984 $\delta^{13}C$ Measurements as Indicators of Carbon Flow in Marine and Freshwater Ecosystems. *Contributions in Marine Science* 27:13–47.

Grove, J. M.
2001 The Initiation of the "Little Ice Age" in Regions Round the North Atlantic. *Climatic Change* 48:53–82.

Hanson, J. M., and M. Lanteigne
2000 Evaluation of Atlantic Cod Predation on American Lobster in the Southern Gulf of St. Lawrence, with Comments on Other Potential Fish Predators. *Transactions of the American Fisheries Society* 129(1):13–29.

Harrison, P. G.
1989 Detrital Processing in Seagrass Systems: A Review of Factors Affecting Decay Rates, Remineralization and Detritivory. *Aquatic Botany* 35:263–288.

Harrison, R. G., and M. A. Katzenberg
2003 Paleodiet Studies Using Stable Carbon Isotopes from Bone Apatite and Collagen: Examples from Southern Ontario and San Nicolas Island, California. *Journal of Anthropological Archaeology* 22: 227–244.

Herrick, F. H.
1911 Natural History of the American Lobster. *Bulletin of the U.S. Fisheries* 1909:149–408.

Hildebrandt, W. R., and T. Jones
2002 Depletion of Prehistoric Pinniped Populations along the California and Oregon Coasts: Were Humans the Cause? In *Wilderness and Political Ecology: Aboriginal Influences and the Original State of Nature* edited by C. E. Kay and R. T. Simmons, pp. 72–110. University of Utah Press, Salt Lake City.

Hoshiko, A., M. J. Sarkar, S. Ishida, Y. Mishima, and N. Takai
2006 Food Web Analysis of an Eelgrass (Zostera marina L.) Meadow in Neighboring Sites in Mitsukuchi Bay (Seto Inland Sea, Japan) Using Carbon and Nitrogen Stable Isotope Ratios. *Aquatic Botany* 85:191–197.

Jackson, J. B. C., M. X. Kirby, W. Berger, K Bjorndahl, L. Botford, B. Bourque, R. Bradbury, R. Cooke, J. Erlandson, J. Estes, T. Hughes, S. Kidwell, C. Lange, H. Lenihan, J. Pandolfi, C. Peterson, R. Steneck, M. Tegner, and R. Warner.
2001 Historical Overfishing and the Recent Collapse of Coastal Ecosystems. *Science* 293: 629–638.

Josefson, A. B., T. L. Forbes, and R. Rosenberg
2002 Fate of Phytodetritus in Marine Sediments: Functional Importance of Macrofaunal Community. *Marine Ecology Progress Series* 230: 71–85.

Kaehler, S., E. A. Pakhomov, and C. D. McQuaid
2000 Trophic Structure of the Marine Food Web at the Prince Edward Islands (Southern Ocean) Determined by $\delta^{13}C$ and $\delta^{15}N$ analysis. *Marine Ecology Progress Series* 208:13–20.

Kurlansky, M.
1997 *Cod: A Biography of the Fish that Changed the World.* Walker, New York.

Langton, R., and R. Bowman
1980 Food of Fifteen Northwest Atlantic Gadiform Fishes. *NOAA Technical Report NMFS SSRF* 740:1–23.

Lesage, V., M. O. Hammill, and K. M. Kovacs
2001 Marine Mammals and the Community Structure of the Estuary and Gulf of St. Lawrence, Canada: Evidence from Stable Isotope Analysis. *Marine Ecology Progress Series* 210:203–221.

Lotze, H. K., and I. Milewski
2004 Two Centuries of Multiple Human Impacts and Successive Changes in a North

Atlantic Food Web. *Ecological Applications* 14:1428–1447.

Lotze, H.K., H.S. Lenihan, B.J. Bourque, R.H. Bradbury, R.G. Cooke, M.C. Kay, S.M. Kidwell, M.X. Kirby, C.H. Petersen, and J.B. Jackson

2006 Deletion, Degradation, and Recovery Potential of Estuaries and Coastal Seas. *Science* 312:1806–1809

McLane, C.B.

1992 *Islands of the Mid-Maine Coast,* Vol. 3: *Muscungus Bay and Monhegan Island.* Tilbury House, Gardiner, Maine.

McMahon, K.W., B.J. Johnson, and W.G. Ambrose, Jr.

2005 Diet and Movement of the Killifish, *Fundulus heteroclitus,* in a Maine Salt Marsh Assessed Using Gut Contents and Stable Isotope Analyses. *Estuaries* 28:996–973.

McMillan, C., P.L. Parker, and B. Fry

1980 $^{13}C/^{12}C$ Ratios in Seagrasses. *Aquatic Botany* 9:237–249.

Mann, K.H.

1973 Seaweeds: Their Productivity and Strategy for Growth. *Science* 182:975–981.

Michener, R.H., and D.M. Schell

1994 Stable Isotope Ratios as Tracers in Marine Aquatic Food Webs. In *Stable Isotopes in Ecology and Environmental Science,* edited by K. Lajtha and R. Michener, pp. 138–157. Blackwell Scientific, London.

Michener, R.H., and L. Kaufman

2007 Stable Isotope Ratios as Tracers in Marine Food Webs: An Update. In *Stable Isotopes in Ecology and Environmental Science,* edited by R.H. Michener and K. Lajtha, pp. 238–282. Blackwell Publishing, Malden, Massachusetts.

Minagawa, M., and E. Wada

1984 Stepwise Enrichment of ^{15}N along Food Chains: Further Evidence and the Relation between (^{15}N and Animal Age. *Geochimica et Cosmochimica Acta* 48:1135–1140.

Myers, R.A., J.A. Hutchings, and N.J. Barrowman

1997 Why Do Fish Stocks Collapse? The Example of Cod in Atlantic Canada. *Ecological Applications* 7:91–106.

O'Leary, W.M.

1996 *Maine Sea Fisheries: The Rise and Fall of a Native Industry, 1830–1890.* Northeastern University Press, Boston.

Paine, R.T.

1980 Food Webs: Linkage, Interaction Strength and Community Infrastructure. *Journal of Animal Ecology* 49:667–685.

Pauly, D., V. Christiansen, J. Dalsgard, R. Froese, and F. Torres, Jr.

1998 Fishing down Marine Food Webs. *Science* 279:860–863.

Perkins, H.C., S.B. Chenoweth, and R.W. Langton

1997 The Gulf of Maine Atlantic Cod Complex, Patterns of Distribution and Movement of the Sheepscot Bay Substock. *Bulletin of the National Research Institute of Aquaculture* 25:101–108.

Phillips, D.L., and P.L. Koch

2002 Incorporating Concentration Dependence in Stable Isotope Mixing Models. *Oecologia* 130:114–125.

Planque, B., and T. Fredou

1999 Temperature and the Recruitment of Atlantic Cod *(Gadus morhua). Canadian Journal of Fisheries and Aquatic Sciences* 56:2069–2077.

Quinn, D.B., and A.M. Quinnn

1983 *The English New England Voyages, 1602–1608.* Hakluyt Society Second Series 161. Hakluyt Society, Cambridge, England.

Redman, C.L.

1999 *Human Impact on Ancient Environments.* University of Arizona Press, Tucson.

Rosenberg, A.A., W.J. Bolster, K.E. Alexander, W.B. Leavenworth, A.B. Cooper, and M.G. McKenzie

2005 History of Ocean Resources: Modeling Cod Biomass Using Historical Records. *Frontiers in Ecology and the Environment* 3:78–84.

Scott, W.B., and M.G. Scott

1988 *Atlantic Fishes of Canada.* Canadian Bulletin of Fisheries and Aquatic Sciences 219. University of Toronto Press, Toronto.

Simenstad, C., J. Estes, and K. Kenyon

1978 Aleuts, Sea Otters, and Alternative Stable-State Communities. *Science* 200:403–411.

Smith, S.I.

1879 The Stalk-Eyed Crustaceans of the Atlantic Coast of North America North of Cape Cod. *Transactions of the Connecticut Academy of Arts and Sciences* 5:27–136.

Spiess, A.E., and R.A. Lewis

2001 *Turner Farm Fauna: 5000 Years of Hunting and Fishing in Penobscot Bay, Maine.* Occasional Publications in Archaeology 11. Maine State Museum and the Maine Historic Preservation Commission, Augusta, Maine.

Steneck, R.S.

1997 Fisheries-Induced Biological Changes to the Structure and Function of the Gulf of Maine Ecosystem. In *Proceedings of the Gulf of Maine Ecosystem Dynamics Scientific Symposium and Workshop,* edited by G.T. Wallace and E.F. Braasch, pp. 151–165. Regional Association for

Research on the Gulf of Maine, Hanover, New Hampshire.

2006 Is the American Lobster, *Homarus americanus*, Overfished? A Review of Overfishing with an Ecologically-Based Perspective. *Bulletin of Marine Sciences* 78:607–632.

Steneck, R. S., and J. T. Carlton

2001 Human Alterations of Marine Communities: Students Beware! In *Marine Community Ecology*, edited by M. Bertness, S. Gaines, and M. Hay, pp. 445–468. Sinauer Press, Sunderland, Massachusetts.

Steneck, R. S., and M. N. Dethier

1994 A Functional Group Approach to the Structure of Algal-dominated Communities. *Oikos* 69:476–498.

Steneck, R. S., and E. Sala

2005 Large Marine Carnivores: Trophic Cascades and Top-down Controls in Coastal Ecosystems Past and Present. In *Large Carnivores and the Conservation of Biodiversity*, edited by J. Ray, K. Redford, R. Steneck, and J. Berger, pp. 110–137. Island Press, Washington, D.C.

Steneck, R. S., and C. J. Wilson

2001 Long-Term and Large-Scale Spatial and Temporal Patterns in Demography and Landings of the American Lobster, *Homarus americanus*, in Maine. *Journal of Marine and Freshwater Research* 52:1302–1319.

Steneck, R. S., M. H. Graham, B. J. Bourque, D. Corbett, J. M. Erlandson, J. A. Estes, and M. J. Tegner

2002 Kelp Forest Ecosystems: Biodiversity, Stability, Resilience and Future. *Environmental Conservation* 29:436–459.

Steneck, R. S., J. Vavrinec, and A. V. Leland

2004 Trophic-Level Dysfunction in Kelp Forest Ecosystems of the Western North Atlantic. *Ecosystems* 7:523–552.

Trzcinski, M. K., R. Mohn, and W. D. Bowen

2006 Continued Decline of an Atlantic Cod Population: How Important is Gray Seal Predation? *Ecological Applications* 16:2276–2292.

Tuross, N., M. L. Fogel, and P. E. Hare

1988 Variability in the Preservation of the Isotopic Composition of Collagen from Fossil Bone. *Geochimica et Cosmochimica Acta* 52:929–935.

Vadas, R. L., and R. S. Steneck

1988 Zonation of Deep Water Benthic Algae in the Gulf of Maine. *Journal of Phycology* 24:338–346.

Vermeij, G.

1991 When Biotas Meet: Understanding Biotic Interchange. *Science* 253:1099–1103.

Wada, E., H. Mizutani, and M. Minagawa

1991 The Use of Stable Isotopes for Food Web Analysis. *Critical Reviews in Food Science and Nutrition* 30:361–371.

Wing, S. R., and E. S. Wing

2001 Prehistoric Fisheries in the Caribbean. *Coral Reefs* 20:1–8.

Witman, J. D., R. J. Etter, and F. Smith

2004 The Relationship between Regional and Local Species Diversity in Marine Benthic Communities: A Global Perspective. *Proceedings of the National Academy of Sciences* 101:15664–15669.

Worm, B. M., and R. A. Myers

2003 Meta-Analysis of Cod-Shrimp Interactions Reveals Top-down Control in Oceanic Food Webs. *Ecology* 84(1):162–173.

9

Codfish and Kings, Seals and Subsistence

NORSE MARINE RESOURCE USE IN THE NORTH ATLANTIC

Sophia Perdikaris and Thomas H. McGovern

IN THE PAST TWO DECADES, the archaeology and paleoecology of the North Atlantic have been transformed by a series of major international, interdisciplinary projects (Barrett et al. 2000; Church et al. 2005; Dockrill et al. 2005; Edwards et al. 2004; McGovern 2001; McGovern et al. 2007; Parker-Pearson and Sharples 1999; Sharples 1998, 2005; Simpson et al. 2001). Most have been inspired by the theoretical framework of historical ecology (Crumley 1994) in their investigation of the complex interactions of climate, landscape change, and human culture in the region, and all have made geoarchaeology, archaeobotany, zooarchaeology, and human osteology part of their fundamental research design. Since 1990, the North Atlantic Biocultural Organization (NABO) has aided these collective efforts by holding coordinating meetings and workshops (New York 1992, Glasgow 1994, St. John's 1997, Glasgow 2001, Akureyri 2002, Copenhagen 2003, Quebec 2006), providing identification manuals and data management tools to aid comparability (Krivogorskaya et al. 2005; McGovern 2004), and by publishing monographs, working group reports, and conference volumes (Bigelow 1991; Arneborg et al. 2006; Housely and Coles 2004; Morris and Rackham 1994; Ogilvie and Jónsson 2001). NABO has sponsored long-running international field schools in Iceland and Shetland, which have drawn students from 26 nations since 1996 (McGovern 2004). Major regional research foci include culture contact, the impact of climatic fluctuations of the Medieval and early modern period, and the varied environmental impacts of imported European agricultural systems upon island ecosystems. The interaction between marine and terrestrial economies is a cross-cutting theme that unites virtually all investigations in the region. This chapter presents some of the results of this long-term international collaboration and makes use of newly available data sets and regional syntheses to provide a broad overview of Norse use of marine resources for subsistence and for exchange in the North Atlantic (Figure 9.1).

VIKING AGE EXPANSION

The dramatic expansion of Scandinavian culture and settlement during the Viking Age

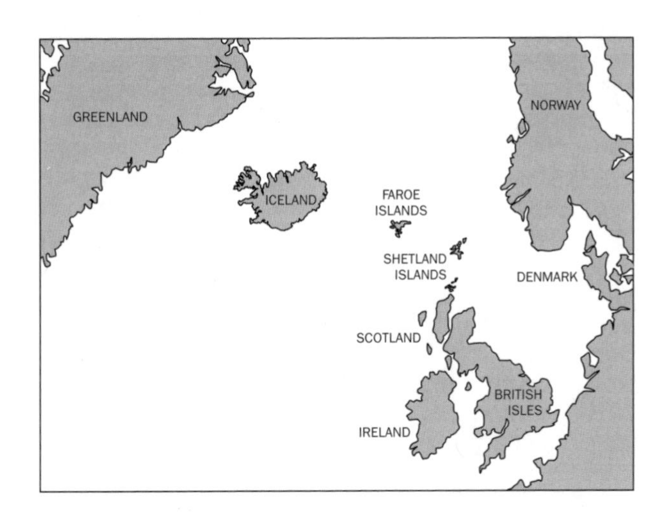

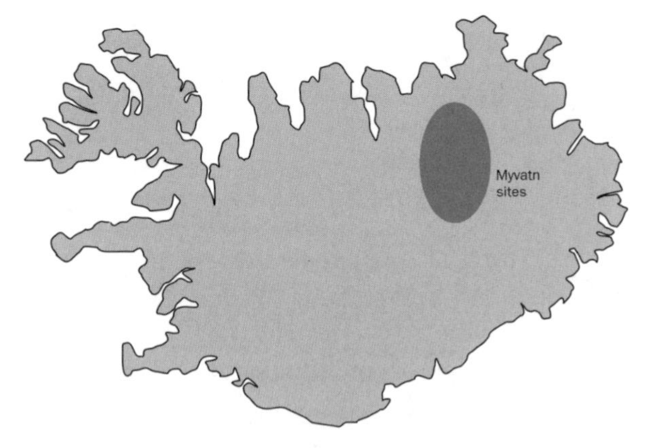

FIGURE 9.1. Map of the North Atlantic study area.

(ca. AD 750–1050) resulted from interconnected factors including climate change, external trade, and technological improvements in ships and navigation that allowed long open-water voyages out of sight of land by AD 750 (Brøgger and Shetelig 1951; Christensen 2000; Crumlin-Pedersen 1997; Nicolaysen 2003; Olsen et al. 1995). All these changes took place in the social context of intense competition among Nordic chiefly elites attempting to consolidate power, undermine social leveling mechanisms, and create an effective transition to state-level organization with themselves as founding monarchs (Hedeager 2000; Jones 1987; Thurston 1999; Thurston and Fisher 2006). The new wealth and expanding opportunities overseas also created tension between established elites and ambitious commoners seeking to become chieftains through successful Viking raids or by establishing new settlements in the islands of the west (Jørgensen et al. 2002). Both routes to power required ships and seafaring skills and the capacity to provision men kept long away from their farms. As raiding and overseas settlement intensified, social conflict also increased among lesser and greater chieftains who made use of both prestige goods and staple goods in their contests for power and followers (Perdikaris and McGovern 2007; Randsborg 1980).

The early Viking period of ca. AD 750–900 saw escalating Nordic piracy and sea-borne raiding along the coasts of northwest Europe and an expansion of settlement into the Faroe Islands (ca. 800–850), Iceland (ca. 875), the Northern Isles (Shetland and Orkney), and the Western Isles (Hebrides) (ca. AD 800–850) (Arge 2005; Vésteinsson 1998, 2000a, 2000b).

The later Viking period of ca. AD 900–1050 encompassed the settlement of Greenland (ca. AD 985) and briefly of Newfoundland (around AD 1000). This was also a period of escalating warfare in Britain, culminating in the establishment of an Anglo-Scandinavian dynasty under Knut the Great (AD 1014–1035) that controlled England, Denmark, and Norway for two decades (Lawson 2004). The end of the Viking period thus saw the transformation of pagan Nordic warlords into Christianized kings using clerics and Latin literacy to administer increasingly unified kingdoms with widespread political and economic contacts in northwest Europe (Adams and Holman 2004). Between ca. AD 1050–1100 and ca. AD 1500 many scholars working in northern Scotland recognize a Medieval "Late Norse" period characterized by increasing political integration within the Earldom of Orkney and the Medieval Norwegian monarchy along with widespread indications of intensified marine fishing (Bigelow 1984; Morris et al. 1995). Iceland and Greenland retained political independence until their integration with the Norwegian crown in 1264. The Greenlandic settlements became extinct by the mid-fifteenth century under what are still somewhat mysterious circumstances (Arneborg 2000; Barlow et al. 1997; Buckland et al. 1996; Diamond 2005; Gulløv 2000; McGovern 2000). The interaction of marine and terrestrial portions of subsistence economy with the changing demands of regional and international trade and changing climate in the later Middle Ages were to have very different impacts on these Scandinavian island communities (Buckland 2000).

CLIMATE CHANGE

The North Atlantic is fortunate in having a steadily developing set of high-resolution, multiproxy climatic indicators with resolution on the scale of decades, years, and occasionally individual seasons (ice, sea, and lake cores; dendroclimatology; historical climatology; see Barlow et al. 1997; Buckland et al. 1996; Jennings and Weiner 1996; Meeker and Mayewski 2002; Ogilvie and Jónsson 2001; Ogilvie and McGovern 2000). These proxy data sets are increasingly being combined with well-validated agroclimatological models, which have transformed our ability to convert these high-resolution temporal series into spatially mapped applications capable of flagging areas of probable maximum climate impact within culturally constrained landscapes (Dugmore et al. 2007; Simpson et al. 2004). In well-studied areas, it is now possible to closely model vegetation responses to both grazing pressure and growing season change on the scale of individual farm holdings and individual seasons (Simpson et al. 2003; Thomson and Simpson 2006, 2007). These steadily improving paleoclimatic data indicate that the old concepts of a uniformly warm "medieval optimum" (ca. AD 700–1300/1400) followed by a uniformly cold "little ice age" (ca. AD 1300/1400–1850) are considerable oversimplifications of actual variability, though the labels are still widely used as shorthand by climatologists (Ogilvie and Jónsson 2001).

For island societies and maritime economies, perhaps the most important climatic variables are the frequency and intensity of storms and the degree of interannual predictability of weather patterns. Storminess and unpredictable weather directly affect the hazards of both local and offshore voyages and raise the human costs of intensified marine resource exploitation. In the western North Atlantic, sea ice also plays a major factor in both marine and terrestrial ecosystems, and variation in sea ice extent in summer and winter has been traced by a combination of ice chemistry in the GISP-2 Greenland ice core (Meeker and Mayewski 2002; Rohling et al. 2003), sea cores in Danmark Strait between Iceland and Greenland (Jennings and Weiner 1996), and documentary evidence (Ogilvie 1997). The amount of sea salt sodium in ice cores has been used to reconstruct winter storm frequency and intensity, and the impact of both changing storm frequency and of catastrophic storms is a major topic of regional research (Bigelow 1991; Dugmore et al. 2007; Meeker and Mayewski 2002). The early Viking Age (ca. AD 700–950) appears in the multiproxy record as a period of relatively low storminess and low sea ice incidence, with a spike in storm frequency AD 975–1025 and a dramatic increase in the frequency and intensity of North Atlantic storms after AD 1425 that continued into the eighteenth century. Based on the sea salt sodium record and the sea core evidence, summer sea ice appears to have increased between Greenland and Iceland around the mid-thirteenth century and definitely saw a major increase after AD 1450. The mid-thirteenth-century Norwegian text *King's Mirror* reports on some of the hazards imposed by the newly arrived summer drift ice:

> It has frequently happened that men have sought to make the land too soon and, as a result, have been caught in the ice floes. Some of those who have been caught have perished; but others have got out again . . . . All those who have been caught in these ice drifts have adopted the same plan: they have taken their small boats and have dragged them up on the ice with them, and in this way have sought to reach land; but the ship and everything else of value had to be abandoned and was lost. Some have had to spend four days or five upon the ice before reaching land, and some even longer.
>
> (Larsen 1917: section XVI)

Sea ice in summer was thus not a regular problem for Viking Age navigators in the western North Atlantic but became a significant hazard in the later Middle Ages, especially between Iceland and Greenland. Interannual variability in storminess seems to have increased markedly after 1425, and there seems to have been a general increase in interannual climate variability after the late tenth century. Taken together, the changes in sea ice distribution, increased frequency of major storms, and decreasing interannual predictability of climate probably produced a significantly more hostile North Atlantic maritime environment for the seafarers of the thirteenth through eighteenth centuries than that experienced by their Viking Age precursors.

CULTURE CONTACT AND MARITIME ADAPTATIONS

In Arctic Norway and Sweden, Norse farmers had long been in contact with northern hunter-fishers today called Saami. By the ninth century AD, these contacts had settled into a regular pattern of interactions between Norse and Saami, which apparently involved a mix of trade and tribute extraction that brought Arctic products (furs, seal oil, walrus ivory, and hide) south, and metal and woolen cloth north (Olsen 2003). A frequently cited account by a North Norwegian chieftain named Ottar was recorded in the court of King Alfred of Wessex in the late ninth century. Ottar describes income from "tribute" collected regularly from the Saami, including reindeer farming, whaling, and walrus hunting (Lund 1984). A wondering Anglo-Saxon scribe noted that this North Norwegian chieftain owned far fewer cattle than any respectable Thane of Wessex but was still "accounted wealthy in his own country." Archaeology confirms that while Ottar's kinsmen above the Arctic Circle in the Lofoten and Vesterålen Islands may have lacked the agrarian resources of southern England, they were far from poor, boasting huge chieftain's halls, rich troves of imported silver, gold, and glassware, and impressive *naust*—the boat houses that protected the seagoing ships that were the source of so much of this wealth and power (Munch et al. 2003). The chieftains of these Arctic islands demonstrated their military and naval power later in the Viking Age, repeatedly intervening in the power struggles associated with the development of the Norwegian state ca. AD 950–1050, with huge fleets filled with well armed supporters.In the eleventh century these northern earls of Hålogaland acquired additional power and moved south to settle near Nidaros (Trondheim), assuming the title "Earls of Hlaði" and became Cnut of England's allies.

After AD 1300, the nature of Norse-Saami contact changed in the North Cape area, as Norway and Novgorod began to contest access to the walrus and fish of Finnmark and the Kola Peninsula. Warfare pitting Norwegian and Saami against the Karelians employed by the Russian cities led to construction of castles and fortified churches in the high Arctic, negotiated territorial divisions, and eventually the development in the late Middle Ages of an ethnically diverse society caught up in market production as well as subsistence (Urbańczyk 1992). Recent work in coastal Finnmark indicates the complexity of economic and social interactions around North Cape in the fourteenth to sixteenth centuries, and has documented settlements aimed at market production of cod as well as settlements mainly engaged in subsistence (Amundsen et al. 2003; Henriksen et al. 2006).

Nordic contacts with seafaring Celtic populations in the Northern and Western isles of Scotland accelerated after AD 800. Scandinavian colonization early in the Viking Age is attested by burials, farm sites, and widespread place name change (especially in the outer Hebrides and the Northern Isles; see Barrett 2003; Graham-Campbell and Batey 1998; Jennings and Kruse 2005; Owen 2004; Richards 2001). The nature of Norse contact with Late Iron Age Celtic populations (Picts, Dalriadic Scots, Britons) in what is now Scotland remains hotly debated but is usually seen as involving a mixture of conflict and assimilation on both sides, now illuminated by ongoing modern and ancient DNA investigations and stable isotope analyses (Barrett 2003; Barrett et al. 2000). The economic aspects of the Iron Age–Viking Age cultural transition in Shetland, Orkney, Hebrides, and Caithness have been investigated by several major projects, and it is now possible to compare pre-Norse, Viking Age, and Late Norse-Medieval marine resource use in some detail for this region (see Barrett 2003; Bond et al. 2005; Dockrill et al. 2005; Owen 2004; Parker-Pearson and Sharples 1999; Sharples 1998, 2005; Sharples et al. 2004; Smith and Mulville 2003).

In the Faroe Islands and Iceland, Norse settlers may have encountered small communities of Christian monks, but these appear to have been driven off or swamped by the later colonists, as historic settlement pattern, house forms, and place names are entirely Scandinavian. (Osteological and DNA evidence suggests that despite Nordic cultural and linguistic dominance, the ninth- to tenth-century settlers of the Faroes, Iceland, and Greenland did include a substantial British Isles component, especially visible in maternal mtDNA of modern Icelanders [Helgason et al. 2000]).

In Greenland, the Dorset Paleoeskimo who had once occupied most of the long coastline apparently survived only in the far northwest, and the Norse settled in the southwestern fjords in what had become an abandoned landscape (Gulløv 2004). Norse-Dorset contacts took place in the Norse northern walrus hunting grounds *(Norðursetur)* in northwest Greenland, and these contacts may have extended into Arctic Canada (McGovern 1985a, 1985b; Schledermann and McCullough 2003). Sometime in the late twelfth to early thirteenth centuries, the Norse Greenlanders came into contact with the Thule Inuit who had migrated across the Canadian Arctic, replacing the earlier Dorset hunters (Gulløv 2000). By the mid-fifteenth century, the Norse Greenlanders were also replaced by the Thule people, and different cultural marine resource exploitation strategies probably played a significant role in the Norse extinction in Greenland (Arneborg 2003; Diamond 2005; McGovern 2000). The North Atlantic thus provides a rich array of case studies of the interaction of cultures, changing economies, and different uses of marine resources. Since space is necessarily limited in this chapter, we will focus on a few themes that may illustrate different aspects of Norse marine resource use and that serve to highlight some recent advances in North Atlantic maritime research.

MARINE PRESTIGE GOODS: NORSE WALRUS HUNTING IN ICELAND AND GREENLAND

As Ottar's account illustrates, Norse marine resource use provided sources of wealth and power unfamiliar to the managers of the primarily agrarian economy of Anglo-Saxon Wessex, generating prestige goods whose low bulk and high value made them attractive items for the elite exchange networks of the early Viking Age. Walrus ivory and hide (used for high-quality ships' line) were clearly in this category, and Norse chieftains like Ottar controlled the main source of walrus products in Arctic Norway before the late ninth century. When Norse voyagers reached southern Iceland ca. AD 870, they encountered resident walrus populations in the area around modern Reykjavik, as bones of both adult and young juvenile walrus have been recovered from ninth-century midden deposits beneath the city streets at site Tjarnargata 4, and walrus place names are found along the nearby Reykjanes peninsula.

In 2001, an early dwelling hall dated by volcanic tephra to just after AD 871 was excavated at Adalstraeti in downtown Reykjavik, producing three adult walrus tusks apparently cached but not recovered from beneath the side benches of the hall (McGovern et al. 2001). Walrus tusks are deeply rooted and difficult to extract without breakage, but these ninth-century tusks showed evidence of careful and expert removal from the dense maxillary bone, suggesting familiarity with walrus hunting among these earliest Icelandic settlers. Bjarne Einarsson (1994) has argued on artifactual grounds for a substantial North Norwegian component among the first Icelanders, which may help to explain the Adalstraeti walrus tusks and the ready exploitation of the local Icelandic walrus colonies. Adolf Fridriksson (Fridriksson et al. 2004) and Christian Keller (personal communication 2006) have suggested a period of initial settlement of both Iceland and Greenland characterized by hunting of walrus and other marine species, which may have preceded the process of chiefly land-claiming and establishment of the farming economy documented in later texts. At present, a combined survey and excavation program directed by Orri Vésteinsson of the University of Iceland is underway to attempt to locate early walrus hunting stations in southwest Iceland. In Iceland, the resident walrus population apparently could not long sustain human predation, and most of our evidence for Norse walrus hunting comes from the Greenlandic colony established from Iceland after AD 985 in two parts of west Greenland (the Western Settlement in modern Nuuk district, and the larger Eastern Settlement to the south in the modern Qaqortoq and Narsaq districts).

Documentary sources for the lost Greenland colony are few and often hard to interpret, but it is clear that walrus ivory and walrus hide rope along with polar bear and seal skins were major exports from Greenland from first settlement down to the end of the colony in the mid-fifteenth century (Arneborg 2003; Larsen 1917; Roesdahl 2005). In 1127, the Greenlandic chieftains are said to have traded the king of Norway a live polar bear for their first bishop, illustrating the value of well-placed arctic prestige gifts (Jones 1987). In 1327, the Norse Greenlanders contributed over 300 kg of walrus ivory to help fund a papal crusade against heretics in southern France, but we cannot determine if this special tithe was the product of multiple hunting seasons or a single massive effort (Gad 1970). Archaeological traces of the Greenlandic Norðursetur walrus hunt have been found in the Disko Bay area (800 km north of the northernmost permanently settled farm), the center of the largest walrus populations in Greenland from at least the eighteenth century(McGovern 1985a, 1985b). These finds support textual references to weeks of rowing in six-oared boats from the farms to the south to the Norðursetur hunting area in the spring (McGovern 1985a, 1985b). The Norðursetur hunters seem to have transported only limited portions of the walrus back to the home farms, as walrus bone finds from both the Western and Eastern settlement areas are made up almost entirely of fragments of the

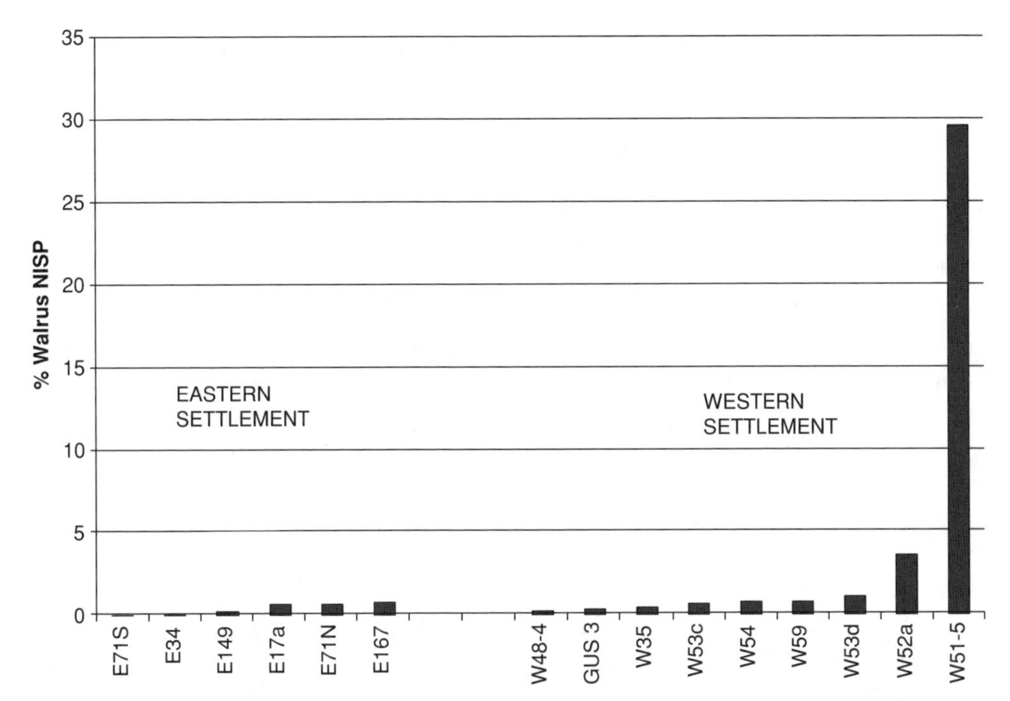

FIGURE 9.2. Relative percentage of walrus bone fragments in the larger Greenlandic archaeofauna (Enghoff 2003; McGovern 1985a, 1985b; McGovern et al. 1996). Note that these are almost all pieces of the dense maxilla around the tusk root. W prefix, Western Settlement; E prefix, Eastern Settlement; NISP, number of identified specimens.

maxilla from around the deep rooted tusks. As Figure 9.2 indicates, walrus bone fragments are found on many Norse sites in both the Western and Eastern Settlements, but their greatest concentration is at the chieftain's farm (W 51) Sandnes in the Western Settlement (McGovern et al. 1996). At Sandnes, the percentage of walrus maxilla rises steadily from the eleventh century to the abandonment of the Western Settlement ca. AD 1350, while the frequency of accidental chipping of the ivory appears to decline over the same period as tusk extraction expertise apparently improved with time (McGovern et al. 1996; Perdikaris and McGovern 2007).

There are also reports of substantial amounts of walrus bone recovered from early twentieth-century excavations at the bishop's manor at Gardar in the Eastern Settlement, perhaps reflecting the Episcopal interest in the walrus ivory trade. The processing of walrus maxillae seems to have been concentrated on coastal elite farms of the Western Settlement, but most Eastern Settlement archaeofauna contain at least a few maxillary fragments, and maxillary fragments are also found at inland farms in both settlement areas (sites E71N, E71S, E167, E34, W35, W53d, W53c, W54, and GUS3 are all more than two hours' walk from the nearest saltwater). There are very few artifacts made of walrus ivory among the many worked bone objects recovered from Norse Greenland. Where walrus tooth was used for buttons or amuletlike figurines (often representing walrus or polar bear, see Fitzhugh and Ward [2000:321]), this came from the apparently unmarketable peglike postcanine teeth. Apart from a few accidentally removed chips and flakes, nearly all the walrus ivory seems to have been efficiently gathered and exported, rather than circulated within Greenlandic society. The ivory was evidently not itself a marker of prestige within Greenland but was instead the means of acquiring status markers from Europe (beer, wine, construction timber, fine garments, church bells, and stained glass all appear archaeologically or in documentary references; Arneborg 2000; Gad 1970). While it appears that boat-owning elites in the settlement area closest to the Norðursetur hunting grounds probably played a leading role in the organization of the hunt and ivory processing, the widespread distribution of walrus maxillary bone chips in virtually every excavated archaeofauna suggests that many, if not all, the Greenlandic households participated at some point in the long-range hunt for walrus.

The walrus bone fragment distribution may be evidence for something like a share system that worked to coordinate scarce labor from inland farms, but it is apparent that "noneconomic" forces were also at work in the form of magic, ritual, and perhaps individual rights of passage. Walrus bacula (the massive penis bone) are found on many Norse farms, including a set of four bacula apparently displayed on the wall of the inland farm W54 in the fourteenth century, and a row of whole walrus and narwhal skulls were found buried within the churchyard wall at the cathedral at Gardar. The dangerous long-distance hunt, probably socially embedded in a matrix of economic, political, and ritual activity, very likely became more dangerous as the frequency of storms increased in the later Middle Ages and as Thule Inuit began to occupy the core of the Norðursetur sometime after AD 1250 (Gulløv 2004). After 1200, taste in Europe began to switch away from ivory, and elite interest in Greenlandic ivory seems to have plummeted at the same time the expanded Norwegian and Russian exploitation of North Cape and Finnmark expanded access to Barents Sea walrus populations (Roesdahl 2005). The majority of the surviving clerical correspondence regarding the walrus ivory crusade tithe of AD 1327 centers on the problem of profitably marketing such a large cargo of unfashionable material without further depressing prices. The Norse Atlantic walrus hunt may well have played a major role in the initial colonization of Iceland and Greenland, and it certainly retained a major place in the moral and political economy of the Medieval Greenlanders. However in the long term, walrus products could not hold their value as an elite prestige item, and Greenland may have been increasingly cut off by summer drift ice from European market centers after 1300.

SEABIRD AND SEAL SUBSISTENCE HUNTING

Seabirds and seals are seasonally accessible nearshore and onshore resources, which have been exploited by a range of northwestern European cultures going back at least to Mesolithic times. As Viking Age colonists moved into the offshore islands of the Faroes and Iceland, they encountered huge populations of seabirds in nesting colonies probably little disturbed by prior human predation. Later Icelandic written sources noted that at the time of settlement, wild animals of all kinds were unwary and easy to catch (Vésteinsson et al. 2002). Figure 9.3 presents percentages of seal and bird bones from a selection of sites from the North Atlantic dating to the Iron Age, Viking Age, and Late Norse-Medieval period.

In most of the collections from all periods, birds form a fairly minor element in the overall archaeofauna (<10 percent), clearly providing a supplement rather than a staple resource. The notable exceptions are in the stratified archaeofauna from Junkarinsfløtti on Sandoy in the Faroes, and two early settlement period sites from southern Iceland (Tjarnargata 4 and Herjolfsdalur). Both the Faroese and early southern Icelandic collections are made up mainly of colonially nesting seabirds (mainly Alcidae, auk family), and the Icelandic Tjarnargata 4 collection includes a few bones of the now extinct great auk *(Pinguinis impennis)*. The two early Icelandic sites appear to reflect the drawdown of the massive natural capital represented by the previously unharvested bird colonies of the south coast at a time when the small herds and flocks of imported domesticates could not yet fully provision the first colonists. After the early settlement period, Icelandic archaeofaunas generally have less than 5 percent bird bone, the

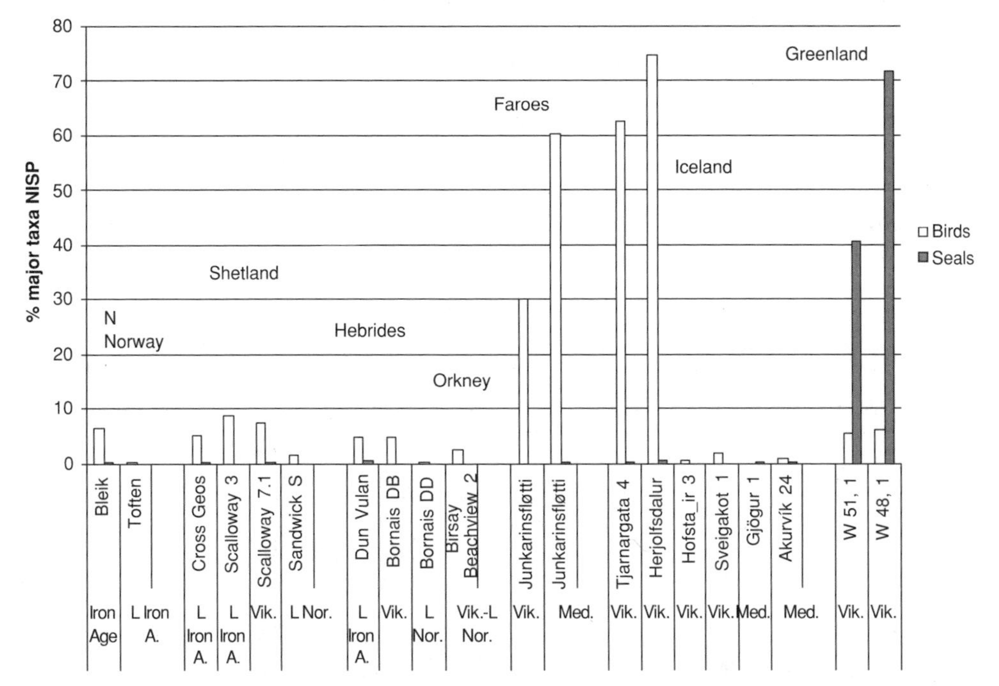

FIGURE 9.3. The relative percent of major taxa made up by birds and seals from a selection of systematically sampled sites across the North Atlantic. NISP, number of identified specimens. For references and data sources see Table 9.1.

only exception being the site of Miðbaer on the small island Flatey in the midst of major seabird colonies in Breiðafjord (Amundsen 2004). In the Faroes, recent excavations led by Simun Arge at the deeply stratified site of Junkarinsfløtti on Sandoy have produced a very large archaeofauna extending from the Viking period to the early thirteenth century (Church et al. 2005). These collections have produced massive numbers of bird bones (mainly puffins, *Fratercula arctica*), whose frequency actually increases relative to domestic mammals and fish in the sample, reaching 60 percent of identified fragments in the Medieval upper layers. Older unsieved collections from the Faroes also show high percentages of puffins extending into the later Middle Ages, indicating that the Junkarinsfløtti archaeofauna is not unique (Gotfredsen 2007). Excavations on Sandoy are ongoing, but results so far suggest a successful long-term management of the Faroese bird colonies for sustainable yield rather than an immediate and rapid drawdown paralleling the south Icelandic pattern. Norse seabird exploitation in the North Atlantic has clearly followed more than one trajectory.

As Figure 9.3 also indicates, seals have been hunted throughout the region, but only in Greenland do seal bones make up a major portion of the bone collections from first settlement onward. Viking Age colonists in the eastern North Atlantic encountered gray seals *(Halichoerus gryphus)* and common or harbor seals *(Phoca vitulina)*, both mainly vulnerable during spring pupping seasons and usually taken by clubbing on shore or netting (Fenton 1978). In Greenland, the Norse encountered a different range of seal species, some seasonally accessible in vast numbers. Colonies of the familiar common seal existed where summer drift ice was not prevalent, but far more numerous were the millions of harp seals *(Pagophilus groenlandicus)* and hooded seals *(Cystophora cristata)* that ride the drift ice from Labrador along the coast of Greenland each spring. These migratory ice-riding Arctic seals were vulnerable

TABLE 9.1

Data Sources for Chapter 9 Figures

PERIOD	REGION	SITE AND PHASE	REFERENCE
Iron Age	N Norway	Bleik	Perdikaris 1999
	N Norway	Toften	Perdikaris 1999
	Orkney	Brough Rd. 3	Rackham 1989
	Orkney	Howe 8	Dockrill et al. 1994; Smith 1995
	Orkney	Pool 6	Nicholson 1998
	Orkney	Saevar Howe 1	Colley 1983; Rowley-Conwy et al. 2004
	S Uist	DunVulan 5	Sharples 2005
	Shetland	Scalloway E 3	Ceron-Carrasco et al. 2005; Sharples 1998
	Shetland	Cross Geos	Bruns and McGovern 1985
Viking Age	Faroes	Junkarinsfløtti 1	Church et al. 2005
	Caithness	Smoo 5	Barrett 1995
	Orkney	Brough Rd. 1 and 2	Rackham 1989
	Orkney	Birsay Beachview	Morris 1996
	Orkney	Pool 7	Nicholson 1998; Bond 2007
	Orkney	Saev. H	Colley 1983
	S Uist	Bornais DB	Sharples 2005
	Shetland	Scalloway L 3	Sharples 1998, 2005
	N Iceland	Sveigakot 1–3	McGovern et al. 2006
	N Iceland	Hofstaðir 1–3	McGovern et al. 2006
	N Iceland	Hrísheimar 2	McGovern et al. 2006
	N Iceland	Granastaðir	Einarsson 1994
	S Iceland	Tjarnargata 4	Amorosi et al. 1996
	S Iceland	Herjolfsdalur	Amorosi et al. 1996
	S Iceland	Adalstraedi	McGovern et al. 2001
	Greenland	W 51-1 Sandnes	McGovern et al. 1996
	Greenland	W 48-1	McGovern 1985a, 1985b
	Greenland	E 17a Narsaq	McGovern 1994
Late Norse–Medieval	N Norway	Storvågan 11th	Perdikaris 1998
	N Norway	Storvågan 13th	Perdikaris 1998
	N Norway	Helgoy 14th	Perdikaris 1998
	Faroes	Junkarinsflotti 3	Church et al. 2005
	S Uist	Bornais DD	Sharples 2005
	Shetland	Sandwick North 3-4	Bigelow 1984
	Caithness	Roberts Haven	Barrett 1997
	Caithness	Freswick	Morris et al. 1995
	NW Iceland	Gjögur 1-2	Amundsen et al. 2005; Krivogorskaya et al. 2005
	NW Iceland	Akurvík 22, 24	Amundsen et al. 2005; Krivogorskaya et al. 2005
	NW Iceland	Finnbogastaðir	Edvardsson et al. 2004
	Greenland	W54	McGovern 1985b
	Greenland	W52a	McGovern 1985b
	Greenland	W35	McGovern 1985b
	Greenland	W53d	McGovern 1985b
	Greenland	W53c	McGovern 1985b
	Greenland	E149	McGovern 1985b
	Greenland	E167	McGovern 1985b
	Greenland	E 71N	McGovern 1985b
	Greenland	E71 S	McGovern 1985b
	Greenland	GUS	Enghoff 2003
	Greenland	E 34	Nyegaard pers. com. 2004

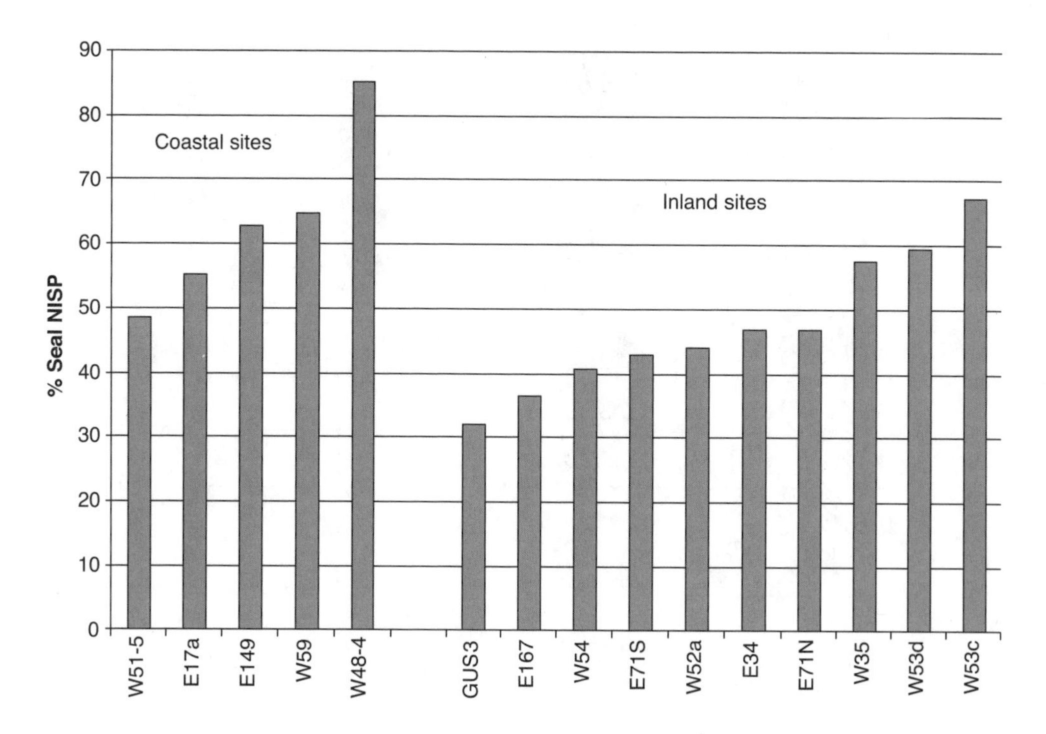

FIGURE 9.4. Seal bones are found in all Norse archaeofauna in large numbers (Enghoff 2003; McGovern 1985, 1985b; McGovern et al. 1996; Nyegaard G., personal communication 2004). These later-phase large collections illustrate a widespread pattern of transport of seal carcasses inland. Bone element distribution study indicates that nearly complete carcasses (gutted, as the baculum is generally missing) were transported up to 30 km from any possible landing point. W prefix, Western Settlement; E prefix, Eastern Settlement; NISP, number of identified specimens.

to clubbing and to entrapment in nets and could be taken effectively with existing Norse technology. Two resident nonmigratory seals also lived in the Greenlandic fjords, the very common ringed seal *(Phoca hispida)* and the rarer but larger bearded seal *(Erignatus barbatus)*. Both of these arctic seals are most reliably taken with toggling harpoons and other ice edge and breathing hole hunting gear, which was part of Inuit and Paleoeskimo, but not Norse, subsistence technology.

Despite their lack of harpoons, Norse hunters in Greenland became expert in taking substantial numbers of the resident and migratory species most vulnerable to netting and clubbing, and seals rapidly became a key subsistence resource in Greenland rather than the occasional supplement sealing represented in the other North Atlantic islands. Figure 9.4 presents the relative percentage of seal bones from the later (fourteenth- to mid-fifteenth-century) archaeofauna from the two settlement areas, divided by coastal or inland location. The effort involved in transporting seal carcasses over rough terrain to upland farmsteads up to 30 km from the sea indicates the importance seals had for the Norse economy. Seasonality studies based on tooth cementum analysis indicates that the harp and common seal were all early spring kills, thus occurring at the low point of the agricultural year when the stored dairy produce and meat from the previous summer would be running low and domestic stock were no longer producing milk. The arrival of the migratory ice-riding harp and hooded seals and the spring pupping concentrations of the common seals would thus have been critical points in the Norse seasonal round in Greenland. The distribution of seal bones so far from any beach also underlines the communal scale

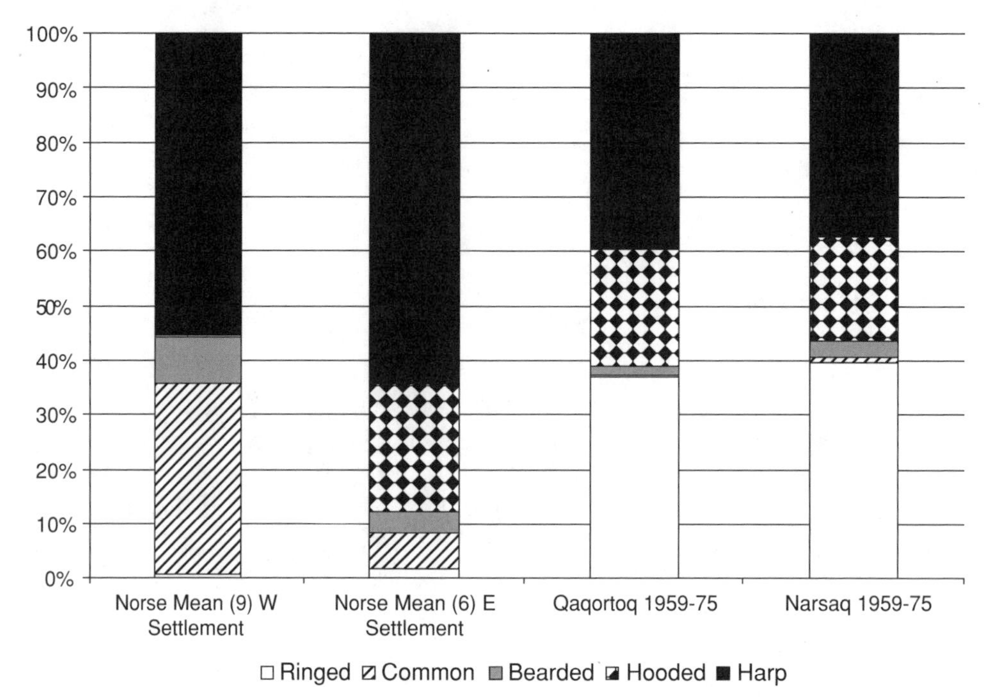

FIGURE 9.5. Comparison of the mean relative percentages of identified seal bones (number of identified specimens) for the 15 substantial archaeofauna available for the later phases of Norse Greenland, with catch statistics gathered for the two modern communities in the former Eastern Settlement (relative percentages of count of seals landed).

of Norse subsistence in Greenland—the individual farm was not the basic unit of survival, and a great many households would be dependent upon a limited number of boats maintained at the larger farms with access to the fjord or coast. Seabird bones (mainly auk family) are also found on inland farms in both settlement areas.

Figure 9.5 presents a comparison of the mean of the relative percentage of seal bones identified to species level in the major Norse archaeofauna, with modern seal catch statistics from Inuit Greenlandic communities in the area of the former Eastern Settlement. The differences between seal species regularly taken by Norse hunters in the Western Settlement and the Eastern Settlement are explained by biogeography: hooded seals seldom reach the Western Settlement area, and common seal colonies are much larger in the Western Settlement area. However, the differences between the seal bones present in the Eastern Settlement Norse archaeofauna and the modern catch statistics for the very same hunting area point to cultural and technological differences. Ringed seals make up the majority of seals taken by traditional Inuit subsistence hunters in winter in most of the eastern Arctic, heavily supplemented by migratory harp seals where available (as in southwest Greenland). The ability of Inuit hunters to take ringed seals in winter at their breathing holes and at the ice edge has meant that these seal hunting communities do not rely so completely upon the timing of the arrival of the migratory seals in spring, and that they have the capacity to reprovision themselves throughout the winter months. The failure of the Norse to acquire Inuit ice-hunting technology despite centuries of contact is remarkable and has been cited repeatedly as a potential root cause for the subsistence failure that ultimately doomed Norse Greenland during a period of climatic change and increased interannual variability (Diamond 2005; McGovern 1985a, 1985b, 1994).

However, it is important to see Norse sealing in Greenland in its economic and social context bound up in a socially embedded seasonal round that involved a nested sequence of activities requiring communal coordination: hay harvest and seabird hunting in late summer followed by caribou drives in autumn and sealing and the Norðursetur hunt in summer. All or most of these activities drew upon the labor of multiple farms and produced the inland distribution of the seal and seabird bones and walrus maxillae that we can observe archaeologically. Inland farmers were dependent upon coastal boat-keepers, who were in turn in need of boat crews and laborers during the short, intensively scheduled Greenlandic summer. Seal hunting by drives into nets and clubbing on shore or on sea ice was a coordinated communal activity, and the sharing of the mass kill was an occasion for carefully calibrated redistribution that reinforced cooperation. Modern Faroese cooperative pilot whaling by mass stranding is carried out today, as much for asserting community solidarity as for the whale meat that is meticulously shared out according to Medieval regulations still enforced. Seal hunting with harpoons at breathing hole or ice edge is an individual or small-group activity, with seals killed singly rather than en masse and thus less suitable for comparable distribution. We do not yet understand the full dimensions of sealing in Norse Greenland, but it seems clear that it represented a powerful social and economic pattern that drew on ancient Nordic seal-hunting techniques, but which developed far more intensively than in any other part of the Norse North Atlantic and played a vital role in subsistence for coastal and inland farmers alike.

BETWEEN LOCAL SUBSISTENCE AND WORLD SYSTEM: FISHING IN THE VIKING AGE

While seals and seabirds contributed to local subsistence, and walrus hunting for a time provided valuable elite prestige goods for trade, none of these marine resources was to have more than local economic importance within the communities of the North Atlantic. By contrast, a series of changes in marine fishing and in the preservation of fish for later consumption that spread throughout the North Atlantic by the close of the Viking Age were to transform the economies of northwest Europe and to contribute to the colonization of North America five hundred years later. Some of the most significant advances in the archaeology and paleoecology of the North Atlantic in recent years have resulted from coordinated work by a number of field projects and analysts aimed at recovering, identifying, and comparing fish bones and other evidence for early fishing (Amundsen et al. 2005; Barrett 2003; Bigelow 1984; Bond et al. 2004; Cerron-Carrasco et al. 2005; Church et al. 2005; Krivogorskaya et al. 2005; McGovern et al. 2007; Morris 1996; Morris et al. 1995; Mulville and Thoms 2005; Nicholson 2004; Parker-Pearson and Sharples 1999; Perdikaris 1996, 1999; Sharples 1998, 2005; Smith and Mulville 2003). Thanks to the work of these collaborating teams we are now able to compare multiple large, fish-rich archaeofauna from well-dated, systematically sampled and sieved contexts that span the period from later Iron Age (sixth to ninth centuries AD) through Viking Age to Late Norse-Medieval period. A key question uniting these investigations has been the causes and origin of the expansion of marine fishing from subsistence production for local consumption into the vast commercial bulk-goods marketing program that brought dried and salted cod to distant inland consumers across Europe after AD 1100.

Historical documents have long indicated the role of preserved cod and herring in the expansion of trade and towns and the strengthening of royal power in Scandinavia by the early twelfth century (Perdikaris 1999). When properly air-dried, low-fat white-fleshed cod family (Gadidae) fish can keep without salting for over five years, providing a light and storable source of high-quality protein that could sustain sailors, armies, and inland populations far from the shore. The two air-dried products of the early Medieval period were *stockfish* (beheaded and dried in the round, hung from racks) and *rotscher* (beheaded and flat dried, sometimes simply spread on cobble beaches). Stockfish production requires winter temperatures within a few degrees of freezing and was best carried out with fish 60 to 110 cm long. Rotscher could be produced under a wider range of temperatures and made use of fish optimally 40 to 70 cm long. Cod became the preferred fish for curing, and methods of fish cutting and preparation became standardized as the older artisanal fisheries were rapidly converted into full-scale commercialization, producing a commodity that could be stored for long periods, widely transported in bulk, and bought and sold in counting houses far from the fishing grounds. By the thirteenth century the Hanseatic League had been formed, with the control and regulation of the codfish trade as a primary collective concern. By the mid-fifteenth century, over 100 cities in Europe had Hansa connections, run from major offices in Novgorod, Bergen, Brugge, and London. Demand continued to expand, and the codfish became a commodity used for futures trading, as collateral for loans, and as a dietary standard in late Medieval cook books (most of which suggest pounding the dried fish with a large hammer for an hour to tenderize it; Fagan 2006). By 1500, the discovery of the Grand Banks fishery off Newfoundland attracted hundreds of vessels from Europe to the New World annually, and the cod fisheries remained an economic engine driving the European colonization of what became New England and the maritime provinces of Canada for the next three centuries (Kurlansky 1997).

"FISH MIDDENS" IN THE NORTH ATLANTIC

The role of preserved cod in the later history of the North Atlantic is thus both dramatic and well documented, but the origins of this phenomenon prior to the beginning of historical records around AD 1100 remains an archaeological problem. The existence of "fish middens" apparently dominated by dense masses of fish bones has been observed in many parts of the North Atlantic for many years, but prior to the late 1970s excavations of these deposits that employed systematic sieving programs were rare, and the analysis and identification of marine fish bones was still in its infancy (Morris et al. 1995). This situation has since changed dramatically, with advances in both excavation and zooarchaeological analysis of fish bone. Systematic sieving and flotation of bone-bearing deposits has become standard, generating substantial, comparably excavated fish bone collections from a wide area, and at the same time significant advances have been made in standards of analysis and data comparability ("number of identified specimens" becoming a standard comparative measure, and species-level identification based on most of the bones of the fish skeleton replacing selected-element approaches). Figure 9.6 presents some of these new data for the relative abundance of fish bones on a sample of comparably excavated major sites in different parts of the North Atlantic (see Amundsen et al. 2005; Barrett et al. 1998, 2000; Bigelow 1991; Krivogorskaya et al. 2005; Perdikaris 1999; McGovern 1985a, 1985b, 1990; McGovern et al. 1996, 2001; Morris et al. 1995; Nicholson 1998, 2004; Parker-Pearson and Sharples 1999; Sharples 1998, 2005).

Relative abundance of marine fish bone in an archaeofauna is a relatively crude indicator of fishing intensity, but it may provide a starting point for comparative discussions (Krivogorskaya et al. 2007; Perdikaris 1999; Perdikaris and McGovern 2007). Figure 9.6 underlines the degree of variation between sites, regions, and periods in the North Atlantic from pre-Viking Iron Age to the Late Norse-Medieval period. The major role of fish in the two Arctic Iron Age Norwegian sites of Bleik (a high-status chieftain's farm on Andoya) and Toften (a middle-ranking Andoya site) is clear. Classic fish middens definitely extend back to

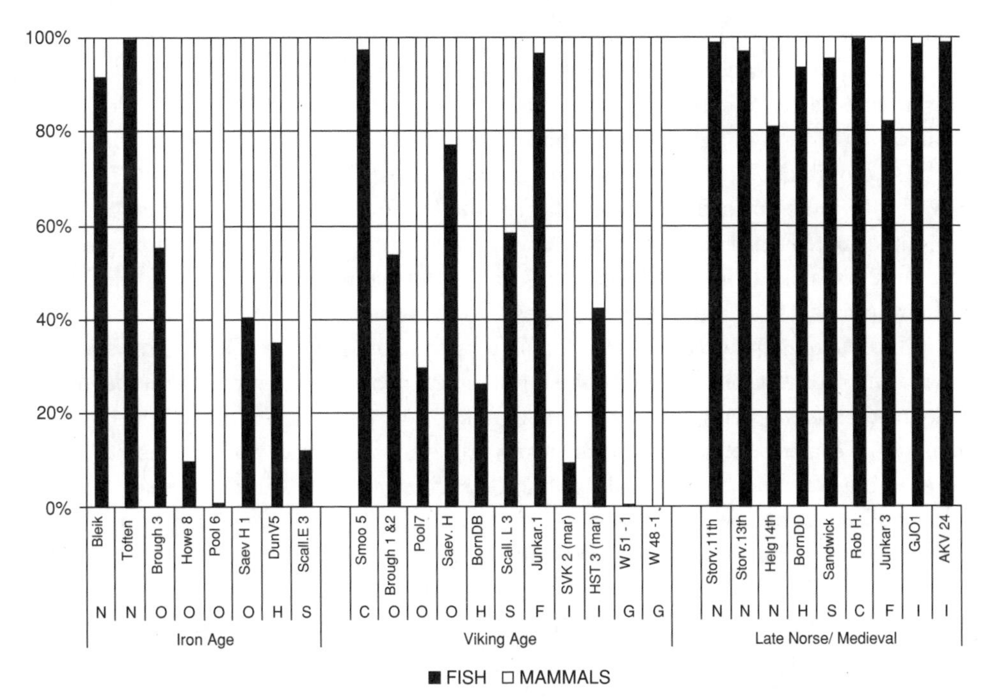

FIGURE 9.6. Relative percentages of marine fish bones and mammal bones from sites dating to the Iron Age, Viking Age, and Late Norse–Medieval period grouped by region. Only sites with systematic sieving are included. N, arctic Norway; O, Orkney; H, Hebrides; S, Shetland; C, Caithness; I, Iceland; G, Greenland. See Table 9.1 for data sources.

the second and third centuries AD in the Lofoten and Vesterålen region of arctic Norway, predating the Viking Age Scandinavian expansion by centuries (Perdikaris 1999). Equally striking is the nearly complete absence of fish from the Viking Age and Medieval Greenland (W51-1 and W 48-1, both in the Western Settlement; McGovern et al. 1996). Despite generally excellent conditions of preservation that regularly allow the recovery of hair, cloth, feathers, and large amounts of mammal bone, and the intensive sieving and whole-soil sampling efforts of multiple projects since 1975, marine and freshwater fish bones have remained exceptionally rare in all Greenlandic archaeofauna (Enghoff 2003; McGovern 1985a, 1985b, 1990; McGovern et al. 1996). Explanations for this shortage of fish remains have ranged from issues of seasonal scheduling and labor shortage (Perdikaris and McGovern 2007) to religious sanction (Diamond 2005). Whatever the cause, it is clear that the Norse Greenlanders made sealing rather than fishing the mainstay of their subsistence economy, and that they never generated the fish middens that create archaeofauna predominately composed of fish bone.

THE PICTISH-NORSE TRANSITION IN NORTHERN SCOTLAND

Between the extremes of apparently intensive marine fishing in Iron Age Arctic Norway and apparently insignificant fishing in Viking and Medieval Greenland are the more complex zooarchaeological patterns of the Northern and Western isles, Faroes, and Iceland. In an important regional synthesis combining the evidence of zooarchaeology, radiocarbon chronology, and stable carbon isotopes in human bone from pre-Norse, Viking Age, and Late Norse contexts, Barrett et al. (2000) presented a strong argument for profound differences between Late Iron Age Pictish and Viking Age diet and economy in the Northern Isles. Pictish fishing was argued to be mainly inshore, aimed at small saithe (1- to 2-year-old "sillocks and piltocks"), which can be caught close inshore, sometimes from rocky cliff-side craig seats (Fenton 1978) rather than from boats. These young saithe are too small to effectively air dry, but they can be smoked for later consumption, and their liver yields oil (in later centuries extensively used for lighting as well as for food). Other marine species taken on Iron Age sites include anadromous species such as eels and salmon, which can be taken in streams or close inshore. Fishing, sealing, and seabird hunting were seen as supplementary to well-developed cereal agriculture (mainly based on barley but including oats in the Late Iron Age; Bond et al. 2005). Available carbon isotopic ratios from Pictish skeletons indicated a largely terrestrial diet (human bone $\delta^{13}C$ values around −20 ‰), and produced a strong contrast with more maritime samples from Viking Age and Late Norse skeletons indicating a stronger participation in the marine food web ($\delta^{13}C$ values around −17 ‰; Barrett et al. 2000:149). Since the 2001 synthesis, important new analyses have been carried out on collections from the sites of Dun Vulan (Late Iron Age; Parker-Pearson and Sharples 1999) and Bornais (Viking to Late Norse; Sharples 2005), approximately 3 km apart on South Uist in the outer Hebrides, and from Viking and Late Iron Age phases at the elite Iron Age site of Scatness in Shetland (Bond et al. 2005; Nicholson 2004). As Figure 9.6 indicates, the Iron Age–Viking Age transition between nearby Dun Vulan and Bornais did not result in an increase in the proportion of fish bones deposited, though there was a shift toward cod and toward larger individuals, as seen in most other transitional contexts. The later pre-Norse Iron Age phases at Scatness in Shetland are producing evidence for intensified fishing and probably storage of dried fish and fish liver oil for elite redistribution prior to the Scandinavian settlement, and the analyst now argues for some significant continuities between Pictish and Viking Age fishing patterns (Nicholson 2004:161). Barley, smoked saithe, and fish liver oil all may have played a role in Late Iron Age Pictish chiefly economies based on resource concentration and redistribution by local elites (Dockrill and Batt 2004). While the Viking Age saw the intensification of flax production (probably in part providing fishing line and sails) and continued soil fertilization and amendment, these patterns are now often seen to represent intensifications of economic patterns begun in the late Iron Age (Bond et al. 2005). Although the clarity of the Pictish-Norse economic transition has blurred slightly in recent years, most workers still model the Pre-Viking Iron Age inhabitants of the Northern and Western isles as primarily farmers who made use of a range of largely inshore marine resources to supplement their terrestrial subsistence economy, in strong contrast to their Pre-Viking Scandinavian contemporaries in the islands of Arctic Norway, whose economies were based primarily upon deep-sea fishing. If the Iron Age peoples of northern Scotland could be characterized as farmers who fished, the contemporary Iron Age North Norwegians were fishermen who also farmed.

While the Norse arrival in northern Scotland ca. AD 800 is very visible in terms of elite burials, change in house forms, and modern place names, it remains far less clear cut in the zooarchaeological and archaeobotanical record. At some sites a strong zooarchaeological case can be built for radical change in degree and type of maritime adaptation after the arrival of the Norse, but in other areas the case for significant continuity can also be made. Some of these differences among archaeofauna may be due to sampling issues (midden vs. house floors, column samples vs. open-area excavations) or issues of chronology (the resolution required to securely date an occupation as before or after the Norse arrival pushes the limits of radiocarbon chronology), but it increasingly appears that this local variability in marine resource use at the Pictish-Viking transition may be signal rather than noise. Different economic patterns in marine resource use may well have existed in the same area, even on the same islands. This economic complexity has contributed to continued debate about the extent of population replacement versus assimilation during the Viking Age contact period (Jennings and Kruse 2005; Owen 2004). Major linguistic and ritual changes can be associated with conquest and replacement of elites, but even scattered continuities in basic subsistence practices may point to a variable blending of local and imported subsistence strategies and technology at the very local (perhaps intra-household) level, and the very discontinuity in the different forms of evidence may carry its own message. The economic and environmental story of Norse-Pictish contact in Scotland is by no means a settled issue and remains a very active research focus for many teams.

As Figure 9.6 indicates, the major jump in fish bone relative percentages in northern Scotland comes not with first contact in the ninth century, but with the Viking Age to Late Norse–Medieval transition around AD 1050–1100 throughout the region. While cod and cod-family gadid fish make up most of the majority of these new fish middens, herring bones provide the jump in fish abundance between Viking and Late Norse contexts at Bornais (Sharples 2005). Continued excavation coupled to a major radiocarbon dating program at the site of Quoygrew in Orkney has clearly demonstrated a break between mammal-dominated Viking Age middens and a classic Late Norse fish-dominated midden that can be closely dated to a generation or so on either side of the year 1000 AD (Barrett 2003; Simpson et al. 2004). As Barrett et al. (2000) have argued, the apparent arrival of a coherent Viking Age economic package involving simultaneous intensification of fishing, grain production, and long-distance dried fish trading breaks down when additional dates on stratified sequences can be applied. These processes may begin with the Norse arrival in northern Scotland around AD 800 but do not take off synergistically until two centuries later. The Late Norse transition has been identified by Bigelow (1984, 1991) as the origin point of the crofter farmer-fisher life ways so evident in the historic and ethnographic record of the Northern and Western isles:

> In fundamental terms, the model proposes that particularly during the twelfth century the majority of Shetland household economies shifted from a predominately subsistence orientation to the mixed subsistence/surplus-for-exchange pattern that in various forms was characteristic of the islands until the twentieth century.
>
> (Bigelow 1991:20)

THE FISH EVENT HORIZON IN BRITAIN

The dating of the fish middens in the Northern and Western isles has become more significant with the documentation of what Barrett et al. (2004) have called the "fish event horizon" (FEH) in Britain (see also Bailey et al., this volume). Derived from a comprehensive data survey of the very large number of archaeofauna reported for Britain from early Iron Age to Medieval periods, the FEH marks a striking transition point in the pattern of marine fish distribution in inland sites in Britain; prior to the temporal horizon of AD 950–1050 there are virtually no marine fish bones in any archaeofauna more than 10 km from the coast. After the FEH, bones of Gadidae and herring become increasingly common on inland sites, as would be expected from the well-documented dried-fish trade after AD 1100. Marine fishing and the intensive marketing of preserved marine fish thus do not date to the Anglo-Saxon period (it is not clear that Anglo-Saxon speakers had a word specifically meaning "cod"; Barrett et al. 2000) nor does it begin with the first Scandinavian settlements in what became the Danelaw in northern England in the late ninth and early tenth centuries. This finding has stimulated a major reevaluation of the evidence for pre-FEH fishing and inland fish distribution in the North Atlantic. We can now be confident that the combination of intensive fishing (producing specialized fish middens in producing areas) and large-scale bulk marketing of preserved marine fish (producing inland fish bone distributions in consuming areas) was not the result of Celtic or Anglo-Saxon cultural traditions, and that (in the British Isles at least) it was also not a product of the early Viking Age arrival of Scandinavian raiders and settlers. We need to look outside the British Isles for the ultimate origins of the FEH and the highly visible Late Norse transition.

EARLY FISHING IN FAROES AND ICELAND

As Figure 9.6 indicates, marine fish appear to have played a very important role in northern Norway, Faroes, and Iceland from Iron Age times into the early Viking Age and Medieval period, both before and after the British FEH. A Scandinavian origin for the FEH seems a clear possibility on these grounds alone, but it is fish preservation technology and fish distribution expertise (through exchange, tribute, or market) that underlies the economic and social changes of the FEH, not simply seafaring skills and level of fishing effort. Since fisher folk everywhere tend to consume fish themselves as well as sell them (often eating the less salable species or size ranges), most coastal site middens tend to reflect a confusing mix of local subsistence and production for market even in later Medieval and early modern situations long after the FEH. This mixture can make unraveling the commercial signature in fish archaeofauna challenging, and clear-cut patterns of species abundance and element representation are not always easy to see in coastal midden deposits (Morris et al. 1995). As has been argued elsewhere, the way forward involves a multiindicator, multiarchaeofauna approach combining species proportions, skeletal element distribution, size and age reconstruction, and the use of later sites of known function as reference points for the understanding of earlier archaeofauna (Amundsen et al. 2005; Krivogorskaya et al. 2005; Perdikaris 1999; Perdikaris and McGovern 2007). A simple comparison of the recovered proportions of different skeletal elements can provide some indications of specialized fish processing.

Figure 9.7 presents a comparison of the relative proportions of two fish skeletal elements (cleithrum around the gill slit, and premaxilla in the upper jaw) for a range of sites of different period and location. In the production of all types of dried gadid fish, the head and jaws (including the premaxilla) are removed in the first stages of fish cutting, and thus premaxillae and other head and jaw bones tend to accumulate where fish are processed. The cleithrum and associated bones in the pectoral arch are normally left in the headless body, as these bones help to keep the fish together and aid drying of the carcass when spread apart. Thus cleithra tend to travel with the prepared body, while premaxillae tend to remain at processing sites. Where subsistence fishers consume their catch on site, middens tend to accumulate cleithra and premaxillae in more or less equal proportions (both elements are robust and easily identified to species level). Where export of preserved stockfish or rotscher takes place, cleithra and premaxilla of the same fish may be deposited thousands of kilometers apart. As Figure 9.7 illustrates, Iron Age sites in Norway (Bleik and Toften in Perdikaris 1999) and at Dun Vulan on South Uist in the Hebrides (Sharples 2005) show some minor fluctuations around the 50:50 natural proportion, with the closest match to the high-status site of Bleik being the eighteenth-century contexts at Finnbogastaðir in northwest Iceland, which is known to derive from a mixed subsistence/market production pattern (Edvardsson et al. 2004). The Beachview (Bv) Birsay Area 3 collection has also been interpreted as primarily domestic consumption refuse probably mixing initial processing and local consumption (Morris et al. 1995). The Viking–Late Norse contexts from Freswick Links Caithness (Phases T, U, V, Area 4; Morris et al. 1995:185–190) by contrast come from a more specialized classic fish midden deposit and show a marked surplus of cod premaxilla (or deficit of cleithra), suggesting processing for distant consumption.

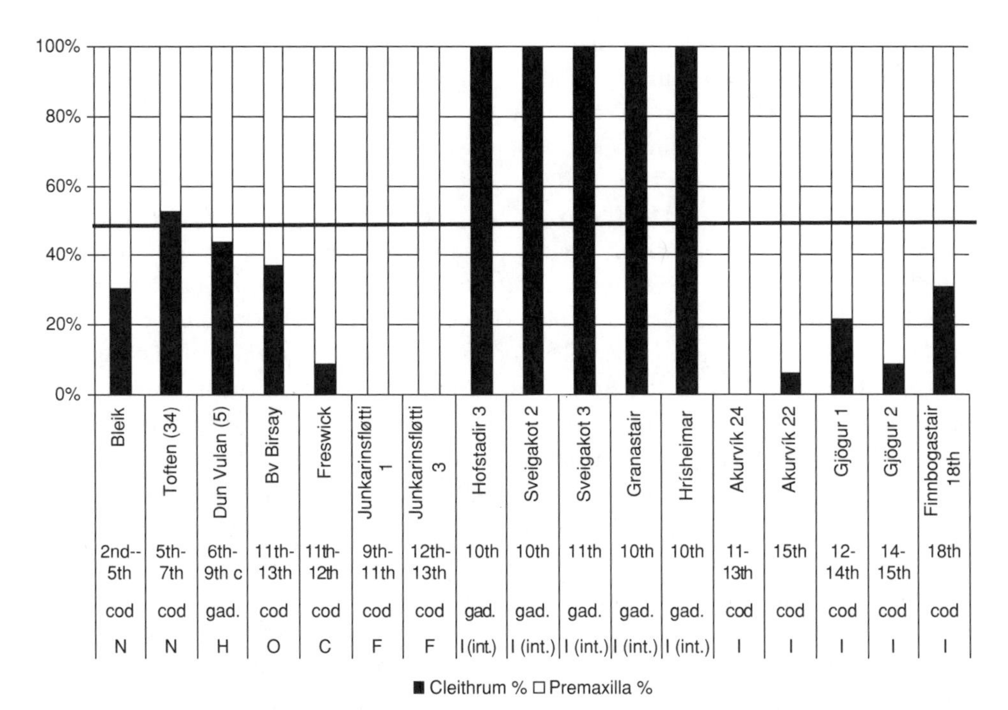

FIGURE 9.7. Relative proportions of cleithrum and premaxilla (number of identified specimens) on selected large archaeofauna. Where possible cod (the major commercial species) has been used for this comparison, but in some cases all gadid fish have been pooled due to sample size issues. The horizontal line indicates the natural proportion of these paired elements in a whole fish. N, arctic Norway; H, Hebrides; O, Orkney; C, Caithness; F, Faroes; I, Iceland. See Table 9.1 for data sources.

The Viking Age and Late Norse-Medieval contexts at Junkarinsfløtti in Sandoy on the Faroes show closely similar patterns of concentration of premaxillae and near absence of cleithra (Church et al. 2005). The site area sampled thus far includes many bird and domestic mammal bones and is thus not simply a fish processing area, but the strong patterning does raise some questions about the role of the site in producing preserved cod for either local or distant markets. Excavation is ongoing at this site, and a multiseason effort now underway should generate very large stratified collections as well as a better understanding of the site as whole.

In Iceland, the current Viking Age coastal archaeofauna available are all from incompletely sieved excavations, from contexts with poor bone preservation, or not yet large enough to effectively quantify. However, large well-preserved modern collections are available from inland sites dating to the Viking Age and from coastal sites in northwest Iceland dating to the twelfth through fifteenth centuries (Amundsen et al. 2005; Krivogorskaya et al. 2005; McGovern et al. 2007). The set of large sieved inland archaeofauna from northern Iceland dating to the late ninth to tenth centuries comes from both inland Eyjafjord (Granastaðir; Einarsson 1994) and from the Lake Mývatn region to the east (Hofstaðir, Sveigakot, Hrisheimar; McGovern et al. 2007). These sites are from 50 to 80 km from the sea and are securely dated through a combination of AMS radiocarbon dates and volcanic tephra. Despite their distance inland, all have produced substantial numbers of marine fish bones datable to before, after, and during the British FEH, indicating the presence of a Viking Age fish distribution system in ninth- to tenth-century Iceland perhaps comparable to the later communal system that transported so many seal carcasses to inland farms in Greenland. The pattern of element distribution signaled by the complete absence of any premaxilla and the abundance of cleithra extends to cod, haddock, saithe, ling, and other Gadidae imported inland; all were brought in as headless preserved fish. Scattered finds of postcranial gadid elements on other early inland sites in both southwestern and eastern Iceland indicate that the Mývatn and Eyjafjord pattern of distribution was widespread, and that inland households throughout Iceland had no problem in provisioning themselves with preserved marine fish early in the Viking Age (McGovern et al. 2007). These Viking Age inland sites not only provide a clear consumer's profile of fish element distribution, but also indicate the existence of a regional-scale network that was capable of producing and distributing substantial amounts of preserved fish over a wide area.

One source for this Viking Age regional fish trade may have been the West Fjords of northwest Iceland. This district has very limited pasture but excellent access to fishing grounds and seems to have supported wealthy elite manors from the Viking Age whose income came from fishing rather than farming (Edvardsson et al. 2004). The sites of Akurvík and Gjögur, which appear in Figure 9.7, are less than 3 km apart on the shores of Nordurfjord in northwest Iceland (Amundsen et al. 2005; Krivogorskaya et al. 2005). Akurvík is a seasonal fishing station composed of briefly occupied sod booths with associated fish middens, while Gjögur is a deeply stratified farm mound. Both produce fish-dominated archaeofauna, and both show strong "producer" signatures in their early and later Medieval contexts. The early phases of both sites predate the historically known mid-fourteenth-century date for the arrival of commercial fishing via England and the Hanseatics and suggest the presence of a still poorly understood precommercial chiefly fish production and exchange network that operated in Iceland (and perhaps in the Faroes and Arctic Norway) before the FEH and the full-scale commercialization of the high Middle Ages.

Another major tool for North Atlantic marine zooarchaeology is the reconstruction of live fish length from measurement of bone fragments (Wheeler and Jones 1989). As Figure 9.8 indicates, dentary and premaxillary measurements tend to produce similar distributions, and these can be used to investigate the nature of the fish product being prepared at fishing stations. While both the Gjögur fishing farm and the Akurvík seasonal fishing station were producing and exporting cod in the Middle Ages, only the Akurvík station seems to have been systematically processing cod fish within the "stockfish window" of ca. 700 to 1,100 mm. Were the stockfish bound for the new overseas markets and the smaller rotscher circulating in local Icelandic exchange systems dating back to the early Viking Age?

Additional measures that cannot be presented here (vertebral element distribution, age-size correlations) provide more tools for the investigation of these early fisheries, but even this limited discussion of the rich data sets now available may serve to illustrate the main results of the investigation (see Amundsen et al. [2005] and Krivogorskaya et al. [2005] for discussion of other indicators).

During the early Viking Age, prior to the FEH in Britain, parts of the Scandinavian world already practiced a combination of fishing, fish preservation, and fish distribution that lay somewhere between the patterns of local subsistence that seems to characterize most fishing in Iron Age Atlantic Scotland and the full-scale proto-world-system commoditization and commercialization of the Hanseatic League. This earlier Nordic fish distribution system connected distant districts in what was probably a social and economic web of obligation and exchange mediated by chiefly managers who, like Ottar, were quite capable of drawing wealth and power as well as food from the sea. It may be significant that the clearest current evidence for such a precommercial production and distribution system comes from those parts of the North Atlantic where Nordic colonists found few or no preexisting Celtic or Anglian populations

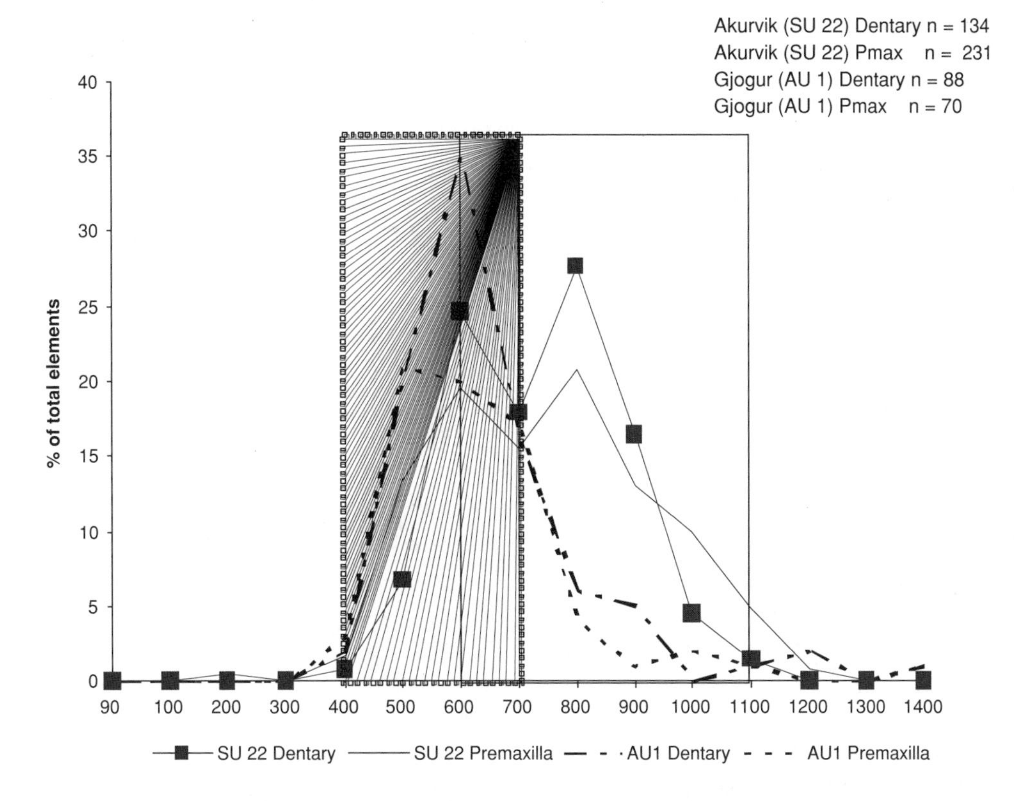

FIGURE 9.8. The distribution of cod live length reconstructions in millimeters for the later Medieval (fourteenth- to fifteenth-century) deposits at the seasonal fishing station Akurvíkand the nearby permanently occupied farm mound Gjögur. The solid line encloses the optimal size range for the production of stockfish, while the dotted line encloses the optimal size range for rotscher production. Akurvík appears to have been actively engaged in both stockfish and klipfish production, while Gjögur seems to have concentrated upon rotscher, perhaps serving different markets.

and were compelled to construct basic economic patterns without reference to preexisting local experience.

ENVIRONMENTAL IMPACT ASSESSMENT

Nordic settlers in the North Atlantic region had varied impacts upon marine and terrestrial ecosystems during the past millennium (Dugmore et al. 2007). In the Northern and Western isles of Britain, Norse settlers entered island ecosystems long exploited by farming cultures who had supplemented barley and stock with varied amounts of fish and sea mammals. Economic changes during the initial ninth- to tenth-century settlement seem to have involved expanded offshore fishing, but substantial fish middens signaling major intensification of fishing and skeletal evidence of substantial numbers of fish being prepared for export seem to date to the eleventh century or later. Human impact on seals, seabirds, and local fish stocks thus probably changed only slightly across the Pictish-Norse transition, and there is little evidence for a significant new impact on these resources. In the Faroes, Iceland, and Greenland, first settlement by ninth- to tenth-century Norse created the potential for significant impact due to the drawdown effect upon accumulated natural capital represented by previously untouched seabird and marine mammal colonies as farmers attempted to provision households while nurturing initially small stocks of imported domesticates. The Faroese evidence is still limited, but ongoing study suggests a heavy but sustainable use of the great nesting cliffs, with no indication of massive seabird depletion after settlement. In Iceland, south coastal sites do show a dramatic "spike and crash" pattern in seabird use, but no species seem to have been driven to local extinction (even the great auk survived to the nineteenth century). Walrus in Iceland did not fare so well, and resident populations seem to have been depleted fairly rapidly, perhaps contributing to the settlement of the Greenland colony in the late tenth century. Nonmigratory seal populations (harbor and gray seals) were also vulnerable to overexploitation but were apparently successfully managed to provide a minor supplementary resource in most areas. Norse walrus hunting in Greenland may well have caused some initial population shifts, but the Norse Norðursetur hunting area of the later Middle Ages was in the same region that retained major walrus populations through the period of intense nineteenth-century exploitation down to the present. In Greenland, seals played a central economic role, with the huge migratory harp seal populations providing a vital seasonal staple resource to subsistence economy throughout the history of the Norse settlements without any credible evidence for depletion. Overall, while the Icelandic walrus colonies may have fallen victim to human impact, there is little evidence for overexploitation of marine resources in the rest of the record, and several indications of successful local level sustainable management of seals and seabird hunting on the millennial scale. While the Atlantic islands' environmental legacy of the Viking Age introduction of farming includes severe erosion and widespread degradation of soils and vegetation, it appears that these chiefly agricultural societies were often effective long-term managers of marine resources. However, in the broader historical view it has recently become clear that the Norse in the North Atlantic bear the mixed responsibility for the origins of western European commercial fisheries that have so drastically altered marine ecosystems worldwide over the past two centuries.

KINGS AND COD: ECONOMIC HERITAGE OF THE LATE VIKING AGE

Although a great deal of work remains to be done in all areas, thanks to the collective efforts of a great many scholars it is now possible to identify the origins of the commercial cod fishery of the North Atlantic with some certainty. The historic fishery seems to have been ultimately a product of prehistoric Scandinavian fishing and fish preservation skills that probably first centered in northern Norway and spread unevenly into the North Atlantic islands colonized after AD 800. In Iceland and possibly in the Faroe Islands, this chiefly fish preservation and distribution experience rapidly translated into economic webs of fish production, curing, and distribution perhaps somewhat similar to those of Iron Age northern Norway. In the Danelaw in northern England and Atlantic Scotland, these networks may have been less easily imposed on local tastes and existing economic structures in the Early Viking Age, and other sources of wealth and power may have been more immediately available through raiding, slave trading, and tribute collection. In the Later Viking Age, the advent of the FEH in Britain seems to have gone unrecorded by the chroniclers, who were far more absorbed in the dramatic story of escalating warfare between an England increasingly unified by Wessex and a succession of ever more powerful Nordic "sea kings" capable of fielding substantial armies of now-professional soldiers. As Randsborg (1980) has argued, the prolonged escalation of military conflict between western Scandinavia and England had social consequences on both sides of the North Sea, promoting centralization, Latinization, and state building on each side. By the time Cnut the Great was briefly able to unite both shores under his rule (AD 1014–1039) he had become a Christian monarch married to a high-ranking Anglo Saxon queen, was a pious patron of monasteries and churches, and was a prolific founder of towns and trading centers in Scandinavia (Lawson 2004). While his dynasty did not long survive his death (and his burial in

Winchester Cathedral near King Alfred), Cnut's rule brought Scandinavian administrators into many inland burghs throughout England and left indications of Anglo-Scandinavian cultural fusion in contemporary sculpture and court poetry. While no mention of un-saga-worthy codfish appear in King Cnut's panegyric poems, it may be significant that his military campaigns have been described as triumphs of logistics rather than of battlefield tactics. While temporal correlation does not in itself prove a connection to the archaeologically visible FEH, it is difficult not to suspect that the strong demand for army rations during the prolonged warfare of AD 950–1012, and the urgent need for funds to sustain postwar economic expansion may have together created an opportunity that only a Scandinavian king could have seized so effectively by introducing a longstanding Nordic resource from the sea to the peoples of Europe.

ACKNOWLEDGMENTS

The research summarized in this chapter is the product of a great many collaborating scholars and institutions in the NABO cooperative, all of whom have our warmest thanks for over a decade of cooperative effort. Special thanks are due to Christian Keller and Andy Dugmore for stimulating suggestions and vital corrections, and to Colin Amundsen, Seth Brewington, George Hambrecht, Ramona Harrison, Yekaterina Krivogorskaya, Konrad Smiarowski, and Jim Woollett for their dedicated work on some of the large archaeofauna reported here. Support for this research was generously provided by the U.S. National Science Foundation (both archaeology and Arctic social sciences programs), the National Geographic Society Committee for Research and Exploration, the Icelandic Science Council, and the Leverhulme Trust.

REFERENCES CITED

Adams, J., and K. Holman (editors)
2004 *Scandinavia and Europe 800–1350: Contact, Conflict, and Coexistence*. Brepols, Turnhout, Belgium.

Amorosi, T., J. Woollett, S. Perdikaris, and T. McGovern
1996 Regional Zooarchaeology and Global Change Research: Problems and Potentials. *World Archaeology* 28(1):126–157.

Amundsen, C. P.
2004 Farming and Maritime Resources at Miðbaer on Flatey in Breiðafjord, North-West Iceland. In *Atlantic Connections and Adaptations: Economies, Environments, and Subsistence in Lands Bordering the North Atlantic,* edited by R. A Housely and G. Coles, 220–231. AEA/NABO Environmental Archaeology Monographs 21. Oxbow Books, Oxford.

Amundsen, C. P., J. Henriksen, E. Myrvoll, B. Olsen, and P. Urbańczyk
2003 Crossing Borders: Multi-room Houses and Inter-ethnic Contacts in Europe's Extreme North. *Fennoscandia Archaeologica* XX: 79–100.

Amundsen, C., S. Perdikaris, T. H. McGovern, Y. Krivogorskaya, M. Brown, K. Smiarowski, S. Storm, S. Modugno, M. Frik, and M. Koczela
2004 Fishing Booths and Fishing Strategies in Medieval Iceland: An Archaeofauna from the of Akurvík, North-West Iceland. *Environmental Archaeology,* 10(2):127–142.

Arge S. V.
2005 Cultural Landscapes and Cultural Environmental Issues in the Faroes. In *Viking and Norse in the North Atlantic: Select Papers from the Proceedings of the 14th Viking Congress, Tórshavn 2001,* edited by A. Mortensen and S. Arge, pp. 22–39. Annales Societatis Scientarium Faeroensis XLIV, Tóshavn, Faroe Islands.

Arneborg, J.
2000 Introduction: From Vikings to Norsemen. In *Vikings: The North Atlantic Saga,* edited by W. W. Fitzhugh and E. I. Ward, pp. 281–284. Smithsonian Institution Press, Washington, D.C.
2003 Norse Greenland: Reflections on Settlement and Depopulation. In *Contact, Continuity, and Collapse: The Norse Colonization of the North Atlantic,* edited by J. Barrett, pp. 163–183. Brepols, Turnhout, Belgium.

Arneborg, J. and B. Grønnow (editors)
2006 Dynamics of Northern Societies: Proceedings of the SILA/NABO Conference on Arctic and North Atlantic Archaeology, Copenhagen, May 10th–14th, 2004. National Museum of Denmark, Copenhagen.

Barlow L., T. Amorosi, P. C. Buckland, A. Dugmore, J. H. Ingimundarsson, T. H. McGovern, A. Ogilvie, and P. Skidmore
1997 Interdisciplinary Investigations of the End of the Norse Western Settlement in Greenland. *The Holocene* 7:489–499.

Barrett, J. H.
1995. *Few Know Earl Fishing-clothes: Fish Middens and the Economy of the Viking Age and Late Norse Earldoms of Orkney and Caithness, Northern Scotland.* Ph.D. thesis, University of Glasgow.

Barrett, J. H.
1997 Fish Trade in Norse Orkney and Caithness: A Zooarchaeological Approach. *Antiquity* 71(273): 616–638.
2003 Culture Contact in Viking Age Scotland. In *Contact, Continuity, and Collapse: The Norse Colonization of the North Atlantic*, edited by James Barrett, pp. 73–113. Brepols, Turnhout, Belgium.
Barrett, J. H., R. Nicholson, and R. Cerron-Carrasco
1997 Fish Trade in Norse Orkney and Caithness: A Zooarchaeological Approach. *Antiquity* 71: 616–638.
Barrett, J. H., R. Beukens, I. A. Simpson, P. Ashmore, S. Poaps, and J. Huntley
2000 What Was the Viking Age and When Did It Happen? A View from Orkney. *Norwegian Archaeological Review* 33:1–39.
Barrett, J. H., A. M. Locker, and C. M. Roberts
2004 Dark Age Economics Revisited: The English Fish Bone Evidence AD 600–1600. *Antiquity* 78(301):618–636.
Bigelow, G.
1984 Subsistence in Late Norse Shetland: An Investigation into a Northern Island Economy of the Middle Ages. Ph.D. thesis, University of Cambridge.
Bigelow, G. F. (editor)
1991 The Norse of the North Atlantic. *Acta Archaeologica* 61.
Bond, J.
2007 The Zooarchaeological Evidence. In *Excavations at Pool, Orkney*, edited by J. Hunter. Oxbow Books, Oxford, in press.
Bond, J. M., E. Guttmann, and I. A. Simpson
2004 Bringing in the Sheaves: Farming Intensification in the Post-Broch Iron Age. In *Atlantic Connections and Adaptations; Economies, Environments and Subsistence in Lands Bordering the North Atlantic*, edited by R. A. Housely and G. Coles, pp. 138–146. AEA/NABO 21. Oxbow Books, Oxford.
Bond, J. M., R. A. Nicholson, and I. Simpson
2005 Living off the Land: Farming and Fishing at Old Scatness. In *Tall Stories? Two Millennia of Brochs*, edited by V. Turner, R. A. Nicholson, S. J. Dockrill, and J. M.Bond, pp. 211–221. Shetland Amenity Trust, Lerwick, UK.
Brøgger, A. W., and H. Shetelig
1951 *The Viking Ships*. Dreyer, Oslo.
Buckland, P. C.
2000 The North Atlantic Environment. In *Vikings: The North Atlantic Saga*, edited by W. W. Fitzhugh and E. I. Ward, pp. 164–153. Smithsonian Institution Press, Washington, D.C.
Buckland, P. C., T. Amorosi, L. K. Barlow, A. J. Dugmore, P. A. Mayewski, T. H. McGovern, A. E. J. Ogilvie, J. P. Sadler, and P. Skidmore
1996 Bioarchaeological and Climatological Evidence for the Fate of the Norse Farmers in Medieval Greenland. *Antiquity* 70(1):88–96.
Bruns, M., and T. H. McGovern
1985 An Archaeofauna from Pictish Deposits at Clibberswick, Shetland Islands. On file Shetland Amenity Trust, Shetland, Scotland, and CUNY Northern Science and Education Center, Hunter College, New York, 15 pp.
Cérron-Carrasco, R. M. Church, and J. Thoms
2005 Towards an Economic Landscape of the Bhaltos Peninsula, Lewis, during the Mid to Late Iron Age. In *Tall Stories? Two Millennia of Brochs*, edited by V. Turner, R. A. Nicholson, S. J. Dockrill, and J. M.Bond, pp. 221–235. Shetland Amenity Trust, Lerwick, UK.
Christensen, A. E.
2000 Ships and Navigation. In *Vikings: the North Atlantic Saga*, edited by W. W. Fitzhugh and E. Ward, pp. 86–99. Smithsonian Institution Press, Washington, D.C.
Church, M. J., S. V. Arge, S. Brewington, T. H. McGovern, J. Woollett, S. Perdikaris, I. T. Lawson, G. T. Cook, C. Amundsen, R. Harrison, K. Krivogorskaya, and E. Dunbar
2005 Puffins, Pigs, Cod, and Barley: Palaeoeconomy at Undir Junkarinsfløtti, Sandoy, Faroe Islands. *Environmental Archaeology* 10:179–197.
Colley, S.
1983 Interpreting Prehistoric Fishing Strategies: An Orkney Case Study, in *Animals and Archaeology: Shell Middens, Fishes, and Birds*, edited by C. Grigson and J. Clutton Brock, pp. 157–171. BAR British Series 183. British Archaeology Reports, Oxford.
Crumley, C. L.
1994 Historical Ecology: A Multidimensional Ecological Orientation. In *Historical Ecology: Cultural Knowledge and Changing Landscapes*, edited by C. Crumley, pp. 1–16. School of American Research, Santa Fe, New Mexico.
Crumlin-Pedersen, O.
1997 Viking-Age Ships and Shipbuilding in Hedeby/Haithabu and Schleswig. In *Ships and Boats of the North*, Vol. 2: *Schleswig and Roskilde*. Archäologisches Landesmuseum der Christian-Albrechts-Universität, Schleswig, Germany.
Diamond, J.
2005 *Collapse: How Societies Choose to Fail or Succeed*. Viking, New York.
Dockrill, S., and C. M. Batt
2004 Power over Time: An Overview of the Old Scatness Broch Excavations. In *Atlantic*

Connections and Adaptations: Economies, Environments and Subsistence in Lands Bordering the North Atlantic, edited by R. A. Housely and G. Coles, pp. 128–138. AEA/NABO 21. Oxbow Books, Oxford.

Dockrill S. J., J. M. Bond, J. Ambers, A. Milles, and I. A. Simpson
1994 Tofts Ness, Sanday Orkney: An Integrated Study of a Buried Orcadian Landscape. In *Whither Environmental Archaeology?* edited by R. M. Luff and P. A. Rowley-Conwy, pp. 115–132. Oxbow Monograph 38. Oxford, Oxbow.

Dockrill S. J., J. M. Bond, and C. M. Batt
2005 Old Scatness: The First Millenium AD. In *Tall Stories? Two Millennia of Brochs*, edited by V. Turner, R. A. Nicholson, S. J. Dockrill, and J. M.Bond, pp. 52–66. Shetland Amenity Trust, Lerwick, UK.

Dugmore, A. J., P. C. Buckland, M. Church, A. Dawson, K. J. Edwards, P. Mayewski, T. H McGovern, K. Mairs, I. A. Simpson, G. Sveinbjarnardóttir
2007 Landscape and Settlement Change in the North Atlantic Islands and the Use of High-Resolution Proxy Climate Records from the Greenland Ice Sheet. *Human Ecology* 35: 169–178.

Edvardsson, R., S. Perdikaris, T. H. McGovern, N. Zagor, and M. Waxman
2004 Coping with Hard Times in North-West Iceland: Zooarchaeology, History, and Landscape Archaeology at Finnbogastaðir in the Eighteenth Century. *Archaeologica Islandica* 3:20–48.

Edwards, K. J., P. C. Buckland, A. J. Dugmore, T. H. McGovern, I. A. Simpson, and G. Sveinbjarnardóttir
2004 Landscapes Circum-Landnám: Viking Settlement in the North Atlantic and Its Human and Ecological Consequences: A Major New Research Programme. In *Atlantic Connections and Adaptations: Economies, Environments, and Subsistence in Lands Bordering the North Atlantic*, edited by R. Housley, and G. M. Coles, pp. 260–271. Oxbow, Oxford.

Einarsson, B. F.
1994 *The Settlement of Iceland: A Critical Approach: Granastadir and the Ecological Heritage.* Series B, Gothenburg Archaeological Theses No. 4. Gothenberg University, Gothenberg, Sweden.

Enghoff, I. B.
2003 Hunting, Fishing, and Animal Husbandry at the Farm Beneath the Sand, Western Greenland: An Archaeozoological Analysis of a Norse Farm in the Western Settlement. *Meddelelser om Grønland Man and Society* 28.

Fagan, B.
2004 *Fish on Friday: Feasting, Fasting and the Discovery of the New World.* Basic Books, New York.

Fenton, A.
1978 *The Northern Isles.* University of Aberdeen Press, Aberdeen.

Fitzhugh W. W., and E. I. Ward
2000 *Vikings: The North Atlantic Saga.* Smithsonian Institution Press, Washington, D.C.

Friðriksson, A., O. Vésteinsson, and T. H. McGovern
2004 Recent Investigations at Hofstaðir, Northern Iceland. In *Atlantic Connections and Adaptations: Economies, Environments, and Subsistence in Lands Bordering the North Atlantic*, edited by R. A Housely and G. Coles, pp. 191–202. AEA/NABO Environmental Archaeology, Monographs 21. Oxbow Books, Oxford.

Gad, F.
1970 *A History of Greenland*, Vol. 1. David Hurst, London.

Gotfredsen, B.
2007 An Archaeofauna from Argisbrekka, Faroe Islands. In *Argisbrekka, A Medieval Shieling Site in the Faroes*, edited by D. Mahler. Munksgaard, Copenhagen, in press.

Graham-Campbell, J., and C. Batey
1998 *Vikings in Scotland.* Edinburgh University Press, Edinburgh.

Gulløv, H. C.
2000 Natives and Norse in Greenland. In *Vikings, the North Atlantic Saga*, edited by W. W. Fitzhugh, and E. Ward, pp. 318–327. Smithsonian Institution Press, Washington, D.C.

Gulløv, H. C. (editor)
2004 *Grønlands Forhistorie.* Gyldendal, Copenhagen.

Hedeager, L.
2000 From Warrior to Trade Economy. In *Vikings, the North Atlantic Saga*, edited by W. W. Fitzhugh, and E. Ward, pp. 84–86. Smithsonian Institution Press, Washington, D.C.

Helgason, A., S. Sigurdardottir, J. R. Gulcher, R. Ward, and K. Stefannson
2000 mtDNA and the Origin of the Icelanders: Deciphering Signals of Recent Population History. *American Journal of Human Genetics* 66(3):999–1016.

Henriksen J., C. Amundsen, E. Myrvoll, and B. Olsen
2006 Culture Contact, Coping, and Commercialization in Arctic Norway AD 1200–1600. Paper presented at the 2006 European Archaeology Conference, Cork, Ireland.

Housley, R., and G. M. Coles (editors)
2004 *Atlantic Connections and Adaptations: Economies, Environments, and Subsistence in Lands Bordering the North Atlantic.* Oxbow Books, Oxford.

Jennings, A., and A. Kruse
2005 An Ethnic Enigma: Norse, Pict, and Gael in the Western Isles. In *Viking and Norse in the North*

Atlantic: Select Papers from the Proceedings of the 14th Viking Congress, Tórshavn 2001, edited by A. Mortensen and S. Arge, pp. 284–297. Annales Societatis Scientarium Faeroensis XLIV, Tóshavn, Faroe Islands.

Jennings A. E., and N. J. Weiner
1996 Environmental Change in Eastern Greenland during the Last 1300 Years: Evidence from Foraminifera and Lithofacies in Nansen Fjord 68N. *The Holocene* 6(2):179–191

Jones, G.
1987 The *Norse Atlantic Saga*. Second edition. Oxford University Press, Oxford.

Jorgensen, T. H., B. Degn, A. G. Wang, M. Vang, H. Gurling, G. Kalsi, A. McQuillin, T. A. Kruse, O. Mors, and H. Ewald
2002 Linkage Disequilibrium and Demographic History of the Isolated Population of the Faroe Islands. *European Journal of Human Genetics* 10:381–387

Krivogorskaya, Y., S. Perdikaris, and T. H McGovern
2005 Fish Bones and Fishermen: The Potential of Zooarchaeology in the Westfjords. Fornleifastofnun Islands, Reykjavik 2005. *Archaeologica Islandica* 4:31–50.

Kurlansky, M.
1997 *Cod: The Biography of the Fish That Changed the World*. Penguin, New York.

Larsen, L. M. (translator)
1917 *The King's Mirror*. American Scandinavian Foundation, New York.

Lawson, M. K.
2004 *Cnut, England's Viking King*. Tempus Press, Stroud, UK.

Lund, N. (editor)
1984 *Two Voyagers at the Court of King Alfred*. William Sessions, York, UK.

McGovern, T. H.
1985a The Arctic Frontier of Norse Greenland. In *The Archaeology of Frontiers and Boundaries*, edited by S. Green and S. Perlman, pp. 275–323. Academic Press, New York.
1985b Contributions to the Paleoeconomy of Norse Greenland. *Acta Archaeologica* 54:73–122.
1990 The Archeology of the Norse North Atlantic. *Annual Review of Anthropology* 19(1): 331–351.
1994 Management for Extinction in Norse Greenland. In *Historical Ecology: Cultural Knowledge and Changing Landscapes*, edited by C. Crumley, pp. 127–154. School of American Research, Santa Fe, New Mexico.
2000 The Demise of Norse Greenland. In *Vikings: The North Atlantic Saga*, edited by W. W. Fitzhugh and E. I. Ward, pp. 327–339. Smithsonian Institution Press, Washington, D.C.
2004 North Atlantic Biocultural Organization (NABO) 10 Years On: Science, Education, and Community. In *Atlantic Connections and Adaptations: Economies, Environments, and Subsistence in Lands Bordering the North Atlantic*, edited by R. Housley, and G. M. Coles, pp. 254–259. Oxbow Books, Oxford.

McGovern T. H., T. Amorosi, S. Perdikaris, and J. W. Woollett
1996 Zooarchaeology of Sandnes V51: Economic Change at a Chieftain's Farm in West Greenland. *Arctic Anthropology* 33(2)94–122.

McGovern T. H., S. Perdikaris, and C. Tinsley
2001 Economy of Landnám: The Evidence of Zooarchaeology. In *Approaches to Vinland, Nordal Institute Studies*, edited by A. Wawn and T. Sigurðardottir, vol. 4, pp. 154–165. Nordal Institute, Reykjavik.

McGovern, T. H., S. Perdikaris, B. F. Einarsson, and J. Sidell
2006 Coastal Connections, Local Fishing, and Sustainable Egg Harvesting: Patterns of Viking Age Inland Wild Resource Use in Myvatn District, Northern Iceland. *Environmental Archaeology* 11(2):187–205.

McGovern, T. H., O. Vesteinsson, A. Fridriksson, M. Church, I. Lawson, I. A. Simpson, A. Einarsson, A. Dugmore, G. Cook, S. Perdikaris, K. Edwards, A. M. Thompson, W. P. Adderley, A. Newton, G. Lucas, and O. Aldred
2007 Landscapes of Settlement in Northern Iceland: Historical Ecology of Human Impact and Climate Fluctuation of the Millennial Scale. *American Anthropologist*, 109:27–51.

Meeker, L. D., and P. A. Mayewski
2002 A 1400 Year Long Record of Atmospheric Circulation over the North Atlantic and Asia. *The Holocene* 12:257–266.

Morris C. D. (editor)
1996 *The Birsay Bay Project*, Vol. 2; *Sites in Birsay Village and on the Brough of Birsay*, Orkney. University of Durham, Department of Archaeology Monograph 2. Alden Press, Oxford.

Morris, C. D., and D. J. Rackham
1994 *Norse and Later Settlement and Subsistence in the North Atlantic*. Archetype Publications, Glasgow.

Morris, C. D., C. Batey, and J. Rackham
1995 *Freswick Links Caithness: Excavation and Survey of a Norse Settlement*. University of Glasgow and NABO. University of Glasgow Press, Glasgow.

Mulville, J., and J. Thoms
2005 Animals and Ambiguity in the Iron Age of the Western Isles. In *Tall Stories? Two Millennia of Brochs*, edited by V. Turner, R. A. Nicholson, S. J. Dockrill, and J. M. Bond, pp. 235–246. Shetland Amenity Trust, Lerwick, UK.

Munch, G. S., O. S. Johansen, and E. Roesdahl (editors)
2003 *Borg in Lofoten: A Chieftain's Farm in North Norway*. Lofoter. Series: Arkeologisk Skriftserie 1.Vikingmuseet på Borg. Tapir Academic Press, Tromso, Norway.

Nicholson, R.
1998 Fishing in the Northern Isles: A Case Study Based on Fish Bone Assemblages from Two Multi-period Sites on Sanday, Orkney. *Environmental Archaeology* 2:15–29.
2004 Iron Age Fishing in the Northern Isles: The Evolution of a Stored Product? In *Atlantic Connections and Adaptations: Economies, Environments and Subsistence in Lands Bordering the North Atlantic*, edited by R. A. Housely and G. Coles, pp. 155–163. AEA/NABO 21. Oxbow Books, Oxford.

Nicolaysen, N.
2003 *The Viking-Ship Discovered at Gokstad in Norway. Christiania 1882*. Norwegian and English Faximile edition. Gregg International, Sandefjord.

Ogilvie, A. E. J.
1997 Fisheries, Climate and Sea Ice in Iceland: An Historical Perspective. In *Marine Resources and Human Societies in the North Atlantic Since 1500*, edited by D. Vickers, pp. 69–87. Institute of Social and Economic Research, Memorial University of Newfoundland, St. Johns, Newfoundland.

Ogilvie, A. E. J., and T. Jónsson
2001 Little Ice Age Research: A Perspective from Iceland. *Climatic Change* 48:9–52.

Ogilvie, A. E. J., and T. H. McGovern
2000 Sagas and Science: Climate and Human Impacts in the North Atlantic. In *Vikings: The North Atlantic Saga*, edited by W. W. Fitzhugh and E. I. Ward, pp. 385–393. Smithsonian Institution Press, Washington, D.C.

Olsen, B.
2003 Belligerent Chieftains and Oppressed Hunters? Changing Conceptions of Interethnic Relationships in Northern Norway during the Iron Age and Early Medieval Period. In *Contact, Continuity, and Collapse: The Norse Colonization of the North Atlantic*, edited by J. Barrett, pp. 9–33. Brepols, Turnhout, Belgium.

Olsen O., J. Skamby Madsen, and R. Flemming (editors)
1995 *Shipshape: Essays for Ole Crumlin-Pedersen. On the Occasion of His 60th Anniversary February 24th 1995*. The Viking Ship Museum, Roskilde, Denmark.

Owen, O.
2004 The Scar Boat Burial and the Missing Decades of the Early Viking Age in Orkney and Shetland. In *Scandinavia and Europe 800–1350 Contact, Conflict, and Coexistence*, edited by J. Adams and K. Holman, pp. 3–35. Brepols, Turnhout, Belgium.

Parker-Pearson, M., and N. Sharples (editors)
1999 *Between Land and Sea: Excavations at Dun Vulan, South Uist*. Sheffield Academic Press, Sheffield, UK.

Perdikaris, S.
1996 Scaly Heads and Tales: Detecting Commercialization in Early Fisheries. Ichthyoarchaeology and the Archaeological Record. Proceedings of the 8th Meeting of the ICAZ Fish Remains Working Group, Madrid, Spain, edited by A. Morales. *Archaeofauna* 5:21–33.
1997 From Chiefly Provisioning to Commercial Fishery: Long Term Economic Change in Arctic Norway. *World Archaeology* 30:388–402.

Perdikaris, S., and T. H. McGovern
2007 Cod Fish, Walrus, and Chieftains: Economic Intensification in the Norse North Atlantic. In *Seeking a Richer Harvest*. Springer US, pp. 193–216.

Rackham, D. J.
1989 Excavations beside the Brough Road: The Biological Assemblage. In *The Birsay Bay Project*, edited by C. D. Morris, Chapter 8, pp. 231–271. Department of Archaeology Monograph 1. University of Durham, Durham, UK.

Randsborg, K.
1980 *Viking Age Denmark: The Formation of a State*. St. Martin's Press, New York.

Richards, J.
2001 *Blood of the Vikings*. Hodder and Stoughton, London.

Roesdahl, E.
2005 Walrus Ivory: Demand, Supply, Workshops, and Greenland. In *Viking and Norse in the North Atlantic: Select Papers from the Proceedings of the 14th Viking Congress, Tórshavn 2001*, edited by A. Mortensen and S. Arge, pp. 182–192. Annales Societatis Scientarium Faeroensis XLIV, Tóshavn, Faroe Islands.

Rohling, E., P. Mayewski, and P. Challenor
2003 On the Timing and Mechanism of Millennial-Scale Climate Variability during the Last Glacial Cycle. *Earth and Environmental Science* 20(2–3):257–267.

Rowley-Conwy, P., P. Arias, M. Budja, D. Gronenborn, A. Jones, L. P. Louwe Kooijmans, P. O. Nielsen, L. G. Straus, and J. Thomas.
2004 How the West Was Lost. *Current Anthropology* 45:S83–S113.

Schledermann P., and K. M. McCullough
2003 Inuit-Norse Contact in the Smith Sound Region. In *Contact, Continuity, and Collapse: The Norse Colonization of the North Atlantic*, edited by

J. Barrett, pp. 183–207. Brepols, Turnhout, Belgium.

Sharples, N. (editor)

1998 *Scalloway: A Broch, Late Iron Age Settlement, and Medieval Cemetery in Shetland.* Oxbow Monograph 82. Oxbow Books, Oxford.

2005 *A Norse Farmstead in the Outer Hebrides, Excavations at Mound 3, Bornais, South Uist.* Oxbow Books, Oxford.

Sharples, N., M. Parker-Pearson, and J. Symonds

2004 The Archaeological Landscape of South Uist. In *Atlantic Connections and Adaptations Economies, Environments, and Subsistence in Lands Bordering the North Atlantic,* edited by R. A. Housely and G. Coles, pp. 28–48. AEA/NABO 21. Oxbow Books, Oxford.

Simpson, I. A., A. J. Dugmore, A. M. Thomson, and O. Vésteinsson

2001 Crossing the Thresholds: Human Ecology and Historical Patterns of Landscape Degradation. *Catena* 42:175–192.

Simpson, I. A., O. Vésteinsson, W. P. Adderley, and T. H. McGovern

2003 Fuel Resources in Landscapes of Settlement. *Journal of Archaeological Science* 30: 1401–1420.

Simpson, I. A., G. Guðmundsson, A. M. Thomson, and J. Cluett

2004 Assessing the Role of Winter Grazing in Historic Land Degradation, Mývatnssveit, Northeast Iceland. *Geoarchaeology* 19:471–503.

Smith, H., and J. Mulville

2002 Resource Management in the Outer Hebrides. In *Atlantic Connections and Adaptations: Economies, Environments, and Subsistence in Lands Bordering the North Atlantic,* edited by R. A. Housely, and G. Coles, pp. 48–65. AEA/NABO 21. Oxbow Books, Oxford.

Smith, K. P.

1995 Landnám: The Settlement of Iceland in Archaeological and Historical Perspective. *World Archaeology* 26:319–347.

Thomson, A. M., and I. A. Simpson

2006 A Grazing Model for Simulating the Impact of Historical Land Management Decisions in Sensitive Landscapes: Model Design and Validation. *Environmental Modelling & Software* 21(8): 1096–1113.

2007 Modeling Historic Rangeland Management and Grazing Pressures, Mývatnssveit Iceland. *Human Ecology*, 35:151–165.

Thurston, T. L.

1999 The Knowable, the Doable, and the Undiscussed: Tradition, Submission, and the Becoming of Rural Landscapes in Denmark's Iron Age. *Antiquity* 73:661–671.

Thurston, T. L., and C. T. Fisher (editors)

2006 *Seeking A Richer Harvest: The Archaeology of Subsistence Intensification, Innovation, and Change.* Springer Science + Business Media, New York.

Urbańczyk, P.

1992 *Medieval Arctic Norway.* Semper, Warsaw, Poland.

Vésteinsson, O.

1998 Patterns of Settlement in Iceland: A Study in Pre-History. *Saga-Book of the Viking Society* 25:1–29.

2000a *The Christianization of Iceland. Priests, Power, and Social Change 1000–1300.* Oxford University Press, Oxford.

2000b The Archaeology of Landnám: Early Settlement in Iceland. In *Vikings: The North Atlantic Saga,* edited by W. W. Fitzhugh and E. Ward, pp. 164–174. Smithsonian Institution Press, Washington, D.C.

Vésteinsson, O., T. H. McGovern, and C. Keller

2002 Enduring Impacts: Social and Environmental Aspects of Viking Age Settlement in Iceland and Greenland. *Archaeologica Islandica* 2: 98–136.

Wheeler, A., and A. K. G. Jones

1989 *Fishes.* Cambridge Manuals in Archaeology. Cambridge University Press, New York.

10

Historical Ecology of the North Sea Basin

AN ARCHAEOLOGICAL PERSPECTIVE AND SOME PROBLEMS OF METHODOLOGY

Geoff Bailey, James Barrett, Oliver Craig, and Nicky Milner

THE NORTH SEA BASIN IS one of the most fertile marine environments in Europe. Its relatively shallow seabed, cool-temperate climate, and winter storms ensure rapid recycling of nutrients, while the presence of land masses on three sides and large rivers draining extensive catchments, such as the Thames, the Rhine, and the Elbe, bring additional inputs of nutrients from land. The geographical limits of the Basin are defined to the west by the coastline of Britain, to the east by the coastlines of southern Norway, western Sweden, and Denmark, and to the south by the coastlines of northern France, the Low Countries, and northwest Germany. To the north, there is a broad opening to the North Atlantic, and to the coastlines of northern Norway and Iceland. To the south there is a much narrower opening through the English Channel to the Bay of Biscay and the southern Atlantic, and to the east a narrow connection between Denmark and Sweden to the progressively more brackish waters of the Baltic (Figure 10.1). With populous countries on every side, the North Sea Basin is also vulnerable to the pressure of human demand on its marine resources. In the past century, and especially in recent decades, it has become a byword for overexploitation of its fish stocks. Historical records suggest that the productivity and abundance of cod *(Gadus morhua)* and herring *(Clupea harengus)* were much greater than today, but the accuracy or wider relevance of these records is unclear (Jackson et al. 2001). Certainly the present-day stocks of some major commercial fish are under serious threat, and a complete ban on fishing for cod has recently been advocated to avoid regional extinction. Given the acute impact of recent human activities on marine ecosystems, knowledge of the frequency and scale of past impacts on marine life is not only historically informative but is also crucial for assessing the current crisis facing ocean fisheries.

The rim of the North Sea Basin has witnessed continuous occupation throughout the last 10,000 years with a succession of communities and cultures who have variously interacted around its perimeter or across an east-west axis through colonization, trade,

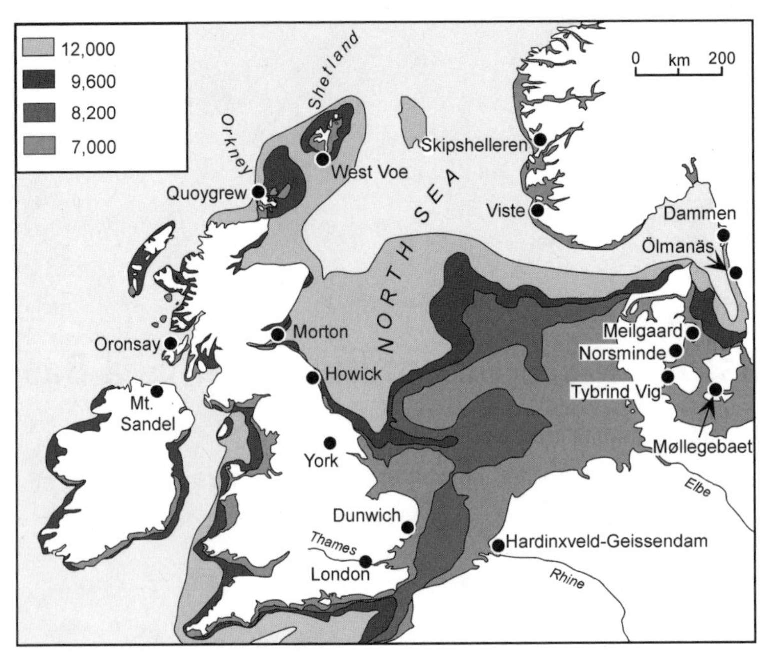

FIGURE 10.1. Map of North Sea region indicating sites and places mentioned in the text and the position of the coastline at different periods. Coastline positions are based on data from Shennan and Andrews (2000), and dates are calibrated radiocarbon dates BC. Drawn by G. Bailey.

conquest, and competition for resources. The "natural" baseline against which to judge present-day impacts is poorly known. Moreover, it was almost certainly an ever-changing baseline, due both to climatic and environmental changes, each amplified in its turn by relatively high latitude and proximity to the Scandinavian ice sheet of the last glacial, and to a changing history of human interest in and exploitation of marine resources over the past 10,000 years.

Because of complex changes in palaeogeography associated with glacial isostasy, the archaeological record of long-term coastal settlement is fragmentary and subject to large uncertainties of differential preservation or visibility. Some of the best-preserved coastal sites, particularly for the Mesolithic period, are on the peripheries of the North Sea proper, in northern and western Scotland, and in the inner waters of Denmark. As far as marine subsistence is concerned, most of the emphasis has been on the Stone Age, and particularly the Mesolithic period (c. 9500–3800 cal BC [10,000–5000 radiocarbon years BP]) (Bailey and Spikins 2008). For the later prehistoric and protohistoric periods (Neolithic, Bronze, and pre-Roman Iron ages, and Roman, Viking, and Medieval periods), the emphasis on archaeological interpretation has been more on issues of trade and culture contact than on marine exploitation and subsistence (cf. Cunliffe 2001). This reflects both the relative rarity of coastal settlement, particularly around the southern rim of the basin, which has undergone long-term submergence and accumulation of marine sediments, and different research agendas in different periods of the archaeological sequence. For the later periods, in contrast to the Mesolithic period, the emphasis has generally been on social and cultural change rather than on palaeoeconomy, and on agriculturally based and urban societies rather than maritime ones.

The questions asked and specific evidence assembled vary tremendously between communities of scholars working in different periods and regions. Seasonality of resource scheduling remains a critical question in the Mesolithic, whereas studies of chiefdoms and states in the Middle Ages are more concerned with issues such as gender and ideology (e.g., Barrett and Richards 2004). Mollusks are the subject of intensive study in Mesolithic contexts but are seldom given equally detailed treatment in later periods. This diversity is exacerbated by the fact that there appear to be only two periods in the postglacial history of the North Sea during which marine resources played a major role in daily routine, economy, and social life. These periods, the Mesolithic and the Medieval (Barrett et al. 2004a; Enghoff 1999, 2000; Milner et al. 2004; Richards et al. 2003; Tauber 1981) occupy temporal extremes of the Holocene. Common ground between them, and those who study them, is hard to find. Thus, it is not possible to construct a single narrative of maritime historical ecology around the North Sea.

Whether this perceived emphasis on marine resources in the Mesolithic and the Medieval periods reflects biases in the archaeological record and the different interests of archaeologists working in different periods is hard to judge. But it is worth noting that in both periods when marine resources seem to be particularly prominent, we are dealing with colonization processes involving the expansion of human populations into new territory—in the case of the Mesolithic, the entry of populations into pristine territory newly exposed after the retreat of the ice sheets, and in the early Medieval period, the westward expansion of the Vikings from Norway and Denmark across the North Sea and around the coasts of Britain involving conquest and occupation of already-populated territory. There are good reasons to suppose that marine resources played an important facilitating role in both cases.

Given these complexities, this chapter will focus on one issue that has been perceived as critical in both periods, and that also unites social and ecological concerns: the intensification of exploitation. We offer a broad overview of the evidence, in order to provide a long-term perspective on the history of human interaction with marine resources. Given the incompleteness of the evidence, we pay particular attention to methodological issues. We examine the large palaeogeographic and climatic changes that have affected the region, and the ways in which these have influenced both the visibility of archaeological evidence for coastal settlement and maritime activity and the ecological characteristics of the marine environment. We emphasize the extreme patchiness of the archaeological record, particularly for the earlier prehistoric period, because of factors of differential visibility and preservation, and focus on two episodes within this longer-term history: changes associated with the introduction of agriculture at the Mesolithic-Neolithic transition at about 4000 cal BC, and the so-called fish event horizon at about AD 1000 associated with the early Medieval and Viking periods (see also Perdikaris and McGovern, this volume).

METHODS

We begin with a brief discussion of methods used to assess changes in the dependency of human populations on marine resources and human impact on marine ecosystems. All the methods we refer to here have been used in discussion of the European archaeological evidence, and all are involved in the examples we discuss later. We distinguish between direct methods, which inform on the ecological impact of human subsistence on the exploited organisms, and indirect methods, which estimate the contribution of marine foods in human palaeodiet.

Direct Methods

The term "direct methods" refers to the morphological, biological, or biochemical characteristics of exploited organisms, which may be sensitive to human impact, and we identify four such indicators. First and most commonly relied on is reduction over time in the mean size of organisms that grow continuously throughout their life span. This indicator is especially popular in studies of mollusks and

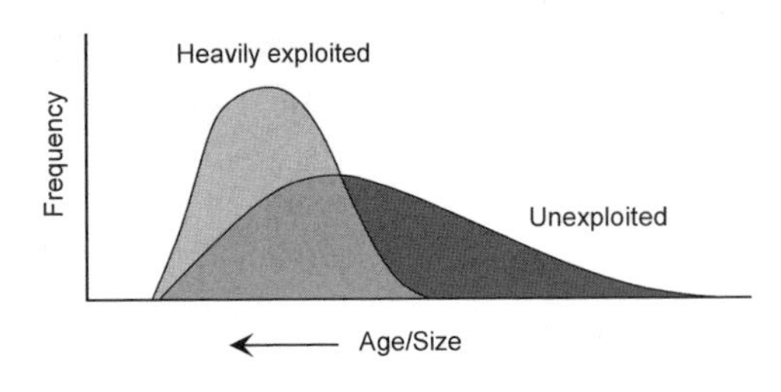

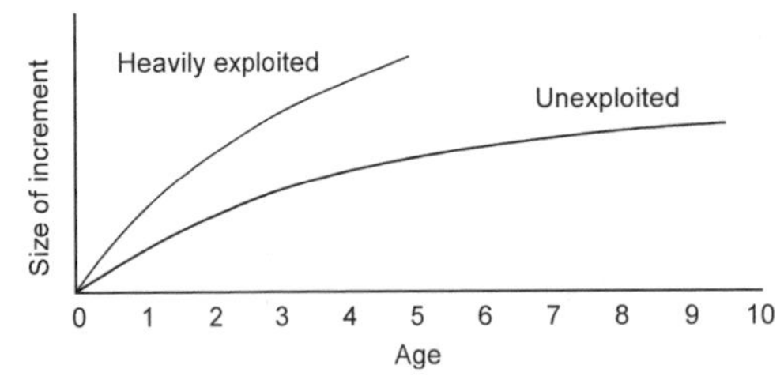

FIGURE 10.2. Effect of human impact and environmental change on age structure and size characteristics of marine organisms. The example given here is for mollusks. Similar effects are likely with all marine organisms that grow continuously throughout their life span such as fish, but are likely to be moderated by differences in feeding areas and behavior patterns of different age cohorts. Drawn by G. Bailey.

assumes that size reduction is due to intensified exploitation that removes the older and larger individuals. However, this measure is highly unreliable in the absence of data on age structure unless environmental changes affecting growth rates can be excluded (Figure 10.2). In other words, shells may undergo long-term reduction in size not because of increased human impact, but because of other environmental changes that slow down rates of growth without affecting the overall age structure. Similar considerations apply to the study of fish bones. Here there is an additional caution, and that is that different age classes of fish may feed in different areas. Moreover, these may change from season to season, so that age and size classes of fish represented in archaeological deposits may reflect differences in the areas fished or the season of activity represented at that particular site. Differences in fishing methods can also affect the size of the caught fish, quite independently of any larger-scale impact of human populations on fishing stocks.

Age at death, our second indicator, can be estimated for many mollusks and vertebrates by measurement of incremental growth structures in the shell or in fish-bone elements such as vertebrae and otoliths and is an essential control on interpretation of changes in mean size.

Another indicator is change in growth rates, which is relevant for many mollusks and fish that are subject to density dependent controls on growth. Under conditions of crowding we should expect slower growth rates, whereas reduced density, for example thinning out by increased predation pressure, would result in increased growth rate. Growth rate information, like age at death, can be obtained from incremental growth structures. However, interpretation is complicated by other environmental changes, which may produce similar results. Change in growth rates, in its turn, is an additional confounding variable that can complicate assessment of size changes.

Finally, changes in the composition of stable nitrogen isotopes in higher-level predators may indicate changes in ecosystem structure and in particular evidence of "fishing down the food web" under conditions of very heavy overall human predation pressure (Hirons et al. 2001; Pauly et al. 1998; Wainwright et al. 1993). We are just beginning to obtain results from this method, but it is clear that measurements on individual organisms need to take account of overall isotope ecology and of geographical variations in isotope ecology.

Indirect Methods

Indirect methods essentially comprise two types of analysis. The traditional method of palaeodietary measurement is to estimate differences in presence/absence or proportion of different food remains in archaeological deposits, using counts of specimens or minimum number of individuals, expressed in terms of flesh weight, calories, protein, or some other measure of relative food value. These provide a measure of change in emphasis on marine versus terrestrial sources of food and, hence, expectations about increased/decreased impact on the marine sector. However, they should be used with extreme caution because of problems of potential bias and confounding variables. For example, the apparent intensification of marine resource exploitation recorded in southern European cave sites from about 17,200 cal BC (16,000 BP) onward and culminating in the concentrations of estuarine open-air shell mounds in Portugal, Denmark, and Scotland from about 6500 cal BC (7000 BP) appears to be largely a function of differential visibility of evidence related to sea level rise, reinforced by stabilization of sea level and development of productive estuarine habitats from about 6500 cal BC onward (cf. Bailey 2004a, 2004b; Bailey and Craighead 2003; Bailey and Milner 2002; Fischer 1995a). Problems of differential preservation of various types of food organisms are also a major source of uncertainty, as are problems of differential discard and differential spatial distribution of food remains, at both intrasite and intersite scales.

Changes in processing methods, particularly of mammals and fish, as indicated by differential frequencies of distinct parts of the skeleton, provide an added refinement. For these may reflect changes in the ways in which the carcasses were prepared in response to increased demand. Systematic overrepresentation of the heads and other unwanted parts of fish, for example, might indicate mass processing of fish for drying or smoking and long-distance export and consumption elsewhere. However, as with other archaeological methods, it is necessary to allow for factors such as differential preservation or differential spatial discard of different parts of the skeleton. Underrepresentation of fish heads, for example, could equally well indicate preparation of fish carcasses for local consumption rather than for long-distance trade.

The second major measure of food consumption is stable carbon and nitrogen isotope measurements on human bone collagen, which can measure changes in the relative proportion of marine and terrestrial protein in the human diet. These techniques offer a powerful insight into human palaeodiet uncomplicated by the biases that afflict archaeological food remains. However, they have uncertainties of their own. Human skeletal material is often confined to limited geographical locations and may not be representative of the wider population. The precision with which the isotope data can measure the relative proportion of marine and terrestrial foods in mixed diets is also subject to controversy (Hedges 2004; Milner et al. 2004, 2006; Richards and Schulting 2006).

In summary, there are a variety of methods that can be deployed, but all are open to various uncertainties. The principal difficulty with direct methods is that all the indicators of increased impact can also result from other sorts of environmental changes not involving human activity, posing major issues of disentangling human and nonhuman effects. The principal difficulty with the indirect methods, quite apart from any uncertainties in the accuracy or reliability of palaeodietary measures, is that variation in the proportion of marine foods that contribute to the diet of coastal people does not necessarily correlate closely or at all with variations in their ecological impact on the exploited organisms.

PALAEOGEOGRAPHY AND PREHISTORIC COASTAL SETTLEMENT

With respect to the earliest period of the sequence, northwestern Europe enjoys several advantages as a region for testing models of initial human colonization of empty landscapes. Foremost is the certainty that humans were colonizing unoccupied landscapes. Northwestern Europe was undoubtedly unoccupied prior to late Pleistocene colonization, since ice sheets thousands of meters thick had covered most of it. The region also has a long history of archaeological study and, hence, a comparatively fine-grained archaeological record, with numerous sites and a well-established chronological framework. However, the dynamics of the retreating ice sheet, rising sea levels in the Baltic and North seas, and isostatic rebound greatly complicate characterization of the landscape and its coastal ecology—and even the extent of available land area and the position of the retreating coastline. Furthermore, the retreating ice simply expanded the area of the European subcontinent that was available for human habitation. No physical barriers (such as seaways, a narrow land isthmus, or ice sheets) precluded two-way movement between the newly available landmass and the Pleistocene refugia for human populations further south in Europe. Also, the differences between the environments of the source region and the colonized area appear to have been minimal. Fauna recolonizing the deglaciated zones would have had a long history of adaptation to (and avoidance of) humans. Colonization processes may have differed from those associated with first occupation of the continents of Australia and the Americas, which were separated from source populations by oceanic or ice barriers, had flora and fauna that differed from those of the colonists' homeland, and included "naïve" animals not used to human predation (Barton et al. 2004:141–142). With these caveats in mind, we examine the archaeological data for colonization of northwestern Europe from about 13,000 years ago onward, following the retreat of the Scandinavian ice sheet.

As on coastlines in most other parts of the world, there is a sharp increase in the evidence of coastal settlement and marine exploitation in the form of shell middens and other indicators of maritime activity from about 6500 cal BC onward, the later part of the Mesolithic period in the European sequence. Most archaeologists now accept that this cannot be taken at face value as evidence of intensification on marine resources, but rather reflects the history of eustatic sea level change and the loss of evidence on now submerged coastlines. In the North Sea, however, this picture is greatly complicated by isostatic effects, and there is evidence both of significantly earlier coastal settlement in the north on elevated shorelines, and significant loss of coastal evidence for later periods in the south because of ongoing submergence and coastal erosion.

The shallowness of the North Sea (~40- to 100-m depth), particularly the southern basin, means that it has a complex history of palaeogeographical change associated with progressive marine inundation by late glacial and early postglacial sea level rise (Coles 1998; Fitch et al. 2005; Shennan and Andrews 2000). That history is further complicated by the proximity of the Scandinavian ice sheet, resulting in variable degrees of isostatic rebound or depression associated with local changes in the loading of sea and ice on the Earth's crust. In general terms, the northerly regions, especially the Norwegian coastline, which had the largest mass of ice over it, are regions of isostatic rebound, which means that late glacial and early postglacial shorelines that elsewhere are now destroyed or submerged by eustatic sea level rise are exposed above modern sea level. This is also the case, though to a much more limited extent, in northern Britain and southern Scandinavia. However, that advantage is offset by the fact that bone and shell preservation over large areas of this territory, and especially in Norway and Scotland, is mostly very poor or nonexistent, due to acid base rocks and soils (Bjerck 1995; Clark 1983; Hardy and Wickham-Jones 2002; Milner et al. 2007a; Waddington et al. 2003).

Conversely, the southern coastlines, particularly along the coast of eastern England and the Low Countries, have undergone progressive submergence, coastal erosion, and loss or burial of earlier coastal settlements by marine transgression or burial under thick deposits of marine and alluvial settlements. These trends have continued into the historical period and in some regions up to the present day. The coastlines of eastern England, particularly the low cliffs of the Yorkshire Coast and parts of East Anglia, are under active erosion, and the collapse of houses and other property into the sea is a regular and dramatic witness to marine encroachment. Dunwich, on the Suffolk Coast, once a thriving fishing town and harbor with a history extending back to at least the eleventh century AD, is famous for its nine churches, all of which are now underwater. Large areas of former coastal settlement are now almost certainly buried under many meters of alluvial and marine sediments. It is only rarely that major engineering works have allowed excavations on a scale that can give an insight into these deeply buried landscapes, as with the discovery of richly preserved Mesolithic settlements at Hardinxveld-Giessendam in the Netherlands, originally located in a coastal wetlands (Louwe Kooijmans 2003).

At the height of the last glacial maximum, at about 20,000 cal BC, the North Sea Basin and its surrounding territories would have been a single landmass of uninhabitable ice and Arctic desert. Whether the British and Scandinavian ice sheets coalesced to form a continuous ice barrier is uncertain, but the North Sea would have comprised at most an inlet of variable extent accessible only from the north (Coles 1998) (Figure 10.1). In the earliest stages of deglaciation and sea level rise, that inlet would have widened somewhat and extended further south in the vicinity of the present-day Norwegian trench. Much of the Norwegian coastline was ice free by 13,500 cal BC (13,000 BP years ago) and perhaps as early as 17,200 cal BC (16,000 BP) (Bjerck 2008). With the onset of the late glacial interstadials at about 13,500 cal BC (13,000 BP), populations of reindeer hunters spread out from glacial refuges in France and southern Germany into the lowland regions of northern Germany and Britain, and onto the still-dry bed of the North Sea. There are numerous dated occurrences of human occupation from about 13,500 cal BC onward, and these include an antler harpoon dredged from the Leman and Ower Bank with an AMS date of 11,950–11,300 cal BC (OxA-1950), and a flint recovered by oil drilling from a submerged beach deposit on the Viking bank (located between the Shetland islands and the Norwegian coastline), which must be at least 11,000 cal BC on geological grounds (Long et al. 1986; Tolan-Smith and Bonsall 1999; Verhart 2004).

The activities of trawler dredges in the southern part of the North Sea have brought up many tons of terrestrial and marine fauna. These include land mammals such as mammoth and reindeer, which represent several different cold stages of the Pleistocene, when the North Sea would have been a dry basin (Glimerveen et al. 2004). Large quantities of marine mammals have also been recovered, including a cold-adapted Pleistocene marine fauna of walrus, harp seal, ringed seal, and various whale species (Van Kolfschoten and Van Essen 2004). These indicate periods of colder climate and submerged shorelines when glacial-period sea level was high enough for marine conditions to penetrate into the southern part of the North Sea Basin.

In the shallow waters of Denmark, there have been extensive and pioneer investigations of underwater archaeology, which have revealed over 2,000 archaeological sites, mostly in the shallow waters of the Kattegat around the islands of Funen and Zealand between the Jutland Peninsula and southern Sweden, with underwater excavations at Tybring Vig, the Argus Bank, Møllegebaet, and Pilhaken (Fischer 1995a, 2004, 2007; Skaarup and Grøn 2004). These sites are in shallow water (~10 m depth) on shorelines that have undergone isostatic submergence, so the majority belong to the Ertebølle culture (c. 5500–4000 cal BC) of the late Mesolithic period and are not significantly earlier than the famous Ertebølle shell mounds on the isostatically elevated shorelines of the northern Jutland Peninsula. However, because of the excellent conditions of preservation underwater, these sites often provide detailed evidence of organic materials, especially wood. At Tybrind Vig there was a dugout canoe, a paddle with elaborate decoration, the remains of a landing stage, and large numbers of wooden stakes arranged in a line to form a fish trap extending out from the shoreline. Similar evidence of large-scale fish traps has been identified on other parts of the submerged shoreline and suggests a major commitment to fishing. It is, however, clear from the faunal and other evidence of food remains, both from underwater sites and from the dry-land shell mounds, that while fish and to a lesser extent shellfish were major food sources, the overall economy was a mixed one that included hunting of terrestrial and marine mammals and some plant food exploitation. How much further back in time this marine-oriented palaeoeconomy can be traced in Denmark remains unclear, though Fischer (1995b) is optimistic about the possibilities of extending underwater exploration to deeper and, hence, earlier underwater locations.

For insight into earlier periods, we must turn to Norway and Sweden. Here, much of the coastline formed at about 13,500 cal BC is now above modern sea level because of isostatic rebound, especially in Norway (Bjerck 1995, 2007, 2008; Larsson 1996; Schaller Åhrberg 2007). That ought to provide a unique insight into the use of marine and intertidal resources at a time when shorelines of equivalent date in other parts of the world are now 40 to 50 m below present sea level. However, there is no indisputable evidence of human presence on these elevated shorelines for another 3,000 years. Then, at or shortly after 9600 cal BC (10,000 BP), numerous sites appear along the full 2,000-km length of coastline as far as the far north of Finnmark, representing an expansion of coastal settlement that must have taken place within a few generations.

One of the difficulties of evaluating these very early coastal settlements is the almost total lack of preservation of bone or other organic material. However, site locations on promontories and offshore islands with good views of the sea and nearby inlets providing easy access for boats are evidence of a strong maritime orientation. It seems probable, but not demonstrable, that fish, seabirds, and sea mammals were all important, most probably in combination with hunting on land. Deep inlets like the Boknafjord in southwest Norway or the Uddevalla and Götaälv Straits on the Bohuslan Coast of Sweden had extremely high levels of marine productivity resulting from the meeting and mixing of nutrient-rich seawater and glacial melt water. Narrow straits also provided good opportunities for trapping sea mammals, and further upstream narrow corridors of land and crossing points were good for intercepting reindeer. The location of the sites, the nature of the resources available, and the need for substantial intake of animal fats for human survival in these northerly regions indicate that marine resources and especially sea mammals must have been a major source of subsistence. Flake adzes are a prominent tool type in the stone industries and would have been well suited for scraping blubber from seal carcasses (Bjerck 2008).

In the British Isles there appears to be a similar pattern, though the earliest sites seem to be slightly later than in Norway, most dating between about 8500 and 8000 cal BC. These are likewise confined to regions of isostatic uplift in northern Britain, on the Northumberland coastline of northeast England, and in parts of Scotland and northern Ireland, the latter requiring the use of boats and skilled navigation to cross the treacherous straits that separate Ireland and southwest Scotland (Bailey 2007; Tolan-Smith 2008). Since the degree of isostatic uplift in northern Britain is much less than in Norway, it is possible that earlier coastal sites have been submerged. However, unequivocal evidence of hinterland occupation in northern Britain is also scarce in earlier periods. Thus, it seems that as in Norway, this early phase of settlement indicates a new phase of population expansion, with heavy reliance on marine resources, perhaps linked as in Norway to changes in ocean currents and ecological conditions associated with the Preboreal period. Here too, however, actual organic remains of marine resources are sparse and limited to some seal bones at Howick in Northumberland (Waddington et al. 2003) and salmon bones at the riverine site of Mount Sandel in northern Ireland (Woodman 1978).

The time lag in the human colonization of these rich marine environments following the exposure of ice-free coastlines can be explained in one of two ways. According to one view, the preexisting human populations of southern Europe had no prior experience of exploiting marine resources. Moreover, successful movement along the Norwegian coastline and around the Western Isles of Scotland would have required seaworthy boats, most probably framed boats covered in animal skins, and investment in the necessary skills for the development of such a technology along with its attendant social costs and risks (Bjerck 1995, 2008), and there may have been a time lag before these skills were developed. The alternative view is that maritime skills and exploitation of marine resources have a much longer history on European coastlines mostly obscured by the loss of evidence associated with lowered sea levels, and that the more northerly coastal regions were too climatically marginal and risky for sustained human settlement until about 9600 cal BC, regardless of any existing technological skills in the exploitation of the sea (Fischer 1996, 2004; Bailey and Spikins 2008). Bjerck (2008), for example, has compared the marine environment on the late glacial Norwegian coastline to present-day Svalbard, with drifting pack ice and frozen fjords, polar bear, large populations of seal and maritime birds, but few fish apart from the small polar cod *(Boreogadus saida)*. It is arguable as to whether this would have been an accessible or viable environment for year-round human survival, or whether marine resources by themselves in the absence of any alternatives on land would have been sufficient to support long-term human settlement. At about 9600 cal BC, the Scandinavian ice sheet, which had persisted in close proximity to the coastline, finally disappeared, opening up the hinterland to herds of reindeer and elk that provided resources to complement those obtained from the sea. The polar front also shifted north to the latitude of Iceland and was replaced by the warmer waters of the Gulf Stream, creating a more productive environment at sea as well as on land, and providing an essentially modern marine fauna that included large stocks of cod and herring, which became key food staples in later periods. Under such conditions, the greater diversity and richness of resources, including land mammals, may have made a critical contribution to the sustainability of human settlement in otherwise quite extreme climatic conditions. Similar considerations apply to northern Britain.

How far back in time one can extend the deeper history of maritime exploitation on the Pleistocene coastlines of southern Europe beyond the immediate influence of the ice sheets is unclear. On all but the narrowest coastlines, the drop in sea level would have taken the shoreline at least 5 to 10 km out beyond the present position, so that any coastal settlements or archaeological evidence of marine exploitation must necessarily now be destroyed or submerged. Certainly, some of the long cave sequences on the north Spanish Coast show the presence of marine shells as food remains in deposits extending back into the last glacial period, and although the quantities of shells are quite small, these are believed to be only the tip of a more intensive pattern of marine exploitation (Bailey and Craighead 2003; Straus 2008; Straus and Clark 1986). In the Mediterranean, Upper Palaeolithic deposits in the site of Cueva de la Nerja in southern Spain, where the offshore shelf is steep and narrow, have yielded bones of seal and fish along with mollusk shells in deposits extending back to about 14,000 BP (Morales et al. 1998; Morales and Roselló 2004). On Gibraltar, the even older sequence of Vanguard Cave (Stringer et al. 2000) extends back to an earlier period of high sea level associated with the last interglacial period—another window of visibility for human use of marine resources—and recent preliminary excavations in these early deposits have produced small quantities of fish bones, a cut-marked bone of a porpoise, and marine shells (C. Finlayson, personal communication 2005). On the shallower continental shelves around the shorelines of northern France and southern Britain, last glacial shorelines would have been much further out to sea, and any archaeological trace of their use less easily visible.

At any rate, as far as the North Sea Basin is concerned, it seems certain that we can trace continuous interest in and exploitation of marine resources back to about 10,000 years ago. Whether an Inuit-like pattern of adaptation to and exploitation of marine resources existed in earlier periods on now-submerged northerly shorelines with Arctic conditions remains unclear.

For the later part of the Mesolithic period following cessation of eustatic sea level rise after about 6500 cal BC, coastal settlements with or without shell middens appear widely throughout much of the study area. Undoubtedly the largest and best-known group is the Ertebølle sites of Denmark and southern Sweden, which comprise some 400 known shell mounds, in which oysters are the most common molluskan species, dated between about 5500 and 4000 cal BC (6500 and 5200 BP) (Andersen 2000, 2007). Most of these are in the northern Jutland Peninsula and on the northern coastline of Zealand, where the slight elevation of the Litorina shoreline of the period above the present sea level has ensured optimal conditions of site visibility and preservation. Some individual mounds are several hundred meters long and 3 to 4 m thick. Meilgaard alone comprises 2,000 m^3 of deposit containing an estimated 2,000,000,000 oyster shells (Bailey 1978), and other mounds are bigger. These seemingly astronomical quantities have sometimes given rise to the notion of human populations who lived primarily on mollusks, and who might thus have imposed considerable pressure on the available mollusk supply. It is, however, clear from a series of analyses carried out in Europe and elsewhere that these quantities do not warrant such a conclusion (Bailey 1975, 1977, 1978; Bailey et al. 1994; Clark 1975). Rather, they are the result of progressive accumulation over many centuries, combined with high levels of visibility and preservation of mollusk shells relative to remains of other food resources. In all those cases where adequate quantitative controls can be established, the marine mollusks appear to represent a relatively minor food supply, but one that was nevertheless of great significance to human subsistence in representing an easily accessible and predictable food resource (cf. Meehan 1982).

In Denmark, there are as many coastal sites again that are not shell middens, and the absence of shells appears to be due to the lack of local shell beds in the immediate vicinity of such sites. This is consistent with the idea that the oysters, though forming an important food supply, were only one of a range of resources exploited by these coastal populations. Some individual sites may have had quite specialized functions for the exploitation of marine birds, whales, or oysters (Rowley-Conwy 1983), but the overall economy comprised a wide range of resources including hunting of terrestrial and sea mammals, shellgathering, fishing, and collecting of plant foods and a high degree of sedentism. Marine resources exploited included, in particular, oysters *(Ostrea edulis)* and cockles *(Cerastoderma edule)*, gray seal *(Halichoerus grypus)*, and cod *(Gadus morhua)*. Freshwater fish are also well represented in some coastal middens (Enghoff 1986).

In Britain, the best-known examples are in the north, particularly in Scotland. These include the Oronsay shell mounds (dominated by *Patella vulgata*), off the west coast of Scotland (5320–3800 cal BC) (Mellars 1987; Mellars and Wilkinson 1980), Morton (dominated by mollusks of *C. edule*) on the Tay Estuary on the east coast (6600–3790 cal BC) (Coles 1971), and West Voe in the Shetlands, with a Mesolithic shell midden deposit with *P. vulgata, O. edulis,* and *Mytilus edulis* dated at 4320–4030 cal BC and a Neolithic midden stratified above it (Melton and Nicholson 2004, 2007). Morton indicates a range of exploited food resources including land mammals, fish (with cod prominent), and plant foods. The Oronsay mounds suggest a more specialized marine economy dominated by fishing for saithe, *Pollachius virens,* like cod a member of the Gadidae family, shellgathering for limpets (possibly used for baiting fishing lines) and hunting of gray seal. Seasonality data from fish otoliths suggest the use of different mounds on the island at different times of year, but it remains unclear whether these represent a specialized maritime community who lived permanently on the island (Mellars 2004) or people who visited the island for its rich marine resources from bases on the mainland or the large islands nearby (Mithen 2000).

In western Sweden and Norway, as noted, preservation conditions are generally poor. Remains of fishing for cod, sometimes of large size, and other gadids are present along with evidence of land-mammal hunting at a scattering of Mesolithic sites, including Skipshelleren and the Viste Cave in Norway, and Ölmanäs and Dammen in western Sweden (Bjerck 2007, 2008; Schaller Åhrberg 2007; Wigfors 1995). Of these, Dammen is of particular interest in preserving quite a large assemblage of bones of cod and herring, substantial quantities of shellfish, particularly periwinkle *(Littorina littorea)*, and a large artifact collection including numerous fishhooks, dated to about 7000 cal BC. Schaller Åhrberg (2007) interprets this as evidence of an organized fishery for cod using baited hooks on lines and for herring using seine nets.

There are few indicators from this scattered body of Mesolithic material that give insight into the nature of human impact on the marine ecosystem or provide us with a clear baseline from which to judge later impacts. It is clear that a wide variety of marine resources were exploited and were of considerable importance as major sources of subsistence over many millennia, including resources such as oyster, cod, and herring that have become of commercial importance in recent centuries. In reviewing the evidence for fish remains from these sites, Pickard and Bonsall (2004) have noted the considerable variation in size and age of specimens recovered in different sites but have concluded that these represent variations in local topographic conditions relative to fish feeding areas and migratory patterns. They also specifically exclude the practice of deep-sea fishing from boats and argue that all the evidence is consistent with use of line fishing from the shore. We conclude that any impacts on the marine ecosystem were at best localized and most probably moderated by the widespread practice of broad-spectrum economies not specialized or overdependent on any one class of food resources, and by the confinement of marine exploitation, for the most part, to shore-based activities or at any rate a boating technology confined to inshore waters.

THE MESOLITHIC TO NEOLITHIC TRANSITION

The transition from the Mesolithic to the Neolithic is associated with the introduction of agriculture in the region, beginning at about 4000 cal BC in Britain and Denmark, and later further north. We might expect this major economic transition to show up in changes in the use of the marine environment, and a number of lines of evidence have been brought to bear on this issue, in particular changes in size and age structure of exploited molluskan populations in midden sequences, and a new body of stable isotope results from human bone collagen.

A number of Danish shell middens contain Neolithic midden deposits stratified above Mesolithic levels, and these sites show a shift from a predominance of oysters in Mesolithic levels to cockles in the Neolithic, a transition especially well documented at the site of Norsminde (Andersen 1989) (Figure 10.3). This mound comprises ~270 m^3 of midden deposit and was occupied from 4500 to 3200 cal BC. The dominant species are oyster *(O. edulis)* and cockle *(C. edule)*, with other mollusk species present in variable but generally low frequencies—mussel *(M. edulis)*, periwinkle *(Littorina littorea)*, carpet shell *(Tapes decussatus)*, and whelk *(Hinia decussata)*. The shift in taxonomic frequencies is well shown at this site, although the change involves a decline in oysters relative to cockles rather than the complete disappearance of oysters. Remains of other food resources are well represented in the Mesolithic levels, including bones of fish, seal, and land

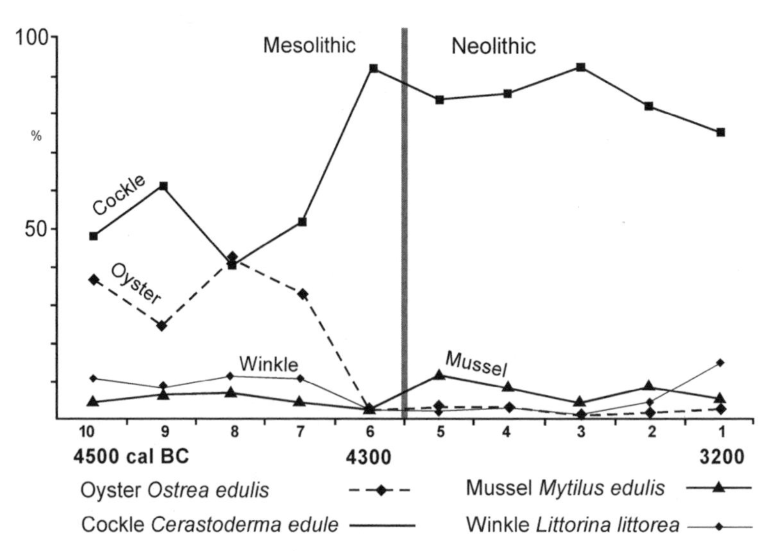

FIGURE 10.3. Change in frequency of molluskan taxa at the Danish shell midden of Norsminde. Data from Bailey and Milner (2007). Drawn by G. Bailey.

mammals, while the Neolithic levels are composed mainly of mollusk shells, suggesting a change in the use of the site from a settlement to a specialized camp used for processing of shellfood by farming people who had their settlements further inland.

Detailed analyses of the size and growth characteristics of the principal mollusk species have been undertaken, informed by control studies of modern populations, to establish limits of accuracy in establishing patterns of growth as indicators of age and season of death. The Mesolithic evidence suggests that after an initial size decline, interpreted as the initial impact of human gathering on a previously unexploited shell bed, there is a very stable pattern of exploitation, with uniform size and age structures of the exploited mollusks, and no evidence that shellgathering activities were making significant inroads into the ecological viability of the mollusk populations (Figures 10.4 and 10.5). In the Neolithic layers, both size and age structure fluctuate, indicating intermittent periods of heavier exploitation pressure, while the shells also appear to have been collected over a longer part of the year (Figure 10.6), in contrast to the narrow season of collection practiced in the Mesolithic period (Bailey and Milner 2007; Milner 2001, 2002). Similar trends have been identified in the cockles as well (Bailey and Milner 2007; Milner and Laurie 2007).

Comparison with stable isotope data from human skeletons suggests a more complicated picture (Figures 10.7 and 10.8). Sample data are now available from quite a large selection of human skeletons from Mesolithic and Neolithic burials throughout western and northern Europe including Britain and Scandinavia (Liden et al. 2004; Lubell et al. 1994; Richards and Hedges 1999a, 1999b; Richards et al. 2003; Schulting and Richards 2002). The results suggest a significant dietary shift from emphasis on marine protein in the Mesolithic to emphasis on terrestrial protein in the Neolithic. Some have gone so far as to suggest an actual avoidance of marine resources by Neolithic people (cf. Thomas 2003), but we do not believe this is warranted by the evidence. Firstly, there is well-known evidence in Neolithic deposits for the continuation of shell-gathering, sea-mammal hunting, and fishing. This is true not only in the Danish sites, but also more widely in the North Sea Basin, notably in Norway, Sweden, and Scotland (Bjerck 2007; Clark 1977; Milner et al. 2004; Wickham-Jones 2007; Wigfors 1995). Secondly, the baseline isotope ecology used to assess human measurements is likely to have varied

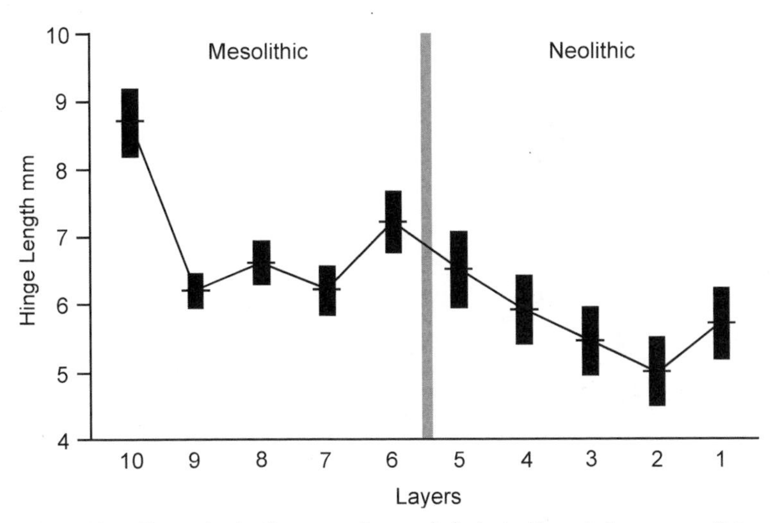

FIGURE 10.4. Change in size frequency of oyster shells in the Norsminde sequence. Data from Bailey and Milner (2007). Drawn by G. Bailey.

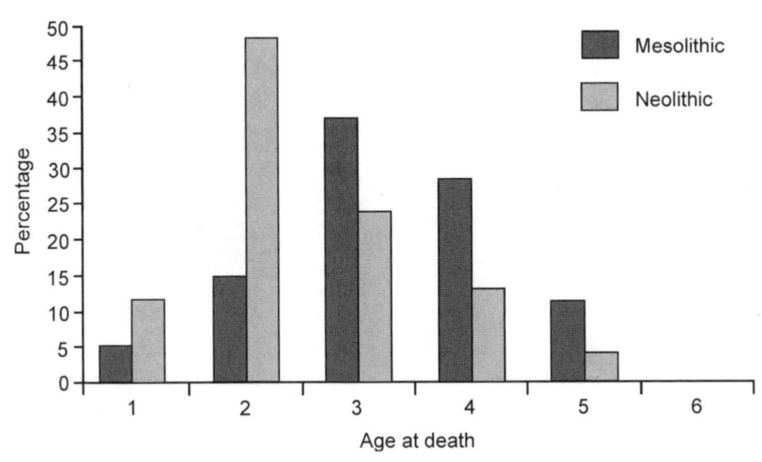

FIGURE 10.5. Change in age structure of oyster shells in the Norsminde sequence. Data from Bailey and Milner (2007). Note the shift from a modal age at death of 3 years in the Mesolithic to 2 years in the Neolithic. Drawn by G. Bailey.

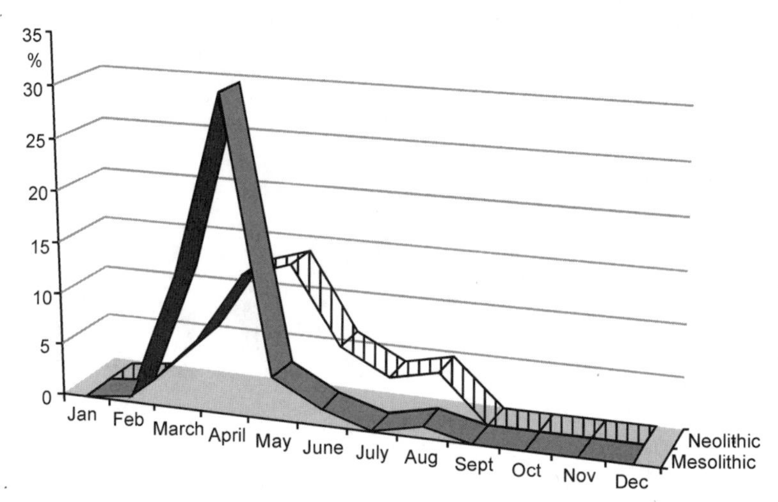

FIGURE 10.6. Seasonality of oyster gathering in the Norsminde sequence. Data from Bailey and Milner (2007). Drawn by G. Bailey.

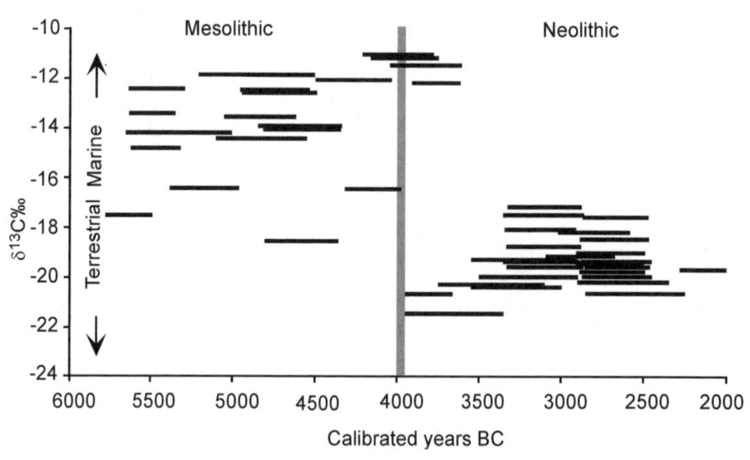

FIGURE 10.7. Stable isotope composition of human skeletal material from Mesolithic and Neolithic contexts in Denmark. Data from Milner et al. (2004). Drawn by G. Bailey.

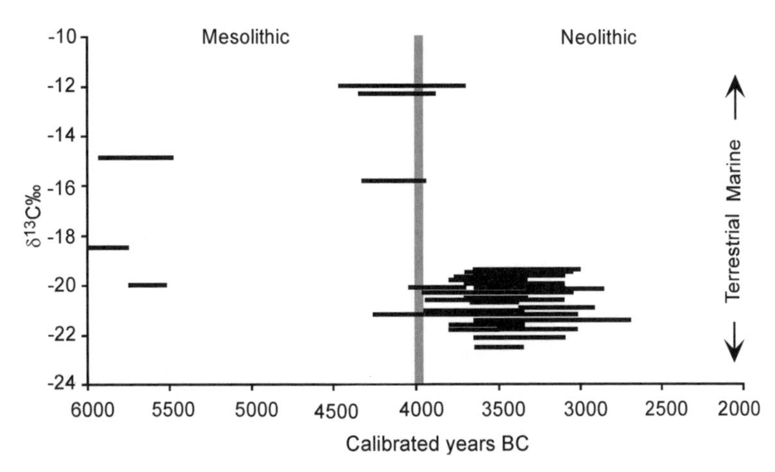

FIGURE 10.8. Stable isotope composition of human skeletal material from Mesolithic and Neolithic contexts in Britain. Data from Milner et al. (2004). Drawn by G. Bailey.

from area to area and period to period. Until appropriate measurements of the isotopic composition of contemporaneous food organisms and environments have been undertaken, the isotope composition of human bone should be interpreted with caution. Thirdly, the stable-isotope technique may be insensitive to differences between diets lacking marine protein and diets in which marine protein makes a relatively small but nonetheless significant contribution (Hedges 2004; Milner et al. 2004, 2006; Richards and Schulting 2006).

Nevertheless, it seems reasonable to infer a general shift over time to greater emphasis, proportionately, on terrestrial protein in the human diet. But this leaves us with the paradoxical result that marine resources appear to have been subject to less pressure from human impact when they formed a larger part of the diet (in the Mesolithic), and more pressure when human populations were apparently less dependent on them (in the Neolithic). That paradox might be resolved in four ways.

Firstly, we might suggest that the results in Figures 10.7 and 10.8 are a biased sample, reflecting individuals that are not representative of all the local populations from which they are drawn. In this regard, it should be noted that there are no stable isotope results on human bone from the Norsminde site, which could conceivably be atypical in relation to the other locations from which human skeletons have been recovered and analyzed. However, it seems unlikely that such a relatively large and diverse sample is systematically biased in this way.

Secondly, it is possible that marine resources declined in abundance because of environmental change and thus became more vulnerable to overexploitation even in the absence of changes in the pattern of human activity. Analysis of sediment cores in the Norsminde Estuary suggests a considerable increase in rate of sediment accumulation during the Neolithic period, quite likely related to agricultural land clearance, which would have improved substrate conditions for cockles at the expense of oysters. In addition, there is independent evidence for minor climatic change at this time, and in particular a reduction in temperature, marking the end of the mid-Holocene climatic optimum. Oysters would have been more sensitive to both these types of environmental change than cockles, and the oyster habitat may well have reduced considerably as a consequence. However, a similar argument cannot be applied to the cockles. On the contrary, the evidence seems to suggest that their habitat expanded, and yet they too came under increased pressure of human exploitation, judging by reductions in size and age of the mollusk shells in the Norsminde midden.

Thirdly, we could argue that Mesolithic populations, being more dependent on marine resources, were more careful to avoid damaging ecological impacts that could have a major effect on human survival and well being than Neolithic populations, who could afford to be less conservation minded, knowing that their livelihood did not depend to the same degree on the continuation of a minor resource. However, we think this unlikely, if only because mollusks appear to have been a relatively minor resource in the Mesolithic period too. Yet they played an important role as a stopgap and an alternative resource at times of the year when other food resources were in short supply, and this applies as much to farming populations as to hunter-gatherer ones (cf. Deith 1988).

Fourthly, it is possible that Neolithic populations, though less dependent on marine resources than their Mesolithic predecessors, were larger overall because of changes in the economic base, including the probable addition of domestic livestock and crops, and thus had a bigger impact on a fixed marine resource base. Thus, changes in economic scheduling of different resources during the course of the year, and in particular a prolongation of the season over which mollusks were collected, as is clearly shown by the seasonality results at Norsminde, exposed the shell beds to increased levels of human impact independently of any other factors.

We cannot at present discriminate with any confidence between these alternatives, not least because analyses of changes in age structure and growth rate need to be applied to a much larger and more diverse sample of food remains in archaeological deposits. It may be that several if not all of these factors have contributed to the patterns visible in the archaeological record. As far as human impact on the molluskan populations is concerned, the main interest of this example is to demonstrate two patterns. The first is an initial reduction in size and age following initial exploitation of a pristine shell bed, followed by a period of stability. The second, apparent in the later part of the Norsminde sequence, is a greater impact on the mollusk population, indicated by episodic reductions in the size and age of the shells with occasional periods of recovery, reflecting wider changes in the organization of the human economy and most probably a larger human population. The details of these changes in human impact are difficult to pursue further in this case, because of relatively small sample sizes, low chronological resolution, and also the ongoing problem of disentangling human impacts on mollusk populations from other environmental changes (Bailey and Milner 2007).

THE MEDIEVAL TO MODERN TRANSITION

A diversity of opinion continues to exist regarding the state of Europe's political economy following the collapse of the Roman Empire (Collins and Gerrard 2004). Nevertheless, it remains clear that the intensity of economic production and market trade declined in the middle centuries of the first millennium AD—the years once referred to as the Dark Ages. Documenting the reemergence of intensive surplus production, market trade, and urbanism (sometimes referred to as the commercial revolution of the Middle Ages) is a well-trodden scholarly path (e.g., Anderton 1999; Hodges 1982; Prestell and Ulmschneider 2003). Perhaps the only significant controversies remaining are the chronology of this development and its causes. Some scholars suggest an early, even eighth-century, development (Hill and Cowie 2001), whereas others would place significant economic changes only around or after the year 1000 (Griffiths 2003).

Within this debate, the North Sea has long been a focus of discussions of trade and its corollary, piracy (e.g., Myhre 1993; Sawyer 1971). The significance of intensified exploitation of marine animals, however, has only been recognized comparatively recently (e.g., Barrett 1997; Benecke 1982; Ervynck and Van Neer 1994; Heinrich 1983; Jones 1981). Hoffmann (1996), in a seminal paper, used historical evidence to hypothesize that Medieval population

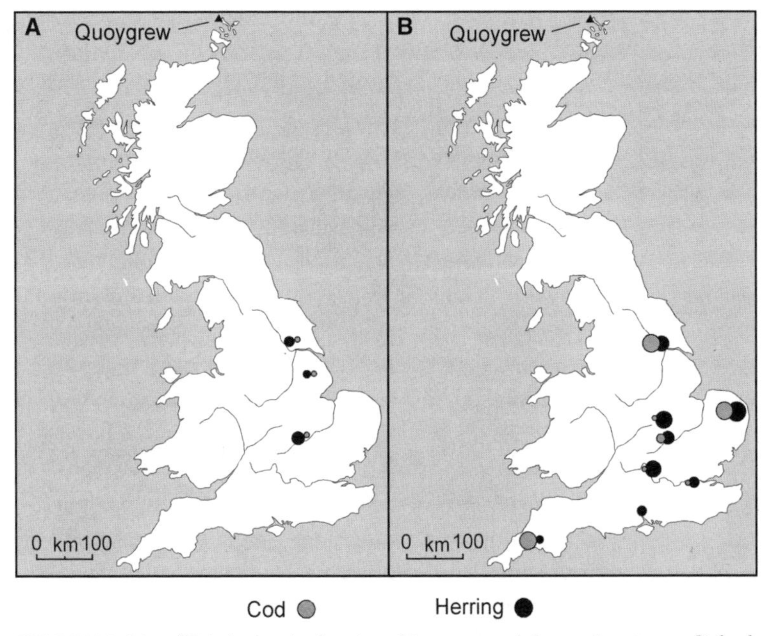

FIGURE 10.9. Map of Britain showing location of Quoygrew and changes in patterns of inland fish consumption: A. ninth and tenth centuries AD; B. eleventh and twelfth centuries AD. Size of circles indicates relative proportion of cod and herring in fish bone assemblages of the two periods. In the earlier period, freshwater fish dominated fish consumption at inland centers. Data from Barrett et al. (2004a). Drawn by G. Bailey. See text for further discussion.

levels, land use practices, and fishing efforts had depleted freshwater fish resources by the early centuries of the second millennium AD. Since then, wider syntheses have convincingly demonstrated that marine fishing increased tremendously around the North Sea at this time, replacing the former dominance of freshwater species in the fish bone record (Barrett et al. 2004a, 2004b; Enghoff 2000). The question of causation remains complex, but Hoffmann's hypothesized shift from freshwater to marine species appears to be real, whether to increase the absolute supply of fish or to replace dwindling freshwater resources. Many of the marine fish are found as bones in inland towns, so it is highly probable that they represent traded commodities (or indirect subsistence, to use Hoffmann's terminology) rather than locally organized provisioning.

To illustrate this trend, one can consider the evidence from different parts of Britain likely to have been net "producers" and "consumers" of traded fish, particularly cod, over the millennium from AD 600 to 1600, namely the Northern Isles of Scotland (Orkney and Shetland) and England (Figure 10.9). The former were the focus of a semiautonomous Scandinavian, then Scottish, earldom (Thomson 2001), whereas the latter transformed from a patchwork of small Anglo-Saxon chiefdoms to the English state over the millennium in question (Hinton 1998).

In the Northern Isles, fish and sea mammals were exploited at very low levels prior to the ninth to tenth centuries. Regardless of recovery method and preservation conditions, cetacean and seal remains are rare finds (e.g., Mulville 2002). Fish bones are more abundant but represent small inshore species easily caught using traps, nets, or lines and with minimal risk or technological investment (Barrett et al. 1999; Colley 1983; Nicholson 1998). Stable isotope analysis of human bone collagen also indicates that marine protein was a negligible component of the Northern Isles diet at this time (Barrett and Richards 2004; Barrett et al. 2001). The most common fish species is saithe, *Pollachius virens*, also the species that dominates the fish remains at Mesolithic Oronsay

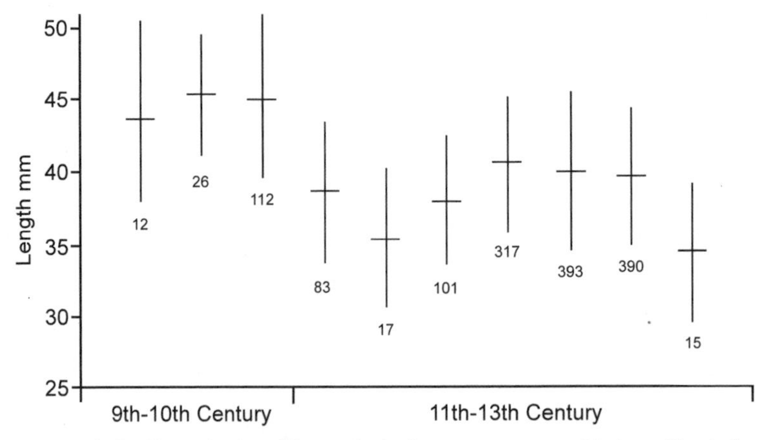

FIGURE 10.10. Change in size of limpets in the Quoygrew sequence. Horizontal bar is the mean, vertical bar is 1 SD. Numbers indicate sample size. Note the sustained size reduction after AD 1000. Data from Milner et al. (2007b). Drawn by G. Bailey.

(see above). Based on ethnohistoric analogy from later centuries, it is likely that prior to the ninth to tenth centuries saithe served principally as a source of lamp oil, although they may also have formed a minor dietary supplement (Nicholson 2004); but it is doubtful that the same analogy can be extended as far back as the Mesolithic period, when it is assumed that saithe were exploited mainly for food.

With the first appearance of burials with Scandinavian grave goods in the Northern Isles (between AD 850 and 950), there was a significant increase in sea fishing, particularly for large cod. This may be evidence for the introduction of a marine-oriented diet, knowledge base, and worldview by Scandinavian migrants (Barrett et al. 1999, 2001; Cerón-Carrasco 2005). Large cod and related species were widely fished in Viking Age Norway (e.g., Barrett et al. 2007; Hufthammer 2003; Perdikaris 1999; see also Perdikaris and McGovern, this volume).

In the eleventh or twelfth centuries, however, the intensity of fishing for cod and related species increased far more. In many zooarchaeological assemblages, fish bones outnumber (and occasionally even outweigh) cattle, caprines, and pigs combined (Barrett et al. 1999, 2001; Harland 2006). Many fish were clearly being eaten locally. However, some sites were producing cured fish based on the evidence of cut marks and element distributions, and these may have been intended for export (Barrett 1997; Barrett et al. 1999; Cerón-Carrasco 2005; Harland 2006). The intensity of bait collection probably also increased, as indicated by a reduction in the size and age of limpet *(Patella vulgata)* shells found in eleventh- to thirteenth-century strata at the site of Quoygrew in Orkney (Figures 10.10 and 10.11) (Milner et al. 2007b). Limpets can be eaten, but isotopic evidence suggests that marine carnivores were preferred (Barrett and Richards 2004). These mollusks were utilized as fish bait in later centuries (Fenton 1978), and the same may be true of the Quoygrew limpets. Stable isotope evidence suggests that some individuals ate more marine protein than is known since the Mesolithic (Barrett and Richards 2004) (Figure 10.12, compare Figure 10.8).

Turning to England, eight taxonomic groups dominate the fish bone record. The marine fish are herring and gadids (cod family fish, in which the closely related hake is included for present purposes). The freshwater fish are cyprinids (carp family) and pike *(Esox lucius)*. The migratory fish are European eel *(Anguilla anguilla)*, salmon and trout (salmonids), smelt *(Osmerus eperlanus)*, and flatfish (including both flounder, *Platichthys flesus*, which enters freshwater, and marine species). When these taxa were compared using correspondence analysis, it became clear that all "catches" from

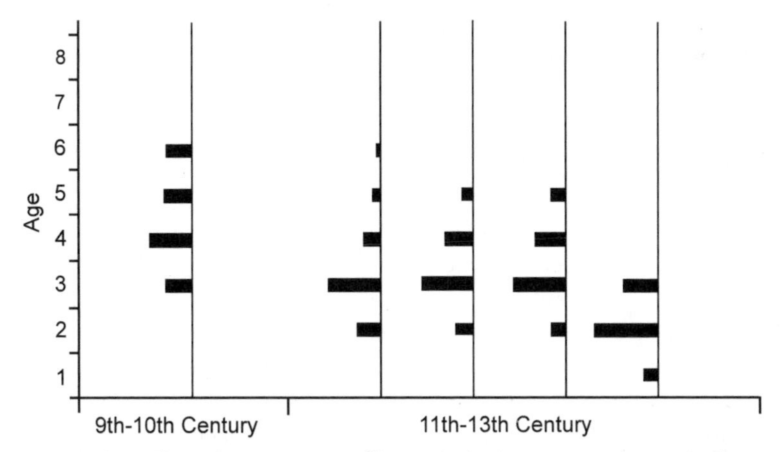

FIGURE 10.11. Change in age structure of limpets in the Quoygrew sequence, using the same horizontal scale as in Figure 10.10. Each age distribution is based on a sample of 30 measurements. Data from Milner et al. (2007b). Drawn by G. Bailey.

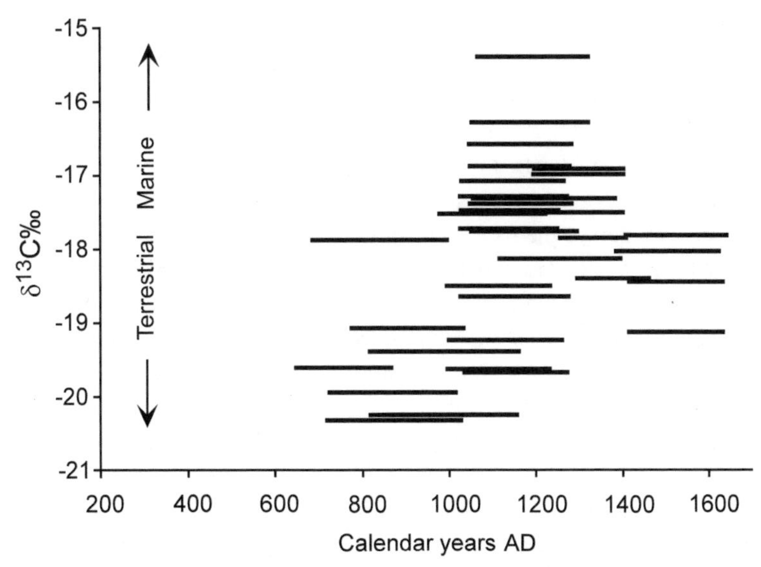

FIGURE 10.12. Stable isotope composition of human skeletal material at Newark Bay, Orkney, showing the greater emphasis on marine food after AD 1000. Data from Barrett and Richards (2004). Drawn by G. Bailey.

the seventh to tenth centuries were dominated by freshwater and migratory species (particularly cyprinids and eels). In the eleventh and later centuries, however, herring and gadids became dominant (Barrett et al. 2004a).

Simply put, the major change through time was an increase in the abundance of herring and gadids. Both became far more abundant in the eleventh to twelfth centuries (Figure 10.9). Herring was found in some earlier sites, particularly proto-urban centers such as York and London (e.g., Jones 1988; Locker 1988), but its importance increased fourfold at this time. For gadids, different species show slightly different patterns. Cod itself first appeared in more than nominal numbers in the eleventh to twelfth centuries. Later, however, its proportion of the total declined as it was joined by related species such as haddock, ling, saithe, and hake. Surprisingly, the few assemblages that can be tightly dated suggest that the initial increase in herring and cod abundance occurred within 50 years of AD 1000. Some corroboration of these observations comes from the limited historical evidence available from this period (see Barrett et al. 2004a, 2004b; Pauley 2004).

Attempts have been made to assess the impact of this early process of increased fishing activity on the dynamics of the fish populations in relation to the more recent impact of heavily industrialized fishing in the twentieth century. Bolle et al. (2004) have used otoliths from archaeological sites of the eleventh century and later to compare age, size, and growth rates of ancient and modern catches of cod, haddock, saithe, and plaice. The expectation is that intensification of fishing activity should show up in a general shift to younger fish and higher growth rates (Figure 10.2). The results are equivocal. Modern data show a clear shift to younger fish and increased growth rates of juveniles in cod and haddock, some change in plaice, and none in saithe. However, interpretation is complicated by two factors. Firstly, there are no data for fish assemblages that precede the fish event horizon to provide a baseline for assessing later changes. Secondly, juvenile fish of some species are subject to predation from adults. Thus, increased predation of adult fish by humans may result in greater overcrowding and reduced growth rate of the juveniles, although this effect will be absent if the nursery grounds for the juvenile fish are separate from the adult feeding areas.

STABLE ISOTOPE ANALYSIS AND FOOD WEB STRUCTURE

Another approach to the investigation of human impact is the analysis of the stable nitrogen isotope composition of the exploited organisms. In the northeast Atlantic fisheries, statistical data on catches over the last 50 years indicate that smaller individuals and species have become increasingly abundant. Essentially, short-lived invertebrates and planktivorous pelagic fish of low trophic level have replaced long-lived carnivorous bottom fish of high trophic level. Pauly et al. (1998) have analyzed these effects in recent fisheries data and describe such disturbances of ecosystem structure and function as fishing down the food web. However, a more complete understanding of this process and particularly its longevity is hampered by the limited data available (Polunin and Pinnegar 2002; Watson and Pauly 2002). Records of fishery landings and fish dietary data rarely extend beyond the past 50 years. Morales and Roselló (2004) have attempted to extend the record further back in time by calculating the mean trophic level of archaeological fish assemblages in a chronostratigraphic sequence that spans thousands of years at the Cueva de la Nerja in southern Spain. However, this approach relies heavily on the availability of extremely well preserved assemblages, which reflect the original, true proportions of fish of different trophic levels, and these are rare in the archaeological record. A technique that offers a more reliable approach is the measurement of the ratio of the stable isotopes of nitrogen (^{15}N and ^{14}N), expressed as $\delta^{15}N$, in archaeological marine fauna, as a means of tracking changes in the diet of marine predators (Hirons et al.

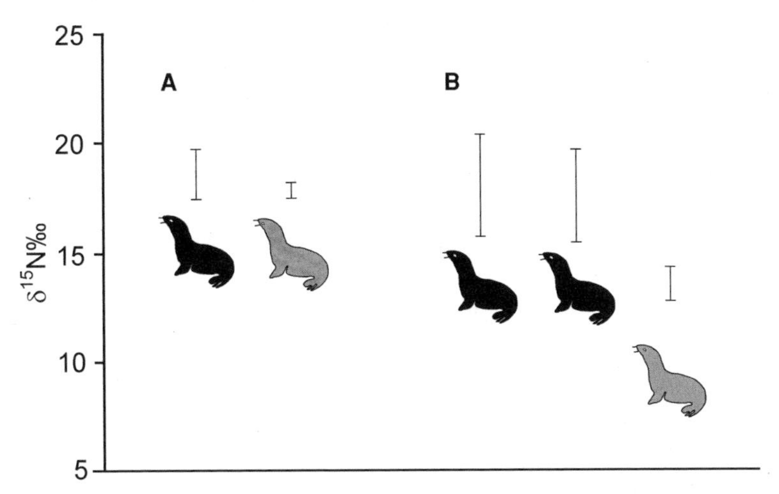

FIGURE 10.13. Changes in $\delta^{15}N$ values of modern (black) and prehistoric (shaded) pinnipeds. Vertical bars indicate the range for each sample. A. Comparison of modern (*n* = 18) and Middle to Late Holocene (*n* = 13) harbor seals from Central California. B. Comparison of modern gray (*n* = 6) and harbor (*n* = 23) seals from the North Sea with prehistoric (ca. 4000 cal BC) gray seals (*n* = 10) from Denmark and southern Sweden. Drawn by G. Bailey.

2001; Wainwright et al. 1993). Nitrogen isotope ratios show clear enrichment of ^{15}N (~2–4‰) in the tissues of a consumer organism relative to its prey, and hence progressive enrichment of ^{15}N with increasing trophic level. Disruptions to the food web such as loss of medium or high trophic level fish should show up as a progressive reduction in $\delta^{15}N$ of top predator species, as they are forced to fish down the food web. Comparing prehistoric, historic, and modern nitrogen isotope data should therefore provide valuable insights into temporal trends in North Sea and northeast Atlantic ecosystem structure in relation to recent human impacts.

Changes in the nitrogen isotope composition of predator species may also be the result of long-term environmental changes that have affected organisms at the base of the food web. In order to demonstrate that shortening of the food web is due to overexploitation, it is essential to compare $\delta^{15}N$ measurements of organisms at different trophic levels and at different time periods and in different geographic locations. We have begun to collect stable isotope data from a number of Danish shell middens of Mesolithic and Neolithic date. An initial comparison of data from pinnipeds (gray and harbor seals) recently beached on the southern coastlines of the North Sea (Das et al. 2003) and archaeological specimens (Craig et al. 2006) shows that the modern specimens actually have higher $\delta^{15}N$ values than their prehistoric counterparts, which is the reverse of what would be expected if recent shortening of the food web had occurred (Figure 10.13). One hypothesis to account for this unexpected result, albeit based on our very small sample sizes, is that there has been an overall shift in the $\delta^{15}N$ values of organisms at the base of the food web due to environmental changes, and we are currently testing that hypothesis by comparing measurements in modern and archaeological mollusk shells. In another example from central California (Figure 10.13), Burton et al. (2001) have shown that there are no significant differences in the mean $\delta^{15}N$ between ancient and modern pinnipeds, suggesting that ecosystem structure has remained stable over the time period in question, although analysis of lower trophic level ancient and modern marine species would be needed to confirm this.

CONCLUSIONS

Human intervention in the North Sea marine ecosystem has been taking place for at least the last 10,000 years and possibly longer. Moreover, there have been major environmental changes affecting marine organisms both at the large scale of the North Sea as a whole, and at a more localized geographical scale, as well as changes in the configuration and size of human populations. There is thus no single baseline against which to judge recent human impact, but rather a constantly shifting baseline subject both to geographical variation and episodic change in time.

There have been at least four periods when intensification of human exploitation of marine resources may be inferred: an early period of late glacial intensification associated with the earliest human colonization; a later postglacial intensification associated with the Mesolithic period; further intensification under the pressure of expanding agricultural populations from the Neolithic period onward; and the fish event horizon of ca. AD 1000, which anticipated the further expansion of the North Sea fisheries to Iceland and Newfoundland in the fifteenth century AD and the industrialization of fishing activity in the twentieth century. There may have been many more such episodes. The evidence of intensification in the prehistoric periods is controversial, especially in the prehistoric periods mentioned, because it is possible that the apparent changes in the archaeological record could equally well be explained by changes in the visibility of coastal archaeological deposits or changes in environmental conditions. The record that we currently have is at best patchy, and we should beware of glossing over the gaps in the record to construct a narrative of long-term progressive and increasing human impact. Equally unwarranted is the assumption that there was no human impact on marine ecosystems before the modern period.

Evidence of increased human pressure on resources is not necessarily evidence of overexploitation of the resource. Resources with limited geographical distributions that are easily accessible (e.g., intertidal mollusks) might seem to be most vulnerable, but even these are likely to be resilient in the face of intensified predation provided that there are reserve stocks on less accessible coastlines or in subtidal locations, capable of replenishing locally depleted shorelines. Moreover, the thinning out of shell beds and the removal of larger specimens may lead to a relaxation of human predation pressure and a switch to other food supplies because of the resulting increase in the cost of locating and collecting edible-sized specimens, long before the risk of local extinction. This is the marine resource that we might expect to have been the earliest to come under pressure of human exploitation, but archaeological records indicate minimal evidence of human impact and none that we would consider threatening to the viability of mollusk populations, not even at a local scale.

Whether increased human impact on marine resources in earlier periods and localized or temporary depletion of specific groups of foods such as mollusks had stressful consequences for the human population is another matter and would have depended on the role played by that resource in the local economy and the availability of alternatives. Claims have been made that local depletion or decline of mollusks was capable of triggering migration in the prehistoric context (Mannino and Thomas 2002), while Rowley-Conwy (1983) has suggested that the decline of oysters at the end of the Ertebølle period (albeit because of environmental change rather than overexploitation) triggered the collapse of the hunter-gatherer economy and the introduction of Neolithic agriculture. Both claims are controversial, the former because it is a hypothetical scenario not currently supported by specific evidence, and the latter because the Norsminde data examined earlier demonstrate that other molluskan taxa continued to be available during the Neolithic period after the decline of the oysters and that the major impact on molluskan size and age structure came after the onset of the Neolithic period rather than before. Again we urge caution in supposing that evidence of increased human impact on marine resources necessarily had stressful consequences for the human population. In particular, in the case of mollusks, we draw attention to Meehan's (1982) detailed and influential study of Aboriginal shellgathering in Australia, which demonstrated the importance of the mollusks as a food resource but also the negligible impact on the local economy when the shell beds were temporarily wiped out by monsoonal storms, a potentially catastrophic event that was dealt with by the local Aboriginal population by switching to alternative resources.

A variety of methods can now be brought to bear on the long-term history of ecological relationships between past human populations and marine resources, including a new generation of stable isotope techniques. However, all of these techniques are subject to uncertainties and potential biases. A major difficulty is that changes that might be taken as evidence of increased human impact might equally well be produced by environmental changes unrelated to human activity (or at best only indirectly so). Changes in the methods used to capture resources can also produce false evidence of intensification in the form of an apparent reduction in size or age structure (e.g., a change in the mesh size of fishing nets or the size of fishing hooks, or a change in the season or location of capture). Nevertheless, it is clear that these techniques, used in combination, open up rich possibilities for tracking long-term changes in human use of and impact on marine resources, subject only to limitations imposed by the availability of archaeological samples and the resources to analyze them.

ACKNOWLEDGMENTS

We acknowledge financial support from the Arts and Humanities Research Council, AHRC UK, to Bailey and Milner through grant B/RG/AN1717/APN14658 from the Leverhulme Trust to Bailey through its Major Research Fellowship scheme and to Barrett for the Medieval Origins of Commercial Sea Fishing Project, and from Historic Scotland to Barrett for the excavations at Quoygrew. We are also grateful for the comments of David Steadman and Steven James.

REFERENCES CITED

Andersen, S. H.

1989 Norsminde: A "Køkkenmødding" with Late Mesolithic and Early Neolithic Occupation. *Journal of Danish Archaeology* 8:13–40.

2000 Kjøkkenmøddinger (Shell Middens) in Denmark: A Survey. *Proceedings of the Prehistoric Society* 66:361–384.

2007 Shell Middens *("Køkkenmøddinger")* in Danish Prehistory as a Reflection of the Marine Environment. In *Shell Middens in Atlantic Europe*, edited by N. Milner, G. Bailey and O. Craig, pp. 31–45. Oxbow, Oxford.

Anderton, M.

1999 *Anglo-Saxon Trading Centres: Beyond the Emporia*. Cruithne Press, Glasgow.

Bailey, G. N.

1975 The Role of Mollusks in Coastal Economies: The Results of Midden Analysis in Australia. *Journal of Archaeological Science* 2:45–62.

1977 Shell Mounds, Shell Middens, and Raised Beaches in the Cape York Peninsula. *Mankind* 11:132–143.

1978 Shell Middens as Indicators of Postglacial Economies: A Territorial Perspective. In *The Early Postglacial Settlement of Northern Europe*, edited by P. A. Mellars, pp. 37–63. Duckworth, London.

2004a The Wider Significance of Submerged Archaeological Sites and Their Relevance to World Prehistory. In *Submarine Prehistoric Archaeology of the North Sea: Research Priorities and Collaboration with Industry*, edited by N. C. Flemming, pp. 3–10. CBA Research Report 141. Council for British Archaeology, London.

2004b World Prehistory from the Margins: The Role of Coastlines in Human Evolution. *Journal of Interdisciplinary Studies in History and Archaeology* 1(1):39–50.

2007 Palaeogeography of the North Sea Basin. In *Mesolithic Studies in the North Sea Basin and Beyond*, edited by C. Waddington and K. Pedersen. Oxbow, Oxford, in press.

Bailey, G. N., and A. Craighead

2003 Late Pleistocene and Early Holocene Coastal Palaeoeconomies: A Reconsideration of the Molluskan Evidence from Northern Spain. *Geoarchaeology: An International Journal* 18(2):175–204.

Bailey, G. N., and N. J. Milner

2002 Coastal Hunters and Gatherers and Social Evolution: Marginal or Central? *Before Farming: the Archaeology of Old World Hunter-Gatherers* 3–4(1):1–15.

2007 The Marine Mollusks from the Mesolithic and Neolithic Deposits of the Norsminde Shell Midden. In *Stone Age Settlement in the Coastal*

Fjord of Norsminde, Jutland, Denmark, edited by S. Andersen, in press.
Bailey, G., and P. Spikins (editors)
2008 *Mesolithic Europe*. Cambridge University Press, Cambridge.
Bailey, G. N., J. Chappell, and R. Cribb
1994 The Origin of Anadara Shell Mounds at Weipa, North Queensland, Australia. *Archaeology in Oceania* 29:69–80.
Barrett, J. H.
1997 Fish trade in Norse Orkney and Caithness: A Zooarchaeological Approach. *Antiquity* 71:616–638.
Barrett, J. H., and M. P. Richards
2004 Identity, Gender, Religion and Economy: New Isotope and Radiocarbon Evidence for Marine Resource Intensification in Early Historic Orkney, Scotland. *European Journal of Archaeology* 7:249–271.
Barrett, J. H., R. A. Nicholson, and R. Cerón-Carrasco
1999 Archaeo-ichthyological Evidence for Long-term Socioeconomic Trends in Northern Scotland: 3500 BC to AD 1500. *Journal of Archaeological Science* 26:353–388.
Barrett, J. H., R. P. Beukens, and R. A. Nicholson
2001 Diet and Ethnicity during the Viking Colonisation of Northern Scotland: Evidence from Fish Bones and Stable Carbon Isotopes. *Antiquity* 75:145–154.
Barrett, J. H., A. M. Locker, and C. M. Roberts
2004a "Dark Age Economics" Revisited: The English Fish Bone Evidence AD 600–1600. *Antiquity* 78(301):618–136.
2004b The Origin of Intensive Marine Fishing in Medieval Europe: The English Evidence. *Proceedings of the Royal Society B* 271:2417–2421.
Barrett, J., A. Hall, C. Johnstone, H. Kenward, T. O'Connor, and S. Ashby
2007 Interpreting the Plant and Animal Remains from Viking Age Kaupang. In *Kaupang in Skiringssal*, Kaupang Excavation Project Publication Series, vol.1, Norske Oldfunn, vol. 22, edited by D. Skre, pp. 283–319. Aarhus University Press, Aarhus, in press.
Barton, C. M., S. Schmich, and S. R. James
2004 The Ecology of Human Colonization in Pristine Landscapes. In *The Settlement of the American Continents: A Multidisciplinary Approach to Human Biogeography*, edited by C. M. Barton, G. A. Clark, D. R. Yesner, and G. A. Pearson, pp. 138–161. University of Arizona Press, Tucson.
Benecke, N.
1982 Zur frühmittelalterlichen Heringsfischerei im südlichen Ostseeraum: Ein archäozoologischer Beitrag. *Zeitschrift für Archäologie* 16: 283–290.
Bjerck, H.
1995 The North Sea Continent and the Pioneer Settlement of Norway. In *Man and Sea in the Mesolithic: Coastal Settlement above and below Present Sea Level*, edited by A. Fischer, pp. 131–144. Oxbow, Oxford.
2007 Mesolithic Coastal Settlements and Shell Middens in Norway. In *Shell Middens in Atlantic Europe*, edited by N. Milner, G. Bailey and O. Craig, pp. 5–30. Oxbow, Oxford.
2008 Norwegian Mesolithic Trends: A Review. In *Mesolithic Europe*, edited by G. Bailey and P. Spikins, pp. 60–106. Cambridge University Press, Cambridge.
Bolle, L. J., A. D. Rijnsdorp, W. van Neer, R. S. Millner, P. I. van Leeuwen, A. Ervynck, R. Ayers, and E. Ongenac
2004 Growth Changes in Plaice, Cod, Haddock, and Saithe in The North Sea: A Comparison of (Post-)Medieval and Present-Day Growth Rates based on Otolith Measurements. *Journal of Sea Research* 51:313–328.
Burton, R. K., J. J. Snodgrass, D. Gifford-Gonzalez, T. Guilderson, T. Brown, and P. L. Koch
2001 Holocene Changes in the Ecology of Northern Fur Seals: Insights from Stable Isotopes and Archaeofauna. *Oecologia* 28:107–115.
Cerón-Carrasco, R.
2005 *Of Fish and Men ("De iasg agus dhaoine"): A Study of the Utilization of Marine Resources as Recovered from Selected Hebridean Archaeological Sites*. BAR British Series 400. Hadrian Books, Oxford.
Clark, J. G. D.
1975 *The Earlier Stone Age Settlement of Scandinavia*. Cambridge University Press, Cambridge.
1977 The Economic Context of Dolmens and Passage Graves in Sweden. In *Ancient Europe: Studies Presented in Honour of Hugh Hencken*, edited by V. Markovic, pp. 35–49. Aris and Phillips, Warminster, UK.
1983 Coastal Settlement in European Prehistory with Special Reference to Fennoscandia. In *Prehistoric Settlement Patterns: Essays in Honor of Gordon R. Willey*, edited by E. von Z. Vogt and R. M. Leventhal, pp. 295–317. University of New Mexico Press, Albuquerque, New Mexico.
Coles, B. J.
1998 Doggerland: A Speculative Survey. *Proceedings of the Prehistoric Society* 64:45–81.
Coles, J. M.
1971 The Early Settlement of Scotland: Excavations at Morton, Fife. *Proceedings of the Prehistoric Society* 37:284–366.

Colley, S. M.
1983 Interpreting Prehistoric Fishing Strategies: An Orkney Case Study. In *Animals and Archaeology 2: Shell Middens, Fishes, and Birds*, edited by C. Grigson and J. Clutton-Brock, pp. 157–171. BAR International Series 183. British Archaeological Reports, Oxford.

Collins, R., and J. Gerrard (editors)
2004 *Debating Late Antiquity in Britain AD 300–700*. BAR British Series 365. Hadrian Books, Oxford.

Craig, O. E., R. Ross, N. Milner, and G. N. Bailey
2006 Sulphur Isotope Variation in Archaeological Marine Fauna from Northern Europe. *Journal of Archaeological Science* 33:1642–1646.

Cunliffe, B.
2001 *Facing the Ocean*. Oxford University Press, Oxford.

Das, K., G. Lepoint, Y. Leroy, and J. M. Bouquegneau
2003 Marine Mammals from the Southern North Sea: Feeding Ecology Data from $\delta^{13}C$ and $\delta^{15}N$ measurements. *Marine Ecology Progress Series* 263:287–298.

Deith, M. R.
1988 A Molluskan Perspective on the Role of Foraging in Neolithic Farming Economies. In *The Archaeology of Prehistoric Coastlines*, edited by G. N. Bailey and J. E. Parkington, pp. 116–124. Cambridge University Press, Cambridge.

Enghoff, I. B.
1986 Freshwater Fishing fromm a Sea-Coast Settlement: The Ertebølle *locus classicus* Revisited. *Journal of Danish Archaeology* 5:62–76.
1999 Fishing in the Baltic Region from the Fifth Century BC to the Sixteenth Century AD: Evidence from Fish Bones. *Archaeofauna* 8:41–85.
2000 Fishing in the Southern North Sea Region from the First to the Sixteenth Century AD: Evidence from Fish Bones. *Archaeofauna* 9:59–132.

Ervynck, A., and W. Van Neer
1994 A Preliminary Survey of Fish Remains in Medieval Castles, Abbeys and Towns of Flanders (Belgium). *Offa* 51:303–307.

Fenton, A.
1978 *The Northern Isles: Orkney and Shetland*. John Donald, Edinburgh.

Fischer, A.
1995b An Entrance to the Mesolithic World below the Ocean: Status of Ten Years' Work on the Danish Sea Floor. In *Man and Sea in the Mesolithic: Coastal Settlement above and below Present Sea Level*, edited by A. Fischer, pp. 371–384. Oxbow, Oxford.
1996 At the Border of Human Habitat: The Late Palaeolithic and Early Mesolithic in Scandinavia. In *The Earliest Settlement of Scandinavia and Its Relationship with Neighbouring Areas*, edited by L. Larsson, pp. 157–176. Acta Archaeologica Lundensia, Series 8:24. Almquist and Wiksell International, Stockholm.
2004 Submerged Stone Age—Danish Examples and North Sea Potential. In *Submarine Prehistoric Archaeology of the North Sea: Research Priorities and Collaboration with Industry*, edited by N. C. Flemming, pp. 23–36. CBA Research Report 141. Council for British Archaeology, London.
2007 Coastal Fishing in Stone Age Denmark—Evidence from below and above the Present Sea Level and from Human Bones. In *Shell Middens in Atlantic Europe*, edited by N. Milner, G. Bailey and O. Craig, pp. 54–69. Oxbow, Oxford.

Fischer, A. (editor)
1995a *Man and Sea in the Mesolithic: Coastal Settlement above and below Present Sea Level*. Oxbow, Oxford.

Fitch, S., K. Thomson, and V. Gaffney
2005 Late Pleistocene and Holocene Depositional Systems and the Palaeogeography of the Dogger Bank, North Sea. *Quaternary Research* 64:185–196.

Glimerveen, J., D. Mol, K. Post, J. W. F. Reumer, H. Van der Plicht, J. De Vos, B. Van Geel, G. Van Reene, and J. P. Pals
2004 The North Sea Project: the First Palaeontological, Palynological, and Archaeological Results. In *Submarine Prehistoric Archaeology of the North Sea: Research Priorities and Collaboration with Industry*, edited by N. C. Flemming, pp. 43–52. CBA Research Report 141. Council for British Archaeology, London.

Griffiths, D.
2003 Exchange, Trade, and Urbanization. In *From the Vikings to the Normans*, edited by W. Davies, pp. 73–106. Oxford University Press, Oxford.

Hardy, K., and C. Wickham-Jones
2002 Scotland's First Settlers: The Mesolithic Seascape of the Inner Sound, Skye and Its Contribution to the Early Prehistory of Scotland. *Antiquity* 76:825–833.

Harland, J.
2006 Zooarchaeology in the Viking Age to Medieval Northern Isles, Scotland: An Investigation of Spatial and Temporal Patterning. Unpublished Ph.D. thesis, Department of Archaeology, University of York, York, UK.

Hedges, R. E. M.
2004 Isotopes and Red Herrings: Comments on Milner et al. and Lidén et al. *Antiquity* 78:34–37.

Heinrich, D.
1983 Temporal Changes in Fishery and Fish Consumption between Early Medieval Haithabu and

Its Successor, Schleswig. In *Animals and Archaeology 2: Shell Middens, Fishes and Birds*, edited by C. Grigson and J. Clutton-Brock, pp. 151–156. BAR International Series 183. British Archaeological Reports, Oxford.

Hill, D., and R. Cowie
2001 *Wics: The Early Medieval Trading Centres of Northern Europe*. Sheffield Academic Press, Sheffield, UK.

Hinton, D. A.
1998 *Archaeology, Economy, and Society: England from the Fifth to the Fifteenth Century*. Routledge, London.

Hirons, A. C., D. M. Schell, and B. P. Finney
2001 Temporal Records of $\delta^{13}C$ and $\delta^{15}N$ in North Pacific Pinnipeds: Inferences Regarding Environmental Change and Diet. *Oecologia* 129(4): 591–601.

Hodges, R.
1982 *Dark Age Economics: The Origins of Towns and Trade AD 600–1000*. Duckworth, London.

Hoffmann, R. C.
1996 Economic Development and Aquatic Ecosystems in Medieval Europe. *The American Historical Review* 101:631–669.

Hufthammer, A. K.
2003 Med kjøtt og fisk på menyen. In *Middelalder-Gården i Trøndelag*, edited by O. Skevik, pp. 182–196. Stiklestad Nasjonale Kultursenter, Stiklestad.

Jackson, J. B. C., M. X. Kirby, W. H. Berger, K. A. Bjorndal, L. W. Botsford, B. J. Bourque, R. H. Bradbury, R. Cooke, J. Erlandson, J. A. Estes, T. P. Hughes, S. Kidwell, C. B. Lange, H. S. Lenihan, J. M. Pandolfi, C. H. Peterson, R. S. Steneck, M. J. Tegner, and R. R. Warner
2001 Historical Overfishing and the Recent Collapse of Coastal Ecosystems. *Science* 293:629–638.

Jones, A. K. G.
1981 Reconstruction of Fishing Techniques from Assemblages of Fish Bones. In *Fish Osteoarchaeology Meeting: Copenhagen 28th–29th August 1981*, edited by I. G. Enghoff, J. Richter and K. Rosenlund, pp. 4–5. Danish Zoological Museum, Copenhagen.
1988 Provisional Remarks on Fish Remains from Archaeological Deposits at York. In *The Exploitation of Wetlands*, edited by P. Murphy and C. French, pp. 113–127. BAR British Series 186. British Archaeological Reports, Oxford.

Larsson, L. (editor)
1996 *The Earliest Settlement of Scandinavia and Its Relationship with Neighbouring Areas*. Acta Archaeologica Lundensia, Series 8:24. Almquist and Wiksell International, Stockholm.

Lidén, K., G. Eriksson, B. Nordqvist, A. Götherstörm, and E. Bendixen
2004 "The Wet and the Wild followed by the Dry and the Tame"or Did They Occur at the Same Time? Diet in Mesolithic-Neolithic Southern Sweden. *Antiquity* 78:23–33.

Locker, A.
1988 The Fish Bones. In *Two Middle Saxon Occupation Sites: Excavations at Jubilee Hall and 21–22 Maiden Lane*, edited by R. Cowie, R. L. Whytehead and L. Blackmore, pp. 149–150. *Transactions of the London and Middlesex Archaeological Society* 39:47–163.

Long, D., C. R. Wickham-Jones, and N. A. Ruckley
1986 A Flint Artefact from the Northern North Sea. In *Studies in the Upper Palaeolithic of Britain and Northwest Europe*, edited by D. A. Roe, pp. 55–62. BAR International Series 296. British Archaeological Reports, Oxford.

Louwe Kooijmans, L. P.
2003 The Hardinxveld Sites in the Rhine/Meuse Delta, the Netherlands, 5500–4500 cal BC. In *Mesolithic on the Move: Papers Presented at the Sixth International Conference on the Mesolithic in Europe, Stockholm 2000*, edited by L. Larsson, H. Kindgren, K. Knutsson, D. Loeffler and A. Åkerlund, pp. 608–624. Oxbow, Oxford.

Lubell, D., M. Jackes, H. Schwarcz, M. Knyf, and C. Meiklejohn
1994 The Mesolithic Neolithic Transition in Portugal: Isotopic and Dental Evidence of Diet. *Journal of Archaeological Science* 21:201–216.

Mannino, M. A., and K. D. Thomas
2002 Depletion of a Resource? The Impact of Prehistoric Human Foraging on Intertidal Mollusk Communities and Its Significance for Human Settlement, Mobility and Dispersal. *World Archaeology* 33:452–474.

Meehan, B.
1982 *Shell Bed to Shell Midden*. Australian Institute of Aboriginal Studies, Canberra.

Mellars, P. A.
1987 *Excavations on Oronsay: Prehistoric Human Ecology on a Small Island*. Edinburgh University Press, Edinburgh.
2004 Mesolithic Scotland, Coastal Occupation, and the Role of the Oronsay Middens. In *Mesolithic Scotland and Its Neighbours*, edited by A. Saville, pp. 167–183. Society of Antiquaries of Scotland, Edinburgh.

Mellars, P. A., and T. Wilkinson
1980 Fish Otoliths as Indicators of Seasonality in Prehistoric Shell Middens: The Evidence from Oronsay (Inner Hebrides). *Proceedings of the Prehistoric Society* 46:19–44.

Melton, N. D., and R. A.,Nicholson
2004 The Mesolithic in the Northern Isles: The Preliminary Evaluation of an Oyster Midden at West Voe, Sumburgh, Shetland, U.K. *Antiquity* 78(299). Electronic document, http://antiquity.ac.uk/ProjGall/nicholson/index.html, accessed July 2007.
2007 A Late Mesolithic–Early Neolithic midden at West Voe, Shetland. In *Shell Middens in Atlantic Europe*, edited by N. Milner, G. Bailey and O. Craig, pp. 94–100. Oxbow, Oxford.

Milner, N.
2001 At the Cutting Edge: Using Thin Sectioning to Determine Season of Death of the European Oyster, *Ostrea edulis*. *Journal of Archaeological Science* 28:861–873.
2002 *Incremental Growth of the European Oyster,* Ostrea edulis*: Seasonality Information from Danish Kitchenmiddens*. BAR International Series 1057. British Archaeological Reports, Oxford.

Milner, N., and E. Laurie
2007 Coastal perspectives on the Mesolithic-Neolithic transition. In *Meso2005*, edited by S. McCartan. Oxford, Oxbow, in press.

Milner, N., O. E. Craig, G. N. Bailey, K. Pedersen, and S. H. Andersen
2004 Something Fishy in the Neolithic? A Re-evaluation of Stable Isotope Analysis of Mesolithic and Neolithic Coastal Populations. *Antiquity* 78:9–22.

Milner, N., O. E. Craig, G. N. Bailey, and S. H. Andersen
2006 A Response to Richards and Schulting. *Antiquity* 80(308):456–468.

Milner, N., G. Bailey, and O. Craig (editors)
2007a *Shell Middens in Atlantic Europe*. Oxbow, Oxford.

Milner, N., J. Barrett, and J. Welsh
2007b Marine Resource Intensification in Viking Age Europe: The Molluskan Evidence from Quoygrew, Orkney. *Journal of Archaeological Science* 34(9):1461–1472.

Mithen, S. (editor)
2000 *Hunter-Gatherer Landscape Archaeology: The Southern Hebrides Mesolithic Project (1988–98)*, Vols. 1 and 2. MacDonald Institute for Archaeological Research, Cambridge.

Morales, A., and E. Roselló
2004 Fishing down the Food Web in Iberian Prehistory? A New Look at the Fishes from Cueva de Nerja (Málaga, Spain). In *Petits Animaux et Sociétés Humaines. Du Complément Alimentaire aux Resources Utilitaires. XXIVe Rencontres Internationales d'Archéologie et d'Histoire d'Antibes*, edited by J.-P. Brugal and J. Desse, pp. 111–123. Éditions APDCA, Antibes.

Morales, A., E. Roselló, and F. Hernández
1998 Late Upper Palaeolithic Subsistence Strategies in Southern Iberia: Tardiglacial Faunas from Cueva de Nerja (Málaga, Spain). *European Journal of Archaeology* 1(1):9–50.

Mulville, J.
2002 The Role of Cetacea in Prehistoric and Historic Atlantic Scotland. *International Journal of Osteoarchaeology* 12:34–48.

Myhre, B.
1993 The Beginning of the Viking Age—Some Current Archaeological Problems. In *Viking Revaluations*, edited by A. Faulkes and R. Perkins, pp. 182–204. Viking Society for Northern Research, London.

Nicholson, R. A.
1998 Fishing in the Northern Isles: A Case Study Based on Fish Bone Assemblages from Two Multi-period Sites on Sanday, Orkney. *Environmental Archaeology: The Journal of Human Palaeoecology* 2:15–28.
2004 Iron-Age Fishing in the Northern Isles: The Evolution of a Stored Product. In *Atlantic Connections and Adaptations: Economies, Environments and Subsistence in Lands Bordering the North Atlantic*, edited by R. A. Housley and G. Coles, pp. 155–162. Oxbow Books, Oxford.

Pauly, D.
2004 Much Rowing for Fish. *Nature* 432:913–914.

Pauly, D., V. Christensen, J. Dalsgaard, R. Froese, and F. Torres, Jr.
1998 Fishing down Marine Food Webs. *Science* 279:860–863.

Perdikaris, S.
1999 From Chiefly Provisioning to Commercial Fishery: Long-Term Economic Change in Arctic Norway. *World Archaeology* 30:388–402.

Pickard, C., and C. Bonsall
2004 Deep-sea Fishing in the European Mesolithic: Fact or Fantasy? *European Journal of Archaeology* 7:273–290.

Prestell, T., and K. Ulmschneider
2003 *Markets in Early Medieval Europe: Trading and "Productive" Sites, 650–850*. Windgather, Macclesfield, UK.

Polunin, N. V. C., and J. K. Pinnegar
2002 In *Handbook of Fish and Fisheries*, Vol. 1, pp 301–320. Blackwell, Oxford.

Richards, M. P., and R. E. M. Hedges
1999a Stable Isotope Evidence for Similarities in the Types of Marine Foods Used by Late Mesolithic Humans on the Atlantic Coast of Europe. *Journal of Archaeological Science* 26:717–722.
1999b A Neolithic Revolution? New evidence of Diet in the British Neolithic. *Antiquity* 73:891–897.

Richards, M. P., and R. J. Schulting
2006 Against the Grain? A Response to Milner et al. *Antiquity* 80(308):444–456.
Richards, M. P., R. J. Schulting, and R. E. M. Hedges
2003 Sharp Shift in Diet at Onset of Neolithic. *Nature* 425:366.
Rowley-Conwy, P.
1983 Sedentary Hunters: The Ertebølle Example. In *Hunter-Gatherer Economy in Prehistory*, edited by G. Bailey, pp. 111–126. Cambridge University Press. Cambridge.
Sawyer, P. H.
1971 *The Age of the Vikings*. Arnold, London.
Schaller Åhrberg, E.
2007 Fishing for Storage: Mesolithic Short Term Fishing for Long Term consumption. In *Shell Middens in Atlantic Europe*, edited by N. Milner, G. Bailey and O. Craig, pp. 46–53. Oxbow, Oxford.
Schulting, R. J., and M. P. Richards
2002 The Wet, the Wild and the Domesticated: the Mesolithic-Neolithic Transition on the West Coast of Scotland. *Journal of European Archaeology* 5:147–158.
Shennan, I., and J. Andrews (editors)
2000 *Holocene Land-Ocean Interaction and Environmental Change around the North Sea*. Special Publications 166. Geological Society, London.
Skaarup, J., and O. Grøn
2004 *Møllegebaet: A Submerged Mesolithic Settlement in Southern Denmark*. BAR International Series 1328. Archaeological Press, Oxford.
Straus, L. G.
2008 The Mesolithic of Atlantic Iberia. In *Mesolithic Europe*, edited by G. Bailey and P. Spikins, pp. 303–327. Cambridge University Press, Cambridge.
Straus, L. G., and G. A. Clark (editors)
1986 *La Riera Cave. Stone Age Hunter-gatherer Adaptations in Northern Spain*. Arizona State University, Tempe.
Stringer, C. B., R. N. E. Barton, and J. C. Finlayson (editors)
2000 *Neanderthals on the Edge: Papers from a Conference Marking the 150th Anniversary of the Forbes' Quarry Discovery, Gibraltar*. Oxbow, Oxford.
Tauber, H.
1981 ^{13}C Evidence for Dietary Habits of Prehistoric Man in Denmark. *Nature* 292:332–333.
Thomas, J.
2003 Thoughts on the "Repacked" Neolithic Revolution. *Antiquity* 77(295):67–75.
Thomson, W. P. L.
2001 *The New History of Orkney*. Mercat, Edinburgh.
Tolan-Smith, C.
2008 Mesolithic Britain. In *Mesolithic Europe*, edited by G. Bailey and P. Spikins, pp. 132–157. Cambridge University Press, Cambridge.
Tolan-Smith, C., and C. Bonsall
1999 Stone Age Studies in the British Isles: The Impact of Accelerator Dating. In *14C et Archéologie. Actes du 3ème Congrès International (Lyon, 6-10 avril 1998)*, edited by J. Evin, C. Oberlin, J-P. Daugas and J-F. Salles, pp. 249–257. Mémoires de la Société Préhistorique Française 26, 1999 et Supplément 1999 de la Revue d'Archéometrie. Société Préhistorique Française, Paris.
Van Kolfschoten, T., and H. Van Essen
2004 Palaeozoological Heritage from the Bottom of the North Sea. In *Submarine Prehistoric Archaeology of the North Sea: Research Priorities and Collaboration with Industry*, edited by N. C. Flemming, pp. 70–80. CBA Research Report 141. Council for British Archaeology, London.
Verhart, L. B. M.
2004 The Implications of Prehistoric Finds on and off the Dutch Coast. In *Submarine Prehistoric Archaeology of the North Sea: Research Priorities and Collaboration with Industry*, edited by N. C. Flemming, pp. 57–61. CBA Research Report 141. Council for British Archaeology, London.
Waddington, C., G. Bailey, A. Bayliss, I. Boomer, N. Milner, R. Shiel, and T. Stevenson
2003 A Mesolithic Settlement Site at Howick, Northumberland: A Preliminary Report. *Archaeologia Aeliana* 32:1–12.
Wainwright, S. C., M. J. Fogarty,R. C. Greenfield, and B. Fry
1993 Long-Term Changes in the Georges Bank Food Web: Trends in Stable Isotopic Compositions of Fish Scales. *Marine Biology* 115:481–493.
Watson, R., and D. Pauly
2002 Systematic Distortions in World Fisheries Catch Trends. *Nature* 414:534–536.
Wickham-Jones, C. R.
2007 Middens in Scottish Prehistory: Time, Space and Relativity. In *Shell Middens in Atlantic Europe*, edited by N. Milner, G. Bailey and O. Craig, pp. 89–93. Oxbow, Oxford.
Wigfors, J.
1995 West Swedish Mesolithic Settlements Containing Faunal Remains: Aspects of the Topography and Economy. In *Man and Sea in the Mesolithic: Coastal Settlement above and below Present Sea Level*, edited by A. Fischer, pp. 197–206. Oxbow, Oxford.
Woodman, P. C.
1978 *The Mesolithic in Ireland: Hunter-Gatherers in an Insular Environment*. Oxford, BAR British Series 58. British Archaeological Reports, Oxford.

11

Twenty Thousand Years of Fishing in the Strait

ARCHAEOLOGICAL FISH AND SHELLFISH ASSEMBLAGES FROM SOUTHERN IBERIA

Arturo Morales-Muñiz and Eufrasia Roselló-Izquierdo

ALONG THE SHORES of southern Iberia, extensive fishing enterprises developed during classical times. Their testimonies are reflected in the many fish factories that dot the present-day coastline and in the thousands of southern Iberian amphorae that distributed fish products throughout the Mediterranean and beyond (Arévalo et al. 2004; Curtis 1991; Étienne and Mayet 2002; Ponsich 1988; Ponsich and Tarradell 1965; Van Neer and Ervynck 2004). From such data one may get the impression that the bounty from the waters on both sides of the Strait of Gibraltar was endless and that the inhabitants of southern Iberia have been making their living from the sea since time immemorial. Indeed, when one reads some of the general works it appears that most of the details concerning these ancient fisheries, the species exploited, and the reasons behind episodes such as the one that in the third century AD wiped out more than half of the fish factories operating along this coast are well known matters (Consejería de Agricultura y Pesca 2004). As it turns out, nothing could be farther from the truth.

Archaeological records of fishing and shellfish collecting from southern Iberia are scarce, patchy, and often of poor quality. This is because until a few years ago there were few attempts to retrieve small faunas in a systematic way and also because most researchers in Spain and Portugal lacked adequate comparative collections for their analysis. Recently, both these problems have been at least partially resolved, and many truly interesting faunal assemblages are currently being studied. Some of the assemblages reported in this chapter appear for the first time in print, although the data and our interpretations are preliminary in more ways than one.

The issues that can be addressed with these assemblages are varied in terms of cultural and environmental questions, thus it is tempting to put forward some hypotheses, however preliminary, to pave the way for future studies. Although the poor and patchy nature of the record makes it impossible to deal with any issue in a detailed way at this moment, we provide an overview of the development of fishing and other marine resource exploitation in southern Iberia as it relates to the following:

1. The earliest evidence of marine fishing.
2. What the taxonomic composition of fish assemblages can tell us about the onset and development of the so-called industrial or commercial fishing of classical antiquity.
3. To what extent the fish assemblages reveal the importance and evolution of fishing in the area and provide evidence of fishing strategies and technologies.
4. Whether overexploitation signatures can be detected and how significant they are.

We have made a conscious decision here to restrict our analyses to faunal data, even though we know that doing so will eliminate some broader context. The complementary evidence on fishing tackle, vessels, salt production, and archaeological contexts of fish factories is not in much better shape as far as southern Iberia is concerned, with too little material evidence and most hypotheses still hotly debated (Alvar 1981; Aura and Pérez 1998; Burgos-Madroñero 2003; Garcia-Vargas 2001; Guerrero 1993; Ladero 1993; Lagóstena 1999; Mederos and Escribano 2005; Muñoz et al. 1988). Under such circumstances we are in no position to judge the various alternatives or to contribute in a meaningful way to the debate. From this perspective, the data we present should be taken more as evidence to enrich the ongoing debate than as an explanation of fishing developments in the area from the restricted perspective of the faunal analyst.

Finally, although the focus of our review is on fish remains, a summary of mollusk assemblages has been included (Table 11.1), and data on mollusks will be incorporated into the discussion to complement some of the patterns evident in the fish samples.

THE COAST AND ITS EVOLUTION

The southern Iberian Coast comprises two different territories separated by the Strait of Gibraltar (Figure 11.1):

1. A western sector bathed by the Atlantic Ocean, the so-called Gulf of Cádiz, which includes the Portuguese Algarve (ca. 170 km) plus some 300 km of the Andalusia Coast.
2. An eastern sector, the Sea of Alborán, which covers about 470 km of the Costa del Sol area at the foot of the Betica Mountains.

Except for the Algarve, the shoreline we see today is a quite recent development that dates back to the Modern Age (Arteaga and Hoffmann 1986; Gavala 1959; Hoffmann 1988, 1994; Hoffmann and Schultz 1988). During the Pleistocene, several periods of lower sea level (up to –100 m on the Sea of Alborán, but of far lower amplitude in the Gulf of Cádiz) resulted in the emergence of extensive continental platforms. During the last glacial maximum around 20,000 BP (uncalibrated radiocarbon years), the global glacio-eustatic low in sea level came to an end, and a transgression started that has, with some minor periods of reversal, lasted to this day—the Mediterranean having gained 5 to 6 m during the last three millennia (Goy et al. 1996; Lario et al. 1993; Margalef 1989; Zazo et al. 1994). As sea level rose in the Mediterranean, the shoreline reached the foot of the Betica Mountains so that by the end of the Pleistocene a previously sandy coastline was transformed into a rocky one, with small beaches and narrow bays along with a wealth of isolated islands (Arteaga and Hoffmann 1986; Aubet et al. 1999). Along the Atlantic Coast, the rising littoral faced no mountain systems, flooding instead the large alluvial plains of rivers such as the Guadiana (dividing Spain and Portugal) and the Guadalquivir (Figure 11.1). The inundation of these lower river valleys created huge estuaries at the end of the Pleistocene, where rising sea levels coupled

TABLE 11.1

Overview of the Archaeological Mollusk Assemblages from Southern Iberia Compared with the Catches of the Andalusian Fishing Sector in the Year 2001

SITE	MAIN TAXA	NUMBER OF IDENTIFIED SPECIMENS	REFERENCE	LOCATION
Nerja (Aurignacian)	Carpet shell (60%), common cockle (30%), limpet (10%)	10	Jorda (1986)	Alboran
Nerja (Magdalenian)	Carpet shell (52–14%), mussel (4–29%), limpet (1–9%)	–	Jorda (1986)	Alboran
Nerja (Mesolithic)	Mussel (55–34%), limpet (9–41%), turban (1–9.5%)	–	Jorda (1986)	Alboran
Vale Santo I (Meso-Neolithic)	Mussel (64–71%), drill (15–18%), limpet (11–13), turban (2–5%)	139 (MNI)	Stiner et al. (2003)	Algarve
Rocha Das Gaivotas (Meso-Neolithic)	Mussel (55–80%), limpet (18–41%), drill (2–21%), turban (2%)	552 (MNI)	Stiner et al. (2003)	Algarve
B. Quebradas II (Meso-Neolithic)	Limpet (29–48%), turban (19–32%), mussel (11–30%), drill (8–13%)	543 (MNI)	Stiner et al. (2003)	Algarve
B. Quebradas I (Meso-Neolithic)	Turban (29–63%), limpet (13–23%), mussel (3–32%), drill (7–19%)	624 (NMI)	Stiner et al. (2003)	Algarve
Nerja (Neolithic)	Limpet (44–64%), mussel (5–25%), turban (5–13%)	–	Jorda (1986)	Alboran
Papauvas (Neolithic-Copper)	Carpet shell (89%), razor clam (6%), oyster (.4%)	10,655	Moreno (1995a)	Gulf of Cádiz
Cueva Frigiliana (Neolithic-Copper)	Limpet (92%), mussel (2.5%)	38	Moreno (1995a)	Alborán
Nerja (Copper)	Turban (15–63%), limpet (21–55%)	–	Jorda (1986)	Alborán
Terrera Ventura (Copper)	Dog cockle (42%), limpet (31.5%)	152	Moreno (1995a)	Alborán (inland)
La Viña (Copper)	Carpet shell (64.5%), turban (24.5%), limpet (3%)	5,024	Moreno (1995a)	Gulf of Cádiz
Almonte (Copper)	Razor clam (70%), carpet shell (21%), oyster (4.5%)	843	Moreno (1995a)	Gulf of Cádiz
Valencina (Copper)	Carpet shell (86%), scallop (2%), limpet (0.5%)	1,694	Moreno (1995a)	Gulf of Cádiz (inland)
Los Millares (Copper)	Limpet (29%), turban (21.5%), dog cockle (21.5%)	1,516	Peters and von den Driesch (1990)	Alborán (inland)
Terrera del Reloj (Bronce)	Oyster (40%), dog cockle (40%), turban (20%)	5	Moreno (1995a)	Alborán (inland)

(continued)

TABLE 11.1 *(continued)*

SITE	MAIN TAXA	NUMBER OF IDENTIFIED SPECIMENS	REFERENCE	LOCATION
Cerro de la Encina (Bronce)	Dentalium (22.5%), dog cockle (47%), rough cockle (14%)	49	Lauk (1976)	Alborán (inland)
Fuente Álamo (Bronce)	Dog cockle (66%), turban (9%), limpet (8%), rough cockle (4%)	550	von den Driesch et al. (1985)	Alborán (inland)
Castillo de Doña Blanca (Phoenician)	Peppery furrow (53%), razor clam (13.5%), carpet shell (13%)	15,919	Moreno (1995a)	Gulf of Cádiz
Toscanos (Phoenician)	Dog cockle (51%), limpet (26%), rough cockle (16.5%)	2,075	Moreno (1995a)	Alborán
Tejada La Vieja (Iron)	Carpet shell (78%), dog cockle (22%)	32	Moreno (1995a)	Gulf of Cádiz (inland)
Puerto 6 (Iron)	Oyster (49.5%), dog cockle (42%)	265	Moreno (1995a)	Gulf of Cádiz
Puerto 29 (Iron)	Cuttlefish (19%), panobe clam (19%), dog cockle (18%)	374	Moreno (1995a)	Gulf of Cádiz
Cabezo de San Pedro (Iron)	Carpet shell (32.5%), dog cockle (18.5%), oyster (18%)	43	Moreno (1995a)	Gulf of Cádiz
La Tiñosa (Iron)	Oyster (37%), carpet shell (29%), pigmy oyster (13%)	117	Moreno (1995a)	Gulf of Cádiz
Setefilla (Iron)	Dog cockle (54%), rough cockle (38%)	13	Moreno (1995a)	Gulf of Cádiz (inland)
Munigua (Roman)	Oyster (66%), drill (33%)	6	Moreno (1995a)	Gulf of Cádiz (inland)
San Nicolás (Roman)	Oyster (68.5%), rough cockle (11%), stripped venus (4%)	1,787	Unpublished data	Alborán
Toscanos (Roman)	Oyster (58.5%), thorny oyster (20%), dog cockle (8%)	65	Moreno (1995a)	Alborán
2001 fishing statistics of the Junta de Andalucia	Stripped venus (45.5%), octopuses (16%), rough cockle (13%, cuttlefish (8%)	10,751,883 kg	Junta de Andalucia (2007)	All Andalusia

NOTE: The abundance values of the Portuguese shell middens refer to minimum number of individuals (MNI), not identified remains.

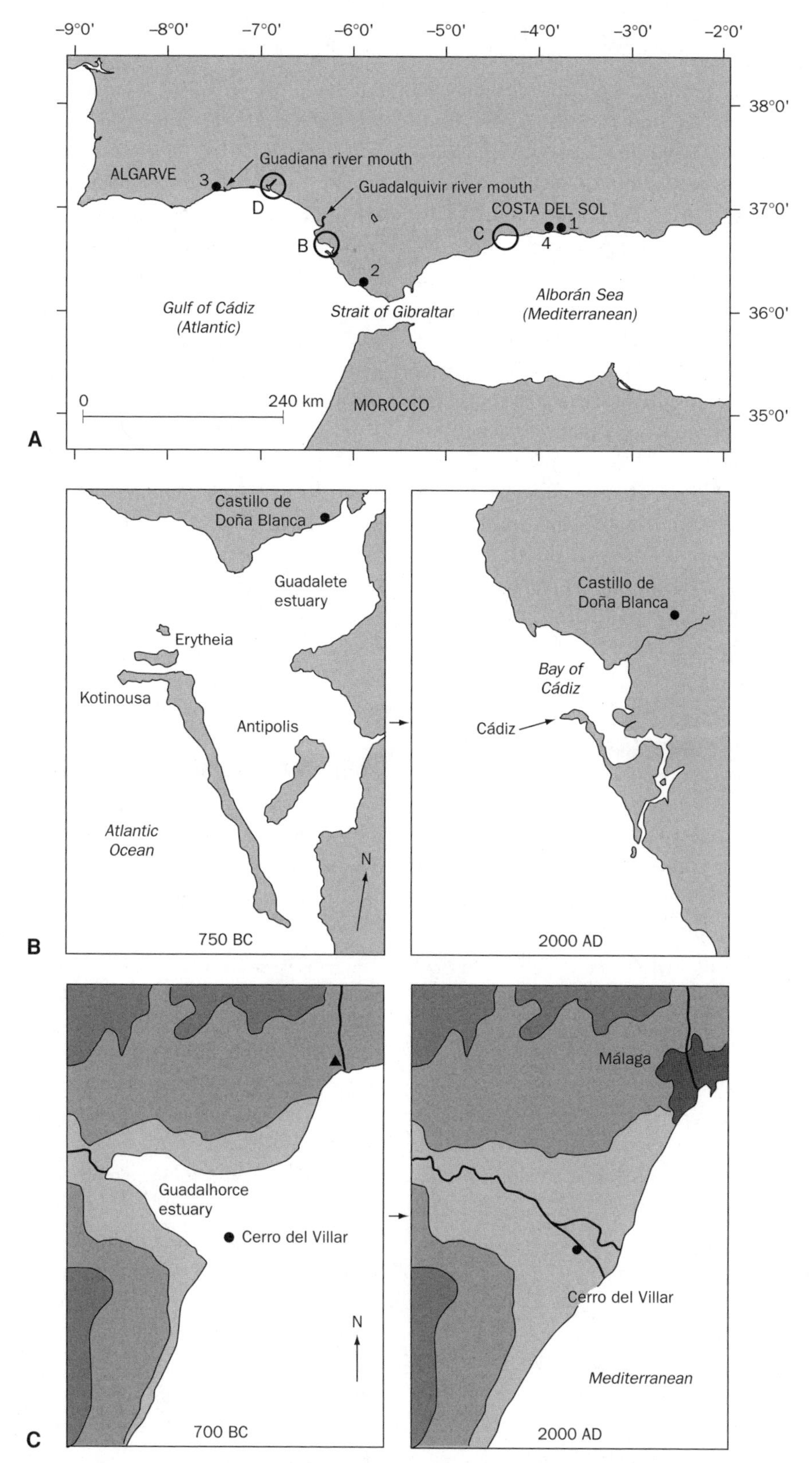

FIGURE 11.1. (A) The southern Iberian coastline with an indication of the major geographical features and sites mentioned in the text. Area *D* is where the sites shown in Figure 11.2 appear. 1, Cueva de Nerja; 2, Baelo Claudia; 3, Castro Marim; 4, Toscanos. The evolution of the coastline around the sites of Doña Blanca (B) and Cerro del Villar (C) is shown. Taken from Aubet et al. (1999) and Roselló and Morales (1994).

with new northeast-southwest littoral currents and the sediments flowing from the rivers formed spit bar systems (Barrera and del Olmo 1994; Goy et al. 1996; Zazo et al. 1994). These spit bar systems gradually closed the mouths of the estuaries, turning the sea into marshland, then into solid ground. Up until the sixteenth century AD this filling process was slow, from a few millimeters to a few centimeters per year. By about AD 1492, when the reconquest fight was concluded and the land given to new Christian landlords, the tempo of events accelerated. The onset of anticyclonic conditions in the area brought about by the Little Ice Age gave rise to strong coastal progradation associated with intensive deforestation due to the introduction of more intensive agricultural practices (Hoffmann 1988; Hoffmann and Schultz 1988; Zazo et al. 1994). Soils were rapidly denuded under a cycle of aridity combined with short episodes of torrential rains, and this produced a fantastic rate of sedimentary flow that by the end of the eighteenth century AD had created a completely new coastline from the Guadiana River to the easternmost corner of the Costa del Sol. This new coastline was sandy and essentially straight and, up until two centuries ago, was still unconsolidated land, difficult to live on and impossible to farm. For several centuries during the Modern Age, the only people occupying it were groups of nomadic fishermen (i.e., the *marengos*), that fished there during the summer and moved inland in the winter (Arbex 1990).

It is within the framework of this highly dynamic prehistoric coast that lasted until the Modern Age, and now features "senseless" coastal towns lying kilometers inland, that the issues of coastal settlement, fishing, and marine adaptations in southern Iberia must be addressed (Figure 11.1).

Additional factors that, on a more local scale, have altered the coast have to do with uplift and subsidence phenomena associated with the numerous faults that cross the littoral (Goy et al. 1996). These are due to the tectonic activity that the collision between the Euroasiatic and African plates, running through the middle of the Strait of Gibraltar on an east-west trajectory, provokes. Occasional side effects of this activity include earthquakes and associated tidal waves that have had restricted, yet often disastrous, consequences documented in the archaeological record (Arteaga and González 2004). It is highly unlikely that tectonic episodes played a major role in human settlement of the coast on a short-term scale, yet their long-term effects on the coastal topography have been important in determining both areas of local fish concentrations and of preferential human settlement.

From an oceanographic perspective, matters have probably been less labile than at the geomorphologic level. The Alborán Sea generates a microtidal coast (i.e., tidal ranges of ca. .5 m), and the absence of major rivers limits the supply of nutrients delivered to its waters. Passive methods of catching fishes based on the tides, such as artificial mounds, common along the Atlantic sector are of little use in the Sea of Alborán (Sáñez-Reguart 1988). The Gulf of Cádiz is a mesotidal coast (with tidal ranges of ca. 2 m) where important supplies of nutrients were once delivered by the major rivers (nowadays most are dammed). Such variables, as well as the absence of any major areas of uplift in the Sea of Alborán and its reliance on eddies to increase the entrance of Atlantic waters and mix surface with deep waters during the winter's anticyclonic conditions, probably have remained stable for millennia (Gil de Sola 1999). Another secular phenomenon is the circulation of deep, hypersaline Mediterranean water. Cascading down the submerged ridges of the Strait of Gibraltar into the Atlantic, this water flows deeper until it meets water of similar density then shifts northward along the western shores of Portugal (Margalef 1989). By doing so, this current interferes with the upwelling of nutrient-rich water in this area, diminishing the potential productivity of the surface waters of the Gulf of Cádiz except around Cape St. Vicente at the westernmost corner of the Algarve (Gil de Sola 1999; Margalef 1989). The potential richness of these waters, however, is not as high as some general works suggest (Consejería de Agricultura y Pesca 2004).

THE FISHES FROM SOUTHERN IBERIA

As expected for a region close to the subtropical belt, the fish fauna from southern Iberia is diverse, featuring some three hundred species, most of which are nowadays of economic interest (Gil de Sola 1999; Whitehead et al. 1984). This number is misleading in the sense that prior to the onset of the Industrial Revolution, the potential number of species available to the local fisherman was far lower, on the order of a hundred (Arbex 1990; Sáñez-Reguart 1993).

The oligotrophic (i.e., low-nutrient) condition dictating the restricted productivity of the Mediterranean combines with its subtropical character to create highly diversified communities, rich in terms of species, but with low levels of dominance and renewal rates (Gil de Sola 1999; Margalef 1989). Only the migratory animals, most of which enter the Mediterranean in spring and leave it during the summer and early autumn, constitute a resource that is concentrated enough in time and space to promote the development of specialized fishing gear and strategies. Most long-distance migrants approach the coast when entering the Mediterranean, so they can often be captured in inshore waters, yet require adequate gear, nets in particular, that is not only expensive to buy and maintain but also needs large numbers of coordinated people to operate. Other migratory species exhibit alternative patterns, either swimming up and down the water column (clupeids) or upstream in rivers (shads, sturgeons). Cooperative mass harvesting may have been difficult to carry out for most of Iberian prehistory, and it is not until a certain threshold of social complexity was reached that a systematic presence of migratory fishes is detected.

For resident fishes, low concentrations and low dominances must have been the rule except under the very specific circumstances of the productive waters at the mouth of the major rivers. From such a standpoint, it may not be a coincidence that most coastal Iron Age sites are concentrated in estuaries (Figures 11.1 and 11.2). The combination of low dominances and low productivities determined a restricted pressure to develop specialized fishing, more so when the low renewal rates of fish communities increased the risk of altering the populations beyond a no-return point, at least during a fisherman's lifetime (Arbex 1990; Gallant 1985; Horden and Purcell 2000; López-Linage and Arbex 1991). Only external agents, in terms of capital investment, would be likely to change the general state of affairs, and this is precisely what the archaeological record documents.

Evaluating the fish assemblages from a cultural perspective requires that one be able to categorize the various taxa in some meaningful way. One traditional system groups fishes into pelagic (i.e., open waters), demersal (living over the bottom), and benthonic (living in contact with or buried in the bottom) categories. Pelagic fishes are often divided into oceanic (i.e., offshore) and neritic or littoral populations, living in inshore waters. This characterization is somewhat misleading because most fishes move actively between zones, often on a circadian (i.e., daily) basis, and also because it does not address features that are also important when evaluating the nature of the exploitation. Among these, size dictates what kinds of fishes are likely to be taken by hooks or spears, and which, whether huge or tiny, will require alternative gear. It is also important to know which fishes in an assemblage are migratory and the extent and nature of those migrations, whether obligatory or facultative, within the water column, and so forth.

Most pelagic fishes have their early development under a stricter control of environmental parameters than is the case for inshore species (Gil de Sola 1999; Margalef 1989). Their planktonic stages are particularly vulnerable to unsuitable water temperatures. For such reasons, pelagic fishes often produce very different recruitments on an annual basis, so that years of glut often alternate with years of dearth. This

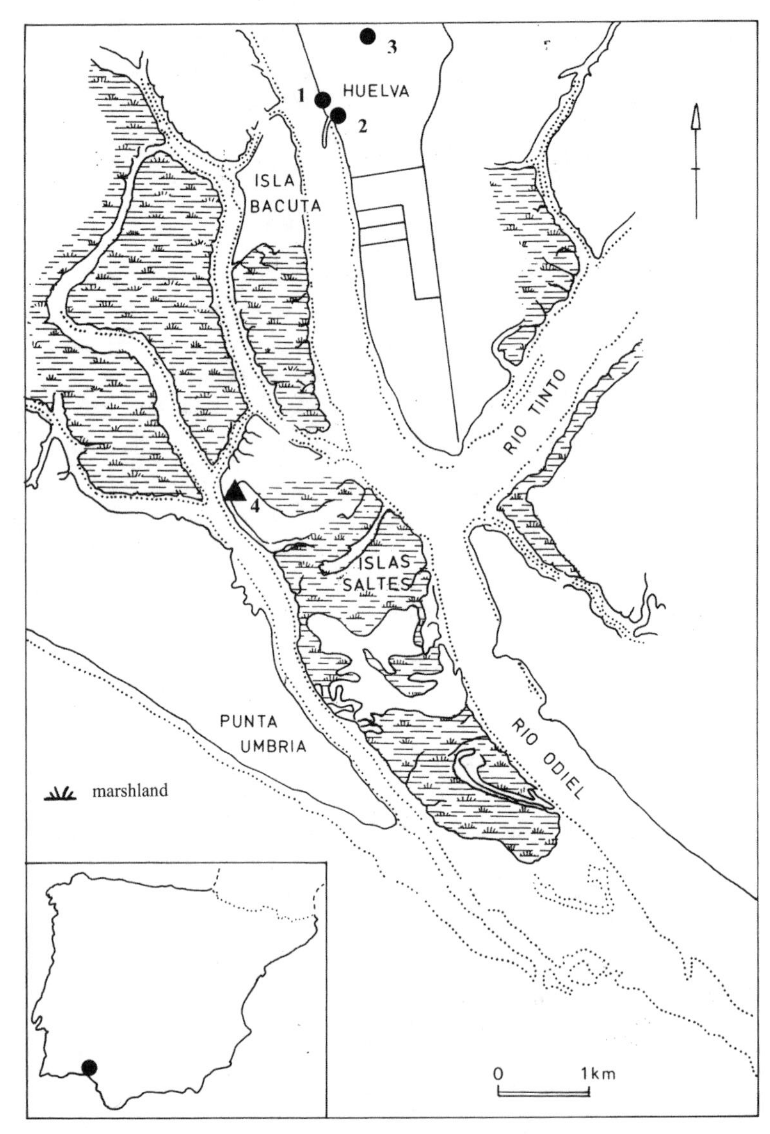

FIGURE 11.2. The estuary and marshlands of the rivers Tinto and Odiel around the city of Huelva with the location of sites mentioned in the text: 1, Puerto 19; 2, Puerto 29; 3, Cabezo de San Pedro; 4, Saltés. Taken, with modifications, from Ervynck and Lentaker (1999).

phenomenon has been documented since ancient times (Gallant 1985; Horden and Purcell 2000). Gulf of Cadiz fishing records from the eighteenth century AD, for instance, reveal highs and lows of devastating consequences for people relying exclusively on these resources (Burgos-Madroñero 2003). For this reason, even when set at far lower average yields, the exploitation of littoral and nonmigratory species was probably a safer bet at all times for the local fishermen of southern Iberia, especially when fishing involved small groups of people and buffering of losses through alternative routes was either restricted or nonexistent (Arbex 1990; Brandt 1984). As a corollary of such limitations, specialized fishing of pelagic taxa must have been a risky choice prior to the existence of exogenous buffers.

A brief introduction to some of the major taxa that appear in the tables that follow will help with their interpretation. The most relevant group of prehistoric and protohistoric littoral fishes in southern Iberia is represented by the seabreams (Sparidae), a family that nowadays includes commercially important genera such as *Sparus* (gilthead), *Pagellus* (pandora), *Pagrus* (porgy), and *Dentex,* but also many others of little economic interest due to their smaller sizes and less gregarious habits. Sparids range from brackish waters to the margins of the continental slope but are common only in inshore waters. They are accompanied there by some demersal sharks (the most common genera being *Mustelus* [smooth-hound], *Galeorhinus* [tope], and *Eugomphodus* [sand shark]) and a variety of bony fishes that includes the demersal red mullets *(Mullus)* and the pelagic gray mullets (Mugilidae, genera *Liza* and *Mugil*), but also the meagre *(Argyrosomus regius),* basses (genus *Dicentrarchus*), and groupers (e.g., *Epinephelus marginatus*). Demersal and littoral taxa can be fished in a variety of ways, but hooks work best for the largest specimens, often slightly below 1 m in the case of the largest sea breams but well beyond that size in some of the sharks and meagre. Nets, both mobile and fixed types, are more adequate for specimens below 40 cm fork length (FL; i.e., from the tip of the snout to the tip of the shortest ray of the caudal fin), particularly when these shoal in large concentrations at the mouth of estuaries or close to the beach.

Within the benthonic taxa, in addition to sharks such as the angel sharks *(Squatina)* and several species of rays (the main genera being *Raja, Dasyatis,* and *Myliobatis*), common taxa include the flatfishes of the family Soleidae (soles), gurnards (Triglidae), scorpionfishes *(Scorpaena),* and, in former times, the sturgeon *(Acipenser sturio).* Other benthonic species include the anguiliforms such as the moray *(Muraena helena)* and conger eels *(Conger conger),* the latter a voracious predator reaching over 80 kg. Most of these benthonic taxa were traditionally taken with hooks, though sturgeons were captured with nets, traps, and also spears when in shallow waters. Soles can be taken only through trawling since their tiny mouths preclude them from taking baited hooks. Nowadays, trawling is the most common method employed in the capture of demersal and benthonic fishes in this area, often with devastating consequences.

In 2001, the combined catches of the Andalusian fishing fleets reached 78,000 tons, fishes representing 82 percent and mollusks 14 percent (Junta de Andalucia 2007). These statistics include data from vessels operating in nonlocal waters, mostly off West Africa, though these constitute a minority of the Andalusian fleet compared to the situation in 2000, when Morocco closed its fishing grounds to fleets from the European Community. The main features of this fishery include the following:

1. For fishes, the main group is represented by the clupeiforms (40 percent), followed by two species of hakes (26 percent), from which the African *Merluccius senegalensis* accounted for 85 percent of the catch. Third were large pelagics, including sharks, tunas, and jacks (12 percent), followed by the blue whiting (*Micromesistius poutassou*) (4.5 percent), a small pelagic gadid. Benthonic taxa comprised 3 percent of the catch, most of the remaining fishes (i.e., 15 percent) being demersal species.
2. Among the mollusks, nearly half of the Andalusian catch consisted of the striped venus clam *(Chamelea gallina),* followed by the octopuses (*Octopus* sp. and *Eledone* sp.) (15 percent), rough cockle (*Acanthocardia tuberculata*) (13 percent), and cuttlefishes (*Sepia* sp.) (8 percent).

These data do not reflect the tremendous changes that the Andalusian fleet underwent during the past 30 years. The temporal trends compiled for the 1985–1999 period show a 33 percent drop in the catch (from 118,000 tons in 1985), along with the collapse of some fisheries and the appearance of new taxa (Junta de Andalucia 2007). The collapse is more evident in the demersal/benthonic fishes as these do not exhibit the dramatic fluctuations that characterize most pelagic species, yet all medium and large pelagic species suffered heavy losses (i.e., the Atlantic pomfret *Brama raii,* declined from 2,000 tons in 1985 to 27 in 2001). New fisheries have focused on underexploited mesopelagic stocks of fishes (the scabbard fish [*Lepidopus caudatus*] now constitutes 4 percent of the total fish catch) and squids.

ARCHAEOLOGICAL SAMPLES: BIASES OF ALL KINDS

The fish assemblages from the most important sites presented below (i.e., Cueva de Nerja, Castro Marim, Castillo de Doña Blanca, Saltés, and Cartuja, as well as the samples deriving from amphorae and salting vats from the fish factories) have been either sieved or floated. These procedures do not eliminate the existence of retrieval biases: at Cueva de Nerja, for example, the smallest mesh size used by one excavation team was 1.5 mm larger than that used by a second team (2.5 mm vs. 1.0 mm). As a result, the differences introduced in the faunal lists were dramatic in the case of perching birds (Morales et al. 1998: Figure 11.4) and may explain why only the 1-mm samples included the remains of clupeid fishes. It now seems clear that even 1 mm is too large a mesh for fishes, significant taxonomic differences having been recorded when using .6-mm mesh instead of the more conventional .8-mm mesh (Bødker-Enghoff 2005). Given that a majority of the collections presented below have not used such small mesh sizes, some having only been collected by hand, retrieval biases render most of the archaeofaunal interpretations that follow questionable (James 1997).

Given the small size of most of the samples, we considered it best to offer the lists of taxa in terms of number of identified specimens (NISP) instead of minimum number of individuals (MNI). The context of most of the assemblages—scattered refuse deposited over prolonged time intervals for the most part—and the similar skeletal profiles of the main taxa should help neutralize those interdependence biases often raised against the use of NISP in zooarchaeological studies (Reitz and Wing 1999).

The overview that follows aims at detecting consistencies in the lists of animal taxa. For that reason, as well as the small size of the samples, our use of statistics is limited to some basic ecological indexes, such as the diversity (H′) and equitability (V′) functions (Table 11.2; Reitz and Wing 1999). Additional parameters of interest for comparative purposes include the trophic level and the contributions of selected faunal groups to the total assemblage, but a detailed discussion of these is beyond the aims of this chapter (Table 11.2; Reitz 2004).

RESULTS

The Origins of Marine Fishing in Europe

The oldest evidences of marine adaptations in southern Iberia are presently restricted to caves and shelters of the eastern, mountainous sector of the Sea of Alborán. These date back to the Mousterian (ca. 200,000–40,000 BP) and are associated with the activities of Neanderthals. The data thus far published for localities such as Vanguard Cave, Gorham's Cave, and Devil's Tower Rockshelter in Gibraltar indicate that shellfish (especially mussels [*Mytilus galloprovincialis*]) contributed regularly to the Neanderthal diet, but not fishes (Barton 2000). In connection with this, one still needs to consider a collection of fish bones including vertebrae from the bluefin tuna *(Thunnus thynnus)*, recovered by Waechter at Gorham's (i.e., stratum E dated to ca. 28,000 BP). These remains are difficult to interpret since the deposits represent "a complex amalgamation . . . that accumulated through a combination of cultural and non-cultural processes" (Erlandson and Moss 2001:424), and their Aurignacian chronology needs to be reconciled with the fact that Neanderthals persisted in Gibraltar until that time (C. Finlayson, personal communication, 2006; Finlayson et al. 2006).

TABLE 11.2

Selected Parameters from the Most Relevant Fish Assemblages from Southern Iberia

SITE	PERIOD	DATE	NISP	H′	S	V′	TL	% PEL	% MIG	% BEN	COMMENTS
Nerja	Solutrean	ca 1.8 ky BP	1,083	2.2121	30	.6497	3.58	9.8	44	7.5	10% migratory if gadids excluded
Nerja	Magdalenian	1.4-1.1 ky BP	4,456	1.82	23	.5804	3.35	27.5	36	5	28% migratory if gadids excluded
Nerja	(epipaleolithic) Mesolithic	ca. 1.08 ky BP	3,645	4.55	27	.5723	3.55	7.6	61	8.5	8% migratory if gadids excluded
Nerja	Neolithic	0.7-0.6 ky BP	210	1.68	15	.6203	3.68	4.2	6	1	
Nerja	Calcolithic	ca. 0.5 ky BP	71	1.41	6	.7869	3.62	–	–	–	Strictly littoral (sea breams + groupers)
Castro Marim	Bronze	X BC?	37	1.85	11	.7715	3.44	30	2.7	11	
Castillo Doña Blanca	Early Iron	VII–VI BC	919	2.93	43	.7790	3.37	5	10	9	See diachronic evolution in text
Toscanos	Early Iron	VII–VI BC	235	2.17	19	.7369	3.54	7.5	4.7	–	
Cerro del Villar	Early Iron	VII–VI BC	(+/−)	–	22	–	–	32	32	–	Only lists of taxa provided
Puerto 29	Early Iron	VII BC	112	2.44	19	.8286	3.41	–	>1	3	
Puerto 10	Late Iron	VII–IV BC	80	2.36	16	.8512	3.47	15	17.5	6	
Cabezo San Pedro	Late Iron	VII–III BC	23	1.97	11	.8215	3.59	26	26	13	Unreliable NISP
Castro Marim	Late Iron	VI–III BC	643	2.61	38	.7175	3.60	31	31.5	8	See diachronic evolution in text
Baelo Claudia	Roman	II–I BC	1,058	.04	6	.0223	3.99	99.5	99.5	–	TL = 3.69 if tunas considered only to genus
Punta Camarinal	Roman	I BC	–	0	1	0	4	100	100		Presently under analysis (TL = 3.7)
Castro Marim	Roman	I AD	116	2.81	25	.8729	3.22	13	14.5	14	
Saltes	Medieval	XII–XIII AD	364	2.38	24	.7488	3.36	14.5	14.5	2.5	
Cartuja	"Modern"	XV–XVI AD	426	2.98	34	.8450	3.12	14	23.5	30	Highest diversity and percentage of benthonic taxa

NOTE: NISP, number of identified remains; H′, Shannon-Weaver index; S, richness; V′, equitability; TL, trophic level; Pel, pelagic; Mig, migratory; Ben, benthonic.

Fish remains associated with *Homo sapiens sapiens* have been retrieved only in the Cave of Nerja in the province of Málaga, despite indirect evidence of fishing at nearby places such as Cueva de la Pileta (rock paintings) and Hoyo de la Mina, where another vertebra from a bluefin tuna has been reported (Figure 11.1; Cortés et al. 1996:74). These assemblages date back to the Magdalenian period (14,000–11,000 BP), yet most are lacking in context, as was the case for the fishes at Gorham's. The same lack of context applies to the "pisciform" depictions in caves, highly questionable in interpretive terms as is the inertia to consider the notochordal canal of fish vertebrae as evidence of human manipulation for ornamental purposes (Cortés and Sanchidrián 1999; Cortés et al. 1996).

Before the Solutrean levels from one of the sectors at Cueva de Nerja (i.e., the Sala del Vestíbulo) were analyzed by us, no finds of fishes earlier than the Magdalenian had been reported in Europe that clearly indicated marine fishing, meaning saltwater taxa and not just those living also in fresh- or brackish waters, as is the case for salmon *(Salmo salar)* and sea trout (*S. trutta trutta;* Cleyet-Merle 1990; Menéndez et al. 1986). At Vestíbulo, three Solutrean levels feature a fairly large spectrum of fish taxa that are for the most part burned and occasionally exhibit cut marks (Table 11.3; Morales and Roselló 2003). There are three uncalibrated ^{14}C dates from the intermediate of these Solutrean strata: 18,420 ± 530 BP (UBAR-158), 17,940 ± 200 BP (UBAR-98), and 15,990 ± 260 BP (UBAR-157).

There are different issues that this fish assemblage reveals besides its coincidence with the last glacial maximum, including a richness and diversity among the highest documented for southern Iberia (Table 11.2). These suggest that the capabilities of Upper Paleolithic people to catch most of the species that would be common in later sites was not impaired by the presumably rudimentary technology at their disposal. Indeed, judging from this wide spectrum of taxa, such technology may not have been rudimentary at all, especially considering that small clupeids such as sardines are reported from the later Magdalenian layers (Table 11.3). At present, however, the only items that might be fishing implements are bone splinters sharpened at each end (with feathering in the middle of the shaft), which have been interpreted as straight hooks (Aura and Pérez 1998). Thus far, these have only been found in the Magdalenian and Mesolithic levels at Sala del Vestíbulo.

Another interesting feature of these Solutrean samples is their taxonomic composition. Along with the typical Mediterranean taxa, all of them incorporate a significant fraction of boreal fishes (i.e., 29 to 35 percent of the NISP). These fishes cannot be further quantified until they are properly studied, but they are all gadids that nowadays are restricted to the waters of the northern Atlantic and that either do not reach the Iberian Peninsula or do so only under exceptional circumstances (Whitehead et al. 1984). Perhaps the most relevant species of this boreal assemblage is the cod, but also revealing is the presence of saithe, haddock, ling, and even pollock (Morales and Roselló 2003). All the boreal fish thus far documented are from fairly large individuals (ranging from 55 to 90 cm FL), despite the fact that smaller fish taxa such as pandora (*Pagellus erythrinus;* 30–45 cm FL) are common in the assemblage (Roselló 2007c). The absence of smaller boreal individuals, if real, could indicate the presence of nonresidential populations entering the Mediterranean during the coldest pulses (winters?) of the Solutrean period. If this idea—currently being tested with stable isotope analyses to determine the latitudinal provenience of the boreal gadids—is confirmed it would suggest that the Sea of Alborán was a feeding rather than a breeding ground. It may also indicate higher productivities in this westernmost sector of the Mediterranean during the late Pleistocene and earlier part of the Holocene, as gadids were the most important group in the later Mesolithic layers at Sala del Vestíbulo (Table 11.3; Aura et al. 2001, 2002). This mixture of boreal and temperate-subtropical species indicates the existence of a

TABLE 11.3

Cueva de Nerja: List of Taxa from the Pleistocene Levels and Sectors

	SOLUTREAN				MAGDALENIAN			
	LX	*LIX*	*LVIII*		*SV*	SM_1	SM_2	*ST*
Acipenser sturio			2	2	31			1
Clupeidae						3		
Salmo sp.		1		1				
Conger conger			1	1				
Belonidae (*Belone belone*)					351	7		2
Gadidae		10	43	53	388	1		
Gadus morhua		8	20	28				
Melanogrammus aeglefinus		1	16	17				
Molva cf. *molva*		5	11	16				
Pollachius pollachius	2	45	193	240				
Pollachius virens		1	2	3				
Pollachius sp.	1		12	13				
Phycis sp.								1
Zeus faber			4	4				
Serranidae					10	35		
Epinephelus marginatus								
Dicentrarchus labrax			1	1				
Carangidae					361	101		
Trachurus cf. *trachurus*	2	24	58	84			4	17
Sciaenidae								
Sparidae	1	20	96	117	1,414	565	17	50
Boops boops			6	6				
Dentex dentex/D. gibbosus		1	1	2			1	1
Diplodus vulgaris								6
Diplodus cf. *sargus*		1	9	10				
Pagellus acarne		1		1				
Pagellus bogaraveo		1		1				
Pagellus erythrinus	2	102	224	328			16	73
Pagrus pagrus	1	2	1	4			1	5
Sparus aurata		6	25	31			5	4
Spondyliosoma cantharus		2	16	18				
Labridae					189	2		
Labrus cf. *bergylta*	1	3	73	77				
Labrus viridis			1	1				2
Sphyraenidae					1			
Mugilidae		1		1	30	358		
Chelon labrosus			1	1				
Liza aurata		1	1	2				
Scombridae					217	160		
Scomber cf. *scombrus*		2	17	19				3
Triglidae						9		
Trigla lucerna			1	1				
Scorpaenidae (*Scorpaena* sp.)					1			3
TOTAL	10	238	835	1,083	2,993	1,241	44	178

NOTE: Sectors: SV, Sala del Vestíbulo; SM, Sala de la Mina; ST, Sala de la Torca. Levels: X, IX, VIII, Solutrean. Data from SV and SM1 taken from Aura et al. (2002); data from the Solutrean samples taken from Morales and Roselló (2003); remaining Magdalenian samples (SM2, ST) taken from combining data from Boessneck and von den Driesch (1980) with data from Roselló et al. (1995).

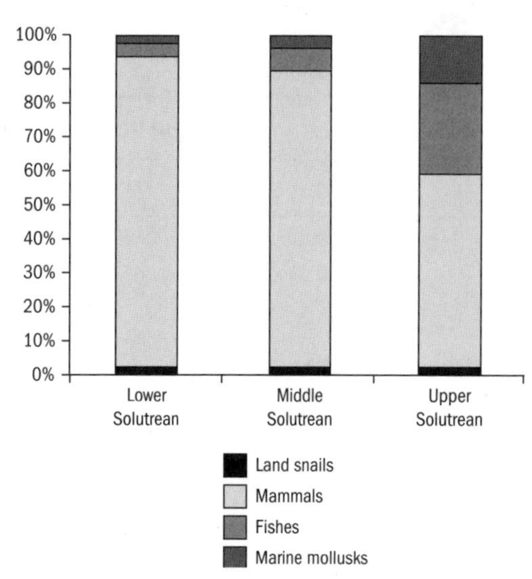

FIGURE 11.3. Cueva de Nerja: the contribution of selected faunal sectors, expressed as percentages of the NISP, in the three Solutrean levels from Sala del Vestíbulo. X, oldest; VIII, youngest. Taken from Morales and Roselló (2003).

former fish community for which, as is the case of the mammalian fauna from the Mammoth steppe, no modern analogues exist.

A third feature of interest is a gradual rise that apparently took place in the contribution of marine resources through the Solutrean (Figure 11.3). Fishes and marine mollusks are marginal contributors to the faunal assemblages in the older levels, but both increase dramatically in the most recent level. This rise does seem to reflect an intensification of marine resource use, for the abundance of both mammals (wild goat in particular) and terrestrial mollusks declines sharply in the last level, yet the shoreline lies at the same distance from the cave as in previous levels (J. Riquelme, unpublished data). On the other hand, given the greater distance of the Solutrean shoreline to the cave (about 4 km vs. 1 km at present), and the assumption that much of the evidence for the exploitation of marine resources must have remained on the beach, the approaching shoreline that the post-Solutrean marine transgression brought about may explain the progressive increase of marine resources documented for the Magdalenian and Mesolithic levels. If such was the case, the intensification of marine resource use through time at Nerja might only be a spurious trend caused by a geomorphological agent (Aura et al. 2001, 2002; Boessneck and von den Driesch 1980; Jordá 1986; Jordá et al. 2003; Morales et al. 1994a, 1998; Pellicer and Morales 1995; Rodrigo 1994; Roselló et al. 1995).

It should be stressed that Cueva de Nerja does not represent a fish midden in any sense. Having never been located on the shore, the overall composition of its faunal assemblages indicates the secondary role played by marine resources at all times (Aura et al. 2001, 2002; Jordá et al. 2003; Pellicer and Morales 1995). If the presence of the boreal gadids coincided with the winter, some other fishes such as the common and horse mackerels (i.e., *Scomber* sp. and *Trachurus* sp.) indicate that fishing also took place during the summer, supporting the idea of a year-round occupation. This hypothesis is reinforced by data being gathered on the size distribution of vertebrae from both pandora and the gadid fishes, whose unimodal behavior has been traditionally taken as proof of nonseasonal fishing (Barrett 1995; Roselló 2007c). Information from contemporaneous sites would help evaluate these hypotheses, but none are currently known from the area for the following millennia.

The Mesolithic, Neolithic, Copper, and Bronze Ages

For 80 percent of the Holocene and except for the Cave of Nerja, fish remains nearly disappear from southern Iberia archaeological sites, with few reported from the Mesolithic, Neolithic, Copper, and Bronze ages (Table 11.4; Amberger 1985; von den Driesch et al. 1985; Hain 1982; Lauk 1976; Morales 1985; Peters and von den Driesch 1990). During the Neolithic, the remains of some cartilaginous fishes appear in the record, but this is probably more a reflection of preservation issues than of changing fishing strategies (Roselló 2007b). The possibility exists that most coastal Early Holocene sites were flooded during the Flandrian transgression or buried under sediments deposited since

TABLE 11.4

Cueva de Nerja: List of Taxa from the Holocene Levels

PERIOD	M	M	M	M	N	N	C	C
Sector	*SV*	*SM1*	*SM2*	*ST*	*ST*	*SM*	*ST*	*SM*
Acipenseridae (*Acipenser sturio*)			1	4	1	1		
Clupeidae	164	11						
Belonidae (*Belone belone*)	6	3		2				
Gadidae	1,947							
Melanogrammus aeglefinus	(30)							
Pollachius pollachius	(+)		4	3		3		
Dicentrarchus sp.			3	1		2		
Epinephelus marginatus			12		22	60	18	4
Serranidae		9						
Carangidae	9	35						
Trachurus trachurus			3	12				
Seriola dumerilii						1		
Sciaenidae		2						
Sparidae	378	153	48	202	8	62	7	3
Dentex dentex + *D. gibbosus*				11	2	7	4	1
Diplodus sp. (+ *D. sargus* + *D. vulgaris*)			2	11				2
Pagellus erythrinus			7	112	4	3	3	
Pagrus pagrus	(+)		4	5	2	17	22	7
Pagrus auriga					1			
Sparus aurata	(+)		5	18	1	4		
Labridae	297	1		3				
Labrus bergylta	(+)							
Labrus merula			1	1				
Mugilidae		108						
Chelon labrosus						1		
Scombridae	30							
Scomber sp.	(+)		1	1		3		
Thunnus thynnus			1			3		
Euthynnus alleteratus						2		
Trygla sp.			1	8				
Scorpaenidae	2							
TOTAL	2,833	322	96	394	41	169	54	17

NOTE: Sectors: SV, Sala del Vestíbulo; SM, Sala de la Mina; ST, Sala de la Torca. Periods: M, Mesolithic; N, Neolithic; C, Copper Age. Data from SV and from SM1 taken from Aura et al. (2002); remaining samples taken from combining data from Boessneck and von den Driesch (1980) with data from Roselló et al. (1995). Parentheses refer to non-quantitative data provided either as approximations or else on a presence–absence basis.

the cycle of intense land erosion started. It is difficult to believe that such an ad hoc hypothesis would be the sole reason for such a state of affairs, however, fish and mollusk remains decline only in the post-Mesolithic levels at Cueva de Nerja despite the progressively approaching coastline (Table 11.4). This trend suggests that the importance of fishing and shellfish gathering probably did not decline within the animal procurement strategies until after the Mesolithic and then not in all places. At Nerja, for example, the Mesolithic from the

Vestibulo sector comes closest to qualifying as a shell midden and testifies to the importance of marine resources, as seems to be general for the coastal Mesolithic throughout Iberia (Aura et al. 2001, 2002; Clark 1987; Pellicer and Morales 1995; Rodrigo 1994; Roselló et al. 1995; Zilhao 1993). The ensuing decline of fishing and shellfish gathering at Nerja may couple with the onset of the Neolithic in the cave's sequence, as has been also evidenced for sites in the Atlantic façade of Portugal (Lentaker 1990–1991; Zilhao 1993). When the Mesolithic is not followed by a clear neolithisation process, such a decline is nowhere evident. In the case of southern Iberia, the four shell middens that bridge the transition from the Mesolithic to the Neolithic suggest stasis instead of change. In their analysis of these sites in the Algarve, Stiner et al. (2003:84) report "little change in the intensity of marine exploitation by humans" between these two periods, concluding that the area may have been occupied by locally persistent forager populations. Whether the phenomenon is valid at a regional scale remains to be seen, but perhaps a sharp decline in marine resource use in a sequence featuring a coastal Mesolithic followed by Neolithic chronologies might be taken as a signal that the shift from hunting and gathering to agriculture and animal husbandry had taken place.

Starting with the Copper Age (i.e., third millennium BC), and mostly along the fertile valley of the Guadalquivir River, progressively more complex societies emerged (Nocete 1994). The subsistence focus in all of them appears to have been on agriculture and husbandry. Under such circumstances, fishing probably became a marginal activity restricted to alternative populations stationed at the coast (the last of the southern Iberian hunter and gatherers?) and unlikely to leave many archaeological traces. That these foragers traded their products to urban centers is indicated by the fact that for both the Copper and Bronze ages most of the records of marine faunas derive from inland sites (Amberger 1985; von den Driesch et al. 1985; Hain 1982; Lauk 1976; Peters and von den Driesch 1990; Table 11.1).

The taxonomic composition of mollusk assemblages hints at faunal spectra dictated by a combination of availability and choice. The former is reflected by the abundance of mussels, limpets (*Patella* sp.), and turban (*Monodonta* sp.) at rocky shore sites, such as the Portuguese shell middens or Cueva de Nerja, and by the dominance of clams at sites along the Gulf of Cádiz shoreline (Table 11.1). Food preference, on the other hand, is suggested by a disproportionate abundance of the carpet shell *(Tapes decussatus)* at various places, including sandy shore sites (e.g., Papauvas, La Viña), but also those lying on cliffs (e.g., Cueva de Nerja) or even inland, as is the case for Valencina. The absence of clams that one would assume to be equally abundant in the same habitats where the carpet shell prospers (e.g., the common cockle, *Cerastroderma edule*), and the fact that the species has been traditionally the most costly of the Iberian clams due to the quality of its meat, suggests that there were more reasons to explain its abundance than mere availability. At the site of Papauvas, the abundance of the razor clam *(Solen marginatus)* may indicate new developments in the cropping of mollusks, as this species burrows deeper in the sediment and lives further offshore than any of the previously recorded taxa.

The Early Iron Age and the Onset of "Commercial" Fishing in Southern Iberia

The Early Iron Age, spanning the tenth to the fifth centuries BC, witnessed a progressive settlement of the coastline. Whether this development was a local phenomenon or somehow triggered by the arrival of trans-Mediterranean colonists is currently being debated, but it may have involved a combination of endogenous and exogenous agents (Aubet 1987; Aubet et al. 1999; Ruíz-Mata 1994). According to Alfredo Mederos (personal communication, 2006), for instance, recent excavations in the city of Huelva have detected a Phoenician presence, including amphorae with fish remains that date back to the start of the first millennium BC. During the Early Iron Age, the economies of local Bronze Age societies, based mostly on mining and cattle herding, were further developed by the Turdetanian kingdoms, many of whose urban centers (including the mythical Tartessos) were established at the mouths of some of the major rivers along the Gulf of Cádiz Coast.

A seventh-century sequence from the site of Castro Marim on the Guadiana River mouth shows a gradual rise in the contributions of fishes from the final moments of the Bronze Age until the third century BC (Levels II–V), indicating that marine fishing preceded the arrival of the Phoenicians by two or perhaps three centuries (Table 11.5). The comparatively high equitability values of this small assemblage indicate that fishing was not focused on any group and that the range of taxa exhibited a fair diversity (Table 11.2). Peculiar to this earliest level were the low frequencies of various species of sea breams and the presence of the John Dory *(Zeus faber)*, an open-water species that classical sources refer to as the favorite fish of the inhabitants of Gades (Cádiz; Roselló 2007a).

The following two Iron Age levels at Castro Marim correspond to moments where the Phoenician presence is well documented in southern Iberia, and both show a gradual increase in the contributions of the fish remains (NISP) accompanied by an increase in the number of taxa (S) and the diversity (H′), but not the equitability (V′) of the samples (Table 11.6). On a qualitative basis, Level III (sixth century BC) witnessed the appearance of many species of sea breams, whereas Level IV, tentatively dated between the sixth and fifth centuries BC (A. Arruda, personal communication, 2005), exhibits a spectacular increase of cartilaginous fishes (Table 11.5). Also revealing is the fact that tunas made their appearance in these early Iron Age levels with modest contributions. Such low frequencies of the bluefin tuna seem to be the rule on all of the Phoenician settlements whose fish faunas have been studied.

At Castillo de Doña Blanca (750–500 BC), tunas were found in low numbers and were exclusively represented by fin rays, a unique skeletal spectrum among the more than 40 documented taxa (Roselló and Morales 1994a, 1994c; Tables 11.2 and 11.7). Given that one of the main products of the later fish factories was *salsamenta* (i.e., salted meat), that in the case of tunas often involved the processing of big chunks of meat devoid of bones (Arevalo et al. 2004; García-Vargas 2001), the fin rays at Doña Blanca suggest that the tunas probably represented processed meat and not necessarily animals fished locally (Roselló and Morales 1994b). In the Mediterranean cities of Toscanos and Cerro del Villar, the very few remains of tunas dated to the latest stages of the occupation (fifth century BC), close to the moment when the Phoenicians were replaced by the Punic (Lepiksaar 1973b; Rodriguez 1999). The only undisputable evidence that the Phoenicians were involved in the commerce of tuna comes from a seventh-century BC Ramón 1.3.1.1-type amphora with "remains of tunas" (sic) found in the site of Acinipo (province of Málaga) some 30 km from the coast (Aguayo et al. 1987). Such evidence is much too restricted to consider the Phoenicians as the promoters of the shift in the local fishery toward large migratory species, although the data suggest that such a shift took place before the first of the fish factories appear in the area (Ruíz-Mata 1994).

As for mollusks, only two Phoenician assemblages have been analyzed from this period, each revealing a different situation (Table 11.1). At Toscanos, the abundance of the dog (*Glycymeris* sp.) and rough *(Acanthocardia tuberculata)* cockles indicates a preferential cropping in the restricted beaches of an otherwise rocky shore. At Castillo de Doña Blanca, on the other hand, the situation appears to be environmentally dictated. As the Bay of Cádiz became isolated from the Atlantic, suspension-feeding species (e.g., the razor clam and carpet shell) that live in clean water are replaced by the detritivorous peppery furrow *(Scrobicularia plana)*, which prospers in muddy waters (Moreno 1995b).

TABLE 11.5

Castro Marim: List of Taxa

PERIOD	BRONZE	IRON AGE			ROMAN	
Level	*II*	*III*	*IV*	*V*	*VI*	*Total*
Unidentified Chondrictian	1		7	9	3	20
Chondricthyan 1	1	5	15	26	16	63
Chondricthyan 2		1	17	29	10	57
Rajidae	2	1	7	4	10	24
Rhinobatidae			1	2		3
Myliobatis sp./Myliobatidae				1	1	2
Sphyrna sp.	1	1				2
Lamnidae			1		1	2
Cf. *Mustelus* sp.			6	4	1	11
Cf. *Galeorhinus galeus*			2		1	3
Squalus sp.			2	2	2	6
Squatina squatina		1	22	2	6	31
Eugomphodus taurus				1		1
Acipenser sturio		1	1	2	3	7
Cyprinus carpio		1				1
Zeus faber	10		2			12
Halobatrachus didactylus	2		4	5		11
Dicentrarchus labrax				1		1
Dicentrarchus punctatus				2		2
Epinephelus cf. *marginatus*		1			1	2
Seriola dumerilii			1			1
Trachurus trachurus					1	1
Plectorhincus mediterraneus					3	3
Dentex sp.		2			2	4
Dentex gibbosus	1	2		6	1	10
Pagellus erythrinus/P. pagrus		2	3	10	4	19
Pagellus erythrinus	1	1	8	9	3	22
Pagrus sp.		2	1	1		4
Pagrus auriga				1		1
Pagrus pagrus			1	3	8	12
Sparus aurata	5	8	28	27	17	85
Sparidae	2	2	2	16	3	25
Sciaenidae		1				1
Argyrosomus regius	11	28	80	8	5	132
Sciaena sp./*Umbrina* sp.			1			1
Scomber japonicus				3	1	4
Sarda sarda				1		1
Scombridae[a]		9	1	43	3	56
Thunnus cf. thynnus		3	2	136	10	151
Mugil sp./*Liza* sp.			2	1		3
TOTAL IDENTIFIED	37	72	217	355	116	797

NOTE: Taken from Roselló (2007a).

[a]Tuna-like fish remains that could not be identified down to genus level.

TABLE 11.6

Castro Marim: Selected Parameters from the Iron Age Levels

LEVEL	NISP	S	H′	V′	TL
III	72	19	2.18	.7403	3.39
IV	217	25	2.28	.7083	3.43
V	355	28	2.30	.6902	3.70

NOTE: NISP, number of identified remains; H′, Shannon-Weaver index; S, richness; V′, equitability; TL, trophic level. Data taken from Roselló (2007c).

The Late Iron Age and the Appearance of Fish Factories

At the end of the fifth century BC the Phoenicians were expelled from Iberia by the Punic. At the same time the first fish salting installations ("factories") appear around the Bay of Cádiz. We exclude here a series of funerary spaces found in the city of Gades, misinterpreted as "factories" (e.g., pools described as salting tanks, etc.; Muñoz and Frutos 2003; Muñoz et al. 1988; Ruíz-Mata, personal communication, 2006). The Las Redes site across the Bay of Cádiz, on the other hand, constitutes the best-preserved case of some two dozen fish salting installations featuring a room with a paved and sloping floor for preparing fish, a storage area for amphorae, and a room with fishing equipment such as fishhooks, line sinkers, and so forth. (Muñoz et al. 1988). Only the fish remains seem to be missing at Las Redes. This factory, as all the others in the area, reached the height of its activity between about 430–325 BC, eventually ceasing operation ca. 200 BC. No Punic fish factories outside the Bay of Cádiz have been found, and only in one of these, Puerto 6, have fish remains been reported (Roselló 1989). It is striking that these represent meagres *(Argyrosomus regius)*, a species never mentioned in classical texts. Besides meagres, tunas were the only fishes reported in Punic amphorae of southern Iberian origin, which have been retrieved as far east as the Forum at Corinth (Arévalo et al. 2004; Williams 1978, but see García-Vargas et al. 2007). Tradition has it that the Iberian amphorae from the Punic amphora building at Corinth included the remains of tunas and porgies (*Pagrus* sp.), but faunal analysis reveals that none of these remains could be attributed to any particular amphora (Williams 1978). Fish scales adhering to the inner face of some of the amphora fragments remain unidentified but do not correspond to those of any scombrid fish (Williams 1978)

Tunas found during the Punic period in southern Iberia came from Turdetanian settlements around the Guadiana River mouth and the city of Huelva (Figure 11.2). At Castro Marim, tunas (including the remains of large scombrids) constitute 30 percent of the identified NISP of Level V, dated between the fifth and fourth centuries BC, but probably reaching into the third century BC (A. Arruda, personal communication, 2005). None of the fish assemblages from the Turdetanian sites evidence the dominance of tunas so characteristic of later Roman factories, and all incorporate an appreciable amount of the littoral taxa documented since the Upper Palaeolithic (Table 11.5; Lepiksaar 1973a; Roselló 1989; Roselló and Morales 1990). In the Punic factories, implements associated with the huge inshore nets called *madragues* indicate that these were already deployed by at least the third century BC (Arévalo et al. 2004).

The mollusk assemblages, though restricted in number and size, likewise hint at important developments within this fishery. At Puerto 29, two previously unheard of taxa, the cuttlefish and the deeper water panobe clam *(Panopea glycymeris)*, reveal a focus on species that are more difficult to crop than any that preceded them. In addition, at most sites oysters *(Ostrea edulis)* appear in significant numbers for the first time. This increasing importance of the oyster marks the beginning of a trend that, as in the case of the bluefin tuna, culminated during the ensuing Roman period.

For over a century after the Punic sites ceased operation, a lack of evidence for commercial

TABLE 11.7

List of Taxa from Selected First Millennium BC Sites

TAXON	DBR	DBM	DBO	DBT	PUERTO 29	PUERTO 10	TOSCANOS	C. S. PEDRO
Scyliorhinus sp.	1	1	1	3	1			
Eugomphodus taurus						1		1
Lamna nasus								3
Isurus oxyrhinchus		1		1		1		
Galeorhinus galeus	3			3			5	3
Mustelus mustelus	24	1	31	56		9		
Sphyrna zygaena								1
Raja sp.	4	2	9	15	1	1		
Dasyatis pastinaca	1			1				
Myliobatis aquila	18	5	1	24	7	1		
Acipenser sturio	37		9	47	1	3		
Muraena helena	13	1		14		1		3
Conger conger	3			3				
Barbus sclateri	26	4	3	33			1	
Halobatrachus didactylus	2		3	5				
Zeus faber							1	
Serranidae			2	2				
Polyprion americanum							6	
Epinephelus marginatus	3	1		4		1	36	
Dicentrarchus labrax	2			2			1	
Dicentrarchus punctatus			4	4	1			
Trachurus trachurus	3		11	14				
Plectorhinchus mediterraneus	42	44	25	112	8	6		
Argyrosomus regius	22	3	33	58	12	12	7	5
Mullus barbatus	1			1				
Sparidae	43	13	84	140	27		25	
Dentex dentex	5	1		6	2	17	66	
Dentex gibbosus	14	3	19	36	4			
Diplodus sp.	7	1	4	12	1			
Diplodus annularis			1	1				
Diplodus sargus			2	2				
Diplodus vulgaris	12		11	23	2			
Lithognathus mormyrus	1		5	6				
Oblada melanura			2	2				
Pagellus acarne	1	6	4	11	3			
Pagellus erythrinus	2		31	33	10	2	14	1
Pagellus centrodontus							4	
Pagrus caeruleostictus	2	2	1	5		3		2
Pagrus pagrus	71	11	64	146	17	5	48	1
Pagrus auriga	5	3		8	3			
Sparus aurata			4	4	10	6	12	1
Mugilidae	11	2	6	19	1		1	
Chelon labrosus	1			1	2			
Liza aurata	1	3	2	6				
Liza ramada	12	1	5	18				
Mugil cephalus	3			3				
Sphyraena sphyraena							5	
Scomber sp.			6	6				

TABLE 11.7 (*continued*)

TAXON	DBR	DBM	DBO	DBT	PUERTO 29	PUERTO 10	TOSCANOS	C. S. PEDRO
Scomber japonicus			1	1				
Scomber scombrus							2	
Sarda sarda							1	
Thunnus cf. thynnus	9	16	1	26		11	3	6
Bothidae	1			1				
TOTAL	406	125	386	919	112	80	235	23

NOTE: DBO, Castillo de Doña Blanca (oldest stage); DBM, Castillo de Doña Blanca (intermediate stage); DBR, Castillo de Doña Blanca (youngest stage); DBT, Castillo de Doña Blanca (total). Data taken from Roselló and Morales (1994b) (Doña Blanca), Roselló (1989) (Puerto 10, Puerto 29), Lepiksaar (1973a) (Cabezo de San Pedro), and Lepiksaar (1973b) (Toscanos).

activities—including fishing—in southern Iberia has been interpreted as a response to extensive geopolitical transformations associated with Roman provincialization. At the end of the second century BC commercial fishing reemerges, first appearing at the factory of Baelo Claudia on the Atlantic entrance to the Strait of Gibraltar (Arévalo and Bernal 1999; Ponsich 1988; Ponsich and Taradell 1965).

The Roman Period and the Collapse of Southern Iberian Fish Factories

A review of the Roman fish factories is not the aim of this chapter, as the subject has been extensively dealt with by many authors (Curtis 1991; Edmonson 1987; Étienne and Mayet 2002; Lagóstena 1999, 2001; Ponsich 1988; Ponsich and Tarradell 1965). Archaeological data reveal that no fewer than 50 factories, with a combined processing capacity in excess of 1,500 m^3 of salting vats, were established along the southern Iberia littoral during the seven centuries of Roman occupation. Most are in extremely poor condition, and for many not even basic data (e.g., capacities, period of operation, etc.) are available. Data on fish remains from these factories are scarce and restricted to five sites: Quinta do Marim and Olhâo in the Algarve, Baelo Claudia in the Gulf of Cádiz, San Nicolás in the Bay of Gibraltar, and Cerro del Mar in the Sea of Alborán. Combined, these cover the entire Roman period, but their data are disconnected, and it is impossible to obtain a precise idea about the fish faunas through time. For this review, the database has been enlarged to incorporate fishes reported in amphorae from the Betica (i.e., Andalusia) and Lusitania (i.e., Portugal) provinces recently compiled by García-Vargas et al. (2007).

On the basis of fish remains, we detected three well-defined stages in the evolution of the Roman fish productions from southern Iberia:

1. From the end of the second century BC to the middle of the first century AD, a "tuna stage" is documented at factories such as Baelo Claudia and Cerro del Mar, as well as amphorae from both factories and shipwrecks (Baelo Claudia, Cerro del Mar, Chiessi, Titan?). This stage constitutes a prolongation of the situation documented during Punic times (Delussu and Wilkens 2000; von den Driesch 1980; Roselló et al. 2003; Tailliez 1961).
2. A "mackerel stage," from the middle of the first to the beginning of the third centuries AD, exhibits a striking overlap with the last moments of the previous stage. Occasional mackerel remains have been found in later contexts, but such finds are scarce and restricted to factories outside the area under review (Bruschi and Wilkens 1996; Delussu and Wilkens 2000; Desse and Desse-Berset 1993; Desse-Berset 1993; Desse-Berset and Desse 2000; von den Driesch 1980). Except for Cerro del Mar, the mackerel

stage is documented only in amphorae found throughout the western Mediterranean and western Europe, testifying to the existence of an intense commerce of this fish. No less than 23 isolated finds of mackerel have also been documented in different European sites dating from the end of the first to the second centuries AD (García-Vargas et al. 2007; Van Neer and Ervynck 2004). The only species systematically documented is the Spanish mackerel *(Scomber japonicus)*, not the nowadays far more common *S. scombrus* (Roselló 1989).

3. From the beginning of the third to the end of the sixth centuries AD, a "clupeid stage" is documented at the Quinta do Marim, Olhão, and San Nicolás factory sites, as well as three Dressel 7 amphorae from Cerro del Mar (Arévalo et al. 2004; Desse-Berset and Desse 2000; von den Driesch 1980). Although clupeid remains in the area date back to the Magdalenian levels from Nerja (Table 11.2) and are also present at the Phoenician site of Cerro del Villar (Rodriguez 1999), this stage marks the moment when sardines and anchovies became the dominant taxa in local assemblages. Most of these specimens are of less than 10 cm FL, indicating a preferential capture of fishes less than one year old (Desse-Berset and Desse 2000; Morales and Roselló 2007a). Apparently all of these tiny fishes went into the production of fish sauces, since no articulated skeletal remains have ever been found.

Of these three groups, only the mackerel appears to exclusively represent salted meat, since both tunas and sardines have been recorded as both *salsamenta* and fish sauces (Bruschi and Wilkens 1996; Delussu and Wilkens 2000; Desse-Berset and Desse 2000; Morales and Roselló 1989; Wheeler and Locker 1985). None of the amphorae carrying salted sardines have been shown to originate in southern Iberia, the Iberian clupeids being exclusively devoted to the production of the lowest-quality sauces. So far, no evidence of whole salted anchovies has been found. Adult anchovies live farther offshore and in deeper water than sardines, so their absence may reinforce the notion that fishing in most places was strictly littoral at that time. The onset of the clupeid stage coincides with the so-called third-century crisis that, together with multiple social developments, witnessed the collapse of most of the southern Iberia fish factories and the development of new ones along the Atlantic façade of Portugal and northern Africa (García-Vargas et al. 2007).

Undoubtedly, there are both profound social and economic reasons that contributed to this state of affairs, and we are not in a position to challenge them (Curtis 1991; Étienne and Mayet 2002; García-Vargas et al. 2007). Our concern remains with the disturbingly tight correspondence between the "taxonomy" and the "chronology" of the fish products. Whether such correlation could reflect phenomena of a more biological nature that have been thus far overlooked because of the priorities given to historical agents sensu stricto is open to debate. Noteworthy also is the progressively shorter time that each of these stages lasts. Tunas were the dominant fishes for some 600 years (one full millennium if Phoenicians are included), whereas mackerels dominated the market for barely two centuries, and clupeids less than that. Could such "acceleration" reflect some kind of disturbance in the exploited populations? A shift in fish products from mainly salted meat to mainly sauces? As usual, there are more questions than answers.

The reasons why fishing industries settled at some particular places resulted from a combination of factors of which fishes were just one element. Two equally decisive variables, the availability of salt and of freshwater, were tightly interconnected, but as of this writing have received little attention (Mederos and Escribano 2005). Any shift toward a drier climate in such a complex system of environmental and economic agents, for example, could have had drastic consequences by reducing the outflow of freshwater from the rivers. Such a change could seriously affect the fish industries by diminishing their ability to keep the facilities clean and not allowing them to provide water for the legions of workers operating their complex fishing gear and installations. From another perspective, changes in the migratory patterns of Atlantic bluefin tuna populations have been detected time and again, although the triggers have yet to be determined. One side effect of such shifts could be the displacement of a previously inshore migratory flux offshore (L. Gil de Sola, personal communication, 2005). For fishermen stationed at the beach, such shifts would have had the same devastating consequences as a de facto depletion of the stocks. If behavioral and geomorphological changes or additional ones combined at some point, they could have set the stage for a collapse of particular fisheries well before any "social crisis" struck, the social agents then merely acting as triggers. In fact, the main shift in the faunal replacement sequence, the "disappearance" of the tunas in the first century AD, does not coincide with any moment of social unrest, and no environmental triggers have ever been invoked in connection with it. The replacement of tunas by mackerels around the middle of the first century AD went unnoticed, possibly because the fishing industries managed to adapt to the change. Whether this was because they could keep on deploying the same gear and people for fishing mackerel remains to be seen. The second replacement episode, coinciding with the third-century AD crisis, was far more drastic since the fishing of clupeids is a low-gear strategy that requires far more modest vessels and tackle, and a profound fragmentation of the fishing crews. Since the shift apparently also implied the abandonment of salted meat production in favor of cheap fish sauces, the resulting enterprises had nothing to do with anything that had preceded them in the area. Were these the reasons for the collapse of many traditional factories? The available data do not allow one to decide whether these constraints or the social crisis were the cause or the consequence, but it seems probable that several agents must have been involved in such a profound rearrangement of the fisheries and fish factories.

In connection with these developments, the mollusk assemblages may offer additional clues despite their apparent homogeneity. It appears that the prevalence of oysters was a significant feature throughout the Roman period, since the three documented sites range from the first to the end of the sixth century AD (Table 11.1). This supply of oysters may have kept going for a totally different reason: oyster culture. At the fish factory of San Nicolás we have found evidence of huge oysters growing in some of the "salting tanks." These oysters are attached to the tiles on which the seed was originally placed, much in the way that historical sources described (Günther 1897). Perhaps such production hides the fact that most of the oyster beds in southern Iberia, restricted to some favorable estuarine areas, were probably depleted during Roman times. A far more convincing sign of resource depletion comes from the virtual absence of the highly esteemed carpet shell at the San Nicolás factory (2 out of 1,787 specimens) and by the first appearance of what has become the main mollusk resource of the present-day Andalusian market: the striped venus clam.

While these changes were taking place at the factories, it appears that local populations stationed at the coast (e.g., Castro Marim) kept practicing the fishing that seems to have been the rule for southern Iberia throughout prehistory (Tables 11.2 and 11.5). From such sparse data, one wonders whether the emphasis on excavating fish factories has seriously biased our perception of the overall picture during those times and whether these "fishing crises" were limited to the industrial entrepreneurs, not the average fishermen who kept on targeting the far less productive yet far more reliable littoral fishes.

Post-Roman Sites

All the archaeological evidence of fish faunas from medieval and modern times in southern

Iberia comes from two sites that offer only a glimpse of the species taken. One of these is Saltés, a Moorish town located in the estuary of the Tinto and Odiel rivers that dates to the Almohad period, from the late twelfth century to early thirteenth century AD (Ervynck and Lentaker 1999). At Saltés, except for the systematic presence of clupeids, nothing indicates major deviations from traditional fish assemblages.

The second site is La Cartuja, a Carthusian monastery located in the city of Seville about 70 km from the present-day mouth of the Guadalquivir River. At Cartuja, fish remains come from a period corresponding to the transition between the Middle and Modern ages (i.e., end of the fifteenth to the beginning of the sixteenth centuries AD; Roselló 1991–1992; Roselló et al. 1994). Cartuja is a completely different site than the others reported here since the context, a cesspit on the prior's quarters where members of the nobility stayed, has little to do with the refuse from the production sites on the coast. For such reason the estimated parameters of diversity, richness, and equitability are not comparable to those from sites located on the coast. Still, the data from Cartuja (Table 11.8) are important because they: (1) document long-distance commerce with the North Atlantic through the presence of processed cod (Roselló et al. 1994); (2) record an unprecedented contribution of strictly benthonic taxa (nearly a quarter of the sample), including some very large conger eels weighing over 50 kg (Roselló 1991–1992); and (3) reveal offshore fishing for the first time in southern Iberia—indicated by the presence of hake *(Merluccius merluccius)* and other midwater species (i.e., the pomfret, *Stromateus fiatola*).

All these changes suggest developments in fishing vessels and gear and may explain the diversity values of Cartuja being the highest recorded in the area (Table 11.2). Most of the information from this period is documentary and ambiguous, the authors not even agreeing on the role played by fishes in the diet (Arié 1974–1975; García-Sánchez 1996; Malpica 1984). The only feature of the fish record in Moorish Iberia that matches historical sources is the existence of long distance commerce with the peninsular interior where pelagic species such as sardine and hake, but not tuna, reportedly played a prominent role and have been documented as early as the tenth century AD in distant inland cities such as Mértola, Calatrava, and Madrid (Morales et al. 1994b; Roselló 1989, 1993; Roselló and Albertini 1997). The need to supply distant markets could be one reason behind the expansion of fisheries beyond the strictly littoral ecosystems.

Overfishing: Sizes and What Else?

Detecting signs that fishing stocks may have become depleted in the distant past is fraught with difficulties, especially when one deals with proxies of the live animals such as bones (Desse-Berset 1993; Jackson et al. 2001). With records as incomplete as the archaeological evidence for fisheries from southern Iberia, one faces further challenges in transforming isolated data into a connected sequence of events.

The fishing strategies inferred from the previous taxonomic lists do not seem to have been of an intensity capable of doing much harm to the exploited populations. In this way, both the high diversity of taxa (Tables 11.3–11.8) and low dominance levels of most assemblages (Table 11.2), together with the apparently marginal importance that fishing represented at most of the coastal sites, convey the impression of a fishing activity incapable of depleting stocks. This hypothesis is reinforced by the fact that the most productive, yet fragile, resources represented by the concentrations of migratory fishes do not seem to have been systematically exploited for most of the time. Only during the millennium from the fifth century BC to the sixth century AD is there reason to believe that this overall state of affairs changed drastically.

At the descriptive level where our research presently stands, the most one can do is to provide data that would allow for a more refined framing of the pertinent questions. The analysis of the size of fishes through time constitutes one such data set.

TABLE 11.8

List of Taxa from Baelo Claudia

TAXON	BAELO	SALTES	CARTUJA
Unidentified chondrichthyans (3 taxa)		1	24
Galeorhinus galeus		6	
Mustelus mustelus		24	
Isurus oxyrhinchus	1		
Eugomphodus taurus		1	
Raja sp.?		7	
Acipenser sturio			39
Alosa fallax + *A. alosa*		4	36
Sardina pilchardus		34	5
Muraena helena			1
Anguilla anguilla			1
Conger conger			30
Barbus cf. *sclateri*			1
Halobatrachus didactylus			23
Merluccius merluccius			29
Gadus morhua			4
Serranus sp.	1	1	
Epinehelus marginatus + *Epinephelus* sp.			4
Dicentrarchus labrax + *D. punctatus*			5
Pomatomus saltratix			11
Parapristipoma sp.			2
Argyrosomus regius		3	59
Chelon labrosus		1	1
Liza aurata + *L. ramada*			16
Mugilidae			12
Sparidae		117	21
Dentex dentex + *D. gibbosus*			7
Dentex macrophalmus + *D. maroccanus*			32
Pagellus acarne		24	8
Pagellus erythrinus	1	17	5
Pagellus bellotti			3
Pagellus sp.		25	
Pagellus sp./*Pagrus* sp.		43	
Pagrus pagrus		2	
Pagrus auriga		1	
Pagrus sp.		6	
Sparus aurata		1	2
Diplodus annularis		18	
Diplodus sp.	3	11	
Stromateus fiatola			1
Scombridae			7
Scomber scombrus + *S. japonicus*		13	
Thunnus thynnus	1,052	2	1
Triglidae			5
Eutrigla gurnardus			8
Aspitrigla cf. *cucculus*			19
Solea sp.		2	2
Pleuronectiformes			1
TOTAL	1,058	364	426

NOTE: Baelo Claudia data taken from Morales and Roselló (2007b); Saltés data taken from Ervynnk and Lentaker (1999); La Cartuja data taken from Roselló (1991–1992).

TABLE 11.9

Bluefin Tuna Size and Weight Estimations from Selected Southern Iberian Sites

SAMPLE	DATE	NISP	RANGE	COMMENTS
Toscanos	VII–VI BC	3	90–120	Two vertebrae from mixed layers; MNI = 1
Puerto 10	VII–IV BC	11	170–200?	Reported as well over 150 cm
Cabezo San Pedro	VII–III BC	4	ca. 200	6–7 years
Castro Marim	V–IV BC	+80	80/ca. 200	Mean FL: ca. 150 cm/63 kg (preliminary)
Baelo Claudia IV	II/I BC	23	117–250	Mean FL: 162 cm/ca. 80 kg
Baelo Claudia III	II/I BC	21	97–263	Mean FL: 164 cm /ca. 81 kg
Baelo Claudia II	II/I BC	18	127–262	Mean FL: 169 cm (180)/ca. 90 kg (120)
Punta Camarinal	I BC	64	138–155	Mean FL: 148 cm/ca. 60 kg
Cerro del Mar	I AD	151	+200	1 specimen ca. 50–60 cm; 150 over 6 years
San Nicolás	VI AD	3	80–150	75/85 cm (2 specimens), 150 cm (1 specimen)

NOTE: NISP, number of identified specimens; MNI, minimum number of individuals; FL, fork length, in centimeters. At Baelo Claudia II, size and weight mean values in parentheses reflect the inclusion of a giant tuna with a very rough FL estimation of 390 cm (See Figure 11.5). Data taken from: Lepiksaar (1973a) Cabezo San Pedro, (1973b) Toscanos; von den Dreisch (1980) Corro del Mar; Roselló and Marales (1990) Puerto 10; Roselló (2007a) Castro Marim; Morales and Roselló (2007a) San Nicholás, (2007b) Baelo Claudia and Punta Camarinal.

Table 11.9 summarizes the data on the sizes of tunas from the Iron Age and Roman periods. Some of the data have been obtained through direct comparisons with recent specimens of known size, while others (e.g., Castro Marim, Baelo Claudia, Punta Camarinal) have been obtained with the use of regression equations of vertebral sizes on FL (Aguado-Jiménez and García-García 2005; Crockford 1997). Despite their limitations, these data suggest some intriguing patterns:

1. The smallest tunas concentrate in the earliest and the latest samples, those from Castro Marim including a couple of animals of around 80 cm FL. If tunas then matured in their third year of life, as most do today (FL: 90–95 cm), these two animals could well have been immature (Magnusson 1994). An 80- to 90-cm FL tuna weighs anywhere from 10 to 20 kg and could, in theory, be taken by hook and line. Some data suggest that tuna growth rates could have been some 30 percent slower in Roman times, however, in which case a fish in its third year of life could have weighed less than 10 kg (Morales and Roselló 2007b).
2. Although tentative, a diffuse trend of increasing size is hinted at with time. In this way, the tunas from pre-Roman, meaning Turdetanian, assemblages never seem to exceed 2 m, whereas a significant fraction of those at the Roman factory of Baelo Claudia do. Although the means do not deviate as much as the ranges, these size differences suggest the deployment of progressively more efficient gear. This hypothesis is based on the very different bulk that the different sizes represent. A 150-cm tuna weighs around 65 kg, but a 200-cm specimen almost trebles that weight (165 kg), and tunas of 2.5 m reach well over 300 kg (Aguado-Jiménez and García-García 2005). All tunas, when captured en masse, will require the very powerful and sophisticated gear exemplified by the madragues, with implements such as double-fork needles and pebble and metal-

FIGURE 11.4. Depiction of the final stages in the lifting of the death chamber in a madrágue of the so-called buche-type, specifically designed for the capture of tuna. Taken from Ponsich (1988:37).

lic sinkers, documented since Punic times as pictorial depictions from the Roman period so eloquently evidence (Figure 11.4; Arévalo et al. 2004:116, 138).

3. At a more subtle level, this shift toward catching a greater proportion of larger tunas through time is also hinted at in a century-long sequence at Baelo Claudia (Figure 11.5). Again, this change is not so much evident in terms of average values as it is on the displacement of the ranges toward the right of the abscissae axes. In this context, the presence of a huge fish in the latest stage of the sequence (Phase II) is noteworthy. The FL estimate here is an extrapolation well beyond the limits of a formula obtained for far smaller animals. This is a giant tuna (1 ton?) whose capture, surely the cause of great celebration, reveals the efficiency of the fishing gear deployed.
4. The Punta Camarinal assemblage, from near the Baelo Claudia factory, is another instance where the leftovers from a short-lived episode of fishing, processing, and rapid burial of remains are documented (Morales and Roselló

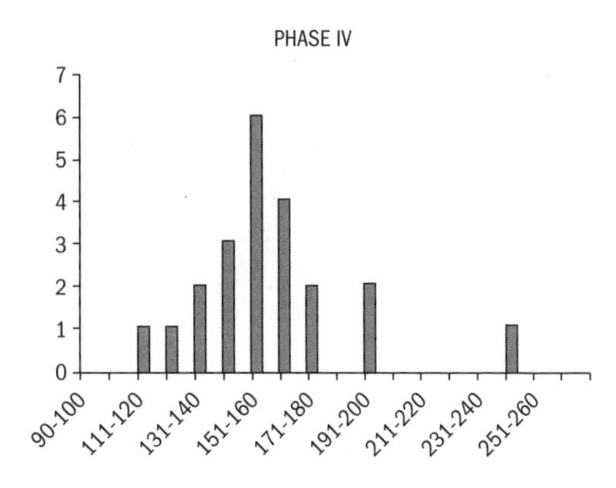

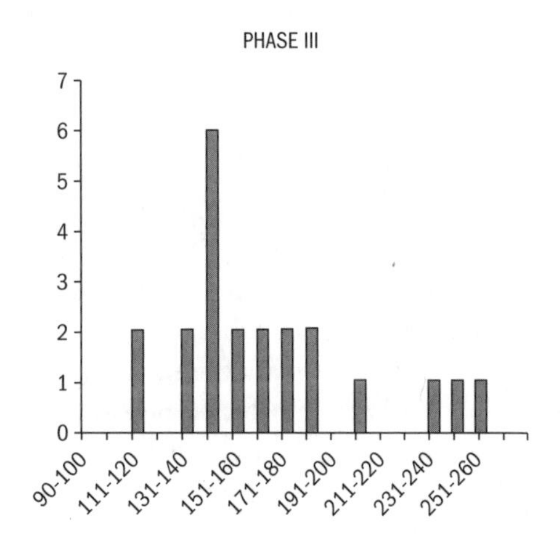

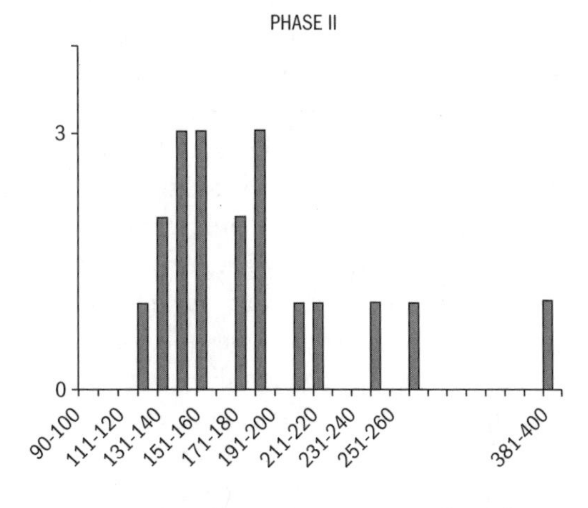

FIGURE 11.5. Distribution of tuna sizes, as inferred from vertebral width values, in the three successive stages of the second to first century BC levels of the fish factory number VI at Baelo Claudia. Data taken from Morales and Roselló (2007b).

2007b). As such, it is not strictly comparable to the Baelo Claudia samples, deposited over a prolonged period of time. Dated at the beginning of the first century AD, the smaller sizes at Punta Camarinal may provide evidence that the larger specimens were already becoming scarcer by that time around Baelo Claudia. The size estimates of larger tunas that von den Driesch provides for Cerro del Mar close to Málaga are too disconnected in time and space to link causally to the Baelo Claudia series (Table 11.9).

Does any of this evidence indicate overfishing? Hardly. For one thing, the data are too scarce, too disconnected, and, above all, too coarse to allow for any truly meaningful comments on the evolution of size of the catches. Size changes could, at best, be taken as an indication of improvements in the fishing gear, but unless one can reliably quantify the number of fish taken at different times, there is no way to make the move from diet to the economy.

In a previous study, Morales et al. (2001) attempted to explain changes in size through time as evidence of overfishing events. In sites throughout Iberia, including the area under consideration here, a recurrent pattern of smaller animals during the Neolithic and larger ones during the Bronze Age first, and the Iron Age later was evident for species such as the gilthead, pandora, porgy, dentex, and grouper at the level of unweighted data plots. That pattern was later proved to be statistically significant at the level of mean values (*t*-tests) once the variances of the samples were shown not to differ significantly among them (*f*-tests; Morales et al. 2001:48).

What these analyses hinted at, when data from all the Iberian sites were pooled, was a gradual rise in the mean size of those species from the Neolithic to the Iron Age. While this happened, the ranges became smaller during the Bronze Age yet expanded during the last period under consideration. Such recurrent patterns coupled with the biology of these littoral species were taken to indicate that the people were fishing very close to shore at the beginning, moved offshore during the Bronze Age to concentrate on the larger fishes from each species, only to return to the shallower waters during the Iron Age without abandoning the offshore zones (Morales et al. 2001).

There are some elements of interest in that model, yet the interpretations offered at the time were too simplistic. For one thing, the increase of the size ranges at the end of the sequence was taken to indicate a shift from fishing to supply local needs to fishing to supply markets far away, yet it now seems clear that long-distance commerce never involved any of these essentially littoral species. A more parsimonious explanation of pre- and protohistoric size trends through time is that fishing technology only gradually allowed people to move to offshore waters. In the beginning (pre-Neolithic and Neolithic), fishes could only be taken from the shore or close to it, where only the juveniles and smaller adults of the seabream species under consideration are frequent. Later on, as sailing technology improved, fishing targeted the larger specimens from these same species that live further away from the coast. However, the number of large specimens, even in a nonexploited population, is restricted. Consequently, there would eventually come a time when targeting only the largest specimens would not produce the minimum required yield. Fishers would then be expected to target fishes of any size, increasing the ranges retrieved in the archaeological deposits. Such nonspecific targeting on sizes does not necessarily imply that a resource had been overharvested, for such problems mainly arise when fishing pressure aims at the base of the population pyramid, not the top. With the current limitations of archaeological data on the history of southern Iberian fishing, it is not possible at this point to connect the various samples causally or to proceed beyond the loose link that the statistical analysis of ranges and means suggested.

By the same token, it would be impossible to leap another five centuries and try to explain the current status of the Andalusian fishery with archaeological data. For both fishes and mollusks, the common species today were rare or nonexistent in the past. In the case of mollusks, more than half of those recorded at sites (i.e., various oysters, limpets, panobe clam, turban, drill and dog cockle) do not appear in modern fishery statistics. Of those that do, almost a third (i.e., carpet shell, razor clam, scallop, mussel) are insignificant items. Much the same applies to the fishes, where one of the most paradigmatic taxa in archaeological samples, the sturgeon *(Acipenser sturio)*, had disappeared by the middle of the twentieth century (Junta de Andalucia 2007). Both fishes and mollusks today are dominated by offshore, often deep-water species that were unreachable until the development of motor boats and sophisticated gear. Today's fishery is not an inshore or subsistence activity, and its multiple problems developed for reasons that, for the most part, are not the ones that caused trouble in the past—although the specter of poor management lurks now as then.

CONCLUSIONS

The taxonomic composition of southern Iberian fish assemblages reveals that except from the millennium that runs from the fifth century BC to the sixth century AD, fishing never focused on specific taxa. The dominance of littoral fishes is probably a reflection of availability, as determined by selective fishing gear and fishing in inshore waters, but also the result of conscious decisions we can only speculate about at this point. Although no material evidence exists during the earlier parts of the sequence, fishing gear and practices were not apparently deployed that allowed for a mass harvesting of resources. Under such circumstances, the scarce number of fish assemblages, as well as the reduced NISPs in most of them, coupled with the scarcity of coastal settlements in the area prior to the first millennium BC, may hint at the secondary role played by fishing in the local economies. It is most doubtful that a marginal activity, not affecting the offshore stocks, could ever lead to the depletion of their counterparts from inshore waters. It remains to be seen whether the size range of the various species has been artificially displaced, the smallest specimens missed due to partial recovery biases.

From the standpoint of seasonality, the evidence conveys the impression of year-round fishing, a shift in tendency being detected only beginning with the fifth century BC. This shift may reflect a change in exploitation strategy, but it may also be the result of a bias toward the preferential excavation of fish factories. The presence of tuna remains in the earlier Iron Age and Phoenician settlements may reflect an improvement of fishing gear more than a shift toward seasonal fishing. Whether changes in fishing gear and targeted taxa were triggered by the colonists remains to be seen, but the involvement of the Phoenicians with the trade of tunas and the Punic fish factories implies that this might be so.

Biases notwithstanding, after the fifth millennium BC changes in fishing and mollusk gathering appear to have been radical. For one thing, the focused seasonal fishing that the data reveal was probably coincident with those times when agricultural activities were at their lowest (i.e., June–July). The change likewise implied the implementation of gear capable of harvesting fishes en masse and the existence of a market capable of absorbing any potential surpluses. After this time, one may entertain the idea of punctuated overfishing events having taken place at a local or regional scale in the area.

The putative size trends of tunas that the available data suggest, and the diachronic faunal replacement that took place within the Roman fish factories through time, however, can hardly qualify as evidence of overfishing. The data at the most reveal that fishing became more efficient with time and that there was a shift to fish smaller-sized pelagics for which no satisfactory explanation exists at this moment. Although this fishing concentrated on migratory species, it did not imply a move to offshore waters—madragues are operative only in shallow waters—so we presume that offshore stocks remained largely undisturbed. More intriguing from the standpoint of stock depletion happen to be the sudden disappearances of some emblematic taxa (e.g., bluefin tuna, Spanish mackerel, carpet shell, etc.) at specific moments during Roman times. At present, all these phenomena rest at a most descriptive level of characterization.

After the Roman period, fish remains become extremely rare in the archaeological record. At Saltés and La Cartuja, there is a return largely to the pre–Iron Age situation of nonfocused inshore fishing. Not until the transition from the Middle to the Modern Age do the first records of foreign fish imports and a move to offshore waters appear, along with evidence of an improved capacity for the capture of both large and small benthonic fishes. The singular nature of these phenomena, however, does not allow one to articulate them within a broader explanatory framework at this point.

A descriptive overview, such as the one attempted here, is bound to end up generating more questions than answers. If the former help promote a more systematic retrieval and study of the archaeological fish and mollusk remains from southern Iberia, we may harbor hope that in a not-too-distant future we will be able to move from the domain of speculation to that of scientific enquiry.

ACKNOWLEDGMENTS

We are very grateful to David Steadman, Steven James, and the editors of this volume for their constructive review of the manuscript and for pointing out data and references with which we were not acquainted. Alfredo Mederos (Madrid), Ana Arruda (Lisbon), Diego Ruiz Mata (Cádiz), Clive Finlayson (Gibraltar), Luis Gil de Sola (Málaga), and José Antonio Riquelme (Granada) are gratefully acknowledged for sharing unpublished data and commenting on preliminary drafts. Angel Luque (Madrid) reviewed the sections on mollusks. This contribution benefited from grants of the Patronato de la Cueva de Nerja (Málaga), the Proyecto MARCAS of the Fundaçao da Ciencia e Tecnologia (FCT, Portugal), and the Spanish National Research Council (CGL-2004-00891).

REFERENCES CITED

Aguado-Jiménez, F., and B. B. Garcia-Garcia
2005 Growth, Food Intake, and Feed Conversion Rates in Captive Atlantic Bluefin Tuna (*T. thynnus* Linnaeus, 1758) under Fattening Conditions. *Aquaculture Research* 36:610–614.

Aguayo, P., M. Carrilero, and G. Martínez
1987 La Presencia Fenicia y el Proceso de Aculturación de las Comunidades del Bronce Final de la Depresión de Ronda (Málaga). In *Atti II Congresso Internazionale Studi Fenici e Punici*, Vol. 2, pp. 559–571. Rome.

Alvar, J.
1981 *La Navegación Prerromana en la Península Ibérica: Colonizadores e Indígenas*. Editorial Universidad Complutense, Madrid.

Amberger, G.
1985 Tierknochenfunde vom Cerro Macareno/Sevilla. *Studien über frühe Tierknochenfunde von der Iberische Halbinsel* 9:76–105.

Arbex, J. C.
1990 *Pescadores Españoles*. Ministerio de Agricultura, Pesca y Alimentación, Madrid.

Arévalo, A., and D. Bernal
1999 La Factoría de Salazones de Baelo Claudia (Tarifa, Cádiz). Balance Historiográfico y Novedades en la Investigación. *Cuadernos de Prehistoria de la UAM* 25(2):75–29.

Arévalo, A., D. Bernal, and A. Torremocha
2004 *Garum y Salazones en el Círculo del Estrecho*. Fundación Municipal de la Cultura, Algeciras, Spain.

Arié, R.
1974–1975 Remarques sur l' Alimentation des Musulmans d'Espagne au cours du Bas Moyen Age. *Cuadernos de Estudios Medievales* 2–3:299–312.

Arteaga, C., and J. A. González
2004 Presencia de Materiales Marinos y Dunares Sobre un Alfar Romano en la Bahía de Algeciras (Cádiz, España). *Libro de Actas. VII Reunión Nacional de Geomorfología*: 393–400.

Arteaga, O., and G. Hoffman
1986 Investigaciones Geológicas y Arqueológicas Sobre los Cambios de la Línea Costera en el Litoral de la Andalucía Mediterránea. In *Anuario Arqueológico de Andalucía*, Vol. 2, pp. 194–195. Junta de Andalucía, Sevilla, Spain.

Aubet, M. E.
1987 *Tiro y Las Colonias Fenicias de Occidente*. Ediciones Bellaterra, Barcelona, Spain.

Aubet, M. E., P. Carmona, E. Curiá, A. Delgado, A. Fernández-Cantos, and M. Párraga (editors)
1999 *Cerro del Villar I: El Asentamiento Fenicio en la Desembocadura del Río Guadalhorce y su Interacción con el Hinterland.* Consejería de Cultura de la Junta de Andalucía, Sevilla, Spain.

Aura, J. E., and C. Pérez
1998 ¿Micropuntas Dobles o Anzuelos? Una Propuesta de Estudio a Partir de los Materiales de la Cueva de Nerja. In *Las Culturas del Pleistoceno Superior en Andalucía,* edited by J. L. Sanchidrián and M. D. Simón, pp. 339–348. Patronato de la Cueva de Nerja, Nerja, Spain.

Aura, J. E., J. F. Jordá, M. Pérez, and M. J. Rodrigo
2001 Sobre Dunas, Playas y Calas. Los Pescadores Prehistóricos de la Cueva de Nerja (Málaga) y su Expresión Arqueológica en el Tránsito Pleistoceno-Holoceno. *Archivo de Prehistoria Levantina* 24:9–39.
2002 The Far South: The Pleistocene-Holocene Transition in Nerja Cave (Andalucia, Spain). *Quaternary International* 93–94:19–30.

Barrera, F., and F. D. Del Olmo
1994 Paleogeografía Post-Flandriense del Litoral de Cádiz. Transformación Protohistórica del Paisaje de Doña Blanca. In *Castillo de Doña Blanca: Archaeoenvironmental Investigations in the Bay of Cadiz, Spain (750–500 BC),* edited by E. Roselló and A. Morales, pp.185–199. BAR International Series 593. British Archaeological Reports, Oxford.

Barrett, J.
1995 *Few Know an Earl in Fishing-Clothes: Fish Middens and the Economy of the Viking Age and Late Norse Earldoms of Orkney and Caithness, Northern Scotland.* Ph.D. dissertation, University of Glasgow, Glasgow.

Barton, N.
2000 Mousterian Hearths and Shellfish: Late Neanderthal Activities on Gibraltar. In *Neanderthals on the Edge,* edited by C. Stringer, N. Barton, and C. Finlayson, pp. 211–220. Oxbow Books, Oxford.

Bødker-Enghoff, I.
2005 Viking Age Freshwater Fishing at Viborg, Denmark. Results from an Interdisciplinary Research Project with Special Focus on Methods of Excavation. Paper presented at the 13th ICAZ Fish Remains Working Group. Basel, October 4–9.

Boessneck, J., and A. von den Driesch
1980 Tierknochenfunde aus vier Südspanischen Höhlen. *Studien über frühe Tierknochenfunde von der Iberische Halbinsel* 4:1–83.

Brandt, A. von
1984 *Fish Catching Methods of the World.* Revised edition. Fishing News Books, Farnham, England.

Bruschi, T., and B. Wilkens
1996 Conserves de Poisson a Partir de Quatre Anfores Romaines. *Archaeofauna* 5:165–169.

Burgos-Madroñero, M.
2003 *Hombres de Mar, Pesca y Embarcaciones en Andalucía. La Matrícula de Mar en los Siglos XVIII y XIX (1700–1850).* Junta de Andalucía, Consejería de Agricultura y Pesca, Sevilla, Spain.

Clark, G. E.
1987 From Mousterian to the Metal Ages: Long-Term Change in the Human Diet of Northern Spain. In *The Pleistocene Old World: Regional Perspectives,* edited by O. Soffer, pp. 293–316. Plenum, New York.

Cleyet-Merle, J.-J.
1990 *Le Prehistorie de la Pêche.* Edition Errance, Paris.

Consejeria De Agricultura y Pesca
2004 *Recursos Pesqueros del Golfo de Cádiz.* Junta de Andalucía, Sevilla, Spain.

Cortés, M., and J. L. Sanchidrián
1999 Dinámica Cultural del Pleistoceno Superior en la Costa de Málaga. *Cuaternario y Geomorfología* 13(1–2):63–77.

Cortés, M., V. E. Muñoz, J. L. Sanchidrián, and M. D. Simón (editors)
1996 *El Paleolítico en Andalucia. La Dinámica de los Grupos Predadores en la Prehistoria Andaluza. Ensayo de Síntesis.* Universidad de Córdoba, Córdoba, Spain.

Crockford, S.
1997 Archaeological Evidence of Large Northern bluefin tuna, *Thunnus thynnus,* in Coastal Waters of British Columbia and Northern Washington. *Fishery Bulletin* 95:11–24.

Curtis, R.
1991 *Garum and Salsamenta: Production and Commerce in Materia Medica.* E. J. Brill, Leiden, The Netherlands.

Delussu, F., and B. Wilkens
2000 Le Conserve di Pesce. Alcuni Dati da Contesti Italiani. *MEFRA* 112:17–18.

Desse, J., and N. Desse-Berset
1993 Pêche et Surpêche en Mediterranée: Le Temoignage de l'os. In *Explotation des Animaux à Travers le Temps,* edited by J. Desse and F. Audoin-Rouzeau, pp. 327–340. Editions APDCA, Juan-les-Pins, France.

Desse-Berset, N.
1993 Contenus d'Amphores et Surpêche: L'Exemple de Sud-Perduto (Bouches de Bonifacio). In *Explotation des Animaux à Travers le Temps,* edited by J. Desse and F. Audoin-Rouzeau, pp. 341–346. Editions APDCA, Juan-les-Pins, France.

Desse-Berset, N., and J. Desse
2000 Salsamenta, Garum et Autres Préparations de Poissons. *MEFRA* 112:73–97.
Edmonson, J. C.
1987 *Two Industries in Roman Lusitania: Mining and Garum Production.* BAR International Series 362. British Archaeological Reports, Oxford.
Erlandson, J. M., and M. L. Moss
2001 Shellfish Feeders, Carrion Eaters, and the Archaeology of Aquatic Adaptations. *American Antiquity* 66(3):413–432
Ervynck, A., and A. Lentaker
1999 The Archaeofauna of the Late Medieval, Islamic Harbour Town of Saltés (Huelva, Spain). *Archaeofauna* 8:141–157.
Étienne, R., and F. Mayet
2002 *Salaisons et Sauces de Poisson Hispaniques.* E. de Boccard, Paris.
Finlayson, C., F. Giles-Pacheco, J. Rodríguez-Vidal, D. A. Fa, J. M. Gutierrez-López, A. Santiago-Pérez, G. Finlayson, E. Allue, J. Baena-Preysler, I. Cáceres, J. S. Carrión, Y. Fernández-Jalvo, C. P. Gleed-Owen, F. J. Jimenez, P. López-Garcia, J. A. López-Sáez, J. A. Riquelme-Cantal, A. Sánchez-Marco, F. Giles-Guzmán, K. Brown, N. Fuentes, C. A. Valarino, A. Villalpando, C. B. Stringer, F. Martinez-Ruiz, and T. Sakamoto
2006 Late survival of Neanderthals at the Southernmost Extreme of Europe. *Nature* 443:850–853.
Gallant, T. W.
1985 *A Fisherman's Tale.* Miscellanea Graeca, Fasciculus 7. Belgian Archaelogical Mission in Greece, Ghent, Belgium.
Garcia-Sánchez, E.
1996 La Alimentación Popular Urbana en Al-Andalus. *Arqueología Medieval* 4:219–235.
Garcia-Vargas, E.
2001 Pesca, Sal y Salazones en las Ciudades Fenicio-Púnicas del Sur de Iberia. *Treballs del Museu Arqueologic D'Eivissa i Formentera* 47:9–66.
García-Vargas, E., E. Roselló, D. Bernal, and A. Morales
2007 Salazones y Salsas de Pescado en la Antigüedad. Un Pprimer Acercamiento a las Evidencias de Paleocontenidos y Depósitos Primarios en el Ámbito Euro-Mediterráneo. In *Las Factorias de Salazón de Traducta. Primeros Resultados de las Excavaciones Arqueológicas en la Calle San Nicolás,* edited by D. Bernal. Publicaciones Universidad de Cádiz, Cádiz, Spain, in press.
Gavala, J.
1959 *La Geología de la Costa y Bahía de Cádiz y el Poema "Ora Marítima" de Aviceno.* Instituto Geológico y Minero de España, Madrid.
Gil de Sola, L.
1999 *Ictiofauna Demersal del Mar de Alborán: Distribución, Abundancia y Espectro de Tamaños.* Ph.D. dissertation, Universidad Autónoma de Madrid, Madrid.
Goy, J. L., C. Zazo, C. Dabrio, J. Lario, F. Borja, F. Sierro, and J. A. Flores
1996 Global and Regional Factors Controlling Changes of Coastlines in Southern Iberia (Spain) during the Holocene. *Quaternary Science Reviews* 14:773–780.
Guerrero, V.
1993 *Navíos y Navegantes en las Rutas de Baleares Durante la Prehistoria.* El Tall, Mallorca, Spain.
Günther, R. T.
1897 The Oyster Culture of the Ancient Romans. *Journal of the Marine Biology Association* 4:360–365.
Hain, F. H.
1982 Kupferzeitliche Tierknochenfunde aus Valencina de la Concepción (Sevilla). *Studien über Frühe Tierknochenfunde von der Iberische Halbinsel* 8:1–184.
Hoffmann, G.
1988 Holozänstratigraphie und Küstenlinienverlagerung an der Andalusischen Mittelmeerküste. *Fachbereich Geowissenschaften der Universität Bremen 2.* Bremen. 1994 Hand Drillings in the Area of Castillo de Doña Blanca. In *Castillo de Doña Blanca: Archaeoenvironmental Investigations in the Bay of Cádiz, Spain (750–500 BC),* edited by E. Roselló and A. Morales, pp. 199–200. BAR International Series 593. British Archaeological Reports, Oxford.
Hoffmann, G., and H. Schultz
1988 Coastline Shifts and Holocene Stratigraphy on the Mediterranean Coast of Andalucía (Southeastern Spain). In *Archaeology of Coastal Changes,* edited by Avner Raban, pp. 53–70. BAR International Series 404. British Archaeological Reports, Oxford.
Horden, P., and N. Purcell
2000 *The Corrupting Sea: A Study of Mediterranean History.* Blackwell, Oxford.
Jackson, J. B. C., M. X. Kirby, W. H. Berger, K. A. Bjorndal, L. W. Botsford, B. J. Bourque, R. H. Bradbury, R. H. Cooke, J. Erlandson, J. A. Estes, T. P. Hughes, S. Kidwell, C. B. Lange, H. S. Lenihan, J. M. Pandolfi, C. H. Peterson, R. S. Steneck, M. J. Tegner, and R. R. Warner
2001 Historical Overfishing and the Recent Collapse of Coastal Ecosystems. *Science* 293: 629–638.
James, S. R.
1997 Methodological Issues Concerning Screen Size Recovery Rates and Their Effects on Archaeofaunal Interpretations. *Journal of Archaeological Science* 24:385–397

Jordá, J. F.
1986 La Fauna Malacológica de la Cueva de Nerja. In *La Prehistoria de la Cueva de Nerja*, edited by J. F. Jordá, pp. 145–177. Patronato de la Cueva de Nerja, Nerja, Spain.

Jordá, J. F., J. E. Aura, M. J. Rodrigo, M. Pérez, and E. Badal
2003 El Registro Paleobiológico Cuaternario del Yacimiento Arqueológico de la Cueva de Nerja (Málaga, España). *Boletín de la Real Sociedad Española de Historia Natural (Sección Geología)* 98(1–4):73–89.

Junta de Andalucia
2007 http://www.juntadeandalucia.es/agricultur aypesca/portal/opencms/portal/portada.jsp.

Ladero, M. A.
1993 Las Almadrabas de Andalucía (siglos XIII–XVI). *Boletín de la Real Academia de la Historia* 190: 345–354.

Lagóstena, L.
1999 *El Litoral Hispánico. Explotación de sus Recursos: Origen y Evolución Histórica hasta el Período Tardorromano.* Ph.D. dissertation, University of Cádiz, Cádiz, Spain.
2001 *La Producción de Salsas y Conservas de Pescado en la Hispania Romana (II a. C- IV d. C.).* Publications Universitat de Barcelona, Barcelona, Spain.

Lario, J., C. Zazo, L. Somoza, J. L. Goy, M. Hoyos, P. G. Silva, and F. J. Hernández-Molina
1993 Los Episodios Marinos Cuaternarios de la Costa de Málaga. *Revista de la Sociedad Geológica de España* 6(3–4):41–46.

Lauk, H. D.
1976 Tierknochenfunde aus Bronzezeitlichen Siedlungen bei Monachil und Purullena (Provinz Granada). *Studien über Frühe Tierknochenfunde von der Iberische Halbinsel* 6:1–111.

Lentaker, A.
1990–1991 *Archeozoölogisch Onderzoek van Laatprehistorische Vindplaatsen uit Portugal.* Ph.D. dissertation, University of Ghent, Ghent, Belgium.

Lepiksaar, J.
1973a Fischreste aus einer Tartessischen Siedlung in Huelva. *Studien über Frühe Ttierknochenfunde von der Iberische Halbinsel* 4:32–34.
1973b Fischknochenfunde aus der Phönizischen Faktorei von Toscanos. *Studien über Frühe Tierknochenfunde von der Iberische Halbinsel* 4:109–119.

López-Linage, J., and J. C. Arbex (editors)
1991 *Pesquerías Tradicionales y Conflictos Ecológicos. 1681–1794. Una Selección de Textos Pioneros.* Ministerio de Agricultura, Pesca y Alimentación, Madrid.

Magnusson, J. (editor)
1994 *An Assessment of Atlantic Bluefin Tuna.* National Academy Press, Washington, D.C.

Malpica, A.
1984 El Pescado en el Reino de Granada a Fines de la Edad Media: Especies y Nivel de Consumo. In *Manger et Boire au Moyen Age*, Vol. 2, edited by D. Menjot, pp. 103–117. Publications de la Faculté des Lettres et Sciences Humaines de Nice 27 (1ère série), France.

Margalef, R. (editor)
1989 *El Mediterráneo Occidental.* Ediciones Omega, Barcelona, Spain.

Mederos, A., and G. Escribano
2005 El Comercio de Sal, Salazones y *Garum* en el Litoral Atlántico Norteafricano Durante la Antigüedad. *Empúries* 55:209–224.

Menéndez de la Hoz, M., L. G. Strauss, and G. A. Clark
1986 The Ichthyology of La Riera Cave. In *La Riera Cave: Stone Age Hunter-Gatherer Adaptations in Northern Spain,* edited by L. G. Strauss and G. A. Clark, pp. 285–288. Anthropological Research Papers 36. Arizona State University, Tempe.

Morales, A.
1985 Análisis Faunístico del Yacimiento de Papauvas (Aljaraque, Huelva). In *Papauvas I. Campañas de 1976 a 1979*, edited by J. C. Martín de la Cruz, pp. 233–257. Excavaciones Arqueológicas en España 136. Ministerio de Cultura Madrid, Spain.

Morales, A., and E. Roselló
1989 Informe Sobre la Fauna Ictiológica Recuperada en una Ánfora Tardo-Romana del Tipus KEAY XXVI (Spatheion). In *Un Abocador del Segle V d.C. en el Forum Provincial de Tarraco,* edited by TED'A, pp. 324–328. Memories d'Excavació 2. Tarragona, Spain.
2003 Ictiofaunas Solutrenses de la Sala del Vestíbulo (Cueva de Nerja, Málaga). Interim Report Laboratorio de Arqueozoologia, Universidad Autonoma de Madrid 2003/12, Spain.
2007a Vertebrados de las Factorías de la Calle San Nicolás (Algeciras, Cádiz) y Reflexiones Zoológicas Sobre las Factorías Romanas de Salazones. In *Las Factorias de Salazón de Traducta. Primeros Resultados de las Excavaciones Arqueológicas en la Calle San Nicolás,* edited by D. Bernal. Publicaciones Universidad de Cádiz, Cádiz, Spain, in press.
2007b Los Atunes de Baelo Claudia y Punta Camarinal (Cádiz, Siglos I/II a.C.): Apuntes Preliminares. In *La Factoría de Salazones de Baelo Claudia*, edited by A. Arévalo and D. Bernal, pp. 489–498. Colección de Arqueología. Junta de Andalucía, Sevilla, Spain, in press.

Morales, A., E. Roselló, and J. M. Cañas
1994a Cueva de Nerja (prov. Málaga): A Close Look at a Twelve Thousand Year Ichthyofaunal Sequence from Southern Spain. In *Fish Exploitation in the Past,* edited by W. Van Neer, pp. 253–262. *Annales du Musée Royale de L'Afrique Centrale, Sciences Zoologiques* 274.

Morales, A., E. Roselló, A. Lentaker, and D. C. Morales
1994b Archaeozoological Research in Medieval Iberia: Fishing and Fish Trade on Almohad Sites. *Trabalhos de Antropologia e Etnologia* 34(1–2):453–475.

Morales, A., E. Roselló, and F. Hernández
1998 Late Upper Paleolithic Subsistence Strategies in Southern Iberia: Tardiglacial Faunas from Cueva de Nerja (Málaga, Spain). *European Journal of Archaeology* 1(1):9–50.

Morales, A., E. Roselló, and M. Ruiz
2001 A Renewable Fish Exploitation Model for Application on Archaeoichthyological Assemblages. In *Animals and Man in the* Past, edited by W. Prummel and H. Buitenhuis, pp. 44–52. ARC Publicaties 41. Groningen, The Netherlands.

Moreno, R.
1995a Catálogo de Malacofaunas de la Península Ibérica. *Archaeofauna* 4:143–272.
1995b Arqueomalacofaunas de la Península Ibérica: Un Ensayo de Síntesis. *Complutum* 6:353–382.

Muñoz A., and G. De Frutos
2003 El Comercio de las Salazones en Època Fenicio-Púnica en la Bahía de Cádiz. Estado Actual de las Investigaciones: Los Registros Arqueológicos. In *XVI Encuentros de Historia y Arqueología de San Fernando. Las Industrias Aalfareras y Conserveras Fenicio-Púnicas de la Bahía de Cádiz, San Fernando, 2000,* pp. 131–167. Publicaciones de la Universidad de Córdoba, Córdoba, Spain.

Muñoz, A., G. de Frutos, and N. Berriatua
1988 Contribución a los Orígenes y Difusión Comercial de la Industria Pesquera y Conservera Gaditana a Través de las Rrecientes Aportaciones de las Factorías de Salazones de la Bahía de Cádiz. In *I Congreso Internacional sobre el Estrecho de Gibraltar,* edited by E. Ripoll, pp. 487–508. UNED, Madrid.

Nocete, F.
1994 *La Formación del Estado en las Campiñas del Alto Guadalquivir (3000–1500 a.n.e.).* Servicio de Publicaciones Universidad de Granada, Granada.

Pellicer, M., and A. Morales
1995 *Fauna de la Cueva de Nerja I. Salas de la Mina y de la Torca, Campañas 1980–1982.* Patronato de la Cueva de Nerja, Nerja, Spain.

Peters, J., and A. von den Driesch
1990 Archäologische Untersuchung der Tierreste aus der Kupferzeitlichen Siedlung von Los Millares (Prov. Granada). *Studien über Frühe Tierknochenfunde von der Iberische Halbinsel* 12:51–110.

Ponsich, M.
1988 *Aceite de Oliva y Salazones de Pesca: Factores Geo-económicos de Bética y Tingitania.* Publicaciones Universidad Complutense de Madrid, Madrid.

Ponsich, M., and M. Tarradell
1965 *Garum et Industries Antiques de Salaison dans la Meditérranée Occidentale.* Bibliothèque de L'Ecole des Hautes Etudes Hispaniques XXXVI, Paris.

Reitz, E.
2004 "Fishing down the Food Web": A Case Study from St. Augustine, Florida, USA. *American Antiquity* 69:63–83.

Reitz, E., and E. Wing
1999 *Zooarchaeology.* Cambridge University Press, Cambridge.

Rodrigo, M. J.
1994 Remains of *Melanogrammus aeglefinus* (Linnaeus, 1758) in the Pleistocene-Holocene Passage in the Cave of Nerja, Málaga/Spain. *Offa* 51:348–351.

Rodriguez, C. G.
1999 La Pesca y la Explotación Marina y Fluvial. In *Cerro del Villar I: el Asentamiento Fenicio en la Desembocadura del Río Guadalhorce y su Interacción con el Hinterland,* edited by M. E. Aubet, P. Carmona, E. Curià, A. Delgado, A. Fernández-Cantos, and M. Párraga, pp. 320–324. Consejería de Cultura de la Junta de Andalucía, Sevilla, Spain.

Roselló, E.
1989 *Arqueoictiofaunas Ibéricas: Aproximación Metodológica y Bio-cultural.* Publicaciones Universidad Autónoma de Madrid, Madrid.
1991–1992 Preliminary Comments on a Late Medieval Fish Assemblage from a Spanish Monastery. *Journal of Human Ecology* 2:371–390.
1993 Análisis de los Peces Recuperados en Mértola. *Arqueología Medieval* 2:277–284.
2007a La Ictiofauna de Castro Marim (Algarve, Portugal). Interim Report Laboratorio de Arqueozoologia, Universidad Autonoma de Madrid 2007/6, Spain.
2007b Fish Remains from Pico Ramos with an Overview of Iberian Neolithic and Chalcolithic Ichthyofaunas. In *Humans on the Basque Coast During the 6th and 5th Millennia BC. The Case of Pico Ramos,* edited by L. Zapata. BAR International Series. British Archaeological Reports, Oxford, in press.

2007c Los Peces de la Sala del Vestíbulo (Cueva de Nerja, Málaga), in preparation.

Roselló, E., and D. Albertini

1997 Análisis Ictioarqueológico de la Plaza de Oriente (Madrid). Interim Report Laboratorio de Arqueozoologia, Universidad Autonoma de Madrid 1997/2, Spain.

Roselló, E., and A. Morales

1990 La Ictiofauna del Yacimiento Tartésico de la Calle del Puerto no.10 (Huelva): Consideraciones Generales. *Espacio, Tiempo y Forma Serie I* 3:291–298.

1994b The Fishes. In *Castillo de Doña Blanca: Archaeoenvironmental Investigations in the Bay of Cadiz, Spain (750–500 BC)*, edited by E. Roselló and A. Morales, pp. 91–142. BAR International Series 593. British Archaeological Reports, Oxford.

1994c Castillo de Doña Blanca: Patterns of Abundance in the Ichthyocenoses of a Phoenician Site from the Iberian Peninsula. *Archaeofauna* 3:131–143.

Roselló, E., and A. Morales (editors)

1994a *Castillo de Doña Blanca: Archaeoenvironmental Investigations in the Bay of Cadiz, Spain (750–500 BC)*. BAR International Series 593. British Archaeological Reports, Oxford.

Roselló, E., A. Morales, and D. C. Morales

1994 La Cartuja/Spain: Anthropogenic Ichthyocenosis of Culinary Nature in a Paleocultural Context. *Offa* 51:323–331.

Roselló, E., A. Morales, and J. M. Cañas

1995 Estudio Ictioarqueológico de la Cueva de Nerja (prov. Málaga): Resultados de las Campañas de 1980 y 1982. In *Fauna de la Cueva de Nerja I. Salas de la Mina y de la Torca, Campañas 1980–1982*, edited by M. Pellicer and A. Morales, pp. 163–217. Patronato de la Cueva de Nerja, Nerja, Spain.

Roselló, E., A. Morales, D. Bernal, and A. Arévalo

2003 Salsas de Pescado de la Factoría Romana de Baelo Claudia (Cádiz, España). In *Presencia de la Arqueoictiología en México*, edited by A. F. Guzmán, O. Polaco and F. Aguilar, pp. 152–157. CONACULTA-INAH, México, D.F.

Ruiz-Mata, D.

1994 El Poblado Fenicio del Castillo de Doña Blanca. Introducción al Yacimiento. In *Castillo de Doña Blanca: Archaeoenvironmental Investigations in the Bay of Cadiz, Spain (750–500 BC)*, edited by E. Roselló and A. Morales, pp. 1–19. BAR International Series 593. British Archaeological Reports, Oxford.

Sánez-Reguart, A.

1988 [1791–1795] *Diccionario Histórico de los Artes de la Pesca Nacional*. Imprenta de la Viuda de Don Joaquín Ibarra, Madrid. 1988 facsimile edition. Ministerio de Agricultura, Pesca y Alimentación, Madrid.

1993 [1796?] *Colección de Producciones de los Mares de España. Tomo I.* Unpublished Manuscript, 1993 facsimile edition. Ministerio de Agricultura, Pesca y Alimentación, Madrid.

Stiner, M. C., N. F. Bicho, J. Lindly, and R. Ferring

2003 Mesolithic to Neolithic Transitions: New Results from Shell-Middens in the Western Algarve, Portugal. *Antiquity* 77(295):75–86

Tailliez, P. H.

1961 Travaux de l'Été 1958 sur l'Épave du Titan a l'Ile du Levant (Toulon). In *Actes du IIe Congres International d'Archeologie sous-marine*, pp. 175–178. Bordiguera.

Van Neer, W., and A. Ervynck

2004 Remains of Traded Fish in Archaeological Sites: Indicators of Status, or Bulk Food? In *Behaviour Behind Bones: The Zooarchaeology of Ritual, Religion, Status and Identity*, edited by S. J. O'Day, W. Van Neer, and A. Ervynck, pp. 203–214. Oxbow Books, Oxford.

von den Driesch, A.

1980 Osteoarchäologische Auswertung von Garumresten des Cerro del Mar (Málaga). *Madrider Mitteilungen* 21:151–154.

von den Driesch, J. Boessneck, M. Kokabi, and J. Schäffer

1985 Tierknochenfunde aus der Bronzezeitlichen Höhensiedlung Fuente Álamo, Provinz Almería. *Studien über frühe Tierknochenfunde von der Iberische Halbinsel* 9:1–75.

Wheeler, A., and A. Locker

1985 The Estimation of Size in Sardines *(Sardina pilchardus)* from Amphorae in a Wreck at Randello, Sicily. *Journal of Archaeological Science* 12:97–100.

Whitehead, P. J. P., M.-L. Bauchot, J.-C. Hureau, J. Nielsen, and E. Tortonese

1984 *Fishes of the North-Eastern Atlantic and the Mediterranean*. UNESCO, Paris.

Williams, C. K.

1978 Corinth, 1977: Forum Southwest. *Hesperia* 47:1–39.

Zazo, C., J.-L. Goy, L. Somoza, C.-J. Dabrio, G. Belluomini, S. Improta, J. Lario, T. Bardají, and P.-G. Silva

1994 Holocene Sequence of Sea-level Fluctuations in Relation to Climatic Trends in the Atlantic-Mediterranean Linkage Coast. *Journal of Coastal Research* 10:933–945.

Zilhao, J.

1993 The Spread of Agro-pastoral Economies across Mediterranean Europe. A View from the Far West. *Journal of Mediterranean Archaeology* 6:5–63.

12

Human Impact on Precolonial West Coast Marine Environments of South Africa

Antonieta Jerardino, George M. Branch, and Rene Navarro

WITH VERY EXTENSIVE, DIVERSE, and productive coastlines, it is no surprise that South Africa offers a superb opportunity to understand how marine ecosystems function, and the effects of people on these environments. This prospect is heightened by the realization of the tremendous time depth of human occupation, including the first emergence of modern humans in Africa and, by default, in the world (Erlandson 2001; Marean et al. 2007). This endeavor is multidisciplinary by necessity: marine ecology and archaeology go hand in hand, along with other closely related specialities such as oceanography, geology, and palaeoenvironmental studies. While ecological studies can provide an understanding of the ecology of present species, their vulnerability to harvesting, and the way they are influenced by physical factors, archaeology and palaeoenvironmental studies offer a window into the past where such ecological relationships and physical variables can be seen changing through time. The result of this multidisciplinary dialogue not only feeds much needed academic debate but also brings new and important dimensions into marine conservation policies and fisheries management. Because of this, the relevance of archaeology to modern society extends beyond its perceived status as a highly specialized and rather esoteric field (Erlandson and Rick, this volume). As demonstrated below, we also propose that this multidisciplinary dialogue benefits the way coastal archaeological sites are studied, by bringing a more exhaustive and systematic approach to answering questions, and by highlighting their main trait as archives of both human and environmental history. Seen in this perspective, the protection, conservation, and management of coastal archaeological sites must be a priority as promoted by the current South African heritage legislation (National Heritage Resources Act, No. 25, 1999).

In this chapter, we explore the relationship that people established with the marine environment on the South African West Coast over several millennia. Such effects have frequently been demonstrated for terrestrial environments worldwide (Redman 1999), but few convincing cases have been made for marine systems. A broad account of the long-term precolonial human settlement of the West Coast of South Africa sets the background for a more focused analysis of selected archaeological sequences in the Lamberts Bay area. An integrated approach

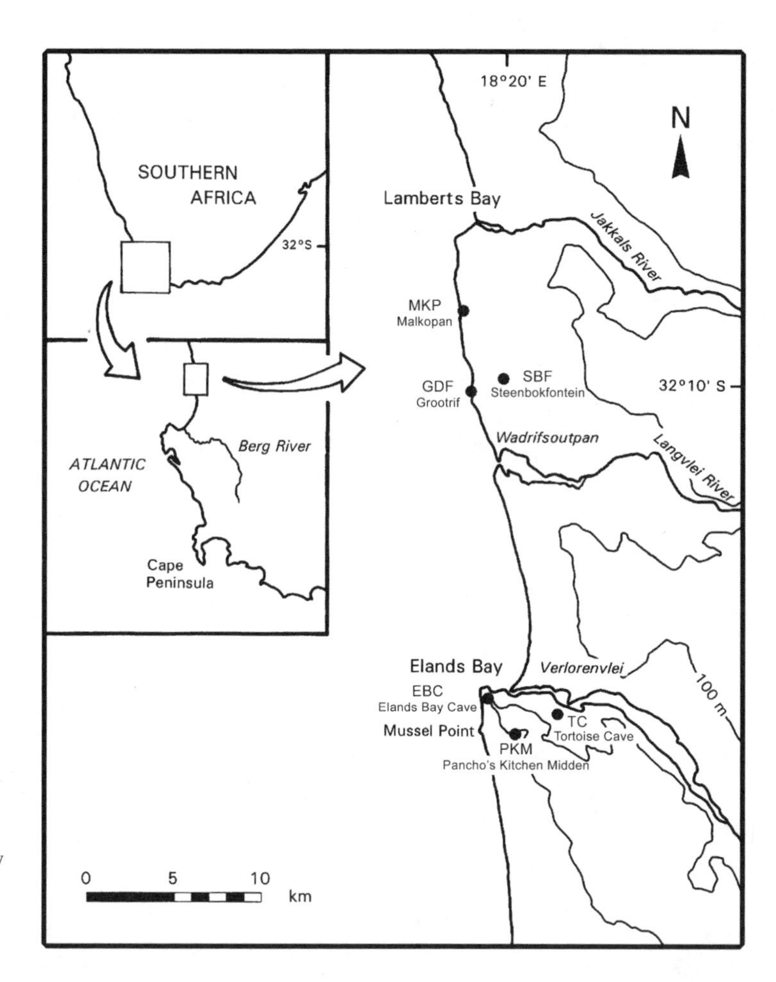

FIGURE 12.1. Map of the study area showing location of sites and places mentioned in the text.

that combines sampling of faunal remains from archaeological sites, current knowledge of the ecology of the species exploited in the past, and detailed palaeoenvironmental reconstructions is employed as a powerful tool for unraveling long-term variability in marine ecosystems and their responses to human intervention. A case is made for human harvesting having been responsible for the local depletion of black mussels and limpets stocks, as reflected by shrinking mean sizes at a time when human population densities were peaking and when reliance on marine resources was greatest. A concurrent decline in rock lobster size for the Lamberts Bay area (Figure 12.1) is interpreted differently, given the likely resilience of this species to subsistence harvesting and its susceptibility to environmental changes. This adds a note of caution in terms of the methodology employed to answer questions regarding human impacts on marine harvested species: not all changes in populations are necessarily related to the same causes. Consequently, all possible lines of evidence and hypotheses deserve serious and equal attention when tackling these types of questions.

PREHISTORIC EXPLOITATION OF MARINE INVERTEBRATES ON THE WEST COAST OF SOUTH AFRICA

Some of the earliest evidence for the exploitation of marine resources in the world is found in South Africa, mostly from deeply stratified sequences along the South Coast dating to between 165,000 and 120,000 years BP (Jacobs et al. 2003a, 2003b; Marean et al. 2007; Thackeray 1988; Vogel 2001). A few Middle Stone

Age (MSA) West Coast sites confirm that subsistence harvesting is very ancient, including reasonably well developed shell middens near the towns of Yzerfontein and Saldanha dated to about 70,000 years ago (Klein et al. 2004; Volman 1978). It is possible that marine resources were collected before this time too, but many of these sites are likely to have been washed out due to a high sea-level stand before the last interglacial period around 127,000 years ago (Klein et al. 2004; Parkington 2003).

The number of marine species present at these early MSA sites is similar to that observed in younger and nearby Later Stone Age (LSA) sites. Only a few species such as rock lobster *(Jasus lalandii)* and fish are either not present in MSA sites or their presence there cannot be attributed to human agency with certainty. The species diversity, however, appears to differ between MSA and LSA sites (Klein et al. 2004). Whether this reflects behavioral or environmental fluctuations is yet to be established, and resolution will require more systematic excavation of MSA sites and analysis of larger shellfish samples. Another pattern emerging from West Coast MSA sites is the significantly larger mean sizes of at least three limpet species when compared to those of LSA sites (Halkett et al. 2003; Parkington 2003; Steele and Klein 2005). No changes in the mean sizes of black mussels *(Choromytilus meridionalis)*, however, are observed in previous comparisons of MSA and LSA sites. Less-intense harvesting of limpets during the MSA due to lower human population levels at that time, and greater ecological resilience of black mussels have been invoked to explain these patterns (Klein et al. 2004; Parkington 2003; Steele and Klein 2005).

Although this scenario seems plausible, it is also important to consider other explanations, such as marked differences in the residential permanence of human groups with similar population levels during both MSA and LSA times. In other words, low exploitation pressure on MSA limpets could well have resulted from shorter visits by people to the coast, while relatively longer visits characterized the latter period. Moreover, there are many factors other than exploitation that influence the size composition of marine invertebrates, including differences or changes in environmental conditions and the intrinsic properties of the species themselves. As shown in this chapter, zonation, exposure to wave action, aquatic productivity, and turbidity can profoundly affect the size composition of limpets and black mussels. It also remains to be explained why black mussel sizes remained relatively constant through time despite the facts that (1) the black mussel is the most abundant species in MSA and LSA sites, and (2) rocky-shore mussels are susceptible to overexploitation as reflected by reductions in their mean sizes caused by modern subsistence harvesting (Lasiak 1992). Data on early shellfish gathering behavior in southern Africa is still preliminary. It needs to be recognized that the number, and at times the size, of available MSA shellfish samples is small when compared to those analyzed from LSA sites (Jerardino 1993, 1997, 2007; Jerardino and Yates 1997; Tonner 2005). Thus, more variables need to be assessed when explaining the observed reductions in limpet sizes between MSA and LSA sites, and more; as well, larger samples of MSA shellfish are also necessary to support any interpretation.

A considerable time gap separates coastal occupation of MSA sites and those of earliest LSA age, probably because most, if not all, former coastal sites dating to this gap were drowned by rising sea levels, and only a few of these may have survived on the Atlantic continental shelf (Miller 1990; van Andel 1989). Given the focused research efforts in the Elands Bay and Lamberts Bay areas for the last 30 years (Figure 12.1), it is not surprising that the earliest LSA radiocarbon-dated evidence for the exploitation of marine invertebrates along the West Coast of South Africa comes from several caves and shelters at these adjacent locations. Shell midden horizons appear for the first time in Elands Bay Cave and Tortoise Cave around 12,000 and 7700 BP (all dates presented in this chapter are uncalibrated), respectively (Parkington 1981;

Robey 1987), and well-developed shell lenses dating to ca. 8400 BP have been excavated from Steenbokfontein Cave (Jerardino and Swanepoel 1999) (Figure 12.1). Bedrock at this latter site is far below these deposits, and further excavations are likely to uncover older shell lenses. Early Holocene evidence for shellfish collection might also be present in Spoegrivier Cave, located about 200 km north of Lamberts Bay, although no radiocarbon dates are yet available for these basal deposits (Webley 2002). Once sea level transgressed to within a few kilometers of these sites during the Early Holocene, a wide range of marine invertebrates (and several vertebrate species) were exploited, including limpets, mussels, whelks, winkles, chitons, and rock lobsters. All of these species would have been collected from rocky shores, although sandy beaches appear to be exploited to a small but significant extent, which was not repeated during subsequent occupations throughout the Late Holocene (A.J., personal observation).

After about 5000 BP, coastal shell middens abound along the West Coast, reflecting a full range of settlement and subsistence choices and new cultural and economic developments (Buchanan 1988; Buchanan et al. 1978; Conard et al. 1999; Jerardino 1996, 2007; Jerardino and Yates 1996, 1997; Parkington et al. 1988; Robertshaw 1978, 1979; Sadr and Smith 1992; Sealy et al. 2004; Smith et al. 1991). With few exceptions, coastal archaeological sites accumulated in close proximity to rocky shores from where much of the marine subsistence was obtained (Jerardino 2003). Although sketchy, differences in the above social and economic variables are apparent among subregions along the West Coast (e.g., Namaqualand, Elands Bay and Lamberts Bay area, Vredenburg Peninsula). These differences seem to have been dictated by environmental variability (rainfall, geomorphology, and availability of resources), changing human population levels, and cultural contact situations. Much work remains to be done in each of these subregions for this emerging and variable picture of coastal harvesting along the West Coast of South Africa to have a more solid empirical foundation. Some of these subregions have been sampled and studied more extensively than others, with the Lamberts Bay and Elands Bay area (hereafter referred to as "the study area") offering more numerous and chronologically deeper sequences. This chapter presents a case study from this particular subregion and attempts to ascertain the nature of the interaction between humans and their marine environment during the Late Holocene. Multiple lines of evidence are used to achieve this goal, including data on population levels and dietary mix, palaeoenvironmental reconstructions, the biology and ecology of rocky shore marine invertebrates, current understanding of the effect of modern harvesting on similar species along other South African shorelines, and statistical analyses on metrical observations of body size for four species of marine invertebrates recovered from several archaeological sites.

SETTLEMENT AND SUBSISTENCE PATTERNS AT LAMBERTS AND ELANDS BAY

The chronological record of the study area shows that relatively few sites were occupied between 8,000 and 4,500 years ago (Figure 12.2). Between 4,500 and 3,000 years ago, human settlement occurred in cave sites and shelters, with volumes of deposit ranging mostly between 1 and 10 m^3. Human impact on marine resources was probably negligible at this time. Around 3500 BP, new sites were being occupied for the first time. Rates of deposition started to increase along with the size of floor areas at sites that had been previously occupied. Longer residential permanence is inferred from higher densities of artifacts that, according to ethnographic observations, were manufactured and/or lost or discarded during longer visits to sites (Jerardino 1995a, 1996; Jerardino and Yates 1996). Volumes of deposits range between 10 and 100 m^3 per site around this time. These are the first signs of population increase in the study area. Subsequently, between 3000 and

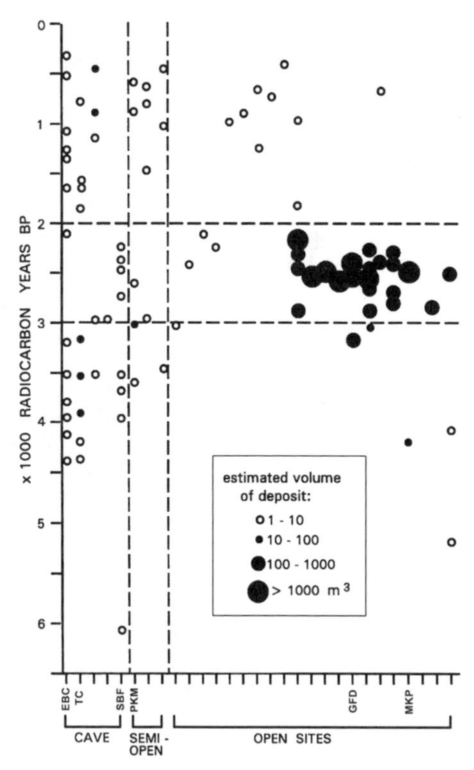

FIGURE 12.2. Intensity of site occupation in the study area (uncalibrated radiocarbon years BP).

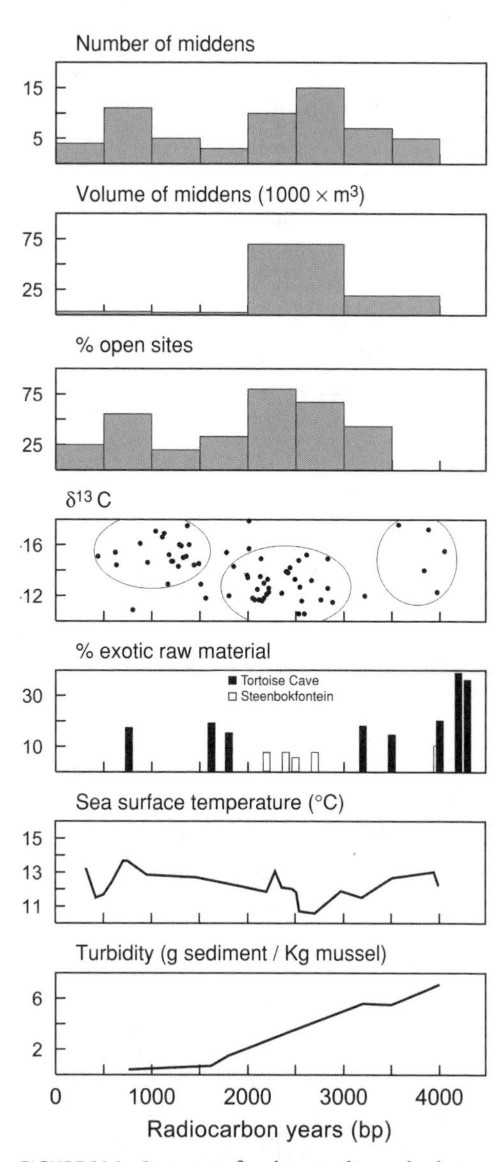

FIGURE 12.3. Summary of settlement, diet, and palaeoenvironmental changes in the study area (uncalibrated radiocarbon years BP).

2000 BP, settlement focused on very large open shell middens situated immediately behind rocky shorelines (Figures 12.2 and 12.3). Tons of marine shell and low densities of artifacts and terrestrial fauna characterize these enormous shell middens (Jerardino and Yates 1997). The term "megamiddens" was coined for these site types when first observed 30 years ago (Parkington 1976). Volumes of deposit range from a hundred to several thousand cubic meters per site. A range of activities was performed at these sites, including the processing of vast quantities of shellfish, possibly for drying and later consumption, stone knapping, and processing of terrestrial animal foods (Henshilwood et al. 1994; Jerardino and Yates 1997). Dating of caves and shelters has also shown that only two such sites were occupied during the megamidden period, namely Steenbokfontein Cave and Pancho's Kitchen Midden (Figure 12.1). Either residential permanence continued to be as extended as before or even longer periods of habitation were involved (Jerardino 1996, 1998; Jerardino and Yates 1996). After 2000 BP, population densities declined sharply as reflected by the overall lower number of sites and reduced volumes of deposits. Settlement also returned to caves and shelters over the last 2,000 years (Figures 12.2 and 12.3), and short visits seem to have characterized this period (Jerardino 1996, 1998).

Changes in hunter-gatherer mobility can be inferred from fluctuations in the frequency of lithic raw materials from Steenbokfontein and Tortoise caves, both with sequences that extend to the Early Holocene. Although never dominant, exotic raw materials such as silcrete and indurated shale (known locally as "hornfels") were more commonly used between 8,000 and 4,000 years ago with frequencies of 13 to 36 percent (Jerardino 1996; new data) (Figure 12.3). Around 3500 BP, these materials were still used regularly, but in a highly variable fashion, with frequencies of 3 to 21 percent. During the subsequent megamidden period, exotic raw materials were rarely used at Steenbokfontein Cave, comprising 6 to 8 percent of the flaked stone assemblage. Slightly smaller frequencies are recorded from contemporary occupations at Pancho's Kitchen Midden (Jerardino 1998). Exotic raw materials continued to be used infrequently over the last 2,000 years as shown by Tortoise Cave and other sites dating to the last 1,000 years (Jerardino 1996, 2000; Orton 2006). From this, a coherent picture of shifting settlement patterns emerge for the study area. As the land became more populated with groups settling for longer periods around 3,500 years ago, and reaching a maximum between 3,000 to 2,000 years ago, hunter-gatherer mobility became increasingly restricted to the coastal margin. Contact with inland areas and beyond the Oliphant River to the north and Berg River to the south was not only rare during the megamidden period, but also after 2000 BP, when human occupation of the study area waned. Lack of contact between the coast and the interior during the megamidden period is also supported by the dearth of radiocarbon dates from hinterland sites (2 out of 42) falling between 3000 and 2000 BP (Jerardino 1996:87). Hence, the suggestion by Henshilwood et al. (1994) that dried shellfish would have been transported from the coast to the interior for their final consumption during this millennium is not supported by the available evidence.

Concomitant changes in subsistence are also reflected in the archaeological record. Isotopic evidence based on carbon isotope measurements on collagen and bone apatite from human skeletons buried along the West Coast shows an increase in marine food intake between 3000 and 2000 BP, the so-called megamidden period (Figure 12.3). Much of the protein and a significant portion of energy-rich foodstuffs were obtained from marine resources during this millennium. In contrast, the last 2,000 years saw a greater contribution of terrestrial proteins and carbohydrates into people's diets (Lee-Thorp et al. 1989; Sealy 1989; Sealy and Van der Merwe 1988). These subsistence trends are closely mirrored by changes in the density of dietary remains that have been preserved in archaeological sites for which we have sufficiently detailed observations. The dietary mix, as reconstructed from density ratios of marine and terrestrial resources from Steenbokfontein Cave and Pancho's Kitchen Midden show an increase in marine foodstuffs during the megamidden period, particularly around 2,600 and 2,500 years ago (Jerardino 1996, 1998; Jerardino and Yates 1996; new data). Thus, both archaeological and isotopic evidence are in agreement, suggesting that hunter-gatherer diet during the megamidden period was more marine than ever before or after. Given the magnitude and nature of these trends in population levels, settlement patterns, and subsistence, it is reasonable to suspect that people would have had their greatest impact on the local marine fauna during the megamidden period.

MEASURING HUMAN IMPACT ON MARINE INVERTEBRATES

Marine ecologists working with invertebrate species have several complementary ways of exploring hypotheses related to human impact on rocky shores (Lasiak 1992; Siegfried et al. 1994). These include (1) the quantification of the volumes or mass of harvested resources and their recruitment levels through time; (2) species composition of catches; and (3) comparison of richness, abundance, and size-frequency distribution of species in areas where human harvesting is prevented, versus areas where harvesting takes place. These observations are evaluated against an understanding of the influence of the physical environment on the collected species, and the biology and community structure of the species under study. Unfortunately, archaeologists have a narrower set of choices when looking for ways to answer similar questions, but species composition of the catches and body-size observations can be retrieved fairly directly from the archaeological record. Proxies can also be established for the physical environment (e.g., degree of exposure of shoreline, water turbidity, and sea surface temperatures). Although archaeologists are able to evaluate their observations through time as marine biologists do, the chronological control allowed by radiocarbon dating does not match that available to researchers working in the present day. Nevertheless, archaeologists have managed over the last 15 years to extricate exciting observations reflecting the ability of people to exert a tangible impact on marine invertebrate species (Jerardino et al. 1992; Spennemann 1986; Swadling 1976).

ARCHAEOLOGICAL OBSERVATIONS FROM THE STUDY AREA

Three molluskan species, namely *Choromytilus meridionalis* (the black mussel), *Cymbula granatina* (the granite limpet), and *Scutellastra granularis* (the granular limpet), and one species of crustacean, *J. lalandii* (the West Coast rock lobster), are considered in this chapter. These species belong to three different trophic levels, and some of them have direct ecological links. Black mussels are filter feeders, the limpets are intertidal grazers, and rock lobsters are top predators with a powerful ability to modify the relative abundance of their prey and that of other species associated with them (Castilla et al. 1994). In particular, rock lobsters consume mussels, urchins, and winkles. Elimination or depletion of these groups by rock lobsters has powerful affects on other elements of the biological community. Urchins harbour juvenile abalone, so any diminishment of urchins results in a decline in juvenile abalone, with repercussions for the adult population. Moreover, consumption of grazers such as winkles allows algae to proliferate (Day and Branch 2002; Mayfield and Branch 2000; Mayfield et al. 2000). As a result, reductions in the abundance of rock lobsters have the capacity to completely alter the nature of benthic communities.

Analyses of shellfish samples from Steenbokfontein Cave, Malkoppan, and Grootrif D megamiddens followed the methodology outlined by Jerardino (1997). Size observations on limpet shells were obtained by measuring the total lengths of unbroken shells. Body sizes of black mussels and rock lobsters were derived, respectively, from measurements of prismatic band widths and calcareous mandibles. These initial observations were then transformed to body-size estimates with the use of morphometric equations (Buchanan 1985; Jerardino et al. 2001). The statistical significance of any changes in the mean sizes of these four invertebrate species was tested using one-way analysis of variance (ANOVA; Zar 1984) followed by Tukey-Kramer multiple comparisons (Stoline 1981).

Black mussels were the most heavily collected shellfish between 3,000 and 2,000 years ago, with relative frequencies of 70 to nearly 100 percent of the weight at sampled megamiddens. Black mussels were also abundant in shellfish samples recovered from Steenbokfontein Cave, with frequencies covering 40 to 90 percent. On the other hand, limpets were almost absent from most megamiddens during this millennium, although important exceptions are Malkoppan and Grootrif D megamiddens (Figure 12.1). Limpets were also collected from Steenbokfontein Cave before and during the megamidden period. Limpet frequencies spanned 5 to 30 percent in both of these megamiddens, and 2 to 38 percent in Steenbokfontein Cave. Of the two limpet species, *Cymbula granatina* was collected more regularly than *S. granularis*, most probably because it produces higher meat yields per individual (Tonner 2005). Rock lobster remains were found in almost all sites that have been systematically excavated (Horwitz 1979; Jerardino 1996, 2000; Jerardino and Navarro 2002). However, large enough samples of mandibles for valid statistical manipulation are present at only two sites where the megamidden period is represented, namely Steenbokfontein Cave and Grootrif D.

In the case of the black mussel, statistically significant changes can be detected through time, particularly reductions in mean shell length during the megamidden period, as illustrated by data for Steenbokfontein Cave (Figure 12.4). Similar changes occurred at Malkoppan as well. Significant reductions in the mean sizes of the granite limpet and the rock lobster were also detected for the megamidden period at Steenbokfontein Cave and Grootrif D megamidden (Figure 12.4). A decreasing trend in shell length during the megamidden period was also detected for *S. granularis* in Steenbokfontein Cave, although this trend was not statistically significant. No significant changes in the mean sizes of *S. granularis* were detected with Grootrif D data either. It is also interesting that the mean sizes of the three quantified mollusk species from Steenbokfontein Cave were smaller around 4000 BP, before the start of the megamidden period (see later).

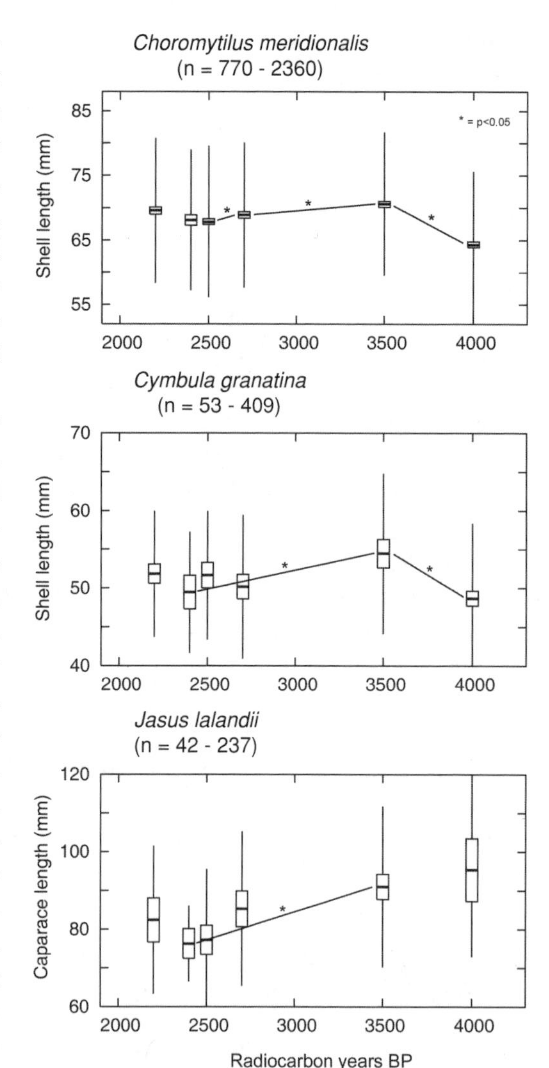

FIGURE 12.4. Changes in mean sizes of black mussels *(Choromytilus meridionalis)*, granite limpet *(Cymbula granatina)*, and rock lobster *(Jasus lalandii)* at Steenbokfontein Cave (uncalibrated radiocarbon years BP).

FAUNAL CHANGES IN THE LIGHT OF ENVIRONMENTAL CONDITIONS

There are convincing changes through time in the size composition and the relative proportions of species in middens, and the immediate temptation is to ascribe this to human harvesting. Support for this view comes from declines in the size of harvested species in middens elsewhere in the world (Jerardino et al. 1992). Additional support comes from the documentation of comparable modern faunal changes that can be linked to subsistence harvesting. Declines in the sizes of limpets and mussels in harvested areas relative to protected areas testifies to this (Branch 1975; Branch and Odendaal 2003; Kyle et al. 1997; Lasiak 1992; Lasiak and Dye 1989). Changes in modern biological community composition have also been recorded, with convergence on a relatively uniform composition in harvested areas (Hockey and Bosman 1994; Lasiak and Field 1995).

Interpretation of the causes of these modern faunal changes benefits from concurrent observations at harvested and protected areas, thus eliminating or at least reducing possible confounding factors that may cloud the influence of human harvesting. No such luxury is possible when deducing the effects of human harvesting from archaeological middens. A range of factors other than human harvesting may influence patterns in size composition, including temperatures, productivity, sea level, turbidity, wave action, storms, red tides, and the inherent relative resilience of different species to harvesting. Without at least a consideration of the potential effects of these factors, it is impossible to be confident that harvesting is the factor that drives faunal changes in middens.

Regional differences in temperature serve as a useful proxy for productivity because sea temperatures are inversely related to nutrient levels (Nielsen and Navarrete 2004). Nutrient levels are in turn correlated with growth rates of primary producers, notably phytoplankton and benthic algae, which underpin the growth of secondary consumers such as mussels that feed on particulate matter and limpets that consume algae. Growth rates are positively correlated with maximum sizes that species achieve. Thus, long-term shifts in temperature, or regional differences in temperature, could plausibly be linked to differences in the sizes attained by mussels or limpets in middens. The West Coast of South Africa as a whole is a region of intense upwelling, and limpet biomass and sizes are larger there than on the South and East coasts, where upwelling is infrequent or absent (Bustamante et al. 1995b). However, the West Coast also has focal points of upwelling at particular sites (Shannon 1985), and growth of mussels and maximum sizes of limpets have both been shown to be greater at upwelling centers than downstream, where upwelling is less marked (Menge et al. 2003; Xavier et al. 2007). The relationship between upwelling and growth (or size) is, however, not necessarily a positive one. Upwelling does enhance nutrient levels, but it also translates into advection of surface waters, so the nutrient-rich upwelled waters are shifted offshore, taking with them any phytoplankton growth that has been spurred by the elevated nutrients. This water is later returned to the shore downstream of the upwelling centers, and particulate food is often more concentrated there than at the upwelling centers (Wieters et al. 2003). Correspondingly, growth of mussels may be expected to be greater downstream than at the focal points of upwelling.

Could temporal differences in productivity have driven the patterns of diminishing sizes recorded in middens over the period 2400–3500 BP, rather than human harvesting? It seems unlikely. First, temperatures were declining over this period (Figure 12.3), indicating elevated rather than diminished productivity; so size should have increased rather than decreased if nutrient-fueled productivity caused the reduction in size. Second, there was a corresponding decrease in temperature during the Little Ice Age (ca. 500 BP), yet mussel sizes *increased* during that period (Jerardino 1997), so opposite responses were recorded during the two periods of temperature decline.

Sea level was at a maximum about 3800 BP, dropped to approximately present levels between 3300 and 2400 BP, and then rose again by about 1 m before declining again (Jerardino 1995b). There is no intrinsic reason why sea level per se would have affected sizes of organisms in middens, but it is possible that associated factors were at work. One option is that turbidity altered with sea level and could have influenced growth rates and size. Patellid limpets are vulnerable to sand cover (Marshall and McQuaid 1989) and are likely to achieve smaller sizes in sand-inundated conditions. Mussels feed less successfully when particulate matter rises above threshold (Stuart et al. 1982) and therefore grow slower and reach smaller sizes. An indirect measure of turbidity can be derived from archaeological middens by quantifying a kind of sediment (water-worn shell and shingle) originally trapped by the byssal threads of mussels and retained later in the middens (see Jerardino 1993: Figure 11.3). The data suggest that turbidity was high around 4000–3000 BP and could explain why sizes of mussels and limpets were low at that time. This does not explain the decreases in size in the subsequent period, when turbidity seemed to be at a minimum. Again, this allows us to reject turbidity as a cause of size changes between 3300 and 2400 BP (Figures 12.3 and 12.4).

Wave action has profound effects on body sizes, as movements of most limpets are inhibited by wave action, and they achieve smaller sizes on wave-exposed than sheltered shores (e.g., Branch and Odendaal 2003). This is particularly true of both *S. granularis* and *C. granatina*, the two species of greatest interest in West Coast middens. Conversely, mussels, which favor wave-exposed shores because wave action suspends and replenishes greater concentrations of particulate food (Bustamante and Branch 1996a, 1996b), attain greater sizes, growth rates, and cover on exposed shores than in sheltered areas (Branch and Steffani 2004; Steffani and Branch 2003). Two lines of thought argue against wave action as an explanation for diminished sizes of limpets and mussels in middens. First, all middens were associated with relatively short outcrops of rocks where the magnitude of wave action is unlikely to have changed in any systematic manner that can be related to sea-level changes over the period when limpet and mussels sizes were declining. Theoretically, changes in sea level could have altered the coastal topography and thus affected wave action, but rises and falls spanning 3 m would probably have had too small an effect on local topography to bring about significant changes in wave action. Second, and more convincingly, even if wave action was altered, we would have predicted opposite responses from limpets and mussels, yet they both declined in size. In short, alterations of wave action seem an implausible cause of body-size declines in mussels and limpets.

Storms could have had more subtle effects on the size composition of mussels and limpets that would have been available for people to harvest. Periodic storms may eliminate large limpets by physically removing them from rocks (Denny et al. 1985) but are a two-edged sword when it comes to mussel size. Storms may remove mussels en masse (Branch and Steffani 2004; Steffani and Branch 2003) but sometimes dump huge quantities of large, subtidal mussels in the intertidal zone where they can take hold, increasing both the quantity and sizes of mussels available to intertidal harvesters (G. M. Branch and S. Eekhout, unpublished data). If the frequency or intensity of storms varied systematically over time, they could have influenced body sizes of both limpets and mussels. However, there is no evidence of such systematic changes over the period when sizes of these animals were declining. There is no easy prehistoric measure of storms, but turbidity serves as a proxy, and it shows no variations over time that clearly correlate with limpet and mussel sizes.

Thus far, we have focused on possible environmental effects that could have influenced body sizes of mussels and limpets, but similar questions can be raised about rock lobsters. To what extent is the decline in rock lobster size over the period 3500–2400 BP likely to reflect environmental conditions rather than harvesting? Two important possibilities exist. The first is that rock lobster growth has declined since the late 1980s (Johnston and Butterworth 2005; Pollock et al. 1997). The causes remain unresolved, but reduced food supply or environmental changes have both been invoked as explanations. Reductions in the rate of growth will result in smaller size and lower productivity, so the question that immediately arises is whether any past variations in growth could have influenced the productivity and size composition of rock lobsters found in shell middens. So far, there is no easy way of testing this possibility, although it seems that prehistoric sizes of rock lobsters were often substantially greater than in modern times (A. Jerardino, unpublished data). Secondly, mass "walk-outs" of rock lobsters have been recorded on the West Coast of South Africa, during which (mostly small) lobsters have moved inshore to avoid oxygen-depleted waters and have ended up becoming stranded on the shore in spectacular quantities—up to 2,000 tons in one episode (Cockcroft 2001). Such events appear to be triggered by upwelling followed by prolonged quiescence, which concentrates phytoplankton inshore in bays, leading to depletion of oxygen as the blooms decay. Ensuing walkouts will influence the availability and size composition of rock lobsters in two ways. They provide a brief bonanza, but they also deplete stocks. There is no way of telling if such walk-outs were more or less frequent in the prehistoric past, but if their frequency has changed it would have powerfully affected the amounts and sizes of lobsters that could have been harvested. In the case of rock lobsters, we are therefore on much less certain ground in attempting to relate declines of size in middens to harvesting pressure. It appears that at least two plausible, but ambiguous, environmental factors may better explain these changes in size.

CHARACTERISTICS OF THE SPECIES

In evaluating the potential effects of human harvesting on the abundance and sizes of target species, consideration also needs to be given to both the relative vulnerabilities of different species to harvesting, and human preferences (Lasiak 1991). Not all species are equally vulnerable to harvesting pressure, and their vulnerability depends on a suite of biological properties.

Accessibility

Species that are confined to the intertidal zone, such as *C. granatina* and *S. granularis*, are readily accessible to harvesters every time the tide recedes sufficiently to expose them. During approximately five days every fortnight the entire shore is exposed during low spring tides, and it is then that human harvesting is most intense. Not all intertidal limpets will, however, be equally vulnerable. *S. granularis* occurs high on the shore and will be most accessible; *C. granatina* occurs lower down but is most abundant on sheltered bays, where it is also easy to harvest; but *S. argenvillei* is characteristically found low on the shore and on wave-beaten shores, where harvesting is more hazardous (Branch and Marsh 1978; Bustamante et al. 1995a). Mussels also occur in the intertidal zone, but they extend down into the subtidal zone. In the midshore they are small but easy to collect, but at the bottom of the shore they are larger although less accessible. In the subtidal, they cannot be harvested without diving. Rock lobsters are even less vulnerable to subsistence harvesting because they live entirely subtidally and extend offshore for several kilometers, into depths of about 80 m (Griffiths and Branch 1997; Heydorn 1969). There is scant osteological evidence that prehistoric harvesting on the West Coast of South Africa involved diving (A. Morris, personal communication, 2006). Fishers may have used simple gourds and twine but would have been limited to very close to the shore. As far as is known, no watercraft or offshore fishing was developed in the region during prehistoric times. Rock lobsters could have been captured in very shallow water only by wading or by luring individuals with bait. This would have meant that only a tiny proportion of the population would have been exposed to harvesting, making it highly unlikely that harvesting could have dented the size composition of the species, in contrast to the impacts that may have been inflicted on intertidal species.

Two key features emerge. First, the apparent absence of watercraft and means of fishing away from the shore would have curbed the capacity of harvesters to influence the population structure of subtidal species. Second, species that have spatial refugia where they cannot be harvested will be relatively less vulnerable to the effects of fishing.

Mobility

Capacity for movement will influence the ability of animals to recover after being harvested. Limpets are sedentary and mussels are sessile. Rock lobsters move inshore and offshore in regular annual migrations (Heydorn 1969), however, and can replenish their shallow-water numbers if harvesting takes place there alone. Again, this points to rock lobsters being relatively immune to the attentions of shore-based harvesters.

Larval Dispersal

Both the mode and frequency of larval dispersal will also influence the vulnerability of species to harvest pressure. Mussels have a widely dispersed planktonic larval stage but often experience intermittent recruitment. Years may pass with little or no recruitment, interspersed with bouts of intense recruitment. Moreover, settlement takes place mainly into existing beds of adult mussels (Harris et al. 1998). No replenishment can take place in years with no recruitment, and even when settlement does take place, recovery is slow where adult beds have been stripped by overharvesting (Dye et al. 1997), thus increasing the chances that harvesting will influence population structure. Species that are most vulnerable to overharvesting are those that have very limited dispersal. A classic example is the solitary ascidian *(Pyura stolonifera)*, colloquially known as "red-bait," which is harvested as a source of food by subsistence fishers on the east coast of South Africa and used as bait for fishing elsewhere. It has a larval stage that lasts only minutes, so its larvae settle within meters of the adults that produce them (Griffiths 1976). As a result, depletion of adults reduces local settlement of larvae, and replenishment by larvae produced afar is impossible. By contrast, the rock lobster has a prolonged larval life lasting 9 to 11 months and is widely dispersed (Silberbauer 1971), so replenishment is possible both by adult movements and by larval settlement.

Growth Rate and Longevity

Species that are fast growing can recover quickly after being depleted, but they tend to be short-lived so that the size-composition of their populations is made up of a small number of year-groups and is inherently unstable from year to year. From a management perspective, there is thus a trade-off: fast growth translates into rapid recovery but high variability in stocks. Even among groups of closely related species, wide differences may exist in growth rate. The limpets *Cymbula oculus* and *C. granatina* grow fast, reaching maturity within 2 or 3 years; but other limpets such as *Scutellastra argenvillei* and *S. cochlear* grow agonizingly slowly, maturing after about 6 years and attaining ages of up to 35 years (Branch 1974; Eekhout et al. 1992). Mussels tend to be fast growing, being harvestable after 1 to 2 years and living for about 5 years, but the rock lobster matures only after 7 to 15 years and lives up to 40 years, making it much more prone to the effects of harvesting and slow to recover (Pollock et al. 2000).

Sex Change

Some species undergo sex change as they age. The result is that older age groups are dominated by one sex. As there are inevitably fewer individuals in these older age groups, this automatically skews the sex ratio. None of the species of central interest here undergoes sex change, but *C. oculus*, a close relative of *C. granatina*, is male during its first 1 to 3 years of life and then becomes female for the rest of its life. Females are consequently not only more rare than males but are prime targets for harvesters because of their larger size. On the east coast of South Africa, where this species is heavily fished by subsistence fishers, it is threatened with extinction because of the depletion of females, which have declined from on average of 36 percent of the sexually mature population in protected areas down to 9 percent in harvested areas (Branch and Odendaal 2003). Clearly, sex change heightens the vulnerability of species to harvesting.

HUMAN PREFERENCES

In addition to the characteristics of the species considered above, human preferences will influence the relative impact of harvesting on different species. Particular species may be harvested more intensely than would be predicted based on their abundance. Factors affecting preference include accessibility, ease of procurement, transportability, relative size, yield in relation to effort, palatability, nutritional value, toxicity, spoilage rate, and desirability. Not all of these aspects can be considered here, but some may have played important roles in determining the rate at which different species were prehistorically harvested on the West Coast of South Africa. For example, the nutritional value of mussels varies on a seasonal basis. Just prior to spawning, mussels are plump and the energy content of the flesh is high due to the buildup of gonads. After spawning, they are scarcely worth collecting. Modern subsistence fishers are well aware of these phenomena and time their harvesting accordingly, often using environmental cues such as the season when particular trees flower as an indication that mussels are "ripe" (Harris et al. 2003).

Another important issue is that periodic blooms of noxious algae on the West Coast of South Africa can result in mussels becoming lethally toxic to humans (Matthews and Pitcher 1996; Pitcher 1999). In Elands Bay Cave, an abrupt hiatus in the harvesting of mussels takes place at about 9500 BP (Parkington 1981), during which harvesting switched for a brief time to focus on much less abundant species, namely, the whelk *Burnupena* and limpets. Although speculative, it is not beyond the bounds of possibility that harvesters were struck by a harmful algal bloom that made mussels toxic, compelling a switch of diet (Parkington 1981; Parkington et al. 1988).

Finally, during the megamidden period of 3000–2000 BP, there is good evidence that mussel were dried and stored for later consumption (Henshilwood et al. 1994). This would have allowed more extended use of mussels, possibly tiding people over periods when the mussels were in poor condition, and overcoming limitations imposed by the greater frequency of toxic algal blooms in summer and storms during winter (Jerardino 1996; Parkington et al. 1988).

INTEGRATING THE EVIDENCE AND CONCLUSIONS

Clearly, the potential impacts of human exploitation on marine resources will depend on a blend of the severity of the impact (e.g., high population pressure), the vulnerability of the species to harvesting, and the extent to which extraneous factors could have influenced the resources. Regarding the first, caution must be exercized when proposing high human population levels in the landscape as the main factor behind declining species' body sizes. Each case needs to be considered and assessed according to its own merits: in some instances population pressure may explain much of the observed variability, but in others, population pressure may be irrelevant. Most important here is the need to present independent evidence for population growth, otherwise, circular logic could feed an argument that ends up presenting population pressure as a self-fulfilling hypothesis.

We have shown that declines in the average sizes of mussels, two species of limpets, and the rock lobster all coincided with a period when human occupation of the study area was intensifying, and when reliance on marine resources was increasing (Sealy and van der Merwe 1988). The coincidence implies that human harvesting was responsible, but this conclusion needs to be evaluated by asking whether these species were sufficiently vulnerable for their populations to have been influenced by harvesting, and whether there were any other factors that could have explained the trends in size.

In the case of both the mussels and the limpets, we could find no convincing evidence that environmental conditions could have caused the declines in size during the megamidden period. Environmental changes did take place over the period when sizes were declining, but none of the factors examined provided a plausible explanation for a diminishment in size in both groups. Limpets and mussels belong to two very different trophic groups, being, respectively, grazers and filter feeders. This alone is significant, because changes in many of the environmental factors examined should yield opposite outcomes for these two trophic groups. Moreover, both mussels and limpets are highly vulnerable to the effects of harvesting because they are accessible, nonmobile as adults, have intermittent larval recruitment, and to a large extent lack refuges beyond the reach of harvesters. We would expect them to be depleted by intense harvesting. Was harvesting intense enough to accomplish this? Rough calculations based on midden sizes and shell densities leads to the conclusion that about 1,666 kg (wet whole mass) would have been removed per kilometer per year (Griffiths and Branch 1997). This compares with a figure of 5,500 kg km^{-1} y^{-1} for highly intense modern subsistence fishing on the southeast coast of South Africa, where severe depletion of stocks has been recorded (Hockey et al. 1988). Given the combination of intense harvesting during the megamidden period from limited available rocky shores, high vulnerability of mussels and limpets, and an absence of alternative explanations for declines in mean sizes of these species, harvesting remains the most parsimonious and robust explanation for the declines in their mean sizes over this period.

However, there is a caveat. Over the same time, there was a significant decline in the sizes of rock lobsters. It is extremely unlikely that this decline can be attributed to shore-based harvesting. It is hard to imagine more than a tiny fraction of the rock lobster population being harvested from the shore. The bulk of the population lives in the subtidal zone, where it would have been inaccessible, and even if the shallow-water sector was harvested, it would have been replenished by movement of adults. In short, rock lobsters would not have been sufficiently vulnerable to shore-based fishing for the population to have been dented sufficiently by harvesting to alter the size composition. Additionally, the availability, productivity, and distribution of rock lobsters all depend strongly on environmental conditions, as demonstrated for modern populations. There are thus good reasons for distrusting prehistoric harvesting as a cause of the decline in rock lobster size, even though it too coincides with the period of intensification of harvesting.

ACKNOWLEDGMENTS

Financial support for archaeological excavations of several sites, processing and dating of the material, and building of the database was received from the History of Marine Animal Populations Project (International Consortium), the National Research Foundation (NRF, South Africa), Swan Fund (Oxford, UK), University Research Fund (UCT), and Wenner-Gren Foundation (Chicago, Illinois). The biological research was funded by the NRF, UCT, and the Andrew Mellon Foundation. Many thanks to J. Erlandson, R. Klein, G. Sampson, C. Marean, and R. Yates for discussions on the subject matter of this chapter, to R. Klein and C. Marean for making bibliographic references available, and to J. Sealy for making isotopic data on West Coast human skeletons available. Thanks are also extended by one of us (A.J.) to J. C. Castilla for introducing me to one of the most fascinating scientific endeavors and for helping me to forge what is needed to follow this path.

REFERENCES CITED

Branch, G. M.

1974 The Ecology of *Patella* Linnaeus from the Cape Peninsula, South Africa. III. Growth Rates. *Transactions of the Royal Society of South Africa* 41:161–193.

1975 Notes on the Ecology of *Patella concolor* and *Cellana capensis*, and the Effects of Human Consumption on Limpet Populations. *Zoologica Africana* 10:75–85.

Branch, G. M., and A. Marsh

1978 Attachment Forces and Shell Shape in *Patella*: Adaptive strategies. *Journal of Experimental Marine Biology and Ecology* 34:111–130.

Branch, G. M., and F. Odendaal

2003 Marine Protected Areas and Wave Action: Impacts on a South African Limpet, *Cymbula oculus. Biological Conservation* 114:255–269.

Branch, G. M., and C. N. Steffani

2004 Can We Predict the Effects of Alien Species? A Case-History of the Invasion of South Africa by *Mytilus galloprovincialis* (Lamarck). *Journal of Experimental Marine Biology and Ecology* 300: 189–215.

Buchanan, W.

1985 Middens and Mussels: An Archaeological Enquiry. *South African Journal of Science* 81: 15–16.

1988 *Shellfish in Prehistoric Diet: Elands Bay, SW Cape Coast, South Africa.* Cambridge Monographs in African Archaeology 31, BAR International Series 455. British Archaeological Reports, Oxford.

Buchanan, W., S. L. Hall, J. Henderson, A. Olivier, J. M. Pettigrew, J. E. Parkington, and R. T. Robertshaw
1978 Coastal Shell Middens in the Paternoster Area, South-western Cape. *South African Archaeological Bulletin* 33:89–93.

Bustamante, R. H., and G. M. Branch
1996a Large Scale Patterns and Trophic Structure of Southern African Rocky Shores: The Roles of Geographic Variation and Wave Exposure. *Journal of Biogeography* 23:339–351.
1996b The Dependence of Intertidal Consumers on Kelp-Derived Organic Matter on the West Coast of South Africa. *Journal of Experimental Marine Biology and Ecology* 196:1–28.

Bustamante, R. H., G. M. Branch, and S. Eekhout
1995a Maintenance of an Exceptional Grazer Biomass in South Africa: Subsidy by Subtidal Kelps. *Ecology* 76:2314–2329.

Bustamante, R. H., G. M. Branch, S. Eekhout, B. Robertson, P. Zoutendyk, M. Schleyer, A. Dye, D. Keats, M. Jurd, and C. D. McQuaid
1995b Gradients of Intertidal Productivity around the Coast of South Africa and Their Relationship with Consumer Biomass. *Oecologia* 102: 189–201.

Castilla, J. C., G. M. Branch, and A. Barkai
1994 Exploitation of Two Critical Predators: The Gastropod *Concholepas concholepas* and the Rock Lobster *Jasus lalandii*. In *Rocky Shores: Exploitation in Chile and South Africa*, edited by R. W. Siegfried, pp. 101–130. Springer-Verlag, Ecological Studies, Berlin.

Cockcroft A. C.
2001 *Jasus lalandii* "Walkouts" or Mass Strandings in South Africa during the 1990s: An Overview. *Marine and Freshwater Research* 52: 1085–1094.

Conard, N. J., T. J. Prindiville, and A. Kandel
1999 The 1998 Fieldwork on the Stone Age Archaeology and Palaeoecology of the Geelbek Dunes, West Coast National Park, South Africa. *Southern African Field Archaeology* 8:35–45.

Day, E., and G. M. Branch
2002 Effects of Sea urchins *(Parechinus angulosus)* on Recruits and Juveniles of Abalone *(Haliotis midae)*. *Ecological Monographs* 72: 133–149.

Denny, M. N., T. L. Daniel, and M. A. R. Koehl
1985 Mechanical Limits to Size in Wave-Swept Organisms. *Ecological Monographs* 55:69–102.

Dye, A. H., T. A. Lasiak, and S. Gabula
1997 Recovery and Recruitment of the Brown Mussel *Perna perna* (L.) in Transkei: Implications for Management. *South African Journal of Zoology* 32:118–123.

Eekhout, S., C. M. Raubenheimer, G. M. Branch, A. L. Bosman, and M. O. Bergh
1992 A Holistic Approach to the Exploitation of Intertidal Stocks: Limpets as a Case History. *South African Journal of Marine Science* 12: 1017–1029.

Erlandson, J. M.
2001 The Archaeology of Aquatic Adaptations: Paradigms for a New Millennium. *Journal of Archaeological Research* 9:287–350.

Griffiths, R. J.
1976 The Larval Development of *Pyura stolonifera* (Tunicata) from the Cape Peninsula. *Transactions of the Royal Society of South Africa* 42:1–9.

Griffiths, C. L., and G. M. Branch
1997 The Exploitation of Coastal Invertebrates and Seaweeds in South Africa: Historical Trends, Ecological Impacts and Implications for Management. *Transactions of the Royal Society of South Africa* 52:121–148.

Halkett, D., T. Hart, R. Yates, T. P. Volman, J. E. Parkington, J. Orton, R. G. Klein, K. Cruz-Uribe, and G. Avery
2003 First Excavation of Intact Middle Stone Age Layers at Ysterfontein, Western Cape Province, South Africa: Implications for Middle Stone Age Ecology. *Journal of Archaeological Science* 30: 955–971.

Harris, J. M., G. M. Branch, B. L. Elliott, B. Currie, A. Dye, C. D. McQuaid, B. Tomalin, and C. Velasquez
1998 Spatial and Temporal Variability in Recruitment of Intertidal Mussels Around the Coast of Southern Africa. *South African Journal of Zoology* 33:1–11

Harris, J., G. M. Branch, C. Sibaya, and K. Bill
2003 The Sokhulu Subsistence Mussel-Harvesting Project: Co-management in Action. In *Waves of Change.Coastal Co-management in Southern Africa*, edited by M. Hauck and M. Sowman, pp. 61–98. University of Cape Town Press, Cape Town.

Henshilwood, C., P. Nilssen, and J. Parkington
1994 Mussel Drying and Food Storage in the Late Holocene, SW Cape, South Africa. *Journal of Field Archaeology* 21:103–109.

Heydorn, A. E. F.
1969 The Rock Lobster of the South African West Coast *Jasus lalandii* (H. Milne-Edwards). 2. Population Studies, Behaviour, Reproduction, Moulting, Growth and Migration. *Investigational Report of the Division of Sea Fisheries, South Africa* 71:1–52.

Hockey, P. A. R., and A. L. Bosman
1994 Man as an Intertidal Predator in Transkei: Disturbance, Community Convergence, and Management of a Natural Food Resource. *Oikos* 46:3–14.

Hockey, P. A. R., A. L. Bosman, and W. R. Siegfried
1988 Patterns and Correlates of Shellfish Exploitation by Coastal People in Transkei: An Enigma of Protein Production. *Journal of Applied Ecology* 25:353–363.

Horwitz, L.
1979 From Materialism to Middens: A Case Study at Eland's Bay, Western Cape, South Africa. Unpublished B.A. (honors) dissertation, Department of Archaeology, University of Cape Town.
Jacobs, Z., G. A. T. Duller, and A. Wintle
2003a Optical Dating of Dune Sand from Blombos Cave, South Africa. II. Single Grain Data. *Journal of Human Evolution* 44:613–625.
Jacobs, Z., A. G. Wintle, and G. A. T. Duller
2003b Optical Dating of Dune Sand from Blombos Cave, South Africa. I. Multiple Grain Data. *Journal of Human Evolution* 44:599–612.
Jerardino, A.
1993 Mid- to Late-Holocene Sea-Level Fluctuations: The Archaeological Evidence at Tortoise Cave, South-western Cape, South Africa. *South African Journal of Science* 89:481–488.
1995a The Problem with Density Values in Archaeological Analysis: A Case Study from Tortoise Cave, Western Cape, South Africa. *South African Archaeological Bulletin* 50:21–27.
1995b Late Holocene Neoglacial Episodes in Southern South America and Southern Africa: A Comparison. *The Holocene* 5:361–368.
1996 Changing Social Landscapes of the Western Cape Coast of Southern Africa over the Last 4500 years. Unpublished Ph.D. dissertation, Department of Archaeology, University of Cape Town.
1997 Changes in Shellfish Species Composition and Mean Shell Size from a Late-Holocene Record of the West Coast of Southern Africa. *Journal of Archaeological Science* 24:1031–1044.
1998 Excavations at Pancho's Kitchen Midden, Western Cape Coast, South Africa: Further Observations into the Megamidden Period. *South African Archaeological Bulletin* 53:17–25.
2003 Precolonial Settlement and Subsistence along Sandy Beaches South of Elands Bay, West Coast, South Africa. *South African Archaeological Bulletin* 58:53–62.
2007 Excavations at a Hunter-Gatherer Site Known as "Grootrif G" Shell Midden, Lambertsbay, Western Cape Province. *South African Archaeological Bulletin* 62:162–170.
Jerardino, J., and R. Navarro
2002 Cape Rock Lobster *(Jasus lalandii)* Remains from South African West Coast Shell Middens: Preservational Factors and Possible Bias. *Journal of Archaeological Science* 29:993–999.
Jerardino, A., and N. Swanepoel
1999 Painted Slabs from Steenbokfontein Cave: The Oldest Known Parietal Art in Southern Africa. *Current Anthropology* 40:542–548.
Jerardino, A., and R. Yates
1996 Preliminary Results from Excavations at Steenbokfontein Cave: Implications for Past and Future Research. *South African Archaeological Bulletin* 51:7–16.
1997 Excavations at Mike Taylor's Midden: A Summary Report and Implications for a Re-characterization of Megamiddens. *South African Archaeological Bulletin* 52:43–51.
Jerardino, A., J. C. Castilla, J. M. Ramírez, and N. Hermosilla
1992 Early Coastal Subsistence Patterns in Central Chile: A Systematic Study of the Marine-Invertebrate Fauna from the Site of Curaumilla-1. *Latin American Antiquity* 3:43–62.
Jeradino, A., R. Navarro, and P. Nilssen
2001 Cape Rock Lobster *(Jasus lalandii)* Exploitation in the Past: Estimating Carapace Length from Mandible Sizes. *South African Journal of Science* 97:59–62.
Johnston, S. J., and D. S. Butterworth
2005 Evolution of Operational Management Procedures for the South African West Coast Rock Lobster *(Jasus lalandii)* Fishery. *New Zealand Journal of Marine and Freshwater Research* 39:687–702.
Klein, R. G., G. Avery, K. Cruz-Uribe, D. Halkett, J. E. Parkington, T. Steele, T. P. Volman, and R. Yates
2004 The Ysterfontein 1 Middle Stone Age site, South Africa, and Early Human Exploitation of Coastal Resources. *Proceedings of the National Academy of Science* 101:5708–5715.
Kyle, R., B. Pearson, P. J. Fielding, and W. D Robertson
1997 Subsistence Shell-fish Harvesting in the Maputaland Marine Reserve in Northern KwaZulu-Natal, South Africa: Rocky Shore Organisms. *Biological Conservation* 82:183–192.
Lasiak, T. A.
1991 The Susceptibility and/or Resilience of Rocky Littoral Mollusks to Stock Depletion by Indigenous Coastal People of Transkei, Southern Africa. *Conservation Biology* 56:245–264.
1992 Contemporary Shellfish-Gathering Practices of Indigenous Coastal People in Transkei: Some Implications for Interpretation of the Archaeological Record. *South African Journal of Science* 88:19–28.
Lasiak T. A., and A. Dye
1989 The Ecology of the Brown Mussel *Perna perna* in Transkei, Southern Africa: Implications for the Management of a Traditional Food Resource. *Biological Conservation* 47:245–257.
Lasiak, T., and J. G. Field
1995 Community-Level Attributes of Exploited and Non-Exploited Rocky Infratidal Macrofaunal

Assemblages in Transkei. *Journal of Experimental Marine Biology and Ecology* 185:33–53.

Lee-Thorp, J., J. Sealy, and N. Van der Merwe
1989 Stable Carbon Isotope Ratio Differences between Bone Collagen and Bone Apatite, and Their Relationship to Diet. *Journal of Archaeological Science* 16:585–599.

Marean, C.W., M. Bar-Mattews, J. Bernatchez, E. Fisher, P. Goldberg, A. Herries, Z. Jacobs, A. Jerardino, P. Karkanas, N. Mercier, M. Minichillo, P.J. Nilssen, E. Thosmpson, C. Teibolo, H. Valladas, and H. Williams
2007 Early Human Use of Marine Resources and Pigment in South Africa during the Middle Pleistocene, *Nature* 449:905–908.

Marshall, D.J., and C.D. McQuaid
1989 The Influence of Respiratory Responses on the Tolerance to Sand Inundation of the Limpet *Patella granularis* L. (Prosobranchia) and *Siphonaria capensis* Q. and G. (Pulmonata). *Journal of Experimental Marine Biology and Ecology* 128:191–201.

Matthews, S.G., and G.C. Pitcher
1996 Worst Recorded Marine Mortality on the South African Coast. In *Harmful and Toxic Algal Blooms,* edited by T. Yasumoto, Y. Oshima and Y. Fukuyo, pp. 89–92. Intergovernmental Oceanographic Commission of UNESCO, Paris, France.

Mayfield, S., and G.M. Branch
2000 Interrelations among Rock Lobsters, Sea Urchins, and Juvenile Abalone: Implications for Community Management. *Canadian Journal of Fisheries and Aquatic Science* 57:2175–2185.

Mayfield, S., G.M. Branch, and A. Cockcroft
2000 Relationships among Diet, Growth Rate, and Food Availability for the South African Rock Lobster, *Jasus lalandii* (Decapoda, Palinuridea). *Crustaceana* 73:8115–834.

Menge, B.A., J. Lubchenco, M.E. S. Bracken, F. Chan, M.M. Foley, T.L. Freidenburg, S.D. Gaines, G. Hudson, C. Krenz, H. Leslie, D.N. L. Menge, R. Russel, and M.S. Webster
2003 Coastal Oceanography Sets the Pace of Rocky Intertidal Community Dynamics. *Proceeding of the National Academy of Science* 100:12229–12234.

Miller, D.
1990 A Southern African Late Quaternary Sea-level Curve. *South African Journal of Science* 86:456–458.

Nielsen, K.J., and S.A. Navarrete
2004 Mesoscale Regulation Comes from the Bottom Up: Intertidal Interactions between Consumers and Upwelling. *Ecology Letters* 7:31–41.

Orton, J
2006. The Later Stone Age Lithic Sequence at Elands Bay, Western Cape, South Africa: Raw Materials, Artefacts and Sporadic Change. *Southern African Humanities* 18:1–28.

Parkington, J.
1976 Coastal Settlement between the Mouths of the Berg and the Olifants Rivers, Cape Province. *South African Archaeological Bulletin* 31:127–140.
1981 The Effects of Environmental Change on the Scheduling of Visits to the Elands Bay Cave, Cape Province, S.A. *In Patterns of the Past,* edited by I. Hodder, G. Isaac, and N. Hammond, pp. 341–359. Cambridge University Press, Cambridge.
1995 *Elands Bay Cave: A View on the Past.* Unpublished Manuscript. University of Cape Town.
2003 Middens and Moderns: Shellfishing and the Middle Stone Age of the Western Cape, *South African Journal of Science* 99:242–247.

Parkington, P., C. Poggenpoel, W. Buchanan, T. Robey, A. Manhire, and J. Sealy
1988 Holocene Coastal Settlement Patterns in the Western Cape. In *The Archaeology of Prehistoric Coastlines,* edited by G. Bailey and J. Parkington, pp. 22–41. Cambridge University Press, Cambridge.

Pitcher, G.C.
1999 *Harmful Algal Blooms of the Benguela Current.* Sea Fisheries Research Institute, Cape Town.

Pollock, D.E., A.C. Cockcroft, and P.C. Goosen
1997 A Note on Reduced Rock Lobster Growth Rates and Related Environmental Anomalies in Southern Benguela. 1988–1995. *South African Journal of Marine Science* 18:287–293.

Pollock, D.E., A.C. Cockcroft, J.C. Groeneveld, and D.S. Schoeman
2000 The Commercial Fisheries for *Jasus* and *Palinurus* Species in the South-East Atlantic and South-West Indian Oceans. In: *Spiny Lobsters: Fisheries and Culture,* edited by B.F. Phillips and J. Kittaka, pp. 105–120. Blackwell, Oxford.

Redman, C.L.
1999 *Human Impact on Ancient Environments.* University of Arizona Press, Tucson.

Robertshaw, P.T.
1978 Archaeological Investigations at Langebaan Lagoon, Cape Province. *South African Archaeological Bulletin* 10:139–148.
1979 Excavations at Duiker Eiland, Vredenburg District, Cape Province. *Annals of the Cape Provincial Museums* 1:1–26.

Robey, T.
1987 The Stratigraphic and Cultural Sequence at Tortoise Cave, Verlorenvlei. In *Papers in the Prehistory of the Western Cape, South Africa,* edited by J. Parkington and M. Hall, pp. 294–325. BAR International Series 332 (ii). British Archaeological Reports, Oxford.

Sadr, K., and A. B. Smith
1992 Final Report of the Vredenburg Peninsula survey 1991/1992. Unpublished report to Anglo-American Chairman's Fund. University of the Witwatersrand, Johannesburg.

Sealy, J.
1989 *Reconstruction of Later Stone Age Diets in the South-western Cape, South Africa: Evaluation and Application of Five Isotopic and Trace Element Techniques*. Unpublished Ph.D. dissertation, Department of Archaeology, University of Cape Town.

Sealy, J., and N. Van der Merwe
1988 Social, Spatial and Chronological Patterning in Marine Food Use as Determined by $\delta^{13}C$ Measurements of Holocene Human Skeletons from the South-western Cape, South Africa. *World Archaeology* 20:87–102.

Sealy, J., T. Maggs, A. Jerardino, and J. Kaplan
2004 Excavations at Melkbosstrand: Variability among Herder Sites on Table Bay. *South African Archaeological Bulletin* 59:17–28.

Shannon, L. V.
1985 The Benguela Ecosystem. Part I. Evolution of the Benguela, Physical Features and Processes. *Oceanography and Marine Biology Annual Review* 23:105–182.

Siegfried, W. R, P. A. R. Hockey, and G. M. Branch
1994 The Exploitation of Intertidal and Subtidal Biotic Resources of Rocky Shores in Chile and South Africa: An Overview. In *Rocky Shores: Exploitation in Chile and South Africa*, edited by R. W. Siegfried, pp. 1–15. Springer-Verlag, Ecological Studies, Berlin.

Silberbauer, B. I.
1971 The Biology of the South African Rock Lobster *Jasus lalandii* (H. Milne-Edwards). I. Development. *Investigational Report of the Division of Sea Fisheries, South Africa* 92:1–10.

Smith, A. B., K. Sadr, J. Gribble, and R. Yates
1991 Excavations in the South-western Cape, South Africa, and the Archaeological Identity of Prehistoric Hunter-Gatherers within the last 2000 years. *South African Archaeological Bulletin* 46:71–91.

Spennemann, D. H. R.
1986 Effects of Human Predation and Changing Environment on Some Mollusk Species on Tongatapu, Tonga. In *The Walking Larder: Patterns of Domestication, Pastoralism, and Predation*, edited by J. Clutton-Brock, pp. 326–335. Unwin Hyman, London.

Steele, T. E., and R. G. Klein
2005 Mollusk and Tortoise Size as Proxies for Stone Age Population Density in South Africa: Implications for the Evolution of Human Cultural Capacity. *Munibe* 57:5–21.

Steffani, C. N., and G. M. Branch
2003 Growth Rate, Condition, and Shell Shape of *Mytilus galloprovincialis*: Responses to Wave Exposure. *Marine Ecology Progress Series* 24:197–209.

Stoline, M. R.
1981 The Status of Multiple Comparisons: Simultaneous Estimation of All Pairwise Comparisons in One Way Anova Designs. *American Statistician* 35:134–141.

Stuart, V, J. G. Field, and R. C. Newell
1982 Evidence for Absorption of Kelp Detritus by the Ribbed Mussel *Aulacomya ater* Using a New ^{51}Cr-Labelled Microsphere Technique. *Marine Ecology Progress Series* 9:263–271.

Swadling, P.
1976 Changes Induced by Human Exploitation in Prehistoric Shellfish Populations. *Mankind* 10: 156–162.

Thackeray, J. F.
1988 Molluskan Fauna from Klasies River, South Africa. *South African Archaeological Bulletin* 43: 27–32.

Tonner, T. W. W.
2005 Later Stone Age Shellfishing Behaviour at Dunefield Midden (Western Cape, South Africa). *Journal of Archaeological Science* 32:1390–1407.

Van Andel, T. H.
1989 Late Pleistocene Sea Levels and the Human Exploitation of the Shore and Shelf of Southern South Africa. *Journal of Field Archaeology* 16:132–153.

Vogel, J. C.
2001 Radiometric Dates for the Middle Stone Age in South Africa. In *Humanity from African Naissance to Coming Millennia*, edited by M. A. Raath, J. Maggi-Cecchi, and G. A. Doyle, pp. 261–268. Florence University Press, Florence, Italy.

Volman, T. P.
1978 Early Archaeological Evidence for Shellfish Collecting. *Science* 201:911–913.

Webley, L.
2002 The Re-excavation of Spoegrivier Cave on the West Coast of South Africa. *Annals of the Eastern Cape Museums* 2:19–49.

Wieters, E. A., D. M. Kaplan, S. A. Navarrete, A. Sotomayor, J. Largier, K. J. Nielsen, and F. Véliz
2003 Alongshore and Temporal Variability in Chlorophyll *a* Concentration in Chilean Nearshore Waters. *Marine Ecology Progress Series* 249:93–105.

Xavier, B. M., G. M. Branch, and E. Wieters
2007 Abundance, Condition, Growth Rate and Recruitment of *Mytilus galloprovincialis* along the West Coast of South Africa in Relation to Upwelling. *Marine Ecology Progress Series* 346:189–201.

Zar, J. H.
1984 *Biostatistical Analysis*. Second edition. Prentice-Hall, Englewood Cliffs, New Jersey.

13

Archaeology, Historical Ecology, and the Future of Ocean Ecosystems

Torben C. Rick and Jon M. Erlandson

> The persistent myth of the oceans as wilderness blinded ecologists to the massive loss of marine ecological diversity caused by human overfishing and human inputs from the land over the past centuries. Until the 1980s, coral reefs, kelp forests, and other coastal habitats were discussed in scientific journals and textbooks as "natural" or "pristine" communities with little or no reference to the pervasive absence of large vertebrates or the widespread effects of pollution. This is because our concept of what is natural today is based on personal experience at the expense of historical perspective.
>
> JACKSON 2001:5411

RAVAGED BY OVERFISHING, pollution, eutrophication, and numerous other processes, fisheries and marine ecosystems around the world are in a state of crisis. Human populations are also growing at a much higher rate along the coast than interior areas, suggesting that the pressure placed on coastal habitats will increase dramatically in the future. Numerous studies have demonstrated that significant steps are needed to restore the world's oceans, including the continued establishment of marine protected areas (e.g., Botsford et al. 1997; Costanza et al. 1998; Dayton et al. 1998; Ellis 2003; Jackson 2001; Jackson et al. 2001; Pauly and Palomares 2005; Pauly et al. 1998, 2005; Pew Oceans Commission 2003; Pitcher 2001; Safina 1997; Woodard 2000; Worm et al. 2006). In many ways, the future of the world's oceans stands at a crossroads, with the balance predicated by our past, present, and future actions.

Archaeologists and anthropologists have recently played an important role in helping to inform contemporary environmental crises by supplying information on the nature of past environments, human influence on past ecosystem structure and function, and human responses to environmental deterioration (e.g., Broughton 2002; Fisher and Feinman 2005; Grayson 2001; Hayashida 2005; Johnson et al. 2005; Kay and Simmons 2002; Kirch 1997, 2004; Kohler 2004; Krech 2005; Lyman 1996; Lyman and Cannon 2004a; Martin 2005; Redman 1999; Redman et al. 2004; Roel et al. 2002; Steadman 2006; van der Leeuw and

Redman 2002). The world's coastal regions and oceans, however, have generally played a minor role in these studies, with much emphasis being placed instead on terrestrial ecosystems. Similar problems are also present in marine ecology as our knowledge of terrestrial environmental degradation has historically been better documented than that of marine ecosystems. Jackson (2001:5416) noted that the problems facing Neotropical forests are well known and widely discussed by governments and the public, for instance, while the problems facing Neotropical coral reefs have been made widely known only relatively recently. Part of the problem stems from the fact that humans are fundamentally terrestrial organisms, and much of the ocean is beyond the realm of daily experience for most people, many of whom consider the sea to be exotic and even dangerous (Earle 1995:xvii).

In a series of case studies from around the world, the contributors to this volume demonstrate that archaeology, anthropology, and history are poised to play an important role in helping understand and shape the future of the world's oceans. These studies of the historical ecology of marine ecosystems in a variety of cultural and environmental settings demonstrate the complex interplay between ancient humans, their demography and technology, climate change, coastal productivity, and the biodiversity inherent in local marine ecosystems. In some cases, ancient people appear to have had relatively minor impacts on nearshore marine habitats or fisheries, while in others human impacts were significant. Linking these past studies to the present is equally complex and often hindered by methodological and epistemological discrepancies between the dynamic modern world and the static realm of archaeological and historical records. Nonetheless, the case studies in this volume provide much-needed context on ancient human impacts on (or interactions with) marine environments, and ultimately a framework on which to build future research.

In this chapter, we provide an overview of the implications of the global case studies presented in this volume. We highlight the contributions archaeologists and other historical researchers can make to understanding contemporary marine communities and the importance of increased collaboration between marine ecologists, biologists, archaeologists, historians, and other researchers, focusing on three primary areas: (1) ancient human impacts on marine environments, (2) methodological promises and problems, and (3) the importance of archaeology for providing baseline data on the structure and nature of marine environments at various points in the past.

ANCIENT HUMAN IMPACTS ON MARINE ECOSYSTEMS

All of the chapters in this book provide important information on ancient human impacts on coastal environments, demonstrating a variety of complex interactions between people and marine ecosystems. In a recent analysis of the ways archaeological and historical data can inform contemporary environmental crises, Jackson et al. (2001:636) noted that "contrary to romantic notions of the oceans as the 'last frontier' and of the supposedly superior ecological wisdom of non-Western and precolonial societies, our analysis demonstrates that overfishing fundamentally altered coastal marine ecosystems during each of the cultural periods we examined. Changes in ecosystem structure and function occurred as early as the late aboriginal and early colonial stages, although these pale in comparison with subsequent events."

These sentiments echo many of the case studies presented in this book. However, archaeological and historical data from the research presented here underscore the variability of ancient human impacts on marine environments. Such impacts and interactions vary through space and time, but they generally appear to have accelerated as human populations grew and more sophisticated fishing technologies were developed. In some cases, precolonial or preindustrial impacts culminate with significant declines in various classes of marine resources or in the structure and function of marine ecosystems. While the perspectives gleaned from this book demonstrate that humans functioned as apex predators in many marine environments (particularly in nearshore habitats), these impacts appear to have been both negative and positive, demonstrating the dynamic nature of human interaction, independent of climate change, with marine ecosystems.

Anderson's (Chapter 2) synthesis of prehistoric and historic occupations of South Polynesian islands demonstrates evidence for significant and fairly rapid impact of Maori, Moriori, and other peoples on seal populations and/or moas. In the case of seals, breeding largely ceased on the north island of New Zealand and dramatically declined on the south island. Although some indications of size reduction in fish and shellfish stocks have been documented, Anderson notes that fish and shellfish populations in these areas do not appear to have been as devastated as marine mammals. Anderson also argues that climate change played a role in these developments, but it is difficult to differentiate climatic versus human-induced changes. While he acknowledges the difficulties of proving this archaeologically, Anderson makes an interesting argument about the possibility of "consumer-prioritized foraging" in the Pacific, noting that although the Maori and others must have seen and understood bird and seal depletion, they may have prioritized their own short-term needs over long-term sustainability. We suspect that such strategies were also prominent elsewhere.

Fitzpatrick, Keegan, and Sullivan Sealey's analysis of the Caribbean record (Chapter 7) documents habitat transformation from agricultural activities (i.e., sea grass habitat to muddy mangrove) and extirpation or reduction in sea turtles, large fishes, shellfish, and birds. Their data demonstrate that prehistoric Caribbean environments were far from "pristine," but like all the case studies in this volume the prehistoric human impact often pales in comparison to modern and historical cases of much larger scale harvest and impact. Bourque, Johnson, and Steneck (Chapter 8) provide similar evidence for localized depletion of fish in the Gulf of Maine, as do Kennett, Voorhies, Wake, and Martinez for the Pacific Coast of Mexico (Chapter 5).

Many of the contributors recognize that people undoubtedly altered local marine environments, but demonstrate that these impacts are often difficult to detect archaeologically, or that human impacts are often juxtaposed with records of considerable continuity through time in the types of resources being used. There is a clear distinction between prehistoric, colonial, and modern impacts, with the latter two tending to be much more significant than the former (see also Jackson et al. 2001). Corbett, Causey, Clementz, Koch, Doroff, Lefevre, and West (Chapter 3), for example, argued that although Aleuts had an impact on nearshore ecosystems and functioned as a top predator in the Aleutian Islands, the underlying current of the regional archaeological record is one of stability rather than change. This is similar to Rick, Erlandson, Braje, Estes, Graham, and Vellanoweth (Chapter 4), who noted that the Chumash and their predecessors along the southern California Coast played an important role in structuring and influencing marine ecosystems, but a similar suite of resources was used across the Holocene (see also Erlandson et al. 2004). For Peru in the Early and Middle Holocene, Reitz, Andrus, and Sandweiss (Chapter 6) present a case for minimal human impacts, emphasizing the profound influence of climatic events (El Niño, etc.). Morales and Roselló (Chapter 11) also explore the role of differing environments, habitats, and seasonality in influencing fishing strategies. Through a series of methodological cautions, Bailey, Barrett, Craig, and Milner (Chapter 10) also suggest that there is a distinct difference in "human pressure" on resources and "overexploitation." They note that while mollusks would be among the most heavily affected, the archaeological record of such cases is often limited.

One of the more profound archaeological findings in recent years is evidence for resource depression or changes in the rank of the prey people used through time (see Broughton 1999; Erlandson and Rick, this volume). Working in a variety of areas, archaeologists have demonstrated that prehistoric peoples had significant impacts on ancient animal populations, greatly reducing their numbers in some instances (see Grayson 2001). This includes coastal regions such as San Francisco Bay (Broughton 1999), the Pacific Northwest (Butler 2000), and New Zealand (Nagaoka 2002, 2005). The archaeological record abounds with cases of people switching to different resources, but as noted above, there is also evidence for long-term stability in the archaeological record of many coastal areas. In a recent review of the evidence for resource depression in the Pacific Northwest of North America, Butler and Campbell (2004) suggested that the record is characterized by stability rather than change. The reasons behind patterns of stability are, of course, perplexing. Are they a result of the technological limitations of ancient peoples, the overwhelming productivity or resilience of certain nearshore ecosystems, or the limited chronological resolution of most archaeological records—or do they show some prehistoric conservation ethic?

As we noted in Chapter 1, evidence for resource depression and overexploitation does not necessarily prove that conservation or sustainable harvests were not practiced by ancient peoples within a given region. Localized impacts may have been managed by periodically moving base camps or villages. Unrestrained or unregulated harvests (and heavy impacts) may have been practiced in periods of abundance, followed by the development of conservation strategies or sustainable harvest practices as resources become scarcer. Over the course of millennia, therefore, we might expect to find evidence for heavy human impacts on some fisheries and marine ecosystems, as well as the adoption of strategies to keep such impacts within manageable and sustainable levels. Whether applied to ancient peoples, historical exploitation patterns, or modern fisheries, these are contentious issues—not easily answered with static archaeological data. What remains clear is that people have influenced the nature of nearshore marine ecosystems for a very long time and that we have much to learn about the success and failures of ancient, historical, and modern human uses and abuses of the oceans and other environments.

Jerardino, Branch, and Navarro's (Chapter 12) South African data show a unique perspective where, despite having one of the longest coastal archaeological sequences in the world, technological limitations led people to subsist largely on shellfish and seabirds, with limited emphasis on marine mammals and fishes. After carefully weighing the role of environmental factors, Jerardino and her colleagues note that human harvest pressure may have caused a reduction in mean shellfish size, but they add that despite this pressure, shellfish harvest in the area persisted for several millennia. These patterns stand in contrast to modern South African data, which show a profound impact on the local marine environment. These studies call into question the important issue of scale, where ancient peoples often had significant impacts on the environments in which they lived, but these are often fundamentally more limited than modern industrial impacts. Such viewpoints have also been noted by marine ecologists (e.g., Jackson et al. 2001). Pauly and Palomares (2005:197), for example, suggested that "while pre-industrial fisheries had the capacity to extirpate some freshwater and coastal fish populations as evidenced in the sub fossil and archaeological records [Jackson et al. 2001; Pitcher 2001], it is only since the advent of industrial fishing that the sequential depletion of coastal, then offshore populations of marine fish has become the standard operating procedure [Ludwig et al. 1993]."

Until recently, however, archaeological data on this subject have been limited. The scale and magnitude of ancient human impacts is something that clearly needs to be gauged on a local rather than global level, as it is clear that impacts varied through space and time, were focused primarily on nearshore fisheries and ecosystems, and were dependent on environmental, cultural, and technological developments.

This variability in ancient and modern records is further echoed by Perdikaris and McGovern's (Chapter 9) research among the Norse in the North Atlantic. Terrestrial land degradation by Viking Age farmers is well documented and includes deforestation, severe erosion, and soil degradation. Perdikaris and McGovern's review of Norse and other use of a variety of marine resources (sea mammals, fishes, and birds) is a data-rich account of Viking Age interactions with marine environments. Despite the impacts of Norse people on terrestrial landscapes, however, Perdikaris and McGovern noted that "these chiefly agricultural societies were often effective long-term managers of marine resources." This includes apparent impacts or reductions in the walrus population of Iceland, but sustainable harvests of birds and other sea mammals. Perdikaris and McGovern caution that the Norse appear to be partly responsible for the origins of western European commercial fisheries that have had lasting ecological consequences for the North Atlantic. This case study points to differences in the ways the Norse and other peoples interacted with and altered terrestrial and marine environments. Although they will undoubtedly vary through space and time, such differences may also be present among some Polynesian agriculturalists (see Kirch 1997). These data provide an important lesson that suggests that even in the wake of heavy environmental impacts, some ancient societies may have used sustainable strategies on other resources or ecosystems.

Another issue raised by contributors to this volume is the possibility that some marine environments are more resilient than others, especially compared to terrestrial ones. In a comparison of the historical ecology of kelp forest ecosystems in Maine, California, and the Aleutians, for instance, Steneck et al. (2002) concluded that the more-diverse food webs and ecosystems of California kelp forests appear to have been more resistant to localized or regional collapse caused by ancient humans than those of the less-diverse kelp forest ecosystems of the Aleutians and the Gulf of Maine. In comparing impacts in terrestrial versus marine ecosystems, the dearth of overwhelming early human impacts on some marine environments may be because terrestrial systems operate in fundamentally different ways, and prior to modern industrial technologies, they were more susceptible to widespread human-induced degradation. After all, many marine environments (e.g., pelagic or deep benthic zones) are difficult to reach and forage in using preindustrial technologies (see Morales and Roselló, Chapter 11). Such distinctions are important and illustrate the need for caution when comparing ancient data with modern data, especially when gauging the impacts of people through time on marine ecosystems. On the other hand, as has been the case with our own work on California's Channel Islands, as archaeologists and ecologists become more sophisticated at recognizing the signatures of human impacts in marine ecosystems, the evidence for such impacts may become more widespread and profound.

MODERN AND HISTORICAL DATA: METHODOLOGICAL PROMISES AND PROBLEMS

All the chapters in this volume demonstrate that traditional zooarchaeological data (number of identified specimens, minimum number of individuals, and weight of shellfish and vertebrates) can greatly inform our understanding of the nature of ancient marine environments and the role of people in influencing their structure and functioning. A significant problem, however, is that these data sets are generally not directly comparable to modern or historical records, creating an imbalance in our understanding of ancient, historical, and modern ecological change. The studies in this volume grapple with these issues, with several providing new and innovative ways to increase our understanding of ancient marine environments and heighten the comparability of archaeological and contemporary datasets.

In their synthesis for the North Sea area, Bailey et al. deal with this issue in perhaps the greatest detail. In surveying the major methods of investigating human impacts on marine environments, they note that the classic methods employed in Europe and beyond include measurement of shell and bone specimens to infer any reduction in size by predation, use of traditional zooarchaeological measures of the relative abundance of various faunal classes identified in a site, and the analysis of stable isotopes in human bones to estimate the importance of various resources in the human diet. Of particular interest is their discussion of the fact that measurements of shell and bone specimens to determine human harvest pressure can be influenced by a variety of environmental factors. Although environmental factors undoubtedly influenced shellfish sizes, studies by Jerardino et al. (Chapter 12), Rick et al. (Chapter 4), and Corbett et al. (Chapter 3) suggest that people probably also had a role in influencing the size and abundance of shellfish species. Determining how much of this was influenced by environmental variables, however, remains an unresolved issue.

The fact that environmental factors often produce similar archaeological signatures as human-induced changes is echoed by Reitz et al. (Chapter 6). For the Peruvian Coast, they emphasize the importance of stable isotope reconstructions of environmental variables but note that El Niño and other phenomena can cause faunal and ecological changes that mirror human-induced impacts. Their conclusions reiterate the variability and complexity of differentiating anthropogenic, climatic, geomorphological, biological, and other sources of ecological change. These problems also plague modern ecological data, where on much shorter time scales (10–20 years) it can be very difficult to separate "top-down" versus "bottom-up" effects on ecological communities, much less human versus climatic influences in ecosystems (see Dayton et al. 1998). Through continued use of stable isotope data and other means of reconstructing marine environmental conditions, the longer time scales offered by archaeology will help improve our understanding of the effects of human versus climatic variables.

Morales and Roselló (Chapter 11) provided one of the few studies with measurements of vertebrate faunal sizes through time. In this case, they measured fish jaw lengths to determine if there were human impacts on the age and size of fishes. Their data were inconclusive, further demonstrating the difficulties in comparing ancient and modern records. Through their analysis of a variety of zooarchaeological data, however, they illustrate the complexities of comparing archaeological and modern data and the need for additional analyses and better comparative techniques.

Perdikaris and McGovern (Chapter 9) demonstrate the utility of zooarchaeological analyses of seal, seabird, and fish remains to document patterns of marine resource exploitation and human impacts by Norse peoples. Their data, including bone element comparisons and changes in the relative abundance of various taxa, demonstrate the changing nature of Norse maritime economies and ultimately how they compare with modern fisheries. They used zooarchaeological data to examine economic and ecological issues, providing an integrated perspective on the past, and noting that environmental issues function within a variety of social, economic, and other factors. Corbett et al.'s detailed reconstructions of changes in sea urchin size through time also show an important contribution of zooarchaeology to our understanding of ancient marine ecosystems (see Chapter 3).

One of the more intriguing analytical tools discussed in several chapters (i.e., Bailey et al., Bourque et al., Fitzpatrick et al., Kennett et al., Morales et al., and Rick et al.) and other recent studies (see Morales and Roselló 2004; Quitmyer and Reitz 2006; Reitz 2004; Wing 2001) involves converting archaeological data into trophic levels that are more comparable to modern ecological measures. Considerable emphasis has been placed in recent years on investigating the possibility that ancient peoples may have "fished down food webs," a practice of overexploiting high trophic level fishes and then moving down to fish at lower trophic levels, something that has happened in numerous modern fisheries around the world (see Erlandson and Rick, this volume; Pauly et al. 1998, 2005).

The archaeological studies in this volume provide interesting trophic level data, but patterns of fishing down the food web have been ambiguous (e.g., Morales and Roselló 2004; Reitz 2004). Using preliminary data from the California Coast, Rick et al. (Chapter 4) suggested that through the Holocene, native peoples in southern California may have fished up the food web, focusing on shellfish and lower trophic level fish for thousands of years before human population growth required the development of more sophisticated fishing technologies and a greater expenditure of energy in pursuit of larger pelagic species. Reitz et al. (Chapter 6) also noted that ancient Peruvians largely targeted low trophic level herrings and anchovetas due to their abundance and ability to be harvested en masse. Morales and Roselló (Chapter 11) described a variety of trophic levels being used throughout most early records from the coast of southern Iberia, and Bailey et al. (Chapter 10) used stable nitrogen isotopes to argue that prehistoric sites reflect lower trophic signatures than modern sites, possibly due to differing nitrogen values in lower food-web organisms due to environmental changes. In contrast, Bourque et al. (Chapter 8) brought a suite of data to bear on this issue, arguing for a case of ancient fishing down the food web in the Gulf of Maine, where climate may have had little influence on changes in fish faunal data and stable isotopes through time. Using a variety of faunal data from the Pacific Coast of Mexico, Kennett et al. (Chapter 5) also argue for an ancient case of fishing down the food web. More research is needed on this topic, but collectively the case studies in this volume provide no evidence of a global trend toward fishing down the food web or the universal overexploitation of high trophic level marine foods by ancient peoples. Even in the cases where evidence for fishing down the food web is fairly convincing, these studies suggest this may have been fairly localized in scale (e.g., Bourque et al., Chapter 8). Due to the technological and labor costs associated with fishing many high trophic level fishes, particularly those generally found in deep waters (tuna, swordfish, etc.), such strategies may have been incongruent with preindustrial fishing technologies and subsistence needs. Such patterns will depend on the environments being studied, however, with estuaries and other bounded environments probably receiving greater impacts and higher probabilities of ancient fishing down food webs.

The use of archaeological data to understand ancient human impacts on the environment is a complicated issue. However, the very fact that archaeologists are pursuing this topic at all gives us hope that our methodological approaches will be greatly improved and expanded in the near future. In our own research, we never would have started to unravel the complex patterns of human impact and influence on the nature of southern California marine environments if we had not started looking for it in the first place. Like many of the chapters in this volume and elsewhere, our research began by reinterpreting and analyzing traditional zooarchaeological data from coastal sites, and then moved into more sophisticated shell measurements, stable isotope studies, and other research. Archaeological analyses of ancient human interactions with the marine environment are sure to change and improve in the future, and the chapters in this book help set us on this important journey.

TIMELINES, BASELINES, AND RESTORATION

Perhaps the most salient contribution of archaeology and other historical sciences to contemporary issues facing our oceans is the long-term perspectives on the structure and function of marine environments provided by archaeological and historical data (see Braje et al. 2006; Jackson 2001; Jackson et al. 2001). Recent research has demonstrated that ecological baselines, benchmarks used to measure the health and status of ecosystems, have "shifted" or changed significantly through time (see Jackson et al. 2001; Pauly 1995). Accurate baseline data, however, are crucial for measuring and maintaining the stability and productivity of our oceans. It is clear that in historic and modern times many ecosystems have been severely degraded and provide flawed perspectives on what the natural state of the ecosystem was (Jackson et al. 2001). The chapters in this volume clearly demonstrate that people had a significant role in influencing the nature of nearshore marine ecosystems for much of the Holocene and earlier. Emerging from these studies it is clear that archaeologists can play a significant role in determining how ecological baselines may have changed and what marine ecosystems may have looked like in deep time. Such paradigms, however, are diverse and variable, and much research is needed to determine the relevance of archaeological data to such studies.

Reitz et al. underscore the complexities of using archaeological data to help restore and remediate modern marine habitats, emphasizing the role of environmental changes (e.g., El Niño) in shaping ancient marine environments. They stress that historical and archaeological data should be viewed as a "complex web" of environmental and cultural variables, suggesting that the prehistoric record of ancient marine environments is dynamic and constantly fluctuating. An important lesson from their research is that modern-day management strategies must also project that future environmental variables, and conditions will be equally (if not more) dynamic (see also Anderson, Bailey et al., and Jerardino et al.). Several marine ecologists have also noted that climatic and human-induced disturbances influence marine ecosystems on long- and short-term cycles (see Graham et al. 2003; Jackson 2001; Steneck et al. 2002). Through several studies in this volume, it is clear that archaeological data can help elucidate these issues, especially the importance of understanding climate change on short- and long-term scales. Accounting for the relationships between climatic and cultural variables in the structure of marine ecosystems will lead to better baseline data and, ultimately, restoration efforts. As Bailey et al., Bourque et al., Corbett et al., and others demonstrate, increased use of stable isotopes from archaeological remains will prove to be a fundamental means of teasing apart the effects of local climate change on human subsistence.

As noted above, many of the case studies in this volume demonstrate that marine ecosystems are relatively resilient in the face of long-term human predation pressure. In the case of southern California, the Mediterranean, the North Sea, and South Africa, such records span at least 10,000 years and in some cases much longer. Although prehistoric peoples influenced marine environments throughout this time, prehistoric marine animal extinction events are rare. Evidence for relative continuity and stability are at odds with the modern record of devastating environmental impacts. Corbett et al.'s data from the Aleutians emphasizes this idea, noting that the structure and function of prehistoric Aleutian ecosystems appears to have operated in a state of relative equilibrium compared to that of the last 50 years. This system also shows changes in its structure, and strong Aleut influence, but the underlying current of stability is very interesting. A fundamental question raised by these and other studies is what the management or restoration baseline should be for any particular fishery or ecosystem. Should the ideal be what the marine environment was like at first human arrival, or should we strive for prehuman arrival? Archaeology shows us that people are part of the natural world and have been a significant force for at least several hundred thousand years. From this perspective and the nature of our current marine crisis, it seems clear that humans must be made part of the baseline equation, but this timeline has also shifted.

Working in concert with the ecological notion of shifting baselines is the archaeological perspective of "shifting timelines" for ancient human use of aquatic foods. As we discussed in Chapter 1, archaeological evidence for the antiquity of human use of coastal resources has been pushed back considerably in many parts of the world. Once thought to have been used primarily during the Holocene, a variety of archaeological data now suggest that some early hominids made at least sporadic use of coastal or aquatic resources, with an intensification of marine resource exploitation after the appearance of anatomically modern humans between roughly 150,000 and 50,000 years ago (see Erlandson 2001). Archaeological research also demonstrates that the development of relatively sophisticated seafaring skills allowed humans to colonize island Southeast Asia, Australia, western Melanesia, and the Ryuku Islands south of Japan between about 55,000 and 35,000 years ago. The early use of coastal resources is also probably much more abundant than the archaeological record currently suggests, as postglacial sea level rise and coastal erosion have submerged or destroyed much of the evidence for the Pleistocene occupation of coastal environments. Nonetheless, these data suggest that human use of coastal areas has great time depth, illustrating that human impact on coastal environments may have begun earlier than previously thought. With such perspectives, we gain greater insight into how complex the shifting baselines phenomenon is, with human influence starting when people first occupied coastal areas and began foraging in them.

CONCLUSIONS

Around the world, coastal shell middens represent a remarkably rich repository of information on the nature of marine ecosystems, the development of early human fishing technologies and economies, and the impacts ancient humans have had on marine fisheries and ecosystems (Erlandson and Fitzpatrick 2006). Increasingly, marine ecologists and fisheries managers are taking note of these treasure troves of data on the deep history of the world's oceans buried in archaeological and historical records. Simultaneously, many archaeologists are making greater use of modern ecological data—or establishing collaborative relationships with biologists, ecologists, and other scientists—to inform archaeological interpretations. Jackson (2001:5416) indicated that "paleoecological, archaeological, and historical reconstructions of coastal marine ecosystems provide the best evidence for predicting ecological consequences of establishing very large scale marine reserves and other forms of rigorous protection of fisheries." While increasing in number, such interdisciplinary studies using archaeological and other historical data to help inform the present are still relatively rare. Much of the problem stems from the fact that these disciplines need to be better integrated, and that traditional intellectual barriers must be broken down. Much of the responsibility falls on the shoulders of archaeologists to make our data accessible, approachable, and applicable to other disciplines. This perspective has recently been echoed by a number of archaeologists (e.g., Hayashida 2005; Kirch 1997; Lyman and Cannon 2005b; Redman 1999; Redman et al. 2004). Lyman and Cannon (2004b:23) suggested that "archaeologists are beginning to argue that they must make their discipline relevant to modern concerns, but they are not always clear about why they should do so other than to note that archaeology provides a time depth to anthropogenically created ecologies [van der Leeuw and Redman 2002]."

The chapters in this volume are a significant step in this direction. Many of them represent collaborative efforts between ecologists, archaeologists, biologists, historians, and other researchers. In those cases, scientists from different disciplines have brought unique perspectives on the past, present, and future of our world's oceans. As the problems facing our oceans, and all environments on our planet, become increasingly dire it is clear that if we are to solve these problems, it will be through increasing interdisciplinary and collaborative research. Archaeology and history alone cannot provide the solutions to these problems, but the deeper time depth of such historical perspectives is needed for supplying baseline information crucial for restoring coastal ecosystems and more effectively managing marine fisheries. The challenge is to increase collaboration and continue to move archaeological data out of the realm of the esoteric and into a field that plays a crucial role in creating a better and more sustainable planet on which to live.

ACKNOWLEDGMENTS

We thank Steven James and David Steadman for comments on an earlier draft of this chapter, Blake Edgar for all his encouragement and help in bringing it to fruition, and all the contributors to this volume for their stimulating discussions of human history and marine ecosystems around the world.

REFERENCES CITED

Botsford, L. W., J. C. Castilla, and C. H. Peterson
1997 The Management of Fisheries and Marine Ecosystems. *Science* 277:509–514.

Braje, T. J., J. M. Erlandson, D. J. Kennett, and T. C. Rick
2006 Archaeology and Marine Conservation. *SAA Archaeological Record* 6(1):14–19.

Broughton, J. M.
1999 *Resource Depression and Intensification during the Late Holocene, San Francisco Bay: Evidence from the Emeryville Shellmound Vertebrate Fauna.* University of California Anthropological Records 32. University of California Press, Berkeley.
2002 Pre Columbian Human Impact on California Vertebrates: Evidence from Old Bones and Implications for Wilderness Policy. In *Wilderness and Political Ecology: Aboriginal Influences and the Original State of Nature,* edited by C. E. Kay and R. T. Simmons, pp. 44–71. University of Utah Press, Salt Lake City.

Butler, V. L.
2000 Resource Depression on the Northwest Coast of North America. *Antiquity* 74:649–661.

Butler, V. L., and S. K. Campbell
2004 Resource Intensification and Resource Depression in the Pacific Northwest of North America: A Zooarchaeological Review. *Journal of World Prehistory* 18:327–405.

Costanza, R., F. Andrade, P. Antunes, M. van den Belt, D. Boersma, D. F. Boesch, F. Catarino, S. Hanna, K. Limburg, B. Low, M. Molitor, J. Pereira, S. Rayner, R. Santos, J. Wilson, and M. Young
1998 Principles for Sustainable Governance of the Oceans. *Science* 281:198–199.

Dayton, P. K., M. J. Tegner, P. B. Edwards, and K. L. Riser
1998 Sliding Baselines, Ghosts, and Reduced Expectations in Kelp Forest Communities. *Ecological Applications* 8:309–322.

Earle, S. A.
1995 *Sea Change: A Message of the Oceans.* Fawcett Columbine, New York.

Ellis, R.
2003 *The Empty Ocean.* Island Press, New York.

Erlandson, J. M.
2001 The Archaeology of Aquatic Adaptations: Paradigms for a New Millennium. *Journal of Archaeological Research* 9:287–350.

Erlandson, J. M., and S. M. Fitzpatrick
2006 Oceans, Islands, and Coasts: Current Perspectives on the Role of the Sea in Human Prehistory. *Journal of Island and Coastal Archaeology* 1:5–32.

Erlandson, J. M., T. C. Rick, and R. L. Vellanoweth
2004 Human Impacts on Ancient Environments: A Case Study from California's Northern Channel Islands. In *Voyages of Discovery: The Archaeology of Islands,* edited by S. M. Fitzpatrick, pp. 51–83. Praeger, New York.

Fisher, C. T., and G. M. Feinman
2005 Introduction to "Landscapes over Time." *American Anthropologist* 107:62–69.

Graham, M. H., P. K. Dayton, and J. M. Erlandson
2003 Ice Ages and Ecological Transitions on Temperate Coasts. *Trends in Ecology and Evolution* 18:33–40.

Grayson, D. K.
2001 The Archaeological Record of Human Impact on Animal Populations. *Journal of World Prehistory* 15:1–68.

Hayashida, F. M.
2005 Archaeology, Ecological History, and Conservation. *Annual Review of Anthropology* 34:43–65.

Jackson, J. B. C.
2001 What Was Natural in the Coastal Oceans? *Proceedings of the National Academy of Sciences* 98:5411–5418.

Jackson, J. B. C., M. Kirby, W. Berger, K. Bjorndal, L. Botsford, B. Bourque, R. Bradbury, R. Cooke, J.

Erlandson, A. James, J. A. Estes, T. Hughes, S. Kidwell, C. Lange, H. Lenihan, J. Pandolfi, C. Peterson, R. Steneck, M. Tegner, and R. Warner
2001 Historical Overfishing and the Recent Collapse of Coastal Ecosystems. *Science* 293:629–637.

Johnson, C. D., T. Kohler, and J. Cowan
2005 Modeling Historical Ecology, Thinking about Contemporary Systems. *American Anthropologist* 107:96–107.

Kay, C. E., and R. T. Simmons (editors)
2002 *Wilderness and Political Ecology: Aboriginal Influences and the Original State of Nature.* University of Utah Press, Salt Lake City.

Kirch, P. V.
1997 Microcosmic Histories: Island Perspectives on "Global" Change. *American Anthropologist* 99:30–42.
2004 Oceanic Islands: Microcosms of "Global Change." In *The Archaeology of Global Change: The Impact of Humans on Their Environment,* edited by C. L. Redman, S. R. James, P. R. Fish, and J. D. Rogers, pp. 13–27. Smithsonian Institution Press, Washington, D.C.

Kohler, T. A.
2004 Pre-Hispanic Impact on Upland North American Southwestern Environments: Evolutionary Ecological Perspectives. In *The Archaeology of Global Change: The Impact of Humans on Their Environment,* edited by C. L. Redman, S. R. James, P. R. Fish, and J. D. Rogers, pp. 224–247. Smithsonian Institution Press, Washington, D.C.

Krech, S., III
2005 Reflections on Conservation, Sustainability, and Environmentalism in Indigenous North America. *American Anthropologist* 107:78–86.

Ludwig, D., R. Hilborn, and C. Walters
1993 Uncertainty, Resource Exploitation, and Conservation: Lessons from History. *Science* 260:17, 36.

Lyman, R. L.
1996 Applied Zooarchaeology: The Relevance of Faunal Analysis to Wildlife Management. *World Archaeology* 28:110–125.

Lyman, R. L., and K. P. Cannon (editors)
2004a *Zooarchaeology and Conservation Biology.* University of Utah Press, Salt Lake City.
2004b Applied Zooarchaeology, Because It Matters. In *Zooarchaeology and Conservation Biology,* edited by R. L. Lyman and K. P. Cannon, pp. 1–24. University of Utah Press, Salt Lake City.

Martin, P. S.
2005 *Twilight of the Mammoths: Ice Age Extinctions and the Rewilding of the Americas.* University of California Press, Berkeley.

Morales, A., and E. Roselló
2004 Fishing down the Food Web in Iberian Prehistory? A New Look at the Fishes from Cueva de Nerja (Malaga, Spain). In *Pettis Animaux Et Societes Humaines Du Complement Alimentaire Aux Ressources Utilitaires XXIV,* edited by J. Brugal and J. Desse, pp. 111–123. APDCA, Antibes.

Nagaoka, L.
2002 The Effects of Resource Depression on Foraging Efficiency, Diet Breadth, and Patch Use in Southern New Zealand. *Journal of Anthropological Archaeology* 21:419–442.
2005 Declining Foraging Efficiency and Moa Carcass Exploitation in Southern New Zealand. *Journal of Archaeological Science* 32: 1328–1338.

Pauly, D.
1995 Anecdotes and the Shifting Baselines Syndrome of Fisheries. *Trends in Ecology and Evolution* 10(10):430.

Pauly, D., and M. L. Palomares
2005 Fishing Down Marine Food Web: It Is Far More Pervasive Than We Thought. *Bulletin of Marine Science* 76(2):197–211.

Pauly, D., V. Christensen, J. Dalsgaard, F. Rainer, and F. Torres
1998 Fishing down Marine Food Webs. *Science* 279:860–863.

Pauly, D., R. Watson, and J. Alder
2005 Global Trends in World Fisheries: Impacts on Marine Ecosystems and Food Security. *Philosophical Transactions of the Royal Society: Biological Transactions* 360:5–12.

Pew Oceans Commission
2003 *Americas Living Oceans: Charting a Course for Sea Change.* Pew Oceans Commission, Arlington, Virginia.

Pitcher, T. J.
2001 Fisheries Managed to Rebuild Ecosystems? Reconstructing the Past to Salvage the Future. *Ecological Applications* 11:601–617.

Quitmyer, I. R., and E. J. Reitz
2006 Marine Trophic Levels Targeted between AD 300 and 1500 on the Georgia Coast, USA. *Journal of Archaeological Science* 33: 806–822.

Redman, C. L.
1999 *Human Impact on Ancient Environments.* University of Arizona Press, Tucson.

Redman, C. L., S. R. James, P. R. Fish, and J. D. Rogers (editors)
2004 *The Archaeology of Global Change: The Impact of Humans on Their Environment.* Smithsonian Institution Press, Washington, D.C.

Reitz, E. J.
2004 "Fishing down the Food Web": A Case Study from St. Augustine, Florida, U.S.A. *American Antiquity* 69:63–83.

Roel, C. G., M. Lauwerier, and I. Plug (editors)
2002 *The Future from the Past.* Oxbow Books, Oxford.

Safina, C.
1997 *Song for the Blue Ocean.* Henry Holt and Company, New York.

Steadman, D. W.
2006 *Extinction and Biogeography of Tropical Pacific Birds.* University of Chicago Press, Chicago.

Steneck, R., M. Graham, B. Bourque, D. Corbett, J. M. Erlandson, and J. Estes
2002 Kelp Forest Ecosystems: Biodiversity, Stability, Resilience, and Their Future. *Environmental Conservation* 29:436–459.

van der Leeuw, S., and C. L. Redman
2002 Placing Archaeology at the Center of Socio-natural Studies. *American Antiquity* 67:597–605.

Wing, E. S.
2001 The Sustainability of Resource Use by Native Americans on Four Caribbean Islands. *International Journal of Osteoarchaeology* 11:112–126.

Woodard, C.
2000 *Ocean's End: Travels through Endangered Seas.* Basic Books, New York.

Worm, B., E. B. Barbier, N. Beaumont, J. E. Duffy, C. Folke, B. S. Halpern, J. B. C. Jackson, H. K. Lotze, F. Micheli, S. R. Palumbi, E. Sala, K. A. Selkoe, J. J. Stachowicz, and R. Watson
2006 Impacts of Biodiversity Loss on Ocean Ecosystem Services. *Science* 314:787–790.

INDEX

Indexer:	Live Oaks Indexing
Composition:	Aptara, Inc.
Text:	9.5/13 Scala
Display:	Scala Sans, Scala Sans Caps
Printer and Binder:	Thomson Shore, Inc.